AF713627

Principles of Uncertainty

CHAPMAN & HALL/CRC
Texts in Statistical Science Series

Joseph K. Blitzstein, *Harvard University, USA*
Julian J. Faraway, *University of Bath, UK*
Martin Tanner, *Northwestern University, USA*
Jim Zidek, *University of British Columbia, Canada*

Recently Published Titles

Introduction to Probability, Second Edition
Joseph K. Blitzstein and Jessica Hwang

Theory of Spatial Statistics
A Concise Introduction
M.N.M van Lieshout

Bayesian Statistical Methods
Brian J. Reich and Sujit K. Ghosh

Sampling
Design and Analysis, Second Edition
Sharon L. Lohr

The Analysis of Time Series
An Introduction with R, Seventh Edition
Chris Chatfield and Haipeng Xing

Time Series
A Data Analysis Approach Using R
Robert H. Shumway and David S. Stoffer

Practical Multivariate Analysis, Sixth Edition
Abdelmonem Afifi, Susanne May, Robin A. Donatello, and Virginia A. Clark

Time Series: A First Course with Bootstrap Starter
Tucker S. McElroy and Dimitris N. Politis

Probability and Bayesian Modeling
Jim Albert and Jingchen Hu

Surrogates
Gaussian Process Modeling, Design, and Optimization for the Applied Sciences
Robert B. Gramacy

Statistical Analysis of Financial Data
With Examples in R
James Gentle

Statistical Rethinking
A Bayesian Course with Examples in R and STAN, Second Edition
Richard McElreath

Statistical Machine Learning
A Model-Based Approach
Richard Golden

Randomization, Bootstrap and Monte Carlo Methods in Biology
Fourth Edition
Bryan F. J. Manly, Jorje A. Navarro Alberto

Principles of Uncertainty
Second Edition
Joseph B. Kadane

An Introduction to Nonparametric Statistics
John Kolossa

For more information about this series, please visit: https://www.crcpress.com/Chapman--Hall-CRC-Texts-in-Statistical-Science/book-series/CHTEXSTASCI

Principles of Uncertainty

Second Edition

Joseph B. Kadane

CRC Press
Taylor & Francis Group
Boca Raton London New York

CRC Press is an imprint of the
Taylor & Francis Group, an informa business
A CHAPMAN & HALL BOOK

Second edition published 2020
by CRC Press
6000 Broken Sound Parkway NW, Suite 300, Boca Raton, FL 33487-2742

and by CRC Press
2 Park Square, Milton Park, Abingdon, Oxon, OX14 4RN

© 2021 Taylor & Francis Group, LLC

First edition published by CRC Press 2011

CRC Press is an imprint of Taylor & Francis Group, LLC

Reasonable efforts have been made to publish reliable data and information, but the author and publisher cannot assume responsibility for the validity of all materials or the consequences of their use. The authors and publishers have attempted to trace the copyright holders of all material reproduced in this publication and apologize to copyright holders if permission to publish in this form has not been obtained. If any copyright material has not been acknowledged please write and let us know so we may rectify in any future reprint.

Except as permitted under U.S. Copyright Law, no part of this book may be reprinted, reproduced, transmitted, or utilized in any form by any electronic, mechanical, or other means, now known or hereafter invented, including photocopying, microfilming, and recording, or in any information storage or retrieval system, without written permission from the publishers.

For permission to photocopy or use material electronically from this work, access www.copyright.com or contact the Copyright Clearance Center, Inc. (CCC), 222 Rosewood Drive, Danvers, MA 01923, 978-750-8400. For works that are not available on CCC please contact mpkbookspermissions@tandf.co.uk

Trademark notice: Product or corporate names may be trademarks or registered trademarks, and are used only for identification and explanation without intent to infringe.

Library of Congress Cataloging-in-Publication Data

ISBN: 978-1-1380-5273-4 (hbk)
ISBN: 978-1-3151-6756-5 (ebk)

Typeset in Computer Modern font
by Cenveo Publisher Services

Dedication

To my teachers, my colleagues and my students.

J. B. K.

Contents

List of Figures

List of Tables

Foreword

With respect to the mathematical parts of this book, I can offer no better advice than (Halmos, 1985, p. 69):

> ...study actively. Don't just read it: fight it! Ask your own questions, look for your own examples, discover your own proofs. Is the hypothesis necessary? Is the converse true? What happens in the classical special case? What about the degenerate cases? Where does the proof use the hypothesis?

In addition, for this book, it is relevant to ask "What does this result mean for understanding uncertainty? If it is a stepping stone, toward what is it a stepping stone? If this result were false, what consequences would that have?"

Preface

Publishing a book is a bit like letting go of a child, now an adult. One is aware of strengths and weaknesses, hopes for the best, and follows the career with interest and concern. But once published, what happens is entirely out of one's control.

Publishing "Principles of Uncertainty" was like that. I had no idea what sort of career the book would have, and no ability to change that career. The book was on its own. As it happened, the world was kind to the book. "Choice," published by the Association of College and Research Libraries, chose it as one of the 100 best technical books published that year. Later, the International Society for Bayesian Analysis chose it for the Morris Degroot Prize for the best book published in a two year interval. I thank both for their consideration and for welcoming Principles. Jim Zidek, a classmate and friend, read several versions and made helpful comments. One principal reviewer was Christian Robert, who liked the book except for Chapter 12, which criticizes the various strands of frequentism. It is certainly true that my perspective is contrary to the modern consensus, which is that polite people don't discuss statistical foundations. I feel I have to, because when I apply statistics, I have to be able to explain why I chose the methods I did. I have expanded Chapter 12 to further explain my views. As I expressed briefly in Chapter 13, in my view the whole purpose of statistical methods is to aid in applications.

That thought led me to write a companion book, "Pragmatics of Uncertainty," which exposes and explores 15 different applied papers of mine. The issue is to what extent was I faithful to "Principles" when I went to use those ideas. Did the application force me to abandon my principles? I invited my readers to judge that for themselves.

The opportunity to revise "Principles" occasions reflections on what to change or add.My first target was the characteristic function approach to the central limit theorem in Chapter 6. While rigorous, it is old-fashioned and clunky. I had hoped that Stein's method would provide a clean way to a simple central limit theorem, but the longer I studied it, the less likely that seemed. Then I found a simple calculus-based method in unpublished work by Pippenger which fit the bill. I also wanted to add material on Bayesian Non-parametrics, which now appears in Chapters 9 and 10. In turn, this material necessitated the establishment of new facts about the Dirichlet distribution, which appears in section 8.9. Chapter 12 also needed some updates, as recent developments indicated, to me at least, that the broader world of applied statistics was coming to some of the same conclusions that I had, about the usefulness of the frequentist statistics I had been taught in graduate school. I thought about the possibility of combining Chapters 1 and 2, since unconditional probability (Chapter 1) is a special case of conditional probability (Chapter 2), but decided that the result would be unintuitive and mysterious, so I dropped that idea. I have added remarks and new problems throughout, and corrected the typos I knew about.

I owe thanks to many. There's hardly anything original in the book. Rather it is my selection of the received wisdom of several centuries of the work of others. The idea is to show how all those pieces fit together, to make a point of view that helps when facing a new applied problem. Once again, my closest companions in understanding the legacy of careful thought about uncertainty have been Teddy Seidenfeld and Mark Schervish, now joined by Robin Gong and Rafael Stern. We still meet once a week, still write papers, and still challenge each other to think more clearly. My assistant, Heidi Sestrich, has been a friend as well as a coworker. My publisher, David Grubbs, has been more patient with me than I deserve. Additionally, Christian Robert and an anonymous reviewer gave me a very useful critique on the revision.

Chapter 1

Probability

"How can I be sure? In a world that's constantly changing, how can I be sure?"
—The Young Rascals

A businessman is exploring a city new to him. He finds a pet store, wanders in, and starts chatting with the owner. After half an hour, the owner says, "I can see you are a discerning gentleman. I have something special to show you," and he brings out a parrot. "This parrot is very smart, and speaks four languages: English, German, French and Spanish," he says. The businessman tries out the parrot in each language, and the parrot answers. "I have to have this parrot," says the businessman, so he buys the parrot, puts it on his shoulder, and leaves the shop.

He goes into a bar. Everyone is curious about the parrot. Nobody believes that the parrot can speak four languages. So the businessman makes bets with everyone in the bar. When all the bets are made, the businessman speaks to the parrot, but the parrot doesn't answer. He tries all four languages, but the parrot is silent. So the businessman has to pay up for all his bets, puts the parrot on his shoulder, and leaves the bar.

When they get to the street, he says to the parrot, "Why wouldn't you say anything in there?" to which the parrot replies, "Listen, stupid, think of all the bets you can make in there tomorrow night!"

1.1 Avoiding being a sure loser

Uncertainty is a fact of life. Indeed we spend much of our waking hours dealing with various forms of uncertainty. The purpose of this chapter is to introduce probability as a fundamental tool for quantifying uncertainty.

Before we begin, I emphasize that the answers you give to the questions I ask you about your uncertainty are yours alone, and need not be the same as what someone else would say, even someone with the same information as you have, and facing the same decisions.

What are you uncertain about? Many things, I suppose, but in order to make progress, I need you to be more specific. You may be uncertain about whether the parrot will speak tomorrow night. But instead, suppose you are uncertain about tomorrow's weather in your home area. In order to speak of the weather, I need you to specify the categories that you will use. For example, you might think that whether it will rain is an important matter. You might also be concerned about the temperature, for example, whether the high temperature for the day will be above 68 degrees Fahrenheit, which is 20 degrees Centigrade or Celsius. Thus you have given four events of interest to you:

A_1: Rain and High above 68 degrees F tomorrow
A_2: Rain and High at or below 68 degrees F tomorrow
A_3: No Rain and High above 68 degrees F tomorrow
A_4: No Rain and High at or below 68 degrees F tomorrow.

Tomorrow, one and only one of these events will occur. In mathematical language, the events are exhaustive (at least one must occur) and disjoint (no more than one can occur). Whatever you are uncertain about, I can ask you to specify a set of disjoint and exhaustive events that describe your categories.

Now I have to ask you about how likely, in your opinion, is each of the events you have specified. I will do this by asking you what price you think is fair for particular tickets I will imagine you will offer to buy or sell. I am going to ask you to name a price at which you would be willing either to sell or to buy such a ticket. You can write such tickets if you are selling them, and I can write them if you are buying them and I am selling them. Tickets are essentially promissory notes. We do not consider the issue of default, that either of us will be unable or unwilling to redeem our promises when the time comes to settle. Consider a ticket that pays \$1 if event A_1 happens and \$0 if A_1 does not happen. A buyer of such a ticket pays the seller the amount p. If the event A occurs, the seller pays the buyer \$1. If the event A does not occur, the seller owes the buyer nothing. (The currency is not important. If you are used to some other currency, change the ticket to the currency you are familiar with.) There is an assumption here that the price at which you offer to buy such a ticket is the same as the price at which you are willing to sell such a ticket. You can count on me to pay if I owe you money after we see tomorrow's weather, and I can count on you similarly. The intuition behind this is that if you are willing to buy or sell a ticket on A_1 for \$0.70, you consider A_1 more likely than if you were willing to buy or sell it for only \$0.10.

Let us suppose that in general your price for a \$1 ticket on A_1 is $Pr\{A_1\}$ (pronounced 'price of A_1'), and in particular you name 30 cents. This means that I can sell you such a ticket for \$0.30 (or buy such a ticket from you for \$0.30). If I sell the ticket to you and it rains tomorrow and the temperature is above 68 degrees Fahrenheit, I would have to pay you \$1. If it does not rain or if the temperature does not rise to be above 68 degrees Fahrenheit, I would not pay you anything. Thus in the first case, you come out \$0.70 ahead, while in the second case I am ahead by \$0.30. Similarly you name prices for A_2, A_3 and A_4, respectively $Pr\{A_2\}, Pr\{A_3\}$ and $Pr\{A_4\}$.

It would be foolish for you to specify prices for tickets for all four events that have the property that I can accept some of your offers and be assured of making money from you, whatever the weather might be tomorrow (*i.e.*, making you a sure loser). So we now study what properties your prices must have so that you are assured of not being a sure loser. But before we do that, I must remind you that avoiding being a sure loser does not make you a winner, or even likely to be a winner. So avoiding sure loss is a weak requirement on what it takes to behave reasonably in the face of uncertainty.

To take the simplest requirement first, suppose you make the mistake of offering a negative price for an event, for example $Pr\{A_1\} = -\$0.05$. This would mean that you offer to sell me ticket A_1 for the price of -\$0.05, (*i.e.*, you will give me the ticket and 5 cents). If event A_1 happens, that is, if it rains and the high temperature is more than 68 degrees Fahrenheit, you owe me \$1, so your total loss is \$1.05. On the other hand, if event A_1 does not happen, you still lose \$0.05. Hence in this case, no matter what happens, you are a sure loser. To avoid this kind of error, your prices cannot be negative, that is, for every event A, you must specify prices satisfying

$$Pr\{A\} \geq 0. \tag{1.1}$$

Now consider the sure event S. In the example we are discussing, S is the same as the event {either A_1 or A_2 or A_3 or A_4}, which is a formal mathematical way of saying either it will rain tomorrow or it will not, and either the high temperature will be above 68 degrees

	Ticket			Net
Outcome	A_1	A_2	$A_1 \cup A_2$	
A_1 but not A_2	1	0	-1	0
A_2 but not A_1	0	1	-1	0
neither A_1 nor A_2	0	0	0	0

Table 1.1: Your gain from each possible outcome, after buying tickets on A_1 and A_2 and selling a ticket on $A_1 \cup A_2$.

Fahrenheit or not. What price should you give to the sure event S? If you give a price below \$1, say \$0.75, I can buy that ticket from you for \$0.75. Since the sure event is sure to happen, tomorrow you will owe me \$1, and you will have lost \$0.25, whatever the weather will be. So you are sure to lose if you offer any price below \$1. Similarly, if you offer a price above \$1 for the sure event S, say \$1.25, I can sell you the ticket for \$1.25. Tomorrow, I will certainly owe you \$1, but I come out ahead by \$0.25 whatever happens. So you can see that the only way to avoid being a sure loser is to have a price of exactly \$1 for S. This is the second requirement to avoid a sure loss, namely,

$$Pr\{S\} = 1. \tag{1.2}$$

Next, let's consider the relationship of the price you would give to each of two disjoint sets A and B to the price you would give to the event that at least one of them happens, which is called the **union** of the events A and B, and is written $A \cup B$. To be specific, let A be the event A_1 above, and B be the event A_2 above. These events are disjoint, that is, they cannot both occur, because it is impossible that the high temperature for the day is both above and below 68 degrees Fahrenheit. The union of A and B in this case is the event that it rains tomorrow.

Suppose, to be specific, that your prices are \$0.20 for A_1, \$0.25 for A_2 and \$0.40 for the union of A_1 and A_2. Then I can sell you a ticket on A_1 for \$0.20, and a ticket on A_2 for \$0.25, and buy from you a ticket on the union for \$0.40. Let's see what happens. Suppose first that it does not rain. Then none of the tickets have to be settled by payment. But you gave me \$0.20 + \$0.25 = \$0.45 for the two tickets you bought, and I gave you \$0.40 for the ticket I bought, so I come out \$0.05 ahead. Now suppose that it does rain. Then one of A_1 and A_2 occurs (but only one. Remember that they are disjoint). So I have to pay you \$1. But the union also occurred, so you have to pay me \$1 as well. In addition I still have the \$0.05 that I gained from the sale and purchase of the tickets to begin with. So in every case, I come out ahead by \$0.05, and you are a sure loser. The problem seems to be that you named too low a price for the ticket on the union. Indeed, any price less than \$0.45 leads to sure loss, as the following argument shows.

To see the general case, suppose $Pr\{A_1\} + Pr\{A_2\} > Pr\{A_1 \cup A_2\}$. Suppose I sell you tickets on A_1 and A_2, and buy from you a ticket on $A_1 \cup A_2$. These purchases and sales cost you $Pr\{A_1\} + Pr\{A_2\} - Pr\{A_1 \cup A_2\} > 0$. There are then only three possible outcomes (remembering that A_1 and A_2 are disjoint, so they cannot both occur). These are listed in Table 1.1.

Therefore the settlement of the tickets leads to a net of zero in each case. Thus, whatever outcome occurs, you lost $Pr\{A_1\} + Pr\{A_2\} - Pr\{A_1 \cup A_2\} > 0$ from buying and selling tickets, and earned nothing from settling tickets after learning the outcome. Hence, all told, you lost $Pr\{A_1\}+Pr\{A_2\}-Pr\{A_1 \cup A_2\}$. In the example above, $Pr\{A_1\} = \$0.20$, $Pr\{A_2\} = \$0.25$ and $Pr\{A_1 \cup A_2\} = \$0.40$, so your sure loss is $Pr\{A_1\} + Pr\{A_2\} - Pr\{A_1 \cup A_2\} = \$0.20 + \$0.25 - \$0.40 = \$0.05$.

So suppose you decide to raise your price for the ticket on the union, say to \$0.60. Now

Outcome	Ticket A_1	A_2	$A_1 \cup A_2$	Net
A_1 but not A_2	-1	0	1	0
A_2 but not A_1	0	-1	1	0
neither A_1 nor A_2	0	0	0	0

Table 1.2: Your gain from each possible outcome, after buying a ticket on $A_1 \cup A_2$ and selling tickets on A_1 and A_2.

suppose I decide to sell you the ticket on the union at your new price, and to buy from you tickets on A_1 and A_2 at the prices you offer, \$0.20 and \$0.25. Now if it does not rain, again no tickets pay off, but you gave me \$0.60 and I spent \$0.20 + \$0.25 = \$0.45, so I am \$0.15 ahead. And if it does rain, again one and only one of A_1 and A_2 pays off, but so does the union, so again we exchange \$1 to settle the tickets, and I am ahead by \$0.15. Once again, you are a sure loser. Here the problem is that you increased the price of the union by too much. The same argument shows that any price greater than \$0.45 leads to sure loss, as the following argument shows.

Now we consider the general case in which $Pr\{A_1 \cup A_2\} > Pr\{A_1\} + Pr\{A_2\}$. Now I do the opposite of what I did before: I buy from you tickets on A_1 and A_2, and sell you a ticket on $A_1 \cup A_2$. From these transactions, you are down $Pr\{A_1 \cup A_2\} - Pr\{A_1\} - Pr\{A_2\} > 0$. Again, one of the same three events must occur, with the consequences shown in Table 1.2. Again, settling the tickets yields no gain or loss for either of us, so your sure loss is $Pr\{A_1 \cup A_2\} - Pr\{A_1\} - Pr\{A_2\} > 0$. In the example, $Pr\{A_1\} = \$0.20, Pr\{A_2\} = \$0.25$ and $Pr\{A_1 \cup A_2\} = \$0.60$. Then your sure loss is $Pr\{A_1 \cup A_2\} - Pr\{A_1\} - Pr\{A_2\} = \$0.60 - \$0.20 - \$0.25 = \$0.15$. The entries in Table 1.2 are the negative of those in Table 1.1, because my purchases and sales if $Pr\{A_1 \cup A_2\} > Pr\{A_1\} + Pr\{A_2\}$ are the opposite of my purchases and sales if $Pr\{A_1\} + Pr\{A_2\} > Pr\{A_1 \cup A_2\}$. Hence if your price for the ticket on the union of the two events is too low or too high, you can be made a sure loser. I hope I have persuaded you that the only way to avoid being a sure loser is for your prices to satisfy

$$Pr\{A \cup B\} = Pr\{A\} + Pr\{B\}, \tag{1.3}$$

when A and B are disjoint.

So far, what I have shown is that unless your prices satisfy (1.1), (1.2) and (1.3), you can be made a sure loser. You will likely be relieved to know that those are the only tricks that can be played on you, that is, that if your prices satisfy equations (1.1), (1.2) and (1.3), you cannot be made a sure loser. To show that will require some more work, which comes later in this chapter. Prices satisfying these equations are said to be **coherent**. The derivations of equations (1.1), (1.2) and (1.3) are constructive, in the sense that I reveal exactly which of your offers I accept to make you a sure loser. Also the beliefs of the opponent are irrelevant to making you a sure loser.

Equations (1.1), (1.2) and (1.3) are the equations that define $Pr\{\cdot\}$ to be a probability (with the possible strengthening of Equation (1.3) to be taken up in Chapter 3). To emphasize that, we will now assume that you have decided not to be a sure loser, and hence to have your prices satisfy equations (1.1), (1.2) and (1.3). I will write $P\{\cdot\}$ instead of $Pr\{\cdot\}$, and think of $P\{A\}$ as your probability of event A.

Although the approach here is called subjective, there are both subjective and objective aspects of it. It is an objective fact, that is, a theorem, that you cannot be made a sure loser if and only if your prices satisfy equations (1.1), (1.2) and (1.3). However, the prices that you assign to tickets on any given set of events are personal, or subjective, in that the theorems do not specify those values. Different people can have different probabilities without violating coherence.

To see why this is natural, consider the following example: Imagine I have a coin that we both regard as fair, that is, it has probability 1/2 of coming up heads. I flip it, but I don't look at it, nor do I show it to you. Reasonably, our probabilities are still 1/2 of a head. Now I look at it, and observe a head, but I don't show it to you. My probability is now 1. Perhaps yours is still 1/2. But perhaps you saw that I raised my left eyebrow when I looked at the coin, and you think I would be more likely to do so if the coin came up heads than tails, and so your probability is now 60%. I now show you the coin, and your probability now rises to 1. The point of this thought-experiment is that probability is a function not only of the coin, but also of the information available to the person whose probability it is. Thus subjectivity occurs, even in the single flip of a fair coin, because each person can have different information and beliefs.

1.1.1 *Interpretation*

What does it mean to give a price $Pr^M\{B\}$, this morning on an event B, and this afternoon give a different price, $Pr^A\{B\}$ for it? Let us suppose that no new information has become available and that inflation of the currency is not an issue. Perhaps you thought about it harder, perhaps you just changed your mind. If this morning you could anticipate whether your price will increase or decrease, then you have opened yourself to a kind of dynamic sure loss. If $Pr^M\{B\} > Pr^A\{B\}$, then you would be willing to buy a ticket on B this morning for $Pr^M\{B\}$, and anticipate selling it back this afternoon for $Pr^A\{B\}$, leading to loss $Pr^M\{B\} - Pr^A\{B\} > 0$. Conversely, if $Pr^M\{B\} < Pr^A\{B\}$, you would be willing to sell a ticket on B in the morning for $Pr^M\{B\}$ and buy it back this afternoon for $Pr^A\{B\}$, leading to loss $Pr^A\{B\} - Pr^M\{B\} > 0$. Thus to avoid dynamic sure loss, your statement that your price in the morning is $Pr^M\{B\}$ is a statement that (absent new information, a complication dealt with in Chapter 2), you anticipate that your probability this afternoon will also be $Pr^M\{B\}$.

A different issue arises in the statement, after an event, of probabilities a person would have given, if asked, before the event. In retrospect, it is easy to exaggerate your probability of what actually occurred. This bias, called hindsight bias (see Fischhoff (1982)), makes whatever happens more likely in retrospect than it was in prospect.

1.1.2 *Notes and other views*

> "We do see things not as they are, but as we are."
>
> —Patrick (1890) *

> "It is generally accepted that...an application of the axioms of probability is inappropriate to questions of truth and belief."
>
> —(Grimmett and Stirzaker, 2001, p. 18)

I think of probability as a language to express uncertainty, and the laws of probability ((1.1), (1.2) and (1.3)) as the grammar of that language. In ordinary English, if you write a sentence fragment without a verb, I am not sure what you mean. Similarly, if your prices are such that you can be made a sure loser, you have contradicted yourself in a sense, and I do not know which of your bets you really mean, and which you would change when confronted with the consequences of your folly. Just as following the rules of English grammar do not restrict the content of your sentences, so too the laws of probability do not restrict the beliefs

*See Crane and Kadane (2008, 2013) for justification of this citation.

you express using them. For additional material on subjective probability, see DeFinetti (1974), Kyburg and Smokler (1964), Press and Tanur (2001), Savage (1954), and Wright and Ayton (1994).

Coherence is a minimal set of requirements on probabilistic opinions. The most extraordinary nonsense can be expressed coherently, such as that the moon is made of green cheese, or that the world will end tomorrow (or ended yesterday). All that coherence does is to ensure a certain kind of consistency among opinions. Thus an author using probabilities to express uncertainty must accept the burden of explaining to potential readers the considerations and reasons leading to the particular choices made. The extent to which the author's conclusions are heeded is likely to depend on the persuasiveness of these arguments, and on the robustness of the conclusions to departures from the assumptions made.

The philosopher Nelson Goodman (1965) has introduced two new colors, "grue" and "bleen." An object is grue if it is green and the date is before Jan. 1, 2100. A grue object is blue after that date. A bleen object simply reverses the colors. Thus empirically all our current data would equally identify objects as grue and green on the one hand, and as bleen and blue on the other. It is our beliefs about the world, and not our data, that lead us to the conclusion that even after Jan. 1, 2100 leaves will be green and the sky blue, not conversely. This thought experiment illustrates just how firmly embedded are our preconceived notions, and how complex, and fraught with possibilities of differing interpretations are our thought processes.

There is a substantial body of psychological research dedicated to finding systematic ways in which the prices that people actually offer for tickets or the equivalent fail to be coherent. See Kahneman et al. (1982), von Winterfeld and Edwards (1986) and Kahneman (2011). Since the techniques of this section show how to make them sure losers, if you can find such people, please share a suitable portion of your gains with an appropriate local charity.

There is a special issue about whether personal probabilities can be zero or one. The implication is that you would bet your entire fortune present and future against a penny on the outcome, which is surely extreme. In the example in section 1.1.1, I propose that when I see that the coin came up heads, my probability is one that it is a head. Could I have misperceived? For the sake of the argument I am willing to set that possibility aside, but I must concede that sometimes I do misperceive, so I can't *really* mean probability one.

The subjective view of probability taken in this section is not the only one possible. There is another view, which purports to be "objective." Generally, proponents of this view say that the probability of an event is the limiting relative frequency with which it appears in an infinite sequence of independent trials. See Feller (1957, p. 5). There are, however, several difficulties with this perspective. I postpone discussion of them until section 2.13 as part of the discussion of the weak law of large numbers. It is very much to be wished that we could find a basis for a valid claim of objectivity, but so far, each such claim has failed. Subjectivity at least acknowledges that people often disagree and does not allow one to claim that his view has a higher claim on truth than another's, without being persuasive as to why.

A second line of argument seeks help from information theory (and in particular entropy) to define and use ideas of ignorance, non-informativeness, reference, etc. Roughly, the idea is that these formulas express what people "ought" to think, and hence how they "ought" to bet. Proponents of this line include Jeffreys (1939), Jaynes (2003), Zellner (1971), Bernardo (1979) and Bayarri and Berger (2004). Unfortunately, this literature does not explain why a person ought to have such opinions and ought to bet accordingly. Some of the difficulties inherent in this approach are considered in Seidenfeld (1979, 1987).

The motivation for both of these attempts to find an "objective" basis for inference seems to be that science in general and statistics in specific would lose credibility and face by giving up a claim of objectivity. If I thought that such a claim could be sustained, I would be in favor of making it. However, anyone familiar with science and other empirical disciplines knows that disagreement is an essential part of scientific discourse, in particular about the matters of current scientific interest. Having a language that recognizes the legitimacy of differing points of view seems essential to helping to advance those discussions. (See Berger and Berry (1988).)

Another treatment of the subject states (1.1), (1.2) and (1.3) (or a countably additive version of (1.3) to be discussed later) as mathematical axioms. See for instance Billingsley (1995). As axioms, they are not to be challenged. However the relationship between those (or any other set of axioms) and the real world is left totally unexplored. This is unsatisfactory when the point is to explain how to deal with a real phenomenon, namely uncertainty. Thus I prefer the treatment given here, which has an explanation of why these are reasonable axioms to explore.

The approach used in this book is sometimes referred to as behavioristic. One limitation of this approach is that you may already have a "position" in the variable in question. For example, if you are an international expert on the ozone hole over Antarctica, you may be subject to one of two influences. If your elicited probability is to be published, you may be inclined to "view with alarm," as you have a personal and financial incentive to do so as your personal services and research efforts would be more valuable if public concern on this issue were increased. On the other hand, if the uses of the elicitation were only private, you might want to use the availability of "tickets" to purchase "insurance" against the possibility that ozone holes are found to be unimportant. For more on biases of these kinds, see Kadane and Winkler (1988). There are markets in which we all, perforce, have a position. What does it mean, for example, to hold a ticket that pays \$1 if a nuclear war occurs? (See Press (1985).) There are other limitations to the set of events to which one might want to apply this theory. I believe that the limitation to bets that can be settled is, while plausible, too stringent. For example, it makes sense to me to speak of opinions about the uses that were made of a particular spot in an archaeological site, despite the fact that a bet on the matter could never be settled. (See Kadane and Hastorf (1988).) Even with these limitations, however, I believe the approach explored here offers a better explanation of probability than its current alternatives.

DeFinetti (1974) is the proponent of the approach taken here. However, he is also (1981) one of its most important critics. The heart of his criticism is that you, in naming your prices, may try to guess what probabilities I may have, and game the system. Thus the act of eliciting your probabilities may change them (shades of Heisenberg's uncertainty principle!). DeFinetti suggests instead the use of proper scoring rules, and explores in DeFinetti (1974) the use of Brier (1950)'s squared-error scoring rule. This was not completely satisfactory either, as he did not address the question of whether different subjective probabilities would be the consequence of different proper scoring rules. Lindley (1982) uses scoring rules to justify the use of personal probability. Following suggestions in Savage (1971), recent work of Predd et al. (2009) and Schervish et al. (2009) relaxes the assumption that the proper scoring rule must be Brier's, and opens the possibility of basing subjective probability on proper scoring rules. However, scoring rules have their own difficulties, as they assume that the decision maker is motivated solely by the scoring rules. By contrast, the "avoid sure loss" approach used here assumes only that the decision-maker prefers \$1 to \$0.

Yet another approach to probability is through the assumptions of Cox (1946, 1961). For commentary, see Halperin (1999a,b) and Jaynes (2003). Cox's approach is not operational, that is, it does not lead to a specification of probabilities of particular events, unlike the approach suggested here.

There are authors who accept the idea of the price of lottery tickets as a way of learning how you feel about how likely various events are, but point out that you might feel uncomfortable having the same price for both buying and selling a ticket. This leads to what is now called the field of imprecise probabilities (for example Walley (1990)). This book concentrates on the simpler theory, supposing that your buying and selling prices are the same for all tickets. Another possibility is that you might refuse to offer bets. While that position prevents you from being a sure loser, it also prevents you from having the benefit of the theory. However, in Chapter 4 there is discussion of reluctance to bet on certain random variables.

An excellent general introduction to uncertainty is Lindley (2006).

1.1.3 *Summary*

Avoiding being a sure loser requires that your prices adhere to the following equations:

(1.1) $Pr\{A\} \geq 0$ for all events A

(1.2) $Pr\{S\} = 1$, where S is the sure event

(1.3) If A and B are disjoint events, then $Pr\{A \cup B\} = Pr\{A\} + Pr\{B\}$.

If your prices satisfy these equations, then they are coherent.

1.1.4 *Exercises*

1. Vocabulary. Explain in your own words the following:
   (a) event
   (b) sure event
   (c) disjoint events
   (d) exhaustive events
   (e) the union of two events
   (f) sure loser
   (g) coherent
   (h) probability
2. Consider the events A_1, A_2, A_3 and A_4 defined in the beginning of section 1.1.1 and as applied to your current geographic area for tomorrow. What prices would you give for the tickets? Explain your reasoning why you would give those prices. Are your prices coherent? Prove your answer. If your prices are not coherent, would you change them to satisfy the equations? Why or why not?
3. (a) Suppose that someone offers to buy or sell tickets on the events A_1 at price \$0.30, on A_2 at price \$0.20, and on the event of rain at price \$0.60. What purchases and sales would you make to ensure a gain for yourself? Show that a sure gain results from your choices. How much can you be sure to gain?
   (b) Answer the same questions if the price on the event of rain is changed from \$0.60 to \$0.40.
4. Think of something you are uncertain about. Define the events that matter to you about it. Are the events you define disjoint? Are they exhaustive? Give your prices for tickets on each of those events. Are your prices coherent? (Show that they are, or are not.) Revise your prices until you are satisfied with them, and explain why you chose to be a sure loser, or chose not to be.
5. Suppose that someone offers to buy or sell tickets at the following prices:

   If the home team wins the soccer (football, outside the U.S. and Canada) match: \$0.75

   If the away team wins: \$0.20

A tie: \$0.10

What purchases and sales would you make to ensure a sure gain for yourself? Show that a sure gain results from your choices. How much can you be sure to gain if you buy or sell no more than four tickets?

1.2 Consequences of the axioms of probability: disjoint events

We now explore some consequences of coherence. Each of these takes the form of showing that if you know certain probabilities (*i.e.*, the price of certain tickets) and do not want to be a sure loser, then you are committed to the price of certain other tickets.

To start, we define the complement of an event A, which we write $\bar{A}$ and pronounce "not A," to be the event that A does not happen. By construction, A and $\bar{A}$ are disjoint, that is; they can't both happen. Hence by equation (1.3),

$$P\{A \cup \bar{A}\} = P\{A\} + P\{\bar{A}\}. \tag{1.4}$$

Now again by construction, either A or $\bar{A}$ must happen: they are exhaustive. Another way of saying this is

$$A \cup \bar{A} = S. \tag{1.5}$$

Therefore, by equation (1.3),

$$P\{A \cup \bar{A}\} = P\{S\} = 1. \tag{1.6}$$

Now from equations (1.4) and (1.6), it follows that

$$P\{\bar{A}\} = 1 - P\{A\} \tag{1.7}$$

for every event A. Equation (1.7) should not come as a surprise. All it says is that whatever you are willing to buy or sell a ticket on A, to avoid sure loss you must be willing to buy or sell a ticket on $\bar{A}$ for \$1 minus what you would be willing to pay for A.

There is a special case of (1.7) that will be useful later. The complement of S is the empty event, which is written often with the Greek letter ϕ (pronounced "fee" in the U.S., and "fie" in the U.K.), although its origin is the Norwegian letter O (Weil, 1992, p. 114). It never occurs, because S always occurs. Using equation (1.3), it follows from (1.7) that

$$P\{\phi\} = 1 - P\{S\} = 1 - 1 = 0. \tag{1.8}$$

Thus you could buy or sell a ticket on ϕ for nothing (*i.e.*, give it away or accept it for free), and be sure of not being a sure loser.

Another consequence of equation (1.7) and equation (1.1) is that, for every event A,

$$P\{A\} = 1 - P\{\bar{A}\} \leq 1. \tag{1.9}$$

Now suppose there are three disjoint events, like the events A_1, A_2 and A_3 in section 1.1. For three sets to be disjoint means that if one occurs, none of the others can. Does the principle of avoiding sure loss commit you to a particular price for the union of those three events, if you have already declared a price for each one separately? Equation (1.3), which applies to the case of two disjoint sets, seems like the logical place to start in addressing this issue.

I can think of the union of the three disjoint events as the union of the union of disjoint events as follows:

$$A_1 \cup A_2 \cup A_3 = (A_1 \cup A_2) \cup A_3. \tag{1.10}$$

Equation (1.10) means that first we consider the union of A_1 with A_2, and then we union that event with A_3. Now because A_1 and A_2 are disjoint (they can't both happen), equation (1.3) applies and says that

$$P\{A_1 \cup A_2\} = P\{A_1\} + P\{A_2\}. \tag{1.11}$$

In order to apply equation (1.3) again, it is necessary to examine whether A_3 is disjoint from $(A_1 \cup A_2)$. But if event A_3 occurs, then neither A_1 nor A_2 can occur, and therefore $(A_1 \cup A_2)$ cannot occur. Thus A_3 is indeed disjoint from $(A_1 \cup A_2)$. Therefore, equation (1.3) can be invoked again, yielding

$$\begin{aligned} P\{A_1 \cup A_2 \cup A_3\} &= P\{(A_1 \cup A_2) \cup A_3\} = P\{A_1 \cup A_2\} + P\{A_3\} \\ &= P\{A_1\} + P\{A_2\} + P\{A_3\}. \end{aligned} \tag{1.12}$$

So now we know from equation (1.3) that the probability of the union of two disjoint events is the sum of their probabilities, and from equation (1.12) that the probability of the union of three disjoint events is the sum of their probabilities. This suggests that perhaps the union of any finite number of disjoint events should be the sum of their probabilities as well.

To see how this might work, let's review how we come to know equations (1.3) and (1.12). Equation (1.3) is an assumption, or axiom, derived from a desire not to be a sure loser. Equation (1.12), however, was shown above to be a consequence of equation (1.3). Thus, we have assumed that the probability of the union of $n = 2$ disjoint events is the sum of their probabilities, and shown that if the statement is true for $n = 2$ disjoint events, it is also true for $n = 3$ disjoint events. Suppose we could show in general that if the statement is true for n disjoint events, then it will be true for $n+1$ disjoint events as well. This would be very convenient. If we wanted the result for, say 21 disjoint events, we have it for $n = 2$, we apply the result to conclude that it is true for $n = 3$ disjoint events, then for $n = 4$, etc. until we get to 21. An argument of this kind is called mathematical induction and is a nice way of proving results for all finite integers.

To apply mathematical induction to this problem, there are just two steps. The first, the basis step, is to establish the result for some small n, here $n = 2$, shown by equation (1.3). Second, in the inductive step, suppose we know that the probability of the union of n disjoint events is the sum of their probabilities, and we want to prove it for $n+1$ events. Let $A_1, A_2, \ldots, A_{n+1}$ be the $n+1$ disjoint events in question. Then, generalizing equation (1.10), I can write the union of all $n+1$ events as follows:

$$A_1 \cup A_2 \cup \ldots A_{n+1} = (A_1 \cup A_2 \cup \ldots A_n) \cup A_{n+1}. \tag{1.13}$$

Now the union in parentheses is the union of n disjoint events, so by the assumption I am allowed to make in mathematical induction, the probability of their union is the sum of their probabilities, which generalizes equation (1.11). Furthermore, the event A_{n+1} is disjoint from that union, because if A_{n+1} occurs, then none of the other A's can occur. This puts all the pieces in place to generalize equation (1.11):

$$\begin{aligned} P\{A_1 \cup A_2 \cup \ldots \cup A_{n+1}\} &= P\{(A_1 \cup A_2 \cup \ldots \cup A_n) \cup A_{n+1}\} \\ &= P\{A_1 \cup A_2 \cup \ldots \cup A_n\} + P\{A_{n+1}\} \\ &= P\{A_1\} + P\{A_2\} + \ldots + P\{A_{n+1}\}, \end{aligned} \tag{1.14}$$

which is the statement for $n+1$. [Check to make sure you can justify each of the equality signs in equation (1.14).] Hence we know that for all finite numbers of disjoint sets, the probability of the union is the sum of the probabilities.

1.2.1 Summary

If A is an event, and $\bar{A}$ its complement, $P\{\bar{A}\} = 1 - P\{A\}$. In particular, $P\{\phi\} = 0$, where ϕ is the empty event. Also $P\{A\} \leq 1$ for all events A.

If $A_1, \ldots, A_n$ are disjoint events, then the probability of their union is the sum of their probabilities.

1.2.2 A supplement on induction

Suppose $S(n)$ is some statement that depends on an integer n. If $S(n)$ can be proved for some (usually small) integer n_0, and if it can be shown that $S(n)$ implies $S(n+1)$ for all n greater than or equal to n_0, then $S(n)$ is proved for all integers greater than n_0. You can think of induction working the way an algorithm would: you start it at $S(n_0)$, which is true. Then $S(n_0)$ implies $S(n_0 + 1)$, which in turn implies $S(n_0 + 2)$, etc.

Take as an example the sum of the first n integers. There are at least three different ways to think about this sum. The first is algebraic: Let T be the sum. Then T can be written as $T = 1 + 2 + 3 + \ldots + n$. However it can also be written as $T = n + \ldots + 3 + 2 + 1$. Add up these two expressions for T by adding the first terms, the second terms, etc. Each pair adds up to $n + 1$, and there are n such pairs. Hence $2T = n(n + 1)$, or $T = n(n + 1)/2$.

A second way to think about T is to imagine a discrete $(n + 1)$ by $(n + 1)$ square like a chess board. Consider the number of points in the square below the diagonal. There are none in the first row, one in the second row, two in the third, .., up to n in the last row, so the number below the diagonal is T. There are equally many above the diagonal, and the diagonal itself has $(n + 1)$ elements. Since the square has a total of $(n + 1)^2$ elements, we have $(n + 1)^2 = 2T + (n + 1)$, from which we conclude that $T = n(n + 1)/2$.

The third way to think about T is by induction. The statement to be proved is $S(n)$; the sum $T(n)$ of the first n integers is $n(n+1)/2$. When $n = 1$, we have $T(1) = 1 \times 2/2 = 1$, so the statement is true for $n = 1$, and we may take n_0 to be 1. Now suppose that $S(n)$ is true, and let's examine $S(n + 1)$. We have $T(n + 1) = 1 + 2 + 3 + \ldots + n + (n + 1) = n(n+1)/2 + (n+1) = (n+1)(n/2+1) = (n+1)(n+2)/2$, which is $S(n+1)$. Therefore we have proved the second step of the induction, and have shown that the sum of the first n integers is $n(n + 1)/2$ for all integers n bigger than or equal to 1. Mathematical induction requires that you already think you know the solution. It is not so useful for finding the right formula in the first place. However often some experimentation and a good guess can help you find a formula which you can then try to prove by induction.

Given how immediate and appealing the first two proofs are, it may seem heavy-handed to apply mathematical induction to this problem. However, I think you will find that the following problems are better solved by induction than by trying to find analogues to the first two proofs:

1. Show that the sum of the first n squares (*i.e.*, $1+4+9+\ldots+n^2$) is $n(n+1)(2n+1)/6$.
2. Show that the sum of the first n cubes (*i.e.*, $1 + 8 + 27 + \ldots + n^3$) is $[n(n + 1)/2]^2$.

You can find an excellent further explanation of mathematical induction in Courant and Robbins (1958).

I anticipate that most readers of this book will be familiar with at least one proof of the fact that $T = n(n + 1)/2$. There are two reasons for discussing it here. The first is to give a simple example of induction. The second is to show that the same mathematical fact may be approached from different directions. All three are valid proofs, and each seems intuitive to different people, depending on their mental proclivities.

1.2.3 A supplement on indexed mathematical expressions

It will become awkward, and at times ambiguous, to continue to use "..." to indicate the continuation of a mathematical process.

For example, consider the expression

$$T = 1 + 2 + 3 + \ldots + n.$$

This can also be written

$$T = \sum_{i=1}^{n} i.$$

Here Σ (capital sigma, a Greek letter) is the symbol for sum. "i" is an index. "$i = 1$" indicates that the index i is to start at 1, and proceed by integers to n. Sometimes, to avoid ambiguity, the symbol above the sigma is written "$i = n$". The i after the sigma indicates what is to be added.

It is important to understand that T does not depend on i, although it does depend on n. Thus,

$$T = \sum_{i=1}^{n} i = \sum_{j=1}^{n} j.$$

Indexed notation can be used flexibly. For example, the sum of the first n squares is

$$1 + 4 + 9 + \ldots + n^2 = \sum_{i=1}^{n} i^2.$$

Other functions can also be used in place of addition. For example, $\cup$ is the symbol for the union of sets and $\prod$ is used for the product of numbers. Thus the result proved by induction using equations (1.13) and (1.14) above can be written as follows:

If $A_1, A_2, \ldots, A_n$ are disjoint sets, then

$$P\{\cup_{i=1}^{n} A_i\} = P\{A_1 \cup A_2 \cup \ldots \cup A_n\} =$$

$$P\{A_1\} + P\{A_2\} + \ldots + P\{A_n\} = \sum_{i=1}^{n} P\{A_i\}.$$

Also there is special notation for the product of the first n integers:

$$\prod_{i=1}^{n} i = (1)(2)(3)\ldots(n) = n!$$

(pronounced n-factorial). Factorials are used extensively in this book.

1.2.4 Intersections of events

If A and B are two events, then the intersection of A and B, written AB, is the event that both A and B happen. For example, if A and B are disjoint (remember that means that they can't both happen), then $AB = \phi$. If you flip two coins, and A is the event that the first coin comes up heads, and B is the event that the second coin comes up heads, then A and B are not disjoint, and AB is the event that both coins come up heads. You can think of "intersection" as corresponding to "and" in the same way that "union" corresponds to "or."

The symbol $\prod$ is used for the intersection of several events. Thus $\prod_{i=1}^{n} A_i$ means the event that $A_1, A_2, \ldots, A_n$ all occur. Thus $\prod$ is used both for events and for arithmetic expressions. This double use should not cause you trouble – just look to see whether what comes after $\prod$ is events or numbers.

Note that $\sum_{i=1}^{n} A_i$, where $A_1, \ldots, A_n$ are sets, is not defined.

1.2.5 *Summary*

The probability of the union of any finite number of disjoint events is the sum of their probabilities.

1.2.6 *Exercises*

1. Vocabulary. Explain in your own words:
   (a) complement of an event
   (b) empty event
   (c) several disjoint events
   (d) the union of several events
   (e) mathematical induction
2. Consider two flips of a coin, and suppose that the following outcomes are equally likely to you: H_1H_2, H_1T_2, T_1H_2 and T_1T_2, where H_i indicates Heads on flip i and similarly for T_i.
   (a) Compute your probability of at least one head.
   (b) Compute your probability of a match, (*i.e.*, both heads or both tails).
   (c) Compute your probability of the simultaneous occurrence of at least one head and a match.
3. Consider a single roll of a die, and suppose that you believe that each of the six sides has the same probability of coming up.
   (a) Find your probability that the roll results in a 3 or higher.
   (b) Find your probability that the roll results in an odd number.
   (c) Find your probability that the roll results in a prime number (*i.e.*, one that can't be expressed as the product of two integers larger than one).
4. (a) Find the sum of the first k even positive integers, as a direct consequence of the formula for the sum of the first k positive integers.
   (b) From the sum of the first $2k$ integers, find by subtraction the sum of the first k odd numbers.
   (c) Prove the result of (b) directly by induction.

1.3 Consequences of the axioms, continued: Events not necessarily disjoint

Think of two events A and B that are not necessarily disjoint. The union of A and B, which is the event that either A happens or B happens or both happen, can be thought of as the union of three events: A happens and B does not, B happens and A does not, and both A and B happen. In symbols, this is

$$A \cup B = A\bar{B} \cup B\bar{A} \cup AB. \tag{1.15}$$

Furthermore, the events $A\bar{B}$, $B\bar{A}$ and AB are disjoint. Therefore applying equation (1.12), we have

$$P\{A \cup B\} = P\{A\bar{B}\} + P\{B\bar{A}\} + P\{AB\}. \tag{1.16}$$

Now it is also true that $A = A\bar{B} \cup AB$. Also the events $A\bar{B}$ and AB are disjoint. Therefore using equation (1.3),

$$P\{A\} = P\{A\bar{B}\} + P\{AB\}. \tag{1.17}$$

Similarly, $B = B\bar{A} \cup AB$, and these sets are disjoint as well. Hence

$$P\{B\} = P\{B\bar{A}\} + P\{AB\}. \tag{1.18}$$

Substituting (1.17) and (1.18) into (1.16) yields the result that

$$\begin{aligned} P\{A \cup B\} &= (P\{A\} - P\{AB\}) + (P\{B\} - P\{AB\}) + P\{AB\} \\ &= P\{A\} + P\{B\} - P\{AB\}. \end{aligned} \tag{1.19}$$

In the special case in which A and B are disjoint, $AB = \phi$, $P\{AB\} = 0$, and (1.19) reduces to (1.3). However, it is quite remarkable that a formula such as (1.3) giving the probability of the union of disjoint sets, implies formula (1.19) specifying the probability of the union of sets without assuming that they are disjoint.

Often it is useful to display geometrically the sets A and B and their subsets. This is done using a picture known as a Venn Diagram, shown below in Figure 1.1. Here the event A is represented by the circle on the left, event B by the circle on the right, the event AB by the shaded area, $A\bar{B}$ by the 3/4 moon-shaped area to the left of the shaded area AB, and $B\bar{A}$ similarly by the 3/4 moon-shaped area to the right of AB.

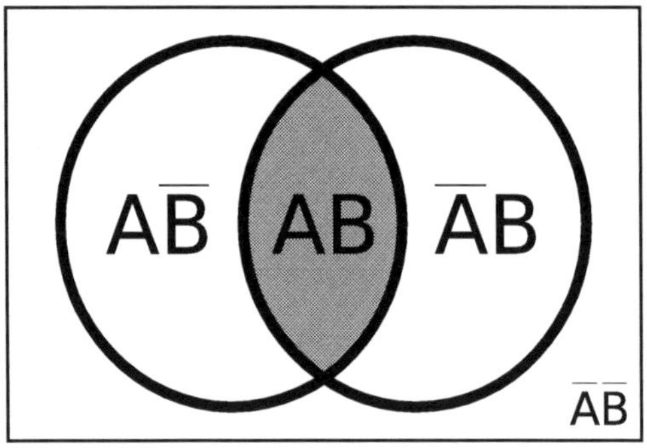

Figure 1.1: A Venn Diagram for two sets A and B.

There is one other implication of the equations above worth noting in passing. Suppose event A cannot occur without event B also occurring. In this case, event B is said to contain event A. This is written $A \subseteq B$. If this is true, $AB = A$. Then equation (1.18) implies

$$P\{B\} = P\{B\bar{A}\} + P\{A\} \geq P\{A\}, \tag{1.20}$$

since $P\{B\bar{A}\} \geq 0$, using (1.1).

1.3.1 A supplement on proofs of set inclusion

Our purpose is to show how to prove facts about whether one set is included in another. Our target is the equality

$$\overline{A \cup B} = \bar{A}\bar{B}. \tag{1.21}$$

In words, this equation says that the elements of $\overline{A \cup B}$ are exactly the elements of $\bar{A}\bar{B}$.

An equality between two such sets is equivalent to two set inclusions:

$$\overline{A \cup B} \subseteq \bar{A}\bar{B} \tag{1.22}$$

and

$$\bar{A}\bar{B} \subseteq \overline{A \cup B}. \tag{1.23}$$

Equation (1.22) says that every element of $\overline{A \cup B}$ is an element of $\bar{A}\bar{B}$, while (1.23) says that every element of $\bar{A}\bar{B}$ is an element of $\overline{A \cup B}$.

To show (1.22), suppose that $x \in \overline{A \cup B}$. Then $x \notin A \cup B$. The notation "$\notin$" means "is not an element of." Then $x \notin A$ and $x \notin B$, so $x \in \bar{A}$ and $x \in \bar{B}$, so $x \in \bar{A}\bar{B}$. Therefore $\overline{A \cup B} \subseteq \bar{A}\bar{B}$, proving (1.22).

To show (1.23), suppose that $x \in \bar{A}\bar{B}$. Then $x \in \bar{A}$ and $x \in \bar{B}$. Therefore $x \notin A$ and $x \notin B$. Therefore $x \notin A \cup B$. And so $x \in \overline{A \cup B}$. Therefore $\bar{A}\bar{B} \subseteq \overline{A \cup B}$, proving (1.23).

Proving (1.22) and (1.23) proves (1.21), since

$$\overline{A \cup B} = \bar{A}\bar{B}.$$

The equality (1.21) is known as DeMorgan's Theorem.

1.3.2 Boole's Inequality

The proof of Boole's Inequality uses (1.19). This inequality is used later in this book. The events in Theorem 1.3.1 need not be disjoint.

Theorem 1.3.1. *(Boole's Inequality) Let $A_1, A_2, \ldots, A_n$ be events. Then*

$$P\{\prod_{i=1}^{n} A_i\} \geq 1 - \sum_{i=1}^{n} P\{\bar{A}_i\}.$$

Proof. By induction on n. When $n = 1$, (1.7) gives the result.

For $n = 2$,

$$\begin{aligned}
P\{A_1A_2\} &= P\{A_1\} + P\{A_2\} - P\{A_1 \cup A_2\} && \text{(uses (1.19))} \\
&= 1 - (1 - P\{A_1\}) - (1 - P\{A_2\}) + (1 - P\{A_1 \cup A_2\}) && \text{(just algebra)} \\
&= 1 - P\{\overline{A}_1\} - P\{\overline{A}_2\} + P\{\overline{A_1 \cup A_2}\} && \text{(uses (1.7))} \\
&\geq 1 - P\{\overline{A}_1\} - P\{\overline{A}_2\} && \text{(uses (1.1))}
\end{aligned}$$

which is the result for $n = 2$.

Now suppose the result is true for $n - 1$, where $n \geq 3$. Then,

$$P\{\prod_{i=1}^{n} A_i\} = P\{A_1 \prod_{i=2}^{n} A_i\}$$

$$\geq 1 - P\{\bar{A}_1\} - P\{\overline{\prod_{i=2}^{n} A_i}\}$$

(uses result at n=2)

$$= 1 - P\{\overline{A}_1\} - 1 + P\{\prod_{i=2}^{n} A_i\}$$

(uses (1.7))

$$\geq 1 - P\{\overline{A}_1\} - 1 + 1 - \sum_{i=2}^{n} P\{\overline{A}_i\}$$

(uses inductive hypothesis at n-1)

$$= 1 - \sum_{i=1}^{n} P\{\overline{A}_i\},$$

which is the statement at n. This completes the proof. □

1.3.3 *Summary*

The probability of the union of two sets is the sum of their probabilities minus the probability of their intersection. Boole's Inequality is also proved.

1.3.4 *Exercises*

1. Vocabulary. State in your words the meaning of:
   (a) the intersection of two events
   (b) Venn Diagram
   (c) subset
   (d) element
   (e) DeMorgan's Theorem
   (f) Boole's Inequality
2. Show that if A and B are any events, $AB = BA$.
3. Show that, if A, B and C are any events, $A(BC) = (AB)C$.
4. Show that, if A, B and C are any events, that $A(B \cup C) = AB \cup AC$.
5. Reconsider the situation of problem 2 in section 1.2.6: Two flips of a coin, and the following outcomes are equally likely to you: H_1H_2, H_1T_2, T_1H_2 and T_1T_2 where H_i indicates heads on flip i and similarly for T_i.
   (a) Find the probability that one or both of the following occur: at least one head and a match. Interpret the result.
   (b) Find the probability that exactly one of at least one head and a match occurs.
6. Consider again the weather example of section 1.1, in which there are four events:
   A_1: Rain and High above 68 degrees F tomorrow.
   A_2: Rain and High at or below 68 degrees F tomorrow.
   A_3: No Rain and High above 68 degrees F tomorrow.
   A_4: No Rain and High at or below 68 degrees F tomorrow.

Suppose that your probabilities for these events are as follows:

$$P(A_1) = 0.1\ ,\ P(A_2) = 0.2\ ,\ P(A_3) = 0.3\ ,\ P(A_4) = 0.4.$$

(a) Check that these probability assignments are coherent.

(b) Check Boole's inequality for these events.

1.4 Random variables, also known as uncertain quantities

The real numbers, that is, numbers like 3, -3.1, $\sqrt{2}$, π, etc., are remarkably useful. A random variable scores the outcome of an event in terms of real numbers. For example, the outcome of a single flip of a coin is an event and can be recorded as H for heads, and T for tails. Since H and T are not real numbers, this scoring is not a random variable. However, instead one could write 1 for a tail and 0 for a head, or 1 for a head and -1 for a tail. Both of these are random variables, since 1, 0, and -1 are real numbers.

One advantage of scoring using random variables is that all of the usual mathematics of real numbers applies to them. For example, consider n flips of a coin, and let $X_i = 1$ if the ith flip is a tail, and let $X_i = 0$ if the ith flip is a head, for $i = 1, \ldots, n$. Then $\sum_{i=1}^{n} X_i$ is a new random variable, taking values $0, 1, \ldots, n$, and is the number of tails that occur in the n flips.

A random variable can be a convenient quantity to express your opinions about in probabilistic terms. For example, if $X = 1$ if a coin flip comes up tails and $X = -1$ if a coin flip comes up heads, then $P\{X = 1\}$ is, in the framework of this book, the worth to you of a ticket that pays \$1 if $X = 1$ occurs (if the coin flip comes up tails). To be coherent, then $1 - P\{X = 1\}$ is the worth to you of a ticket that pays \$1 if $X = -1$ occurs (if the coin flip comes up heads). The probabilities you give to a random variable comprise your **distribution** of the random variable.

Random variables that take only the values zero and one have a special name, indicators. Thus the indicator for an event A, which is written I_A, is a random variable that takes the value 1 if A occurs, and 0 otherwise. Consider the roll of a fair die, having faces 1, 2, 3, 4, 5 and 6. Suppose A is the set of even outcomes, that is, $A = \{2, 4, 6\}$. Then $I_A\{3\} = 0$, but $I_A\{4\} = 1$. Indicators turn out to be very useful. Several examples of solving problems using indicators are given later.

1.4.1 Summary

A random variable scores the outcome of a random event in terms of real numbers. An indicator is a random variable taking only the values 0 and 1.

1.4.2 Exercises

1. Vocabulary. Explain in your own words:
   (a) random variable
   (b) indicator
   (c) distribution of a random variable
2. What is the indicator of
   (a) ϕ
   (b) S
3. Suppose A and B are events, with indicators respectively I_A and I_B. Find expressions in terms of I_A and I_B for

(a) I_{AB}
(b) $I_{A \cup B}$
(c) $I_{\overline{A \cup B}}$

4. Prove $\overline{\cup_{i=1}^{n} A_i} = \prod_{i=1}^{n} \overline{A}_i$ using the methods of set inclusion.
5. Prove the result of exercise 4 by induction on n.

1.5 Expectation for a random variable taking a finite number of values

Suppose that Z is a random variable that takes at most a finite number of values. This is a major constraint on the random variables considered, to be relaxed starting in Chapter 3. Under the assumption that Z takes only finitely many values, there is a sequence of real numbers z_i and an associated sequence of probabilities p_i so that

$$P\{Z = z_i\} = p_i, \; i = 1, \dots, n$$

and $\sum_{i=1}^{n} p_i = 1$. We want a number that will, in some sense, represent a summary of Z. While many such summaries are possible (and used), the one we choose to study first is a weighted average of the values of Z, where the weights are the probabilities. Thus we define the expectation of Z, written $E(Z)$, to be

$$E(Z) = \sum_{i=1}^{n} z_i p_i. \tag{1.24}$$

For those who are physically inclined, consider putting weight p_i at position z_i on a weightless beam. If so, $E(Z)$ is the position on the beam where it balances.

Thus suppose Z a random variable taking the value 2 with probability $\frac{1}{4}$ and 6 with probability $\frac{3}{4}$. Then $z_1 = 2, p_1 = \frac{1}{4}, z_2 = 6, p_2 = \frac{3}{4}$ and

$$\begin{aligned} E(Z) &= z_1 p_1 + z_2 p_2 \\ &= (2)(\frac{1}{4}) + 6(\frac{3}{4}) = \frac{2+18}{4} = \frac{20}{4} = 5. \end{aligned}$$

As a second example of expectation, consider a set A to which you assign probability p, so, to you, $P\{A\} = p$. The indicator of A, I_A, has the following expectation:

$$\begin{aligned} E(I_A) &= 1P\{I_A = 1\} + 0P\{I_A = 0\} \\ &= P\{A\} = p. \end{aligned} \tag{1.25}$$

This relationship between expectation and indicators comes up many times in what follows. If some outcome has probability zero, it has no effect on the expectation of Z.

We now explore some of the most important properties of expectation. The first is quite simple, relating to the expectation of a random variable multiplied by a constant and added to another constant. Again suppose that Z is a random variable taking values z_i with probability p_i (for $i = 1, \dots, n$), where $\sum_{i=1}^{n} p_i = 1$. Let k and b be any real numbers. Then $kZ + b$ is a new random variable taking values $kz_i + b$ with probability p_i. Then expectation of $kZ + b$ is

$$E(kZ + b) = \sum_{i=1}^{n} (kz_i + b)p_i = k \sum_{i=1}^{n} z_i p_i + b \sum_{i=1}^{n} p_i = kE(Z) + b. \tag{1.26}$$

Now let X and Y be two random variables, and we wish to study $E(X + Y)$. To establish notation, let $p_{i,j} = P\{(X = x_i) \cap (Y = y_j)\}$ be your probability that X takes the value $x_i (1 \leq i \leq n)$ and Y takes the value $y_j (1 \leq j \leq m)$. The event $((X = x_i) \cap (Y = y_j))$ can be written more briefly as $(X = x_i, Y = y_j)$. We now find the relationship between the numbers $p_{i,j}$ and the probability that X takes the value x_i. To do so, we use the properties of set inclusion as follows:

$$P\{X = x_i\} = P\{X = x_i, Y = y_1\} + P\{X = x_i, Y = y_2\} + \ldots + P\{X = x_i, Y = y_m\} \tag{1.27}$$

$$= p_{i,1} + p_{i,2} + \ldots + p_{i,m} = \sum_{j=1}^{m} p_{i,j} \text{ for } i = 1, \ldots, n. \tag{1.28}$$

It is convenient to have special notation for the latter sum, and we use

$$p_{i,+} = \sum_{j=1}^{m} p_{i,j}.$$

Therefore we may write

$$P\{X = x_i\} = p_{i,+} \text{ for } i = 1, \ldots, n.$$

Similarly, reversing the roles of X and Y, we have

$$P\{Y = y_j\} = \sum_{i=1}^{n} p_{i,j} = p_{+,j} \text{ for } j = 1, \ldots, m. \tag{1.29}$$

Then

$$\begin{aligned} E(X + Y) &= \sum_{i=1}^{n} \sum_{j=1}^{m} P\{X + Y = x_i + y_j\}(x_i + y_j) \\ &= \sum_{i=1}^{n} \sum_{j=1}^{m} p_{i,j}(x_i + y_j) \\ &= \sum_{i=1}^{n} \sum_{j=1}^{m} p_{i,j} x_i + \sum_{i=1}^{n} \sum_{j=1}^{m} p_{i,j} y_j = \sum_{i=1}^{n} p_{i,+} x_i + \sum_{j=1}^{m} p_{+,j} y_j \\ &= E(X) + E(Y). \end{aligned} \tag{1.30}$$

Formula (1.30) holds regardless of the relationship between X and Y. Of course, by induction

$$E(X_1 + \ldots + X_k) = E(X_1) + E(X_2) + \ldots + E(X_k). \tag{1.31}$$

As an example of the usefulness of indicators, I now derive a formula for the union of many events that need not be disjoint.

We already know that $I_{AB} = I_A I_B$, that $I_{\bar{A}} = 1 - I_A$ and that $\overline{A \cup B} = \bar{A}\bar{B}$.

Therefore we find

$$\begin{aligned} I_{A \cup B} &= 1 - I_{\overline{A \cup B}} = 1 - I_{\bar{A}\bar{B}} = 1 - I_{\bar{A}} I_{\bar{B}} = 1 - (1 - I_A)(1 - I_B) \\ &= I_A + I_B - I_{AB}. \end{aligned}$$

This expression gives a relationship between the random variables $I_{A \cup B}$, I_A, I_B and I_{AB}. Since the random variables on both sides are equal, their expectations are equal. Then using the additivity of expectation proved above, we can write

$$\begin{aligned} P\{A \cup B\} &= E(I_A + I_B - I_{AB}) = E(I_A) + E(I_B) - E(I_{AB}) \\ &= P\{A\} + P\{B\} - P\{AB\}. \end{aligned}$$

When A and B are disjoint, $P\{AB\} = 0$ and the result reduces to (1.3).

This argument can be extended to any number of events as follows: Suppose $A_1, A_2, \ldots, A_n$ are n events. We wish to find an expression for the probability of the not-necessarily-disjoint union of these events in terms of intersections of them. Recall that $\prod_{i=1}^n A_i$ means the event that $A_1, A_2, \ldots,$ and A_n all occur.

We have

$$\overline{\cup_{i=1}^n A_i} = \prod_{i=1}^n \bar{A}_i. \tag{1.32}$$

Therefore

$$\begin{aligned} I_{\cup_{i=1}^n A_i} &= 1 - I_{\overline{\cup_{i=1}^n A_i}} = 1 - I_{\prod_{i=1}^n \bar{A}_i} \\ &= 1 - \prod_{i=1}^n I_{\bar{A}_i} = 1 - \prod_{i=1}^n (1 - I_{A_i}) \\ &= 1 - (1 - I_{A_1})(1 - I_{A_2}) \ldots (1 - I_{A_n}) \\ &= \sum_{i=1}^n I_{A_i} - \sum_{i \neq j} I_{A_i} I_{A_j} + \sum_{i,j,k \text{ not equal}} I_{A_i} I_{A_j} I_{A_k} \ldots \end{aligned}$$

Therefore

$$P\{\cup_{i=1}^n A_i\} = \sum_{i=1}^n P\{A_i\} - \sum_{i \neq j} P\{A_i A_j\} + \sum_{i,j,k \text{ not equal}} P\{A_i A_j A_k\} \ldots \tag{1.33}$$

Thus, when $n = 3$, we have

$$\begin{aligned} P\{A_1 \cup A_2 \cup A_3\} = [P\{A_1\} + P\{A_2\} + P\{A_3\}] - \\ [P\{A_1 A_2\} + P\{A_1 A_3\} + P\{A_2 A_3\}] + [P\{A_1 A_2 A_3\}] \end{aligned}$$

and when $n = 4$, (1.33) is

$$\begin{aligned} P\{A_1 \cup A_2 \cup A_3 \cup A_4\} =& [P\{A_1\} + P\{A_2\} + P\{A_3\} + P\{A_4\}] \\ & - [P\{A_1 A_2\} + P\{A_1 A_3\} + P\{A_1 A_4\} + P\{A_2 A_3\} \\ & + P\{A_2 A_4\} + P\{A_3 A_4\}] \\ & + [P\{A_1 A_2 A_3\} + P\{A_1 A_2 A_4\} + P\{A_1 A_3 A_4\} \\ & + P\{A_2 A_3 A_4\}] \\ & - [P\{A_1 A_2 A_3 A_4\}]. \end{aligned}$$

Example: Letters and envelopes. Consider the following problem. There are n letters to distinct people, and n addressed envelopes. Envelopes and letters are matched at random (*i.e.*, with equal probability). What is the probability $P_{0,n}$ that no letter gets matched to the correct envelope?

Let I_i be the indicator for the event A_i that letter i is correctly matched. Then we seek

$$P_{0,n} = 1 - P\{\cup_{i=1}^{n} A_i\} = P\left\{\prod_{i=1}^{n} \bar{A}_i\right\} = E\left(I_{\prod_{i=1}^{n} \bar{A}_i}\right) =$$

$$E\left[\prod_{i=1}^{n} I_{\bar{A}_i}\right] = E\left[\prod_{i=1}^{n}(1 - I_{A_i})\right]$$

$$= E\left[1 - \sum_{i=1}^{n} I_{A_i} + \sum_{i \neq j} I_{A_i A_j} - \sum_{i,j,k \text{ not equal}} I_{A_i A_j A_k} + \ldots\right].$$

Now $EI_{A_i} = P(A_i) = 1/n$, so $E\left[\sum_{i=1}^{n} I_{A_i}\right] = n(1/n) = 1$.

Similarly if $i \neq j$, $E(I_{A_i A_j}) = P(A_i A_j) = \frac{1}{n} \cdot \frac{1}{n-1}$. Then

$$E[\sum_{i \neq j} I_{A_i A_j}] = \frac{n(n-1)}{2} \frac{1}{n(n-1)} = \frac{1}{2}.$$

In general, for r distinct indices, $EI_{A_{i_1} A_{i_2} \ldots A_{i_r}} = P(A_{i_1} A_{i_2} \ldots A_{i_r}) = \frac{1}{n} \cdot \frac{1}{n-1} \cdots \frac{1}{n-r+1} = (n-r)!/n!$.

How many ways are there of choosing j distinct indices from n possibilities? Suppose we have n items that we wish to divide into two groups, with j in the first group, and therefore $n - j$ in the second. How many ways can this be done? We know that there are $n!$ ways of ordering all the items in the group, so we could just take any one of those orders, and use the first j items to divide the n items into the two groups of the needed size. But we can scramble up the first j items any way we like without changing the group, and similarly the last $(n - j)$ items. Thus the number of ways of dividing the n items into one group of size j and another group of size $n - j$ is $n!/j!(n - j)!$, which I write as

$$\binom{n}{j, n-j}, \text{ but others sometimes write as } \binom{n}{j}.$$

It is pronounced "n choose j and $n - j$" in the first case, and "n choose j" in the second. Both are called binomial coefficients, for reasons that will be evident later in the book (section 2.9). The notation I prefer has the advantage of maintaining the symmetry between the groups, and makes it easier to understand the generalization to many groups instead of just two. (Section 2.9 shows how my notation helps with the generalization.)

How many ways are there to choose, out of n letters, which r will be correctly matched to envelopes and which $n - r$ will not? Exactly $\binom{n}{r, n-r} = \frac{n!}{r!(n-r)!}$ ways. Hence the term in the sum for r matches is

$$\binom{n}{r, n-r} \cdot \frac{(n-r)!}{n!} = \frac{n!}{r!(n-r)!} \cdot \frac{(n-r)!}{n!} = \frac{1}{r!} \quad ,$$

and we have

$$P_{0,n} = 1 - 1 + \frac{1}{2!} - \frac{1}{3!} + \frac{1}{4!} - \ldots + (-1)^n \frac{1}{n!}$$

This is a famous series in applied mathematics.

Taylor approximations are used to study the behavior of a function $f(x)$ in a neighborhood around a point x_0. The approximation is

$$f(x) \approx f(x_0) + (x - x_0)f'(x_0) + \frac{(x - x_0)^2}{2!} f''(x_0) + \ldots .$$

The accuracy of the approximation depends on the function, how far from x_0 one wants to use the approximation, and how many terms are taken.

Recall that $\frac{d}{dx}e^x = e^x$ and $e^0 = 1$. Then expanding e^x around $x_0 = 0$ in a Taylor series,

$$e^x \approx 1 + x + \frac{x^2}{2!} + \frac{x^3}{3!} + \ldots . \tag{1.34}$$

That this series converges for all x is a consequence of the ratio test, since the absolute value of the ratio of the $(n+1)^{st}$ term to the nth term is

$$\left| \frac{x^{n+1}}{(n+1)!} / x^n/n! \right| = \frac{|x|}{n+1}$$

which is less than 1 for large n (see Rudin (1976, p. 66)). Indeed (1.34) is sometimes taken to be the definition of e^x.

Substituting $x = -1$,

$$e^{-1} = 1 - 1 + \frac{1}{2!} - \frac{1}{3!} + \frac{1}{4!} + \ldots .$$

Hence $P_{0,n} \to e^{-1} \approx .368$ as $n \to \infty$. This is a remarkable fact, that as the number of letters and envelopes gets large, the probability that none match approaches .368, and hence the probability of at least one match approaches .632. □

1.5.1 *Summary*

The expectation of a random variable W taking values w_i with probability p_i is $E(W) = \sum_{i=1}^{n} w_i p_i$. The expectation of the indicator of an event is the probability of the event. The expectation of a finite sum of random variables is the sum of the expectations.

1.5.2 *Exercises*

1. Vocabulary. State in your own words what the expectation of a random variable is.
2. Suppose you flip two coins, and let X be the number of tails that result. Also suppose that there is some number p, $0 \leq p \leq 1$, such that

$$P\{X = 0\} = (1-p)^2$$
$$P\{X = 1\} = 2p(1-p)$$
$$P\{X = 2\} = p^2.$$

   (a) Check that, for any such p, these specifications are coherent.
   (b) Find $E(X)$.
3. In the simplest form of the Pennsylvania Lottery, called "Pick Three," a contestant chooses a three-digit number, that is, a number between (000 and 999), good for a single drawing. In each drawing a number is chosen at random. If the contestant's number matches the number drawn at random, the contestant wins \$600. (Each ticket costs \$1.) What is the expected winnings of such a lottery ticket?
4. Consider the first n integers written down in random order. What is the probability that at least one will be in its proper place, so that integer i will be the ith integer in the random order? [Hint: think about letters and envelopes.]

5. (a) Let X and Y be random variables and let a and b be constants. Prove

$$E(aX + bY) = aE(X) + bE(Y).$$

(b) Let $X_1, \ldots, X_n$ be random variables, and let $a_1, \ldots, a_n$ be constants. Prove $E(\sum_{i=1}^{n} a_i X_i) = \sum_{i=1}^{n} a_i E(X_i)$.

1.6 Other properties of expectation

For the next property of the expectation of Z it is now necessary to limit ourselves to indices i such that $p_i > 0$. Suppose those indices are renumbered so that $x_1 < x_2 < \ldots < x_n$, where $\sum_{i=1}^{n} p_i = 1$ and $p_i > 0$ for all $i = 1, \ldots, n$. Let a random variable X be defined as trivial if there is some number c such that $P\{X = c\} = 1$, and non-trivial otherwise. Then a trivial random variable is characterized by $n = 1$ and a non-trivial one by $n \geq 2$. Then the following result obtains:

Theorem 1.6.1. *Suppose X is a non-trivial random variable. Then*

$$\min X = x_1 < E(X) < \max X = x_n.$$

Proof.

$$\min X = x_1 = \sum_{i=1}^{n} p_i x_1 < \sum_{i=1}^{n} p_i x_i = E(X) < \sum_{i=1}^{n} p_i x_n = x_n = \max X$$

□

Corollary 1.6.2. *If X is non-trivial, there is some positive probability $\epsilon_1 > 0$ that X exceeds its expectation $E(X)$ by a fixed amount $\eta_1 > 0$, and positive probability $\epsilon_2 > 0$ that $E(X)$ exceeds X by a fixed amount $\eta_2 > 0$.*

Proof. For the first statement, let $\eta_1 = x_n - E(X) > 0$ and $\epsilon_1 = p_n$. For the second, let $\eta_2 = E(X) - x_1$ and $\epsilon_2 = p_1$. □

This is the key result for the next section.

Example: Letters and envelopes, continued. Let's pause here to consider a classic probability problem, and to show the power of indicators and expectations to solve the problem. Reconsider the envelope and letter matching problem, but now ask, what is the expected number of correct matches? That is, what is the expected number of letters put in the correct envelopes?

If $n = 1$, there is only one letter and one envelope, so the letter and envelope are sure to match. Thus the expected number of correct matches is one. Now consider $n = 2$. There can be only zero or two matched, and each has probability $1/2$. Thus the expected number of correct matches is $\frac{1}{2} \cdot 0 + \frac{1}{2} \cdot 2 = 1$. The expectation takes a value, 1, which is not a possible outcome in this example.

To do this problem for $n = 3$, or more generally, in this way seems unpromising, as there are many possibilities that must be kept track of. So let's use some of the machinery we have developed. Let I_i be the indicator for the event that the ith letter is in the correct envelope. Then the number of letters in the correct envelope is $I = \sum_{i=1}^{n} I_i$. Since we are asked for the expectation of I, we write:

$$E(I) = E(\sum_{i=1}^{n} I_i) = \sum_{i=1}^{n} E(I_i).$$

Now each letter has probability $1/n$ of being in the right envelope. Thus $E(I_i) = 1/n$ for each i. Then

$$E(I) = nE(I_i) = n \cdot 1/n = 1$$

for all n. This is quite simple, considering the large number of possible ways envelopes and letters might be matched. □

Finally, we give a result that is so intuitive to statisticians that it is sometimes called the *Law of the Unconscious Statistician*. Its proof uses expectations of indicator functions.

Theorem 1.6.3. *Let X be a random variable whose possible values are $x_1, \ldots, x_N$. Let $Y = g(X)$. Then the expectation of the random variable Y is given by*

$$E(Y) = E[g(X)] = \sum_{k=1}^{N} g(x_k)P\{X = x_k\}.$$

Proof. Let the possible values of Y be $y_1, \ldots, y_M$. Let I_{kj} be an indicator for the event $X = x_k$ and $Y = y_j = g(x_k)$ for $k = 1, \ldots, N$ and $j = 1, \ldots, M$. With these definitions, $y_j I_{kj} = g(x_k)I_{kj}$.

Then

$$\begin{aligned}
E(Y) &= \sum_{j=1}^{M} y_j P\{Y = y_j\} && \text{(definition of expectation)}\\
&= \sum_{j=1}^{M} y_j E \sum_{k=1}^{N} I_{kj} && \text{(uses (1.25) and (1.14))}\\
&= E \sum_{j=1}^{M} \sum_{k=1}^{N} y_j I_{kj} && \text{(rearranges sum)}\\
&= E \sum_{j=1}^{M} \sum_{k=1}^{N} g(x_k) I_{kj} && \text{(by substitution)}\\
&= \sum_{k=1}^{N} g(x_k) E \sum_{j=1}^{M} I_{kj} && \text{(rearranges sum)}\\
&= \sum_{k=1}^{N} g(x_k) P\{X = x_k\}. && \text{(uses (1.25) and (1.14))}
\end{aligned}$$

□

Theorem 1.6.3 says that if $Y = g(X)$, then $E(Y)$ can be computed in either of two ways, either as $\sum_{j=1}^{M} y_j P\{Y = y_j\}$ or as

$$\sum_{i=1}^{N} g(x_i)P\{X = x_i\}.$$

1.6.1 *Summary*

Expectation has the following properties:

1. Let k be any constant. Then $E(kX) = kE(X)$.

2. Let $X_1, X_2, \ldots, X_k$ be any random variables. Then

$$E(X_1 + X_2 + \ldots + X_k) = E(X_1) + E(X_2) + \ldots + E(X_k).$$

3. $\min X \leq E(X) \leq \max X$. Equality holds here if and only if X is trivial.
4. If $E(X) = c$, and X is not trivial, then there are positive numbers ϵ_1 and η_1, such that the probability is at least ϵ_1 that $X > c + \eta_1$ and positive numbers ϵ_2 and η_2 such that the probability is at least ϵ_2 that $X < c - \eta_2$.
5. Let g be a real-valued function. Then $Y = g(X)$ has expectation

$$E(Y) = \sum_{k=1}^{N} g(x_k) P\{X = x_k\},$$

where $x_1, \ldots, x_N$ are the possible values of X.

The first two were proved in section 1.5, the latter three in this section.

1.6.2 Exercises

1. Vocabulary. Explain in your own words what a trivial random variable is.
2. Write out a direct argument for the expectation in the letters and envelopes matching problem for $n = 3$.
3. Let $P_{k,n}$ be the probability that exactly k letters get matched to the correct envelopes. Prove that $P_{n-1,n} = 0$ for all $n \geq 1$.
4. Suppose there are n flips of a coin each with probability p of coming up tails. Let $X_i = 1$ if the ith flip results in a tail and $X_i = 0$ if the ith flip results in a head. Let $X = \sum_{i=1}^{n} X_i$ be the number of flips that result in tails.
   (a) Find $E(X_i)$.
   (b) Find $E(X)$ using (1.31).

1.7 Coherence implies not a sure loser

Now we return to the choices you announced in section 1.1, to show that if your choices are coherent, you cannot be made a sure loser. So we suppose that your prices are coherent.

Suppose first that you announce price p for a ticket on event A. If you buy such a ticket it will cost you p, but you will gain \$1 if A occurs, and nothing otherwise. Thus your gain from the transaction is exactly $I_A - p$. If you sell such a ticket, your gain is $p - I_A$. Both of these can be represented by saying that your gain is $\alpha(I_A - p)$ where α is the number of tickets you buy. If α is negative, you sell $-\alpha$ tickets. With many such offers your total gain is

$$W = \sum_{i=1}^{n} \alpha_i (I_{A_i} - p_i) \tag{1.35}$$

where your price on event A_i is p_i. The numbers α_i may be positive or negative, but are not in your control. But whatever choices of α's I make, positive or negative, W is the random variable that represents your gain, and it takes a finite number of values. Now we compute the expectation of W:

$$\begin{aligned} E(W) &= E(\textstyle\sum_{i=1}^{n} \alpha_i (I_{A_i} - p_i)) && \text{(by substitution)} \\ &= \textstyle\sum_{i=1}^{n} E(\alpha_i (I_{A_i} - p_i)) && \text{(uses (1.31))} \\ &= \textstyle\sum_{i=1}^{n} \alpha_i E(I_{A_i} - p_i) && \text{(uses (1.26))} \\ &= 0. && \text{(uses (1.25))} \end{aligned}$$

Then we can conclude that one of two statements is true about W, using the corollary to Theorem 1.6.1.

Either

(a) W is trivial (*i.e.*, $W = 0$ with probability 1), so there are no bets and you are certainly not a sure loser,

or

(b) there is positive probability ϵ that you will gain at least a positive amount η. This means that there is positive probability that you will gain from the transaction, and therefore you are not a sure loser.

Therefore we have shown that if your prices satisfy (1.1), (1.2) and (1.3) you cannot be made a sure loser. So we can pull together these results with those of section 1.1 into the following theorem referred to in this book as the Fundamental Theorem of Coherence:

Your prices $Pr(A)$ at which you would buy or sell tickets on A cannot make you a sure loser if and only if they satisfy (1.1), (1.2) and (1.3), or, in other words, if and only if they are coherent.

1.7.1 Summary

The Fundamental Theorem says it all.

1.7.2 Exercises

1. Vocabulary. Explain in your own words:

   Fundamental Theorem of Probability
2. Why is the Fundamental Theorem important?
3. The proof in section 1.7 that coherence implies you can't be made a sure loser rests on the properties of expectation. Where does each of (1.1), (1.2) and (1.3) get used in the proof of those properties?

1.8 Expectations and limits

(This section could be postponed on a first reading.)

Suppose that $X_1, X_2, \ldots$ is an infinite sequence of random variables each taking only finitely many values. Thus, let

$$P\{X_n = a_{ni}\} = p_i \quad , n = 1, \ldots, i = 1, \ldots, I.$$

Suppose

$$\lim_{n \to \infty} a_{ni} = b_i \qquad \text{for } i = 1, \ldots, I. \tag{1.36}$$

Let X be a random variable that takes the value b_i with probability p_i. Then is it true that

$$\lim_{n \to \infty} E[X_n] = E[X] \quad ? \tag{1.37}$$

We pause to analyze this question here, because it constitutes a theme that recurs in Chapter 3 (concerning random variables taking a countable number of values) and Chapter 4 (concerning random variables on a continuous space). To begin, it is necessary to be precise about what is meant by a limit, which is addressed in the following supplement.

1.8.1 A supplement on limits

What does it mean to write that the sequence of numbers $a_1, a_2, \ldots$ has the limit a? Roughly the idea is that a_n gets closer and closer to the number a as n gets large. Consider, for example, the sequence $a_n = 1/n$. This is a sequence of positive numbers, getting closer and closer to 0 as n gets large. It never gets to 0, but it does get arbitrarily close to 0. Here I seek to give a precise meaning to the statement that the sequence $a_n = 1/n$ has the limit 0.

Since the sequence never gets to 0, we have to allow some slack. For this purpose, it is traditional to use the Greek letter ϵ (pronounced "epsilon"). And we assume that $\epsilon > 0$ is positive. Can we find a number N such that, for all values of the sequence index n greater than or equal to N, a_n is within ϵ of the number a? If we can do this for every positive ϵ, no matter how small, then we say that the limit of a_n as n gets large, is a.

Let's see how this works for the sequence $a_n = 1/n$, with the limit $a = 0$. The question is whether we can find a number N such that for all n larger than or equal to N, we have

$$| a_n - a |=| 1/n - 0 |= 1/n$$

less than ϵ. But to write $\epsilon > 1/n$ is the same as to write $n > 1/\epsilon$. Therefore, if we take N to be any integer greater than $1/\epsilon$, the criterion is satisfied for the sequence $a_n = 1/n$ and the limit $a = 0$.

Thus in general we write that the sequence a_n has the limit a provided, for every $\epsilon > 0$, there is an N (finite) such that, for every $n \geq N$,

$$| a_n - a |< \epsilon.$$

If this is the case, we write

$$\text{``} \lim_{n\to\infty} a_n = a\text{'' or ``}a_n \to a \text{ as } n \to \infty.\text{''}$$

Another way of understanding what limits are about is to notice that the criterion is equivalent to the following: for every positive ϵ, no matter how small, $| a_n - a |< \epsilon$ is violated for at most a finite number of values of n (namely, possibly, $1, 2, \ldots, N - 1$).

Yet another way of phrasing the criterion is that every interval I, centered at a and with width 2ϵ, that is, the interval $(a - \epsilon, a + \epsilon)$, excludes only finitely many a_n's.

A property of limits that is used extensively in the materials that follow is the following:
Lemma: Suppose $\lim_{n\to\infty} a_n = a$ and $\lim_{n\to\infty} b_n = b$. Then the sequence $c_n = a_n + b_n$ converges, and has limit $a + b$.

Proof. Let $\epsilon > 0$ be given. Since $\lim_{n\to\infty} a_n = a$, there is some N_1 such that, for all $n \leq N_1$

$$| a_n - a |< \epsilon/2.$$

Similarly, since $\lim_{n\to\infty} b_n = b$, there is some N_2 such that, for all $n \geq N_2$

$$| b_n - b |< \epsilon/2.$$

Let $N = \max\{N_1, N_2\}$. Then for all $n \geq N$,

$$| (a_n + b_n) - (a + b) |\leq| a_n - a | + | b_n - b |< \epsilon/2 + \epsilon/2 = \epsilon.$$

Therefore $\lim_{n\to\infty}(a_n + b_n)$ exists and equals $a + b$. □

It is easy to see that this lemma can be extended to the sum of finitely many convergent sequences.

1.8.2 *Resuming the discussion of expectations and limits*

We now resume our discussion of (1.37), and prove the following:

Theorem 1.8.1. *Under the assumption that* (1.36) *holds,* (1.37) *holds.*

Proof. Let $\epsilon > 0$ be given. According to (1.36), for each i, $i = 1, \ldots, I$, there is an N_i such that, for all $n \geq N_i$, $\mid a_{in} - b_i \mid < \epsilon$. Let $N = \max\{N_1, N_2, \ldots, N_I\}$. Then for all $n \geq N, \mid a_{in} - b_i \mid < \epsilon$. Therefore, for all $n \geq N$,

$$\left|\sum_{i=1}^{I} p_i a_{in} - \sum_{i=1}^{I} p_i b_i\right| \leq \sum_{i=1}^{I} p_i \left|a_{in} - b_i\right| < \sum_{i=1}^{I} p_i \epsilon = \epsilon.$$

Hence

$$\lim_{n\to\infty} E[X_n] = \lim_{n\to\infty} \sum_{i=1}^{I} p_i a_{in} = \sum_{i=1}^{I} p_i b_i = E[X].$$

□

1.8.3 *Reference*

A friendly introduction to limits can be found in Courant and Robbins (1958, pp. 289-295).

1.8.4 *Exercises*

1. Vocabulary. Explain in your own words what the limit of a sequence of numbers is.
2. Do all sequences of numbers have a limit? Prove your answer.
3. Let $a_n = 1/n^2$. Prove $\lim_{n\to\infty} a_n = 0$.
4. Let $a_n = 0$ if n is odd, and $a_n = 1$ if n is even. Does a_n have a limit? Prove your answer.
5. Let $a_n = (n + 1)/n$. Does a_n have a limit? If so, what is it? Prove your answer.

Chapter 2

Conditional Probability and Bayes Theorem

2.1 Conditional probability

We now turn to exploring what is meant by the probability of an event A conditional on the occurrence of an event B. What makes this particularly important is that the comparison of this conditional probability to the probability of A gives a quantitative view of how much the occurrence of B has changed your view of the probability of A. We'll come back to this after exploring what constraints are put on conditional probabilities by avoidance of sure loss.

To make the exposition clearer, I ask your indulgence to allow tickets to be bought and sold not only in integer amounts, as was done in Chapter 1, but now in non-integer amounts. For example, if you buy half a ticket on the event A, it costs you half as much, and if A occurs, you win half as much as you would have with a full ticket. This extension is later shown not to be necessary for the result to be shown next, but it does make the argument simpler.

So let $\Pr\{A|B\}$ (pronounced "A given B") be the price at which you would buy or sell a ticket that pays \$1 if A and B occur, \$0 if B occurs but A does not, and is called off if B does not occur. Thus if B were not to occur, there are no financial consequences to either party.

To explain what is meant by a called-off bet, consider the difference between a ticket on the event $A|B$ and one on the event AB. Suppose you bought one each of such tickets. If A and B occur, you would win a dollar on each ticket. If $\overline{A}$ and B occur, you would win \$0 on each ticket. But if $\overline{B}$ occurs, you would have your purchase price refunded for the ticket on $A|B$ but not on the ticket on AB. The situation is summarized in the following table:

	Ticket	
Outcome	$A\|B$	AB
AB	\$1	\$1
$\overline{A}B$	\$0	\$0
$\overline{B}$	purchase price refunded	no refund (\$0)

Table 2.1: Consequences of tickets bought on $A|B$ and AB.

Table 2.1 makes it clear that a ticket on $A|B$ will be at least as valuable as a ticket on AB, and in general more valuable. The next set of results establishes how much more valuable a ticket on $A|B$ is compared to a ticket on AB.

Theorem 2.1.1. *Either*

$$\Pr\{AB\} = \Pr\{A|B\}\Pr\{B\}$$

or you can be made a sure loser.

Proof. Let $x = Pr\{B\}, y = Pr\{A|B\}$ and $z = Pr\{AB\}$. To show that $z = yx$ is required

	Ticket			Net
Outcome	$A\|B$	AB	B	
$\overline{B}$	y	0	0	y
$\overline{A}B$	0	0	y	y
AB	1	-1	y	y

Table 2.2: Your gains, as a function of the outcome, when tickets are settled, when $xy > z$.

to avoid being a sure loser, this proof shows first that $xy > z$ leads to sure loss, and then that $xy < z$ leads to sure loss.

Suppose first that $xy > z$. I choose to sell you a ticket on $A|B$, buy from you a ticket on AB, and sell you y tickets on B. (Note that $0 \leq y \leq 1$, so you are buying from me a partial ticket on B.) There are three disjoint and exhaustive outcomes, $\overline{B}, \overline{A}B$ and AB. Call them case 1, case 2 and case 3, respectively. We investigate each of these cases in turn.

Case 1: If $\overline{B}$ occurs, the ticket on $A|B$ is called off. You sold me a ticket on AB, which gains you z and bought from me y tickets on B, which cost you xy. Hence your net gain here is $z - xy < 0$, which means a loss for you.

Case 2: Next consider the consequence if $\overline{A}B$ occurs. In addition to your gain of $z - xy$ for the tickets on AB and B, you owe me y for the ticket I sold you on $A|B$, so your gain is $z - xy - y$. When we settle tickets, the y tickets you own on B pay off, so your gain in this case is $z - xy - y + y = z - xy < 0$. Again, you lost.

Case 3: Finally, if AB occurs, the purchase and sale of tickets result in a net gain to you of $z-xy-y$. All three kinds of tickets now pay off, and your net gain is $z-xy-y+y+1-1 = z - xy < 0$. So in this third case, you lost as well.

Since you lost in all three possible outcomes when $xy > z$, you are a sure loser.

The fact that you lost the same amount in each case is not essential to the proof.

It is useful to summarize these transactions as follows: when $xy > z$, I sell you a ticket on $A|B$, which costs you y. I buy a ticket from you on AB, for which I pay you z. Finally, I sell you a fraction y of a ticket on B, which costs you xy. Thus your total costs for these transactions are $xy + y - z$. If B does not occur, I am obliged to return to you the cost, y, of the ticket on $A|B$, so that this bet is called off. Then Table 2.2 shows the consequences of each possible outcome: you gain y.

Thus your total cost is $xy + y - z - y = xy - z > 0$ whatever the random outcome is. You are therefore a sure loser.

Now we move to the second part of the proof, where $xy < z$. Now I choose to buy from you a ticket on $A|B$, sell you a ticket on AB, and buy from you y tickets on B. Again there are the same three disjoint and exclusive events to consider, $\overline{B}, \overline{A}B$ and AB.

You can now follow the pattern of the argument above, showing that in each of these three cases, you have a gain of $xy - z < 0$, which means you lose! Since you lose no matter which of $\overline{B}, \overline{A}B$ and AB happens, you are a sure loser if $xy < z$.

Since you are a sure loser if $xy > z$ and a sure loser if $xy < z$, the only possible way to avoid sure loss is $xy = z$, as claimed. This completes the proof of the theorem. □

It is somewhat remarkable that the principle of avoiding sure loss requires a unique value for the price for which you would offer to buy or sell a called-off ticket except when $\Pr\{B\} = 0$. Again, this treatment is constructive, in that I show exactly which of your offers I accept to make you a sure loser.

I promised some remarks on the case in which you insist that tickets be bought and sold as integers. If the numbers of tickets bought and sold are all multiplied by the same number, the analysis above applies, with each loss being multiplied by the same number. Thus if y is a rational number, that is, it can be written as p/q, where p and q are integers, then when

$xy > z$, I can imagine selling you q tickets on $A|B$, buying from you q tickets on AB, and selling you p tickets on B. Exactly the argument above applies. Similarly, when $xy < z$, I can imagine buying from you p tickets on $A|B$, selling you p tickets on AB, and buying from you q tickets on B. Again the argument applies. Since every real number y can be approximated arbitrarily closely by rational numbers, it can be shown that Theorem 2.1.1 holds for all real y without resorting to non-integer numbers of tickets. If this paragraph is more mathematics than is to your taste, don't worry about it, and just go with the idea of buying and selling y tickets, where y is not an integer.

The argument above shows that if $P\{AB\} \neq Pr\{A|B\}P\{B\}$ then you can be made a sure loser. We now show the converse, that if $P\{AB\} = P\{A|B\}P\{B\}$ you cannot be made a sure loser.

To do so, we need a new random variable to describe the outcome of the ticket that pays \$1 if A and B occur, \$0 if B occurs and A does not, and is called off otherwise. The first two possible outcomes can be modelled with an indicator function, taking the value \$1 if AB occurs, and 0 if $\overline{A}B$ occurs. But what if $\overline{B}$ occurs? You are willing to buy this ticket for $p = Pr\{A|B\}$. For the bet to be called off if $\overline{B}$ occurs means that no money changes hands in this case, which is the same as having your money, $Pr\{A|B\}$, returned. Thus the random variable defined as

$$I_B(I_A - p)$$

properly expresses the consequences of each of the three possible outcomes.

Recall from section 1.7 that the gain from selling for price p a ticket that pays \$1 if A occurs and 0 when it does not is $I_A - p$, and for buying such a ticket is $p - I_A$. Both of these can be represented by $\alpha(I_A - p)$ where α is the number of tickets you buy. Negative α's are interpreted as sales.

With this definition, the payoff from bets on A, B, AB and $A|B$ can be expressed as

$$W = \alpha_1(I_A - P\{A\}) + \alpha_2(I_B - P\{B\}) + \alpha_3(I_{AB} - P\{AB\}) + \alpha_4(I_B(I_A - p)),$$

where the α's are chosen by an "opponent" to try to make you a sure loser. The argument of section 1.7 shows that if your probabilities are coherent, every choice of α_1, α_2 and α_3 leads to

$$E(W') = \alpha_1(I_A - P\{A\}) + \alpha_2(I_B - P\{B\}) + \alpha_3(I_{AB} - P\{AB\}) = 0.$$

Thus I concentrate on the fourth term

$$\begin{aligned} E(W) &= \alpha_4 E[I_B(I_A - p)] \\ &= \alpha_4[E(I_{AB}) - pE(I_B)] \\ &= \alpha_4[P\{AB\} - pP\{B\}]. \end{aligned}$$

Therefore, under the assumption that $P\{AB\} = Pr\{A|B\}P\{B\}$, we have $E(W) = 0$ for all choices of $\alpha_1, \alpha_2, \alpha_3$ and α_4. Again, we can conclude that either

(a) W is trivial (*i.e.*, $W = 0$ with probability 1), so there are no bets and you are certainly not a sure loser,

or

(b) there is positive probability ϵ that you will gain at least a positive amount η. And, therefore, as in section 1.7, you are not a sure loser.

Thus we may conclude

Theorem 2.1.2. *Your price $Pr\{A \mid B\}$ for the called-off bet on A given that B occurs, cannot make you a sure loser if and only if*

$$P\{AB\} = Pr\{A \mid B\}P\{B\}. \tag{2.1}$$

Since I am supposing that you have decided not to be a sure loser, I now equate $\Pr\{A|B\}$ with $P\{A|B\}$, and suppose therefore when $P\{B\} > 0$, that

$$P\{A|B\} = P\{AB\}/P\{B\}. \tag{2.2}$$

It is very important to notice from the outset that the conditional probability of A given B is NOT the same as the conditional probability of B given A. At the time of this writing (2005), there are 14 women and 86 men in the United States Senate. Then the conditional probability of a person being male, given that he is a Senator, is 86%, but the probability that he is a Senator, given that he is male, is very small.

For a second, and perhaps somewhat silly example, the probability that a person has a cold, given that the person has two ears, is, fortunately, substantially less than 1. However the probability that a person has two ears, given that the person has a cold, is virtually 1.

When $P\{B\} = 0$, since $B = AB \cup \overline{A}B$ and this is a disjoint union, we have

$$0 = P\{B\} = P\{AB\} + P\{\overline{A}B\}.$$

Application of (1.1) now yields

$$P\{AB\} = 0.$$

In the context of (2.1), this implies that $Pr\{A|B\}$ is unconstrained, and can take any value including values less than 0 and greater than 1. An exploration of a method to define conditional probability when conditioning on a set of probability 0 is given by Coletti and Scozzafava (2002).

Probability as developed in section 1.1 can be regarded as probability conditional on S, since $P\{A|S\} = P\{AS\}/P\{S\} = P\{A\}$. Indeed, conditioning on a set B can be regarded as shrinking the sure event from S to B, as exercise 4 in section 2.1.2 below justifies.

What happens if there are three events in question, A, B and C? We can write

$$P\{ABC\} = P\{A|BC\}P\{BC\} = P\{A|BC\}P\{B|C\}P\{C\}. \tag{2.3}$$

Indeed there are six ways of rewriting $P\{ABC\}$, since there are three ways to choose the first set, and for each of them, two ways to choose the second, and only one way to then choose the third. Each of these six ways of rewriting $P\{ABC\}$ is correct, but selecting which one is most useful in an applied setting takes some experience.

Equation (2.3) can be generalized as follows:

$$P\{A_1A_2\ldots A_n\} = P\{A_1|A_2\ldots A_n\}P\{A_2|A_3\ldots A_n\}\ldots P\{A_n\}. \tag{2.4}$$

How many ways are there of rewriting the left-hand side of (2.4)? Now there are n ways of choosing the first set, for each of them $(n-1)$ ways of choosing the second, etc. Hence the number of ways is $n!$

2.1.1 Summary

Avoiding sure loss requires that your price $\Pr\{A|B\}$ for a ticket on A conditional on B satisfies $\Pr\{A|B\}\Pr\{B\} = \Pr\{AB\}$.

2.1.2 Exercises

1. Vocabulary. Explain in you own words what is meant by the conditional probability of A given B.
2. Write out the argument for the case $xy < z$ in the proof of Theorem 2.1.1.
3. Make your own example to show that $P\{A|B\}$ and $P\{B|A\}$ need not be the same.

4. Let B be an event such that $P\{B\} > 0$. Show that $P\{\cdot\ |B\}$ satisfies (1.1), (1.2) and (1.3), which means to show that
   (i) $P\{A|B\} \geq 0$ for all events A.
   (ii) $P\{S|B\} = P\{B|B\} = 1$.
   (iii) Let AB and CB be disjoint events. Then
$$P\{A \cup C|B\} = P\{A|B\} + P\{C|B\}.$$

5. Suppose that my probability of having a fever is .01 on any given day, and my probability of having both a cold and a fever on any given day is .001. Given that I have a fever, what is my conditional probability of having a cold?

2.2 The birthday problem

The birthday problem is an interesting application of conditional probability. By "birthday" in this problem, I mean the day a person is born, not the day and the year. Suppose there are k people who compare birthdays, and we want to know the probability $s_{k,n}$ that at least two of them have the same birthday where there are n possible birthdays. For this calculation, assume that nobody is born on February 29 (which obviously isn't true), so that there are $n = 365$ possible birthdays. Also suppose that people have the same probability of being born on any of them. (This is not quite true. There are seasonal variations in birthdays.) Also let $t_{k,n} = 1 - s_{k,n}$ be the probability that no two people have the same birthday. It turns out that this is the easier event to work with.

Now let's look at $t_{1,n}$. Since there is only one person, $t_{1,n} = 1$ because overlap is not possible. Then what about $t_{2,n}$? Well, $t_{2,n} = (\frac{n-1}{n})t_{1,n}$, since the first person occupies one birthday, so the second person has probability $(\frac{n-1}{n})$ of missing it.

Let $E_k = j$ be the event that the kth person has birthday j. Let $E_{\overline{k}} = \overline{j}$ be the event that persons $\overline{k} = (1, 2, \ldots, k-1)$ have birthdays $\overline{j} = (j_1, \ldots, j_{k-1})$, all different.

Then

$$\begin{aligned} t_{k,n} &= \sum_{\overline{j}} \sum_{j \notin \overline{j}} P\{E_k = j | E_{\overline{k}} = \overline{j}\} P\{E_{\overline{k}} = \overline{j}\} \\ &= \sum_{\overline{j}} \left[\frac{n-(k-1)}{n}\right] P\{E_{\overline{k}} = \overline{j}\} \\ &= \left[\frac{n-(k-1)}{n}\right] t_{k-1,n}. \end{aligned}$$

Therefore

$$\begin{aligned} t_{k,n} &= \frac{n-(k-1)}{n} t_{k-1,n} = \left(\frac{n-(k-1)}{n}\right)\left(\frac{n-(k-2)}{n}\right) t_{k-2,n} = \ldots \\ &= \prod_{i=1}^{k-1} (1 - i/n), \text{ if } k > 1 \end{aligned}$$

and $t_{1,n} = 1$.

For any given k and n, this number can be computed simply, but the formula doesn't give much idea of what these numbers look like. Obviously if k grows for fixed n, $t_{k,n}$ decreases, and if n grows for fixed k, $t_{k,n}$ increases.

To approximate $t_{k,n}$, we'll take its logarithm as follows:

$$\log t_{k,n} = \sum_{i=1}^{k-1} \log(1 - i/n).$$

We now apply a Taylor approximation to $f(x) = \log(1 + x)$ in the neighborhood of $x_0 = 0$. (If you have forgotten about Taylor approximations, there is a brief introduction to them in section 1.5.) Since $\log(1) = 0, f(x_0) = \log(1 + 0) = \log 1 = 0$. Also $f'(x) = \frac{1}{1+x}$, so $f'(x_0) = f'(0) = 1$.

Hence for x close to 0, the Taylor approximation to $\log(1 + x)$ is as follows:

$$\log(1 + x) = 0 + x + HOT$$

where HOT stands for "Higher Order Terms." Applying the approximation and neglecting HOT, we have

$$\log(1 - i/n) \simeq -i/n.$$

Therefore

$$\log t_{k,n} = \sum_{i=1}^{k-1} \log(1 - i/n) \simeq \sum_{i=1}^{k-1}(-i/n) = \frac{-1}{n}\sum_{i=1}^{k-1} i = -\frac{1}{2n}k(k-1),$$

using the formula for the sum of the first $k - 1$ integers, as found in section 1.2.2.

Therefore

$$t_{k,n} \simeq e^{-\frac{k(k-1)}{2n}}.$$

Now suppose we want to find k such that $t_{k,n} = 1/2$ (approximately). We know there won't necessarily be an integer k that solves this equation exactly. However, there will be a largest k such that $t_{k,n} \leq 1/2$, and, for that k, $t_{k+1,n} \geq 1/2$.

Thus we want to find the solution k to the equation

$$\frac{1}{2} = e^{-\frac{k(k-1)}{2n}},$$

and we'll accept any real number, not necessarily an integer, as the solution. Taking logarithms again, we have

$$\log\left(\frac{1}{2}\right) = -\frac{k(k-1)}{2n}, \text{ or}$$

$$2n \log 2 = k^2 - k.$$

So we have the quadratic equation in k to solve for k. One way to solve this equation is to complete the square by noticing that the equation is, except for a constant, of the form

$$(k - a)^2 = k^2 - 2ak + a^2.$$

We have to match the linear term, so $-2a = -1$, or $a = 1/2$. So we can re-express the equation, adding $a^2 = 1/4$ to both sides, as

$$2n \log 2 + 1/4 = k^2 - k + 1/4 = (k - 1/2)^2.$$

Hence $k - 1/2 = \pm\sqrt{2n \log 2 + 1/4}$.

Here only the positive square root makes sense, so we find

$$k = 1/2 + \sqrt{2n \log 2 + 1/4}.$$

When $n = 365$, I get $k = 22.99$. Thus with 23 people, half the time there will be a common birthday between some pair of them.

This is a surprisingly small number. The reason why it works is that each person in the group can have a common birthday with each other member of the group, so there is quadratic behavior at the heart of the problem.

Because the Taylor approximation is justified as a limiting argument, it applies in the limit as $x \to 0$. Now that we know that $k = 23$ with $n = 365$, we see that the Taylor's Theorem approximation is being applied around $x_0 = 0$ when x is no larger than $23/365 = .063$. It is therefore plausible that the Taylor's Theorem approximation is accurate.

Many people find this result surprising. Warren Weaver (1963, p. 135) reports

> In World War II, I mentioned these facts at a dinner attended by a group of high-ranking officers of the Army and Navy. Most of them thought it incredible that there was an even chance with only 22 or 23 persons. Noticing that there were exactly 22 at the table, someone proposed we run a test. We got all the way around the table without a duplicate birthday. At which point a waitress remarked, "Excuse me. But I am the 23rd person in the room, and my birthday is May 17, just like the General's over there." I admit that this story is almost too good to be true (for, after all, the test should succeed only half of the time when the odds are even); but you can take my word for it.

An interesting website on the birthday problem is Weisstein (2005).

2.2.1 *Exercises*

1. The length of the Martian year is 669 Martian days. How many Martians would it take to have at least a 50% probability that two Martians would have the same birthday?
2. Do the same problem for Jovians, whose year is 10,503 Jovian days.
3. For each of the three planets, Earth, Mars and Jupiter, how many inhabitants would it take to have a 2/3 probability of having two people with the same birthday?

2.2.2 *A supplement on computing*

Computation is an essential skill for using the methods suggested in this book. While there are many platforms and packages available with which to do statistics, most are limited to doing only the computations anticipated by the package writers. The notable exception is the open-ware package `R`. The spirit of `R` is that it is more like a convenient computer language than like a package. Given the freedom of opinion allowed in the view of probability adopted here, the ability to compute what you want is critical.

Currently `R` can be downloaded (at no charge) from the following website `http://www.r-project.org`. Please do so now. `R` is an interpretive language, which means that it compiles each command line, interactively, as it is given. This makes `R` excellent for exploration of data and figuring out what you want to compute. But this same quality makes it slow for large data-sets and for programs that involve many steps. For computing of this kind, programs are commonly written in Python, and run in that environment. This need not be a concern now.

The first command in `R` is the command that assigns a number to a variable, pronounced "gets" and written as "=" Thus

```
n = 365
```

assigns to `n` the value 365. An alternative is to type $n < -365$. Please type one of these into the console window of `R`. If you now type

```
n
```

R will respond with

```
365.
```

Hence at any time in a computation you can find out the value of a variable simply by typing it. You can also use the print command to find the value of an object. Another feature of R that many users find helpful is the use of up and down arrows to reuse a line that has previously been typed.

R works most conveniently with vectors, and somewhat less efficiently with "do loops." The computations in this section take advantage of this. The goal is to assess the accuracy of the approximation of $1 - \prod_{i=1}^{k-1}(1 - i/n)$ by $1 - e^{-\frac{1}{2}\frac{k(k-1)}{n}}$. While calculus can suggest that this approximation is close, and sometimes derive upper bounds on the error, those bounds tend to exaggerate the extent of error. Computing is an excellent way to find out what the error really is like.

We have already taken n to be 365. Which k's are we interested in? Since the calculation above suggests that k's in the neighborhood of 23 are interesting to us, we'll take all k's up to 30 as being of interest. Therefore, using ul to stand for upper limit, typing

```
ul = 30
```

specifies its value.

Now we need to explore some vectors. A convenient way to get some useful vectors is with the colon command. Try typing

```
1:3
```

You should get the response

```
1 2 3
```

Thus in general l:u, where l (for "lower") and u (for "upper") are integers, gives you a vector of integers starting at l and ending at u. Do you want to find out what would happen if you try the colon command when u is less than l, or if l and u are not integers? Try it. You can't harm anything, and it will give you the right exploratory attitude toward this type of computing.

Using the colon function, then, we'd like to compute the value of $e^{-\frac{k(k-1)}{2n}}$ for each k from 1 to 30, and, since we're thinking in terms of vectors, we'd like a vector of length 30 to do this. We can build this up in stages.

Now we can create a vector of integers from 1 to 30 as

```
1:ul
```

We could also type 1:30 with the same result, but if we think we might want to change the upper limit later to another value, it helps to have a symbol for upper limit.

The next step is to create a vector of length ul whose k^{th} value is k times $(k-1)$. Now if we wrote $k(k-1)$, R would think that k is a function, to be evaluated at the value $k-1$. R returns an error message for this. To express multiplication, * is used. Hence we write

```
k = 1:ul
k*(k-1)
```

to get our intended vector. R does an interesting thing: it knows that what is meant by $(k-1)$ is to subtract the number 1 from each of the elements of k, so that $(k-1)$ is the same, in this case, as 0:(ul-1). To complete the approximation, we add

```
approx = 1 - exp ( k * (k-1)/((-2) * n))
```

This yields a vector of length 30 giving $1 - e^{-\frac{k(k-1)}{2n}}$ for each value of k from 1 to 30.

The key steps are:

```
n = 365
ul = 30
k = 1:ul
approx = 1 - exp (k * (k-1) /((-2)*n))
print (approx)
```

For greater flexibility in changing `n` and `ul` without having to reenter everything, `R` allows us to define `approx` as a function of `n` and `ul`, as follows:

```
approx = function (n, ul) {
   k = 1:ul
   return(1 - exp (k * (k-1) /((-2) * n)))
         }
```

Having entered this function into `R`,

```
print (approx (365, 30))
```

produces the same vector we got before.

Now let's work on the exact calculation, using the same ideas. Fortunately, `R` provides some special tricks. If you type

```
cumsum (1:3)
```

`R` responds with `(1,3,6)`. [Try it!] This is the cumulative sums of the vector `(1,2,3)`. Similarly, `cumprod (1:3)` yields `(1,2,6)`. Hence, `cumprod` is the cumulative product. How convenient! This is just what's needed to compute $\prod_{i=1}^{k}(1-i/n)$, for each k between 1 and `ul`, as follows:

```
k = 1:ul
cumprod (1 - k/n).
```

This is helpful as a step toward what we want, but isn't quite right yet, for two reasons. First, the first number should be 0 (since when there is only one person, there can't be a coincidence of birthdays). Second, the formula we want to compute for $k > 1$ is

$$1-\prod_{i=1}^{k-1}(1-i/n),$$

not

$$1-\prod_{i=1}^{k}(1-i/n).$$

To address the first, we use the function $\texttt{c}(\cdot)$, which permits one to create vectors by inserting elements. For example

```
c(1,3,5)
```

will return `1 3 5`

The second is addressed because using `c` to put a 0 in front of the `cumprod` function automatically shifts each element of the vector to the right by one index. Hence the only adjustment needed is to subtract 1 from `ul`, so that the resulting vector has exactly `ul` elements.

Therefore our computation for the exact probabilities is

```
exact = function (n, ul){
k=1:(ul - 1)
return(c(0,1- cumprod (1-k/n)))
}
```

With this function entered in R,

```
print (exact (365, 30))
```

produces the exact probabilities.

Now, it would be nice to compare the answers obtained to see how close the approximation is. One way to do this is to examine the two vectors that have been calculated, for example by computing the difference. While some checks can be performed visually, it is inconvenient and difficult to see the big picture. Some plots would be nice.

The simplest kind of plot is accomplished with the command

```
plot(approx (365, 30))
```

which gives a picture like Figure 2.1.

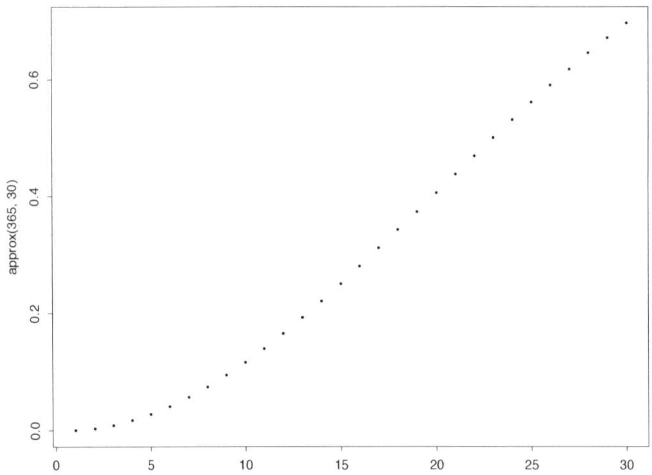

Figure 2.1: Approx plotted against k.

Command: plot (approx (365,30))

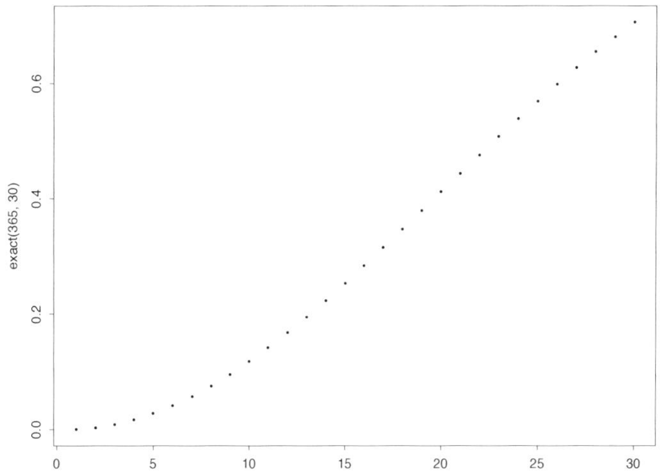

Figure 2.2: Exact plotted against k.

Command: `plot (exact (365,30))`

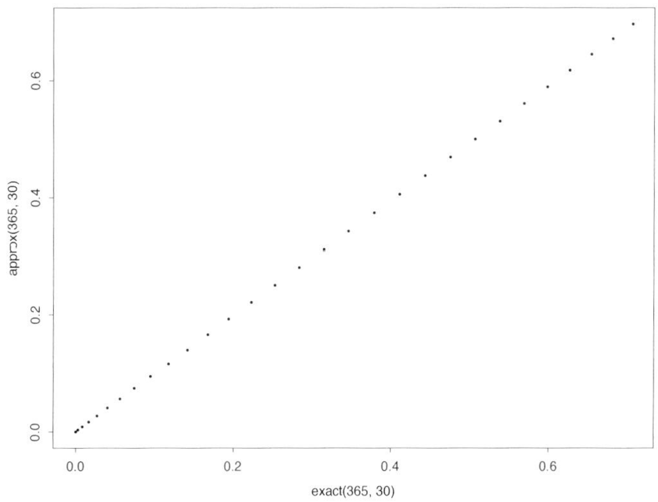

Figure 2.3: Approx plotted against exact.

Command: `plot (exact (365,30)), (approx (365,30))`

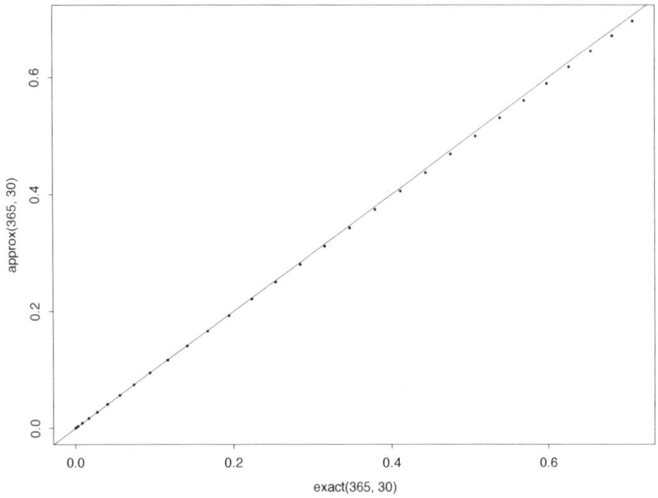

Figure 2.4: Approx plotted against exact, with the line of equality added.

```
Command: abline (0,1)
```

R automatically used k as the second argument, found nice points for the axes, labeled the y axis, but not the x axis, and chose a reasonable plotting character for the points. (Some systems choose other default plotting characters.)

Similarly, the command

```
plot(exact (365,30))
```

gives Figure 2.2.

While these graphs look roughly similar, it would be nice to have them on the same graph. One way to do this is to plot them against each other, for example, by using

```
plot(exact(365,30), approx(365,30))
```

which yields Figure 2.3.

This is a bit more helpful, but it would be nice to see the line $y = x$ put in here, as it would give a visual way of seeing the extent to which the approximation deviates from the exact.

This is accomplished by typing

```
abline(0,1)
```

which gives Figure 2.4. Here the "0" gives the intercept, and the "1" gives the slope. Implicitly the line is being thought of as $y = a + bx$, hence the (somewhat unfortunate) name "abline."

Now we can actually see something, namely that the approximation is a bit too low for larger values of k. Using square brackets to designate the coordinates of a vector, when we examine the exact and approximate calculation in the neighborhood of $k = 23$ we find

```
exact(365,30)[21] = .44369  approx(365,30)[21] = .43749
exact(365,30)[22] = .47570  approx(365,30)[22] = .46893
exact(365,30)[23] = .50730  approx(365,30)[23] = .50000
```

Hence it appears that 23 people are enough to have a 1/2 or more probability of at least one coincident birthday.

Was our Taylor approximation a success? On the one hand, it told us accurately that the number we sought was roughly 23, so 18 is too low and 28 too high. On the other hand, it was not quite accurate, as the approximation could leave serious doubt about whether the correct answer is 23 or 24. Should we be satisfied or not?

There is an art to finding good approximations, and also for appreciating how large the error is likely to be in a given instance. It is learned mostly by comparing exact and approximate results, but there are also some helpful mathematics that can bound errors, or give rates at which errors go to zero, etc. We'll count the first-order Taylor approximation to the birthday problem a qualified success. How useful approximations are depends a lot on the accuracy required for the use you plan to make of the result.

A more precise approximation might be found by taking another term in the Taylor approximation. This would involve adding the first n squares of integers, which you know how to do (see section 1.2.2).

2.2.3 References

There are many fine books on graphics, for example Tufte's volumes (Tufte (1990, 1997, 2001, 2006)) and Cleveland (1993, 1994). An interesting comparative review of five books on graphics is given by Kosslyn (1985).

There are also many excellent books on R. At an introductory level, there's Krause and Olson (1997). At a more advanced level, the book of Venables and Ripley (2002) is widely used. Two more recent books are Wickham and Grolemund (2016) and Ekstrom (2012). There are now an enormous number of books on R. With more coming all the time, I expect this section of my book will be the first to be obsolete. On-line help and links are available as part of R.

For more on the birthday problem, see Mosteller (1962).

2.2.4 Exercises

1. Extend the approximation by calculating the next-order term in the Taylor expansion. Compare the resulting approximation to the approximation discussed above. Is the new approximation more accurate?
2. Compare the approximate and exact solutions to the birthday problem for Martians, both computationally and graphically (see section 2.2.1, exercise 1).
3. Try it for Jovians. Can your computer handle vectors of the lengths required?

2.3 Simpson's Paradox

Imagine two routes to the summit of a mountain, a difficult route D and an easier route $\overline{D}$. Imagine also two groups of climbers: amateurs A, and experienced climbers, $\overline{A}$. Suppose that a person has probabilities of reaching the summit R, as a function of the route and the experience of the climber as follows:

$$P\{R|\overline{D},\overline{A}\} = 0.8 \qquad P\{R|\overline{D},A\} = 0.7$$
$$P\{R|D,\overline{A}\} = 0.4 \qquad P\{R|D,A\} = 0.3$$

Thus experienced climbers are more likely to reach the summit whichever route they take, and both groups are less likely to reach the summit using the more difficult route.

Further suppose also that the experienced climbers are believed to be more likely to take the more difficult route:

$$P\{D|\overline{A}\} = 0.75 \qquad P\{D|A\} = 0.35.$$

Now let's see what the consequences of the choices are for the probability that an experienced climber reaches the summit.

The events $R\overline{D}$ and RD are disjoint, and their union is R. Therefore

$$\begin{aligned} P\{R|\overline{A}\} &= P\{R\overline{D}|\overline{A}\} + P\{RD|\overline{A}\} \\ &= P\{R|\overline{D}, \overline{A}\}P\{\overline{D}|\overline{A}\} + P\{R|D, \overline{A}\}P\{D|\overline{A}\} \\ &= (0.8)(0.25) + (0.4)(0.75) = 0.5. \end{aligned}$$

Similarly

$$\begin{aligned} P\{R|A\} &= P\{R|\overline{D}, A\}P\{\overline{D}|A\} + P\{R|D, A\}P\{D|A\} \\ &= (0.7)(0.65) + (0.3)(0.35) = 0.56. \end{aligned}$$

Thus amateur climbers have a greater chance of reaching the summit (0.56) than do experienced climbers (0.5), although for each route they have a smaller chance.

This is an example of Simpson's Paradox. It may seem paradoxical that the better climbers reach the summit less often. However the tool of conditional probability is useful to see the logic of this apparent contradiction. The amateurs have less chance of reaching the summit than the experienced climbers whichever route they take, but have a greater chance of reaching the summit because more of them take the easier route.

The first point to make is that these choices of conditional probabilities are coherent. Thus there is no way to make a sure loser out of a person who holds these beliefs. Second, if it were the case that $P(D|\overline{A}) = P(D|A)$, so if the rate of taking the more difficult route were regarded as the same regardless of the experience of the climber, the "paradox" would disappear (see problem 2 in section 2.3.2). Indeed, Simpson's paradox is a conundrum, but actually is simply an unexpected consequence of coherence.

Now suppose we had gathered data on the skill of climbers and their success in reaching the summit, but neglected to gather data on what route they chose. This would lead us to the wrong conclusion that amateurs are better climbers.

Instead of mountain climbers, consider an observational study that compares the success rates of two medical treatments. The two treatments are like the two kinds of climbers, and success of the treatment is like reaching the summit. Unmeasured covariates, such as genetics, smoking, diet or exercise may play the role of the route.

This example illustrates why biostatisticians are very concerned to ensure randomization of treatment assignment of patients in a clinical trial. The purpose of randomization is to ensure $P\{D|\overline{A}\} = P\{D|A\}$. (Chapters 7 and 11 of this book return to this topic in greater depth.) For example, consider that a general result of clinical studies is that patients who are sicker don't do as well as patients who are less sick no matter what treatment they have. Left to their own devices, physicians might assign one treatment to sicker patients and the other to less sick ones. Thus an examination of the raw results would not be informative about which treatment is better. Nonetheless, enthusiasts for data mining sometimes propose exactly such an analysis (Mitchell (1997)).

As an example of Simpson's Paradox in practice, consider the data in Table 2.3. To give you some background, the ancestors of the present-day Maori were the indigenous people living in New Zealand at the time when European settlers arrived. As such, they are analogous to the Native Americans in North America, the Inuit of the Arctic, and the Aboriginal People of Australia. In all these places, there are issues of whether these descendants of the original inhabitants are being fairly treated. The data in Table 2.3 were gathered to see whether the Maori were represented in juries in New Zealand in proportion to their numbers in the population. The results show that overall Maoris comprise 9.5% of the population and 10.1% of the jury pool. However, when broken down by geography, Maoris are underrepresented in each district!

Percentage Maori ethnic group			
District	Eligible population (aged 20-64)	jury pool	shortfall
Whangarei	17.0	16.8	.2
Auckland	9.2	9.0	.2
Hamilton	13.5	11.5	2.0
Rotorua	27.0	23.4	3.6
Gisborne	32.2	29.5	2.7
Napier	15.5	12.4	3.1
New Plymouth	8.9	4.1	4.8
Palmerston N	8.9	4.3	4.6
Wellington	8.7	7.5	1.2
Nelson	3.9	1.7	2.2
Christchurch	4.5	3.3	1.2
Dunedin	3.3	2.4	.9
Invercargill	8.4	4.8	3.6
All districts	9.5	10.1	-.6

Table 2.3: The paradox: the Maori, overall, appear to be over-represented, yet in every district they are underrepresented.

2.3.1 *Notes*

The data for Table 2.3 come from Westbrooke (1998). Other real examples of Simpson's Paradox are given by Appleton et al. (1996), by Cohen and Nagel (1934, p. 449), by Bickel et al. (1975), by Morrell (1999), by Knapp (1985) and by Wagner (1982).

Simpson's Paradox is a name used for several different phenomena. It was popularized by Blythe (1972, 1973), after a paper by Simpson (1951). However, the basic idea goes back at least to Pearson et al. (1899, p. 277) and Yule (1903). This is an example of Stigler's Rule, which says that when a statistical procedure is named for someone, someone else did it earlier. Stigler (1980) applies his rule to Stigler's Rule as well. See also Good and Mittal (1987).

2.3.2 *Exercises*

1. Explain in your own words what Simpson's Paradox is. In your view, is it a paradox?
2. Prove the following:
   If
   $$\begin{aligned} &1. \quad P\{D|\overline{A}\} = P\{D|A\} \\ &2. \quad P\{S|\overline{D},\overline{A}\} > P\{S|\overline{D},A\} \end{aligned}$$
   and
   $$3. \quad P\{S|D,\overline{A}\} > P\{S|D,A\},$$
   then
   $$P\{S|\overline{A}\} > P\{S|A\}.$$
3. Suppose that the probabilities for the climbers are as follows, instead of those given in section 2.3:
   $$\begin{array}{ll} P\{R|D,A\} = 0.7 & P\{R|\overline{D},A\} = 0.6 \\ P\{R|D,\overline{A}\} = 0.5 & P\{R|D,A\} = 0.4 \\ P\{D|\overline{A}\} = 0.6 & P\{D|A\} = 0.5 \end{array}$$

Does this lead to Simpson's Paradox? Why or why not?

4. Create a setting and give numbers to probabilities that lead to Simpson's Paradox.
5. In your judgment, do the data in Table 2.3 indicate underrepresentation of Maoris on New Zealand juries? Does your answer depend on whether New Zealand juries are chosen to represent the entire population of New Zealand, or chosen within districts?
6. In section 2.3, Simpson's paradox is introduced in terms of an unmeasured variable (in the example, the expertise of the climber). What is the equivalent variable in Table 2.3? How is it possible for Maoris to be underrepresented in each district, but overrepresented when the districts are put together? Explain your answer.

2.4 Bayes Theorem

There's no theorem like Bayes theorem,
there's no theorem I know.
Everything about it is appealing,
everything about it is a wow!

Box (1980a)

The purpose of this section is to derive several forms of a theorem relating conditional probabilities to each other. The result, Bayes Theorem, is a fundamental tool for the rest of the book. It explains how to respond coherently to data, and forms the mathematical basis for a theory of changing your mind coherently.

Observe that $P\{AB\} = P\{BA\}$, but that (2.2) is asymmetric in A and B. Therefore there are two ways to express $P\{AB\}$, namely

$$P\{A|B\}P\{B\} = P\{B|A\}P\{A\}. \tag{2.5}$$

Supposing $P\{B\} > 0$ and dividing by $P\{B\}$ yields

$$\boxed{P\{A|B\} = \frac{P\{B|A\}P\{A\}}{P\{B\}},} \tag{2.6}$$

which is the first form of Bayes Theorem. Looking at (2.6) might make it clear why $P\{A|B\}$ and $P\{B|A\}$ are not the same.

The event B in (2.6) can be decomposed as follows: $B = AB \cup \overline{A}B$. Furthermore AB and $\overline{A}B$ are disjoint. Therefore using (1.3),

$$P\{B\} = P\{AB\} + P\{\overline{A}B\}.$$

Now each of $P\{AB\}$ and $P\{\overline{A}B\}$ can be rewritten using (2.2) so that

$$P\{B\} = P\{B|A\}P\{A\} + P\{B|\overline{A}\}P\{\overline{A}\}. \tag{2.7}$$

Substituting (2.7) into (2.6) yields

$$\boxed{P\{A|B\} = \frac{P\{B|A\}P\{A\}}{P\{B|A\}P\{A\} + P\{B|\overline{A}\}P\{\overline{A}\}},} \tag{2.8}$$

which is the second form of Bayes Theorem.

Now suppose that instead of A and $\overline{A}$ we have a set of events $A_1, A_2, \ldots, A_n$, that are mutually exclusive (remember that means that no more than one can occur) and exhaustive (at least one must occur). Therefore, exactly one occurs. Then B can be written as

$$B = BA_1 \cup BA_2 \cup \ldots \cup BA_n = \cup_{i=1}^{n} BA_i.$$

Furthermore, the BA_i's are disjoint, so $P\{B\} = \sum_{i=1}^{n} P\{BA_i\}$.

Again each of these can be rewritten using (2.2), yielding

$$P\{B\} = \sum_{i=1}^{n} P\{B|A_i\}P\{A_i\}. \tag{2.9}$$

Now substituting (2.9) into (2.6) and replacing A by A_j yields

$$\boxed{P\{A_j|B\} = \frac{P\{B|A_j\}P\{A_j\}}{\sum_{i=1}^{n} P\{B|A_i\}P\{A_i\}}} \tag{2.10}$$

which is the third and final form of Bayes Theorem. It is important to notice that the second form is a special case of the third form, in which the mutually exclusive and exhaustive sequence consists of the two events A and $\overline{A}$.

Let's see how (2.8) works in practice. Suppose A is the event that a person has some specific disease, and B represents their symptoms. A doctor wishes to assess $P\{A|B\}$, her probability that the person has the disease, given the symptoms the person exhibits. The medical literature is generally organized in terms of $P\{B|A\}$, the probability of various symptoms given diseases a person might have. The bridge between the literature and the desired conclusion is built using Bayes Theorem. To use it (in the second form) requires the doctor to make a judgement about $P\{A\}$. Now $P\{A\}$ is the doctor's probability that the person has the disease before knowing about symptoms B. Depending on what disease we're talking about, she might want to know about the person's medical history, the medical history of the family, what travels the person has recently made, or other information. All of this might go into her belief $P\{A\}$. This belief can be understood in terms of what price she would give to buy or sell a ticket that pays \$1 if the person does indeed have disease A, and nothing otherwise. Additionally she has to assess what she thinks about $P\{B|A\}$ and $P\{B|\overline{A}\}$. These are respectively her probability of the symptoms if the person has the disease and does not. To take a ridiculous example again, suppose B represents "has two ears." Then $P\{B|A\}$ and $P\{B|\overline{A}\}$ are both reasonably taken to be 1, and (2.8) reduces to $P\{A|B\} = P\{A\}$, so the symptom "has two ears" was uninformative.

The case in which the doctor has n disease-states in mind instead of just two (has or has not the disease) is addressed by the third form of Bayes Theorem, equation (2.10).

Formula (2.6) applies to $\bar{A}$ as well, yielding

$$P\{\bar{A}|B\} = \frac{P\{B|\bar{A}\}P\{\bar{A}\}}{P\{B\}} \tag{2.11}$$

Dividing (2.6) by (2.7) yields

$$\frac{P\{A|B\}}{P\{\bar{A}|B\}} = \frac{P\{B|A\}P\{A\}}{P\{B|\bar{A}\}P\{\bar{A}\}}. \tag{2.12}$$

I now introduce the *odds* of an event G, defined as

$$0(G) = P\{G\}/P\{\bar{G}\}. \tag{2.13}$$

Then (2.8) can be rewritten as

$$0\{A|B\} = \frac{P\{B|A\}}{P\{B|\bar{A}\}} 0(A) \tag{2.14}$$

The quantity $\frac{P\{B|A\}}{P\{B|\bar{A}\}}$ is called the *Bayes* Factor.

Then (2.14) can be put into words as follows: The posterior odds of A given B equals the Bayes Factor times the prior odds of A.

Note that Bayes Factors apply only to binary situations (A and $\bar{A}$), but not to the case of more than two possible outcomes, as in (2.10).

2.4.1 *Notes and other views*

Bayes Theorem is a simple consequence of the axioms of probability, and is therefore accepted as valid by all. However some who challenge the use of personal probability reject certain applications of Bayes Theorem. For instance, in the context of the medical example, they sometimes view $P\{B|A\}$ and $P\{B|\overline{A}\}$ as reliably given by the medical literature and therefore "objective," but $P\{A\}$ as "subjective" and therefore not a legitimate probability (Fisher (1959b)). However, this view does not help a doctor treat her patients.

2.4.2 *Exercises*

1. What are the differences among the three forms of Bayes Theorem?
2. Suppose A_1, A_2, A_3 and A_4 are four mutually exclusive and exhaustive events. Also suppose

   $$P\{A_1\} = 0.1$$
   $$P\{A_2\} = 0.2$$
   $$P\{A_3\} = 0.3$$
   $$P\{A_4\} = 0.4.$$

   Let B be an event such that

   $$P\{A_1B\} = 0.05$$
   $$P\{A_2B\} = 0.15$$
   $$P\{A_3B\} = 0.25$$
   $$P\{A_4B\} = 0.3$$

   Compute $P\{A_i|B\}$ for $i = 1, 2, 3, 4$.
3. An Elisa test is a standard test for HIV. Suppose a physician assesses the probability of HIV in a patient who engages in risky behavior (unprotected sex with multiple partners of either sex, or sharing injection drug needles) as .002, and the probability of HIV in a patient who does not engage in those risky behaviors as .0001. Also suppose the Elisa test has a sensitivity (probability of having a positive reading if the patient has HIV) of .99, and a specificity (probability of having a negative reading if the patient does not have HIV) of .99 and does not depend on whether the patient has engaged in risky behavior. Let E stand for "engages in risky behavior," H stand for "has HIV," and R stand for "positive Elisa result." Use Bayes Theorem to compute each of the following:
   (a) $P\{H|E, R\}$
   (b) $P\{H|\overline{E}, R\}$
   (c) $P\{H|E, \overline{R}\}$
   (d) $P\{H|\overline{E}, \overline{R}\}$.

   The low probabilities even after a positive test led to the development of a more expensive but higher-specificity follow up test, which is used after a positive Elisa test before the results are given to patients.
4. In the following problem, choice "at random" means equally likely among the alternatives.

   Suppose there are three boxes, A, B and C, each of which contains two coins. Box A has two pennies, Box B one penny and one nickel, and Box C two nickels. A box is chosen at random, and then a coin is chosen at random from that box. The coin chosen turns out to be a nickel. What is the probability that the other coin in the chosen box is also a nickel? Show each step in your argument.

5. Phenylketonuria (PKU) is a genetic disorder that affects infants and can lead to mental retardation unless treated. It affects about 1 in 10 thousand newborn infants. Suppose that the test has a sensitivity of 99.99% and a specificity of 99%. What is the probability that a baby has PKU if the test is positive?

6. Gamma-glutamyl Transpeptidase (GGTP) is a test for liver problems. Among walking, apparently healthy persons, approximately 98.6% have no liver problems, 1% are binge drinkers, 0.2% have a hepatic drug reaction, and 0.2% have some serious liver disease such as hepatitis, liver cancer, gall stones, metastatic cancer, etc. Suppose the probability of having a positive test in a person with no liver problems is 5%, in a binge drinker 50%, in those with a drug reaction 80% and among those with serious liver disease 95%. Suppose a walking, apparently healthy person has a positive test. What is the probability that such a person has
   (a) no liver problems
   (b) is a binge drinker
   (c) has a hepatic drug reaction
   (d) has a serious liver disease?

   Do the numbers you have computed in (a) to (d) add up to 1? Why or why not?

2.5 Independence of events

Suppose two events A and B have the relationship that learning A does not affect how you would bet on B, that is,

$$P\{B\} = P\{B|A\} = P\{AB\}/P\{A\},$$

or, equivalently (and symmetrically),

$$P\{AB\} = P\{A\}P\{B\}. \tag{2.15}$$

Such events are defined to be independent. The events S (the sure event) and ϕ (the empty event) are independent of every other event. Also if A and B are independent, then A and $\overline{B}$ are independent as well, since $P\{\overline{B}\} = 1 - P\{B\} = 1 - P\{B|A\} = P\{\overline{B}|A\}$.

Consider flipping two coins. Then one way to think of the possible outcomes is $\{H_1H_2, H_1T_2, T_1H_2, T_1T_2\}$. If the coins are fair, it is natural to think that each of these possibilities has equal probability, namely 1/4. In this case, the probability of H_1 is given by $P\{H_1\} = P\{H_1H_2\} + P\{H_1T_2\} = 1/4 + 1/4 = 1/2$. Similarly $P\{H_2\} = 1/2$. The events H_1 and H_2 are independent, since $1/4 = P\{H_1H_2\} = P\{H_1\}P\{H_2\} = (1/2)(1/2)$.

However, suppose someone else decides to code the outcomes by the number of heads, 0, 1 or 2, and believes these are equally likely, so each has probability 1/3.

Can the outcomes of these two flips be regarded as independent? Thus, suppose that $P\{H_1H_2\} = P\{T_1T_2\} = 1/3$. Then for some z, we must have $P\{H_1T_2\} = (1/3)z$ and $P\{T_1H_2\} = (1/3)(1 - z)$. As a consequence

$$\begin{aligned} P\{H_1\} &= P\{H_1H_2\} + P\{H_1T_2\} \\ &= 1/3 + (1/3)z = (1/3)(1 + z) \end{aligned}$$

$$\begin{aligned} P\{H_2\} &= P\{H_1H_2\} + P\{T_1H_2\} \\ &= 1/3 + (1/3)(1 - z) \\ &= (1/3)(2 - z). \end{aligned}$$

Independence then requires

$$\begin{aligned} 1/3 = P\{H_1H_2\} &= P\{H_1\}P\{H_2\} \\ &= (1/3)(1+z)\cdot(1/3)(2-z), \text{ or} \\ 3 = (1+z)(2-z) &= 2+z-z^2. \end{aligned}$$

Thus z must satisfy

$$h(z) = z^2 - z + 1 = 0. \tag{2.16}$$

This function goes to infinity as $z \to \infty$ and $-\infty$. Its minimum occurs at $2z - 1 = 0$, or $z = 1/2$. At $z = 1/2, h(1/2) = 1/4 - 1/2 + 1 = 3/4 > 0$. Therefore there are no real numbers z satisfying (2.16).

Hence the outcome of two flips in this example cannot be regarded as independent.

Can this person be made a sure loser? Provided the person will buy or sell tickets on any two numbers of heads out of the set $\{0, 1, 2\}$ for $2/3$, and on all three possibilities for 1, equations (1.1), (1.2) and (1.3) are satisfied. Thus the person cannot be made a sure loser. What's going on here? This example is a reminder that avoidance of sure loss is a very mild condition on beliefs. Many quite unreasonable beliefs can avoid sure loss. At the same time, it is also useful to be reminded that the idea that flips of coins (the same one or different ones) are independent involves an assumption. This assumption may or may not be natural in the applied context, but it deserves to be justified when it is used.

How might independence be extended to more than two sets? The idea to be captured is that learning that any number of them has occurred does not alter the probabilities of the others. One thought is to apply the definition of independence in (2.15) pairwise; that is, to suppose that (2.15) applies to each pair. However, this idea fails to meet our goal. Consider the following example:

Suppose that there are four possible outcomes of a random variable X, which we'll number 1, 2, 3 and 4. Suppose each outcome is equally likely to you. Let $A_1 = \{1, 2\}$, $A_2 = \{1, 3\}$ and $A_3 = \{2, 3\}$. Then A_1 occurs if and only if $X = 1$ or $X = 2$. Then, by construction, each A has probability $1/2$. Also each pair of A's has one outcome in common, and therefore has probability $1/4$. For example $P\{A_1A_2\} = P\{X = 1\} = 1/4$. Thus the A's are pairwise independent. However the intersection of the three A's is ϕ, the empty set, which has probability zero, which is not the product of the probabilities of all three A's, which is $1/8$. Hence, for example, learning that A_1 and A_2 have occurred, means that the outcome is known to be $X = 1$, and so A_3 cannot occur. Thus $P\{A_3|A_1, A_2\} = 0 \neq P\{A_3\}$.

Hence pairwise independence is not sufficient to capture the idea that probabilities are not altered by learning independent events. As a consequence, we define a set of events $A_1, A_2, \ldots A_n$ to be independent if the probability of the intersection of every subset of them is the product of their probabilities. Formally, this is expressed by writing that if $\{A_{i_1}, A_{i_2}, \ldots, A_{i_j}\}$ is a subset of $\{A_1, A_2, \ldots, A_n\}$, then

$$P\{A_{i_1}A_{i_2}\ldots A_{i_j}\} = P\{A_{i_1}\}P\{A_{i_2}\}\ldots P\{A_{i_j}\}.$$

Events that are not independent are said to be dependent. Independence turns out to be a very important concept in applications of probability.

Having discussed conditional probability in Section 2.1 and independence in this section, it is now possible to move on to conditional independence. It should come as no surprise that events $A_1, A_2, \ldots, A_n$ are defined to be conditionally independent given an event B if every subset $\{A_{i_1}, A_{i_2}, \ldots, A_{i_j}\}$ of $\{A_1, A_2, \ldots, A_n\}$ satisfies

$$P\{A_{i_1}A_{i_2}\ldots A_{i_j}|B\} = P\{A_{i_1}|B\}P\{A_{i_2}|B\}\ldots P\{A_{i_j}|B\}. \tag{2.17}$$

Consider the following experiment: I choose one of two coins, and flip it twice. There are eight possible outcomes, which I label in the following way: $\{C_1, T_1, H_2\}$ means that I chose coin 1, the first flip resulted in tails, the second in heads. Provided I give probabilities for these eight events that are non-negative and sum to 1, equations (1.1) and (1.2) are satisfied. If, in addition, I agree that the probability of any event is to be the sum of the probabilities of the events that comprise it, equation (1.3) is satisfied as well. Thus, whatever my choices, I cannot be made a sure loser.

My choices are as follows:

$$P\{C_1T_1T_2\} = P\{C_2H_1H_2\} = 9/32$$
$$P\{C_1H_1H_2\} = P\{C_2T_1T_2\} = 1/32$$

Each of the other four possibilities, namely $\{C_1T_1H_2\}$, $\{C_1H_1T_2\}$, $\{C_2T_1H_2\}$ and $\{C_2H_1T_2\}$, is to have probability 3/32. To check that these probabilities satisfy (1.2), note that $2(9/32) + 2(1/32) + 4(3/32) = 1$. Since these are all non-negative, they satisfy (1.1) as well. Thus I have satisfied the conditions I set in the previous paragraph. Now let's examine some consequences of these choices.

The probability of choosing the first coin can be found by addition as follows:

$$\begin{aligned} P\{C_1\} &= P\{C_1T_1T_2\} + P\{C_1T_1H_2\} + P\{C_1H_1T_2\} + P\{C_1H_1H_2\} \\ &= 9/32 + 3/32 + 3/32 + 1/32 = 16/32 = 1/2. \end{aligned}$$

Then $P\{C_2\} = 1 - P\{C_1\} = 1/2$, using (1.7).

We can also calculate the probability that the first flip is a tail:

$$\begin{aligned} P\{T_1\} &= P\{C_1T_1T_2\} + P\{C_1T_1H_2\} + P\{C_2T_1T_2\} + P\{C_2T_1H_2\} \\ &= 9/32 + 3/32 + 1/32 + 3/32 = 1/2. \end{aligned}$$

Similarly the calculation for a tail on the second flip gives $P\{T_2\} = 1/2$. Now let's calculate the probability that both flips result in tails, thus:

$$\begin{aligned} P\{T_1T_2\} &= P\{C_1T_1T_2\} + P\{C_2T_1T_2\} \\ &= 9/32 + 1/32 = 10/32 = 5/16. \end{aligned}$$

Now we can examine whether the T_1 and T_2 are independent. We have

$$5/16 = P\{T_1T_2\} \neq P\{T_1\}P\{T_2\} = (1/2)(1/2) = 1/4.$$

Therefore T_1 and T_2 are dependent.

That is not the whole story, however. Let's compute the probability that the first two flips are tails, given that the first coin is chosen.

$$\begin{aligned} P\{T_1T_2|C_1\} &= P\{T_1T_2C_1\}/P\{C_1\} \\ &= (9/32)/(1/2) = 9/16. \end{aligned}$$

Also let's look at $P\{T_1|C_1\}$, which can be calculated as follows:

$$P\{T_1|C_1\} = P\{T_1C_1\}/P\{C_1\} = (9/32 + 3/32)/(1/2) = 24/32 = 3/4.$$

But by a similar calculation, $P\{T_2|C_1\} = 3/4$ as well.

Therefore T_1 and T_2 are conditionally independent given C_1, since

$$9/16 = P\{T_1T_2|C_1\} = P\{T_1|C_1\}P\{T_2|C_1\} = (3/4)(3/4).$$

In fact, one process that would yield the choices of probabilities I made is to think of the process in two parts. Think of each coin as equally likely to be chosen. Conditional on coin 1 being chosen, there are two independent flips of coin 1, each of which has probability 3/4 of coming up tails. If coin 2 were chosen, the flips are again independent, with probability 1/4 of tails. (Now you know why I chose the particular numbers I did.)

2.5.1 *Summary*

Events $A_1, A_2, \ldots, A_n$ are conditionally independent given an event B if every subset of them satisfies (2.17). Events $A_1, A_2, \ldots, A_n$ are independent if they are conditionally independent given S.

2.5.2 *Exercises*

1. Vocabulary. Explain in your own words what it means for a set of events to be independent, and to be conditionally independent given a third event.
2. Make your own example to show that pairwise independence of events does not imply independence.
3. Suppose A and B are two independent and disjoint events. Suppose $P\{B\} = 1/2$. What is $P\{A\}$? Prove your answer.
4. In the example in section 2.5, carefully write out the calculations for $P\{T_2\}$ and $P\{T_2|C_1\}$. Justify each step you make by reference to one of the numbered equations in the book.
5. Suppose you observe two tails in the example just above, but you do not know what coin was used. Apply Bayes Theorem to find the conditional probability that the coin was coin 1.
6. Suppose someone regards 0, 1 and 2 heads as being equally likely in two flips of the same coin and, in the case of exactly one head, considers heads on the first flip to be as probable as heads on the second flip. Compute the conditional probability of heads on the second flip given heads on the first flip. What would such a person have to believe about the coin and the person flipping the coin to sustain these beliefs? Discuss circumstances under which such beliefs might be plausible.
7. Show that S and ϕ are independent of every other event.
8. (a) Suppose you flip two independent fair coins. If at least one head results, what is the probability of two heads?
   (b) Suppose the sexes of children in a family are independent, and that boys and girls are equally likely. If a family with two children has at least one girl, what is the probability they have two girls?
   (c) Again suppose the sexes of children in a family are independent and that boys and girls are equally likely. Imagine a family with two children who are not twins. Suppose that the older child is a girl. What is the probability that they have two girls?
9. Suppose A and B are conditionally independent events given a third event C. Does this imply that A and B are conditionally independent given $\overline{C}$? Either prove that it does, or give a counterexample.
10. Suppose $A \subseteq B$. Find necessary and sufficient conditions on the pair of probabilities $(P(\{A\}, P\{B\})$ for A and B to be independent.
11. Imagine three boxes, each of which has three slips of paper in it, each slip with a number marked on it. The numbers for box A are 2, 4 and 9, for box B 1, 6 and 8, and for box C 3, 5 and 7. One slip is drawn, independently and with equal probability, from each box.
    (a) Compute

$$P\{A \text{ slip } > B \text{ slip}\}$$
$$P\{B \text{ slip } > C \text{ slip}\}$$
$$P\{C \text{ slip } > A \text{ slip}\}$$

(b) Is there anything peculiar about these answers? Discuss the implications.

12. Suppose that events A and B are that people have diseases a and b, respectively. Suppose that having either disease leads to hospitalization $H = A \cup B$. If A and B are believed to be independent events, show that

$$P(A|BH) < P(A|H).$$

Thus if hospital populations are compared, a spurious negative association between A and B might be found. This is called Berkson's Paradox (Berkson (1946)).

13. (a) If A and B are independent events, prove that $\overline{A}$ and $\overline{B}$ are independent.

(b) Suppose $A_1, A_2, \ldots, A_n$ are independent. Prove by induction on n that $\overline{A}_1, \overline{A}_2, \ldots, \overline{A}_n$ are independent.

(c) Suppose $A_1, A_2, \ldots, A_n$ are independent. Prove that $A_1, \ldots, A_k, \overline{A}_{k+1}, \ldots, \overline{A}_n$ are independent, for all $k, 1 \leq k \leq n$ and all n.

2.6 The Monty Hall problem

This problem comes from a popular U.S. television show called "Let's Make a Deal." The show host, Monty Hall, would hide a valuable prize, say a car, behind one of three curtains. Both of the other two curtains are empty or have a non-prize, such as a goat. The contestant is invited to choose one of the curtains. Both of the other curtains are empty if the contestant's initial guess is correct. Monty Hall (with great flourish) opens, and, we'll assume, always opens, one of the remaining two curtains, showing it to be empty. He then asks the contestant whether he or she wishes to exchange the curtain originally chosen for the remaining one. Is it in the interest of the contestant to switch?

We have to specify what Monty Hall does when the contestant correctly chooses the curtain that hides the car. In this case, we'll suppose that Monty chooses with equal probability which of the two remaining curtains to open, neither of which contains the car. Therefore the identity of the unopened and unchosen curtain is irrelevant.

It is natural to suppose, because Monty Hall always opens a curtain which does not conceal the prize, that no information has been conveyed. Thus there would be two equally likely curtains, and the contestant can switch or not, with probability $1/2$ of winning either way. This line of reasoning, while plausible, is wrong.

Suppose that the contestant views the three curtains as equally likely to contain the prize. Then the contestant's first choice has probability $1/3$ of being correct. If his strategy is not to switch, the contestant wins only in the case that this initial choice was correct, which continues to have probability $1/3$. When the contestant chooses whether to switch, the choice is between two curtains, of which one contains the prize. The contestant wins either by switching or by not switching. The probability of winning by switching is then $2/3$. Intuitively, by switching, the contestant gets the probability content of both of the curtains not initially chosen.

Perhaps the point is clearer if expressed in mathematical notation. Define two random variables, C, an indicator that the curtain you chose initially had the prize, and W, an indicator that you win the prize. We want to find $P\{W = 1\}$ for both strategies, "switch" and "don't switch." In both cases, the analysis proceeds by conditioning on C, as follows:

$$\begin{aligned} P\{W = 1\} = & P\{(W = 1)|(C = 1)\}P\{C = 1\} \\ & + P\{(W = 1)|(C = 0)\}P\{C = 0\}, \end{aligned}$$

using 2.9.

Under the "don't switch" strategy, check that $P\{(W = 1)|(C = 1)\} = 1$ and $P\{(W =$

$1)|(C = 0)\} = 0$. Since $P\{C = 1\} = 1/3$ and $P\{C = 0\} = 2/3$, by substitution $P\{W = 1\} = 1 \cdot 1/3 + 0 \cdot 2/3 = 1/3$.

Now under the "switch" strategy, the consequences change as follows: $P\{(W = 1)|(C = 1)\} = 0$ and $P\{(W = 1)|(C = 0)\} = 1$. Because $P\{C = 1\} = 1/3$, $P\{W = 1\} = 0 \cdot 1/3 + 1 \cdot 2/3 = 2/3$. We conclude therefore that switching is the better choice.

This point is perhaps even clearer if we consider a more general problem. Imagine that curtains 1, 2 and 3 have probabilities, respectively, of p_1, p_2 and p_3 of having the car. By necessity $p_1 + p_2 + p_3 = 1$. Suppose the contestant's strategy is to choose a curtain i and not switch. With this strategy the contestant has probability p_i of success, so the best that can be done is $\max\{p_1, p_2, p_3\}$. However, if the contestant chooses curtain i and then switches, his probability of success is $1 - p_i$. The best curtain to choose maximizes $\{1-p_1, 1-p_2, 1-p_3\}$, and therefore is the *least* probable curtain. For example, suppose that the contestant observes that one of the curtains, say curtain 1, does not contain the car, so $p_1 = 0$. Wisely, the contestant chooses curtain 1, is shown that one of the other curtains is empty, switches, and wins for sure! Since $\max\{p_1, p_2, p_3\} < \max\{1 - p_1, 1 - p_2, 1 - p_3\}$ unless some $p_i = 1$, it always pays to choose the least likely curtain, and switch.

The Monty Hall problem became popular after being discussed in a newspaper column by Marilyn Vos Savant. It generated considerable mail, including letters from Ph.D. mathematicians eager to prove that her (correct) solution was wrong! (See Tierney (1991).)

2.6.1 Exercises

1. State in your own words what the Monty Hall problem is.
2. Suppose there are three prisoners. It is announced that two will be executed tomorrow, and one set free. But the prisoners do not know who will be executed and who will be set free. Prisoner A asks the jailer to tell him the name of one prisoner (B or C) who will be executed, arguing that this will not tell him his own fate. The jailer agrees, and says that prisoner B is to be executed.
   Prisoner A reasons that before he had probability 1/3 of being freed, and now he has probability 1/2. The jailer reasons that nothing has changed, and Prisoner A's probability of surviving is still 1/3.
   Who is correct, and why? In what ways is this problem similar to, or different from, the Monty Hall problem?
3. Do a simulation in R to study the Monty Hall problem. Run it long enough to satisfy yourself about the probability of success with the "switch" and "don't switch" strategies.
4. Reconsider the simpler version of the Monty Hall problem, assuming $p_1 = p_2 = p_3 = 1/3$. Suppose that you have chosen box 1. If the prize is in box 2, Monty Hall must open box 3 and show you that it is empty. Similarly, if the prize is in box 3, Monty Hall must show you that box 2 is empty. But if the prize is in box 1 (so your initial choice is correct), Monty Hall has a choice of whether to show you box 2 or box 3. Suppose in this case you have probability $q_{2,3}$ that he chooses box 2, and you have probability $q_{3,2} = 1 - q_{2,3}$ that he chooses box 3.
   (a) What is your optimal strategy as a function of $q_{2,3}$?
   (b) What is your probability of getting the prize using your optimal strategy?
   (c) Show that when $q_{2,3} = q_{3,2} = 1/2$, your optimal strategy and resulting probability of getting the prize coincide with those found in the text for the case $p_1 = p_2 = p_3 = 1/3$.
5. Now consider the general case, where it is not necessarily assumed that $p_1 = p_2 = p_3 = 1/3$. If you choose box i and the prize is in box i, Monty Hall has a choice between showing you that box $j \neq i$ is empty and showing you that box $k \neq i$ is empty (where $j \neq k$).

Suppose you have probability $q_{j,k}$ that he chooses box j, and probability $q_{k,j} = 1 - q_{j,k}$ that he chooses box k.

(a) As a function of $p_1, p_2, p_3, q_{1,2}, q_{1,3}$ and $q_{2,3}$ find your optimal strategy.

(b) What is your probability of getting the prize following your optimal strategy?

(c) Show that when $q_{1,2} = q_{1,3} = q_{2,3} = 1/2$, your optimal strategy and probability of getting the prize are those found in the text.

2.7 Gambler's Ruin problem

Imagine two players, A and B. A starts with i dollars, and B starts with $n - i$ dollars. They play many independent sessions. A wins a session with probability p and gains a dollar from B. Otherwise A pays a dollar to B with probability $q = 1 - p$. They play until one or the other has zero dollars, which means this player is ruined.

Let a_i be the probability that A ruins B, if A starts with i dollars. Then the numbers a_i satisfy the following:

$$a_0 = 0 \tag{2.18}$$

$$a_n = 1 \tag{2.19}$$

$$a_i = pa_{i+1} + qa_{i-1}, \qquad 1 \leq i \leq n-1. \tag{2.20}$$

Equation (2.20) is justified by the following argument: Suppose A starts with i dollars. If he wins a session, which he will with probability p, his fortune becomes $i + 1$ dollars. On the other hand, if he loses a session, which he will with probability $q = 1 - p$, his fortune becomes $i - 1$ dollars. In both cases, with his new fortune his chance of winning the game is the same as if he started with his new fortune.

This reasoning is related to (2.7) as follows: Let R_i be the event that A ruins B starting with i dollars, so $a_i = P\{R_i\}$ for $i = 1, \ldots, n-1$. Let S be the event that A wins the next session, so $P\{S\} = p$ and $P\{\overline{S}\} = q$. The event that A ruins B starting with i dollars given a success on the next session is exactly the event that A ruins B starting with $i+1$ dollars. Thus $P\{R_i \mid S\} = P\{R_{i+1}\} = a_{i+1}$. Similarly $P\{R_i \mid \overline{S}\} = P\{R_{i-1}\} = a_{i-1}$. Then (2.7) applies, yielding

$$\begin{aligned} a_{i+1} = P\{R_i\} &= P\{R_i \mid S\}P\{S\} + P\{R_i \mid \overline{S}\}P\{\overline{S}\} \\ &= a_{i+1}p + a_{i-1}q, \end{aligned}$$

which is (2.20).

Subtracting a_i from both sides of (2.20) yields $0 = p(a_{i+1} - a_i) + q(a_{i-1} - a_i)$, or, reorganized

$$a_{i+1} - a_i = r(a_i - a_{i-1}), \text{ for } 1 \leq i \leq n-1, \tag{2.21}$$

where $r = q/p$.

Writing out instances of (2.21), we have

$$\begin{aligned} a_2 - a_1 &= ra_1 \\ a_3 - a_2 &= r(a_2 - a_1) = r^2 a_1, \end{aligned}$$

etc., and, in general $a_i - a_{i-1} = r^{i-1}a_1, 1 \leq i \leq n$. Adding these together to create a telescoping series,

$$\begin{aligned} a_i - a_1 &= (r^{i-1} + r^{i-2} + \ldots + r)a_1, \text{ or} \\ a_i &= (r^{i-1} + \ldots + r + 1)a_1 = \left(\sum_{j=0}^{i-1} r^j\right) a_1. \end{aligned}$$

In particular,

$$
\begin{aligned}
1 = a_n &= \left(\sum_{j=0}^{n-1} r^j \right) a_1, \text{ so} \\
a_1 &= \frac{1}{\sum_{j=0}^{n-1} r^j}.
\end{aligned}
$$

Therefore

$$
a_i = \frac{\sum_{j=0}^{i-1} r^j}{\sum_{j=0}^{n-1} r^j} \quad i = 0, \ldots, n. \tag{2.22}
$$

When $r = 1, a_i = i/n$ for $i = 0, \ldots, n$. When $r \neq 1$, a neater form is available for a_i. In order to use it, a short digression is necessary. A series is called a geometric series if it is the sum of successive powers of a number.

Both the numerator and denominator of (2.22) are in the form of a geometric series

$$
1 + r + r^2 + \ldots + r^k = G. \tag{2.23}
$$

I multiply G by $(1 - r)$, but will write the result in a special way to make cancellations obvious:

$$
\begin{aligned}
G(1-r) &= 1 + r + r^2 + \ldots + r^k \\
&\quad\;\; - r - r^2 - \ldots - r^k - r^{k+1} \\
&= 1 \qquad\qquad\qquad\qquad\quad - r^{k+1}.
\end{aligned}
$$

Thus $G(1 - r) = 1 - r^{k+1}$, or

$$
G = \frac{1 - r^{k+1}}{1 - r}. \tag{2.24}
$$

Applying (2.24) to (2.22) yields

$$
a_i = \frac{(1 - r^i)/(1 - r)}{(1 - r^n)/(1 - r)} = \frac{1 - r^i}{1 - r^n} = \frac{r^i - 1}{r^n - 1} \; i = 0, 1, \ldots, n, \tag{2.25}
$$

provided $r \neq 1$.

Formula (2.24) has been derived under the assumption that $r \neq 1$. This assumption is necessary in order to avoid dividing by zero in (2.25). However, it is reasonable to hope that as r approaches 1, a_i approaches i/n as an inspection of (2.22) suggests. Let's see if this is the case.

As r approaches 1 (written $r \to 1$), $r^i - 1 \to 0$, as does $r^n - 1$. Therefore both the numerator and denominator in (2.25) approach 0. There is a special technique in calculus to handle this situation, known as L'Hôpital's Rule.

In general, suppose we want to evaluate

$$
\lim_{x \to x_0} \frac{f(x)}{g(x)} \tag{2.26}
$$

where $\lim_{x \to x_0} f(x) = 0$ and $\lim_{x \to x_0} g(x) = 0$. For instance in the Gambler's Ruin example, $x = r, x_0 = 1, f(x) = r^i - 1$ and $g(x) = r^n - 1$.

We will suppose that $f(x)$ and $g(x)$ are continuous and differentiable at x_0. Now

$$
\lim_{x \to x_0} \frac{f(x)}{g(x)} = \lim_{x \to x_0} \frac{f(x) - f(x_0)}{g(x) - g(x_0)} = \lim_{x \to x_0} \frac{\frac{f(x) - f(x_0)}{x - x_0}}{\frac{g(x) - g(x_0)}{x - x_0}} = \frac{\lim_{x \to x_0} \frac{f(x) - f(x_0)}{x - x_0}}{\lim_{x \to x_0} \frac{g(x) - g(x_0)}{x - x_0}} = \frac{f'(x_0)}{g'(x_0)}.
$$

The first step is justified because zero is being subtracted from the numerator and denominator, the second step because the numerator and denominator are being divided by the same quantity, $x - x_0$. The third step is a property of the limit of ratios, and the last step comes from the definition of the derivative.

In our application, $f'(x_0) = \frac{d}{dr}(r^i - 1)\Big|_{r=1} = ir^{i-1}\Big|_{r=1} = i$. Similarly, $g'(x_0) = \frac{d}{dr}(r^n - 1)\Big|_{r=1} = nr^{n-1}\Big|_{r=1} = n$. Hence

$$\lim_{r\to 1} \frac{r^i - 1}{r^n - 1} = i/n, \tag{2.27}$$

which is the result sought. Hence with the understanding that L'Hôpital's Rule applies, we can write

$$a_i = \frac{r^i - 1}{r^n - 1} = \frac{(q/p)^i - 1}{(q/p)^n - 1} \quad i = 0, \ldots, n \tag{2.28}$$

without restriction on $r = q/p$.

Let's see how this result works in an example. Imagine a gambler who has \$98, against a "house" with only \$2. However, the house has the advantage in the game: the gambler has probability 0.4 of winning a session, while the house has probability 0.6. What is the gambler's probability of winning the house's \$2 before he goes broke?

Here $i = 98, n = 98 + 2 = 100, p = 0.4$ and $q = 0.6$. Then $r = q/p = 1.5$, and (2.28) yields

$$a_{98} = \frac{(1.5)^{98} - 1}{(1.5)^{100} - 1} \simeq \frac{(1.5)^{98}}{(1.5)^{100}} = \frac{1}{(1.5)^2} = (2/3)^2 = 4/9. \tag{2.29}$$

Thus, despite the gambler's enormously greater initial stake, he has less than a 50% chance of winning the house's \$2 before losing his \$98 to the house!

2.7.1 *Changing stakes*

Return now to the general Gambler's Ruin problem, but suppose now that instead of playing for \$1 each time, the two gamblers play instead for \$0.50. Then gambler A starts with $2i$ \$0.50 pieces, and needs to win a net of $2(n - i)$ \$0.50 pieces to ruin gambler B. Then in this new game with smaller stakes, gambler A's probability of ruining gambler B is $(r^{2i} - 1)/(r^{2n} - 1)$. In greater generality, if dollars are divided into k parts, gambler A has probability

$$a_i(k) = \frac{r^{ki} - 1}{r^{kn} - 1} \tag{2.30}$$

of ruining gambler B.

To show how this works, reconsider the example discussed at the end of section 2.7. There $p = 0.4$ and $q = 0.6$. Again we take $i = \$98$ and $n - i = \$2$, but now suppose $k = 2$. Then applying (2.30)

$$a_{98}(2) = \frac{(1.5)^{196} - 1}{(1.5)^{200} - 1} \approx \frac{(1.5)^{196}}{(1.5)^{200}} = (2/3)^4. \tag{2.31}$$

Hence the shift to lower stakes, \$0.50 instead of \$1.00, has substantially reduced gambler A's probability of winning. The purpose of this subsection is to explore why this occurs.

In keeping with the example, suppose that gambler A is the less skilled player, so $q > p$ and $r > 1$. Supposing k to be large, we have

$$a_i(k) = \frac{r^{ki} - 1}{r^{kn} - 1} \approx \frac{r^{ki}}{r^{kn}} = r^{k(i-n)}. \tag{2.32}$$

Since $i < n$, $\lim_{k\to\infty} a_i(k) = 0$. This shows that if the stakes are very small, the less skilled gambler is almost sure to lose.

Can a similar analysis apply to the case when the stakes are larger? Returning to our familiar example, suppose now the players are betting \$2 in each session. Then we have

$$a_{98}(.5) = \frac{(1.5)^{49} - 1}{(1.5)^{50} - 1} \approx \frac{(1.5)^{49}}{(1.5)^{50}} = 2/3. \tag{2.33}$$

Thus shifting to higher stakes has substantially improved the chances for the less skilled player A.

Although the exact interpretation of the limiting sequence is problematic, the limiting behavior of (2.30) as $k \to 0$ can be investigated, as follows. By inspection, as $k \to 0$ both the numerator and the denominator of (2.30) approach 0. Therefore we must apply L'Hôpital's Rule. Taking the derivative of the numerator yields

$$\frac{d}{dk}\left\{r^{ki} - 1\right\} = \frac{d}{dk}\left\{e^{ki\log r}\right\} = e^{ki\log r}(i\log r) = r^{ki}(i\log r). \tag{2.34}$$

Similarly, the derivative of the denominator is

$$\frac{d}{dk}\{r^{kn} - 1\} = r^{kn}(n\log r). \tag{2.35}$$

Therefore

$$\lim_{k\to 0} a_i(k) = \frac{r^{ki}(i\log r)|_{k=0}}{r^{kn}(n\log r)|_{k=0}} = i/n. \tag{2.36}$$

Remarkably, then, in this theoretical limit as the stakes get very large, the skill advantage of the more skilled player disappears, and the less-skilled player has the same probability of being successful, i/n, as would be the case if $p = q = 1/2$!

To understand further the behavior of $a_i(k)$ it would be good to check that it decreases as k increases. This is done as follows:

Using (2.34) and (2.35), I have

$$\begin{aligned}\frac{d}{dk}a_i(k) &= \frac{d}{dk}\left\{\frac{r^{ki}-1}{r^{kn}-1}\right\} = \frac{(r^{kn}-1)r^{ki}\cdot i\log r - (r^{ki}-1)r^{kn}\cdot n\log r}{(r^{kn}-1)^2} \\ &= \left(\frac{ir^{ki}}{r^{ki}-1} - \frac{nr^{kn}}{r^{kn}-1}\right)\left\{\left(\frac{r^{ki}-1}{r^{kn}-1}\right)\log r\right\}.\end{aligned} \tag{2.37}$$

The second factor, in curly brackets, is positive. Hence I study the sign of the first factor. The first factor is negative if the function

$$f(i) = \frac{ix^i}{x^i - 1} \tag{2.38}$$

is decreasing in i, for fixed $x = r^k > 1$. To examine this, let

$$\begin{aligned}\Delta f(i) =& f(i) - f(i+1) \\ =& \frac{ix^i}{x^i - 1} - \frac{(i+1)(x^{i+1})}{(x^{i+1}-1)} \\ =& \frac{1}{(x^i-1)(x^{i+1}-1)}\{ix^i(x^{i+1}-1) - x^{i+1}(x^i-1)(i+1)\} \\ =& K\{ix^{2i+1} - ix^i - (i+1)x^{2i+1} + (i+1)x^{i+1}\} \\ =& K\{-ix^i - x^{2i+1} + (i+1)x^{i+1}\} \\ =& x^i K[-i + x^{i+1} + (i+1)x] \\ \text{where } K =& 1/(x^i-1)(x^{i+1}-1) > 0.\end{aligned} \tag{2.39}$$

If it can be shown that the function

$$g(x) = -i - x^{i+1} + (i+1)x \tag{2.40}$$

is negative for all i and all $x > 1$, then (2.37) will be shown to be negative.

$$\text{Now } g(1) = -i - 1 + (i+1) = 0 \tag{2.41}$$

for all i. Furthermore

$$g'(x) = -(i+1)x^i + (i+1) = (i+1)(1-x^i) < 0 \tag{2.42}$$

for all i and all $x > 1$. Hence

$$g(x) < 0 \tag{2.43}$$

for all i and $x > 1$, as was to be shown. Thus we have

$$\frac{d}{dk}a_i(k) < 0. \tag{2.44}$$

As the stakes decrease (k increases), the weaker player's probability of winning, $a_i(k)$ decreases, from $a_i(0) = i/n$ to $a_i(\infty) = 0$.

Figure 2.5 shows a plot of $a_i(k)$ for the example.

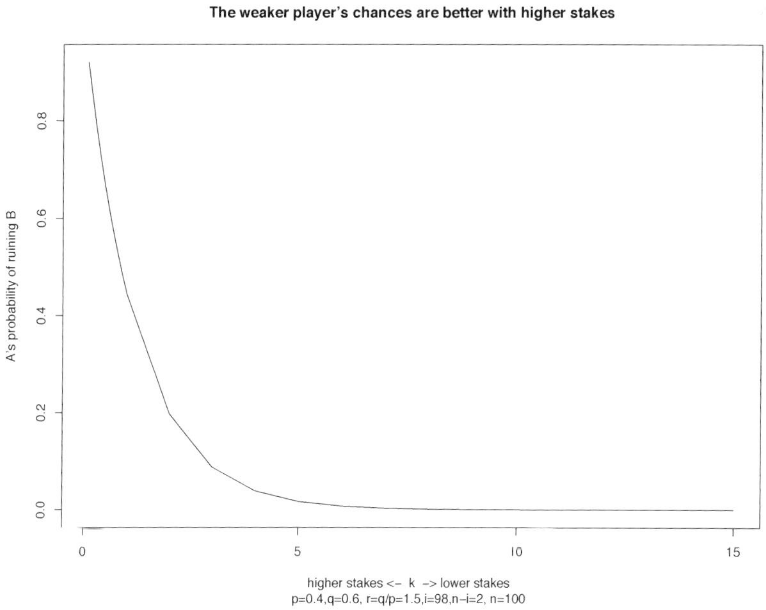

Figure 2.5: The probability of the weaker player winning as a function of the stakes in the example.

```
Commands:  k=c(seq(.1,.9,.1),1:15)
a=(((1.5)**(k*98))-1)/(((1.5)**(k*100))-1)
plot(k,a,xlab="higher stakes <-  k  -> lower stakes",
type="l",ylab="A's probability of ruining B",
main="The weaker player's chances are better
with higher stakes",
sub="p=0.4,q=0.6, r=q/p=1.5,i=98,n-i=2, n=100")
```

This finding is qualitatively similar to the finding that in roulette, where a player has a 1/38 probability of gaining 36 times the amount bet, bold play is optimal in having the best chance of achieving a fixed goal (see Dubins and Savage (1965), Smith (1967) and Dubins (1968)).

2.7.2 *Summary*

Gambler A, who starts with i dollars, plays against Gambler B, with $n - i$ dollars, until one or the other has no money left. A wins a session and a dollar with probability p and loses the session and a dollar with probability $q = 1 - p$. A's probability of ruining B is

$$a_i = \frac{(q/p)^i - 1}{(q/p)^n - 1}.$$

This formula is to be understood, when $q = p$, as interpreted by L'Hôpital's Rule.

The less skilled player has a greater chance of success if the stakes are large than if the stakes are small.

2.7.3 *References*

Two fine books on combinatorial probability that contain lots of entertaining examples are Feller (1957) and Andel (2001).

2.7.4 *Exercises*

1. Vocabulary. Explain in your own words:
   (a) Gambler's Ruin
   (b) Geometric Series
   (c) L'Hôpital's Rule
2. When $p = 0.45$, $i = 90$ and $n = 100$, find a_i.
3. Suppose there is probability p that A wins a session, q that B wins, and t that a tie results, with no exchange of money, where $p + q + t = 1$. Find a general expression for a_i, and explain the result.
4. Now suppose that the probability that A wins a session is p_i if he has a current fortune of a_i, and the probability that B wins is $q_i = 1 - p_i$. Again, find a general expression for a_i as a function of the p's and q's.
5. Use R to check the accuracy of the approximation in (2.29).
6. Consider the Gambler's Ruin problem from B's perspective. B starts with a fortune of $n - i$, and has probability q of winning a session, and hence $p = 1 - q$ of losing a session. Let b_{n-i} be the probability that B, starting with a fortune of $n - i$, ruins A. Then

$$b_{n-i} = \frac{(r')^{n-i} - 1}{(r')^n - 1}, \text{ where } r' = p/q = 1/r.$$

   Prove that $a_i + b_{n-i} = 1$ for all integers $i \leq n$, and all positive p and q satisfying $p + q = 1$. Interpret this result.

2.8 Iterated expectations and independence of random variables

This section introduces two essential tools for dealing with more than one random variable; iterated expectations and independence. We begin with iterated expectations.

Suppose X and Y are two random independence variables taking only a finite number of values each. Using the same notation as in section 1.5, let

$$P\{X = x_i, Y = y_j\} = p_{i,j},$$

where

$$\sum_{i=1}^{n} p_{i,j} = p_{+,j} > 0 \quad j = 1, \ldots, m, \sum_{j=1}^{m} p_{i,j} = p_{i,+} > 0 \quad i = 1, \ldots, n,$$

and

$$\sum_{j=1}^{m} p_{+,j} = \sum_{i=1}^{n} p_{i,+} = 1.$$

Now the conditional probability that $X = x_i$, given $Y = y_j$, is

$$P\{X = x_i | Y = y_j\} = \frac{P\{X = x_i, Y = y_j\}}{P\{Y = y_j\}} = \frac{p_{i,j}}{p_{+,j}}. \tag{2.45}$$

Because this equation gives a probability for each possible value of X provided $Y = y_j$, we can think of it as a random variable, written $X|Y = y_j$. This random variable takes the value x_i with probability $p_{i,j}/p_{+,j}$. Hence this random variable has an expectation, which is written

$$E[X|Y = y_j] = \sum_i x_i p_{i,j}/p_{+,j}.$$

Now for various values of y_j, this conditional expectation can itself be regarded as a random variable, taking the value $\sum_i x_i p_{i,j}/p_{+,j}$ with probability $p_{+,j}$.

In turn, its expectation is written as

$$\begin{aligned} E\{E[X|Y]\} &= \sum_{j=1}^{m} p_{+,j} \sum_{i=1}^{n} x_i p_{i,j}/p_{+,j} \\ &= \sum_{j=1}^{m} \sum_{i=1}^{n} x_i p_{i,j} \\ &= E[X]. \end{aligned} \tag{2.46}$$

This is the law of iterated expectations. It plays a crucial role in the next chapter.

To see how the law of iterated expectations works in practice, consider the special case in which X and Y are the indicator functions of two events, A and B, respectively. To evaluate the double expectation, one has to start with the inner expectation, $E[X|Y]$. (I remind you that what $E[X|Y]$ means is the expectation of X conditional on each value of Y.) Then

$$\begin{aligned} E[X|Y = 1] &= E[I_A | I_B = 1] \\ &= 1P\{I_A = 1 | I_B = 1\} + 0\ P\{I_A = 0 | I_B = 1\} \\ &= P\{I_A = 1 | I_B = 1\} = P\{A|B\}. \end{aligned}$$

Similarly,

$$E[X|Y = 0] = E[I_A | I_B = 0] = E[I_A | I_{\overline{B}} = 1] = P\{A|\overline{B}\}.$$

Now I can evaluate the outer expectation, which is the expectation of $E[X|Y]$ over the possible values of Y, as follows:

$$\begin{aligned} E[E[X|Y]] &= E[E[I_A|I_B]] \\ &= P\{A|B\}P\{B\} + P\{A|\overline{B}\}P\{\overline{B}\} \\ &= P\{AB\} + P\{A\overline{B}\} \\ &= P\{A\} = E[I_A] = E[X]. \end{aligned}$$

The second topic of this section is independence of random variables. Recall from section 2.5 that events A and B are independent if learning that A has occurred does not change your probability for B. The same idea is applied to random variables, as follows:

When the distribution of $X|Y = y_j$ does not depend on j, we have

$$P\{X = x_i|Y = y_j\} = \frac{P\{X = x_i, Y = y_j\}}{p_{+,j}} = \frac{p_{i,j}}{p_{+,j}}$$

must not depend on j, but of course can still depend on i. So denote $p_{i,j}/p_{+,j} = k_i$ for some numbers k_i. Now

$$p_{i,+} = \sum_{j=1}^{m} p_{i,j} = \sum_{j=1}^{m} k_i p_{+,j} = k_i \sum_{j=1}^{m} p_{+,j} = k_i.$$

Hence we have

$$P\{X = x_i|Y = y_j\} = \frac{p_{i,j}}{p_{+,j}} = p_{i,+} = P\{X = x_i\} \text{ for all } j.$$

In this case the random variables X and Y are said to be independent.

If X and Y are independent, and A and B are any two sets of real numbers, the events $X \in A$ and $Y \in B$ are independent events. This can be taken as another definition of what it means for X and Y to be independent.

Intuitively, the idea behind independence is that learning the value of the random variable $Y = y_j$ does not change the probabilities you assign to $X = x_i$, as expressed by the formula

$$P\{X = x_i|Y = y_j\} = P\{X = x_i\}. \tag{2.47}$$

An important property of independent random variables is as follows: If g and h are real-valued functions and X and Y are independent, then

$$\begin{aligned} E[g(X)h(Y)] &= \sum_{i=1}^{n}\sum_{j=1}^{m} g(x_i)h(y_j)p_{i,j} = \sum_{i=1}^{n}\sum_{j=1}^{m} g(x_i)h(y_j)p_{i,+}p_{+,j} \\ &= \sum_{i=1}^{n} g(x_i)p_{i,+} \sum_{j=1}^{m} h(y_j)p_{+,j} = E[g(X)]E[h(Y)]. \end{aligned} \tag{2.48}$$

When X and Y are independent, (2.48) permits certain expectations to be calculated efficiently. This will be used in section 2.11 of this chapter, and reappears as a standard tool used throughout the rest of the book.

When the random variables are not independent, we get as far as the first equality, but cannot use the relation $p_{i,j} = p_{i,+}p_{+,j}$ to go further.

The issue of how to define independence for a set of more than two random variables is similar to the issue of how to define independence for a set of more than two events. For the same reason as discussed in section 2.5, a definition based on pairwise independence does not suffice. Consequently we define a set of random variables $X_1, \ldots, X_n$ as independent if for every choice of sets of real numbers $A_1, A_2, \ldots, A_n$, the events $X_1 \in A_1$, $X_2 \in A_2, \ldots, X_n \in A_n$ are independent events.

Finally, we address the question of a definition for conditional independence. Conditional independence is a crucial tool in the construction of statistical models. Indeed much of statistical modeling can be seen as defining what variables W must be conditioned upon to make the observations $X_1, \ldots, X_n$ conditionally independent given W.

Two random variables X and Y are said to be conditionally independent given a third random variable W if $X|W$ is independent of $Y|W$ for each possible value of W. This relationship is denoted

$$X \perp\!\!\!\perp Y|W.$$

Again, a set of random variables $X_1, \ldots, X_n$ are said to be conditionally independent given W if and only if $X_1|W$, $X_2|W, \ldots, X_n|W$ are independent for each possible value of W.

2.8.1 *Summary*

When X and Y take only finitely many values, the law of iterated expectations applies, and says that

$$E\{E[X|Y]\} = E(X).$$

Random variables $X_1, \ldots, X_n$ are said to be independent if and only if the events $X_1 \in A_1$, $X_2 \in A_2, \ldots, X_n \in A_n$ are independent events for every choice of the sets of real numbers $A_1, A_2, \ldots, A_n$. Random variables $X_1, \ldots, X_n$ are said to be conditionally independent given W if and only if the random variables $X_1|W$, $X_2|W, \ldots, X_n|W$ are independent for each possible value of the random variable W.

2.8.2 *Exercises*

1. Vocabulary. Explain in your own words:
   (a) independence of random variables
   (b) iterated expectations
2. Show that if X and Y are random variables, and X is independent of Y, then Y is independent of X.
3. Show that if A and B are independent events, then I_A and I_B are independent random variables.
4. Show the converse of problem 3: if I_A and I_B are independent indicator random variables, then A and B are independent events.
5. Consider random variables X and Y having the following joint distribution:
   $P\{X = 1, Y = 1\} = 1/8$
   $P\{X = 1, Y = 2\} = 1/4$
   $P\{X = 2, Y = 1\} = 3/8$
   $P\{X = 2, Y = 2\} = 1/4.$
   Are X and Y independent? Prove your answer.
6. For the same random variables as in the previous problem, compute
   a) $E\{X|Y = 1\}$
   b) $E\{Y|X = 2\}$
7. Suppose

$$P\{X = 1, Y = 1\} = x, \ P\{X = 1, Y = 2\} = y, \text{ and}$$
$$P\{X = 2, Y = 1\} = z,$$

where x, y and z are three numbers satisfying

$$x + y + z = 1, \quad x > 0, \quad y > 0, \quad z > 0.$$

Are there values of x, y and z such that the random variables X and Y are independent? Prove your answer.

8. Suppose $X_1, \ldots, X_n$ are independent random variables. Let $m < n$, so that $X_1, \ldots, X_m$ are a subset of $X_1, \ldots, X_n$. Show that $X_1, \ldots, X_m$ are independent.

2.9 The binomial and multinomial distributions

The binomial distribution is the distribution of the number of successes (and failures) in n independent trials, each of which has the same probability p of success. Thus the outcomes of the trials are separated into two categories, success and failure. The multinomial distribution is a generalization of the binomial distribution in which each trial can have one of several outcomes, not just two, again assuming independence and constancy of probability.

Recall from section 1.5 the numbers $\binom{n}{j,n-j} = \frac{n!}{j!(n-j)!}$. We here study these numbers further. Consider the expression $(x+y)^n = (x+y)(x+y)\ldots(x+y)$, where there are n factors. This can be written as the sum of $n+1$ terms of the form $a_j x^j y^{(n-j)}$. The question is what the coefficients a_j are that multiply these powers of x and y. To contribute to the coefficient of the term $x^j y^{(n-j)}$ there must be j factors that contribute an x and $(n-j)$ that contribute a y. Thus we need the number of ways of dividing the n factors into one group of size j (which contribute an x), and another group of size $(n-j)$, (which contribute a y). This is exactly the number we discussed above, n choose j and $(n-j)$. Therefore

$$(x+y)^n = \sum_{j=0}^{n} \binom{n}{j, n-j} x^j y^{n-j},$$

which is known as the binomial theorem.

Next, consider the following array of numbers, known as Pascal's triangle:

$$\begin{array}{ccccccccc} & & & & 1 & & & & \\ & & & 1 & & 1 & & & \\ & & 1 & & 2 & & 1 & & \\ & 1 & & 3 & & 3 & & 1 & \\ 1 & & 4 & & 6 & & 4 & & 1 \end{array}$$

Can you write down the next line? What rule did you use to do so?

The number in Pascal's triangle located on row $n+1$ and at horizontal position $j+1$ from the left and $n-j+1$ from the right is exactly the number $\binom{n}{j,n-j}$. We need the "+1's" because n and j start from zero.

Pascal's triangle can be built by putting 1's on the two edges, and using the relationship

$$\binom{n-1}{j-1, n-j} + \binom{n-1}{j, n-j-1} = \binom{n}{j, n-j} \tag{2.49}$$

to fill in the rest of row n. (You are invited to prove (2.49) in section 2.9.4, exercise 1.) This equation is analogous to the way differential equations are thought of (see, for example, Courant and Hilbert (1989)). Here the relation $\binom{n}{0,n} = 1$ is like a boundary condition, and (2.49) is like a law of motion, moving from the $(n-1)^{st}$ row to the nth row of Pascal's triangle.

Finally, consider n independent flips of a coin with constant probability p of tails and

$1-p$ of heads. Each specific pattern of j tails and $n-j$ heads has probability $p^j(1-p)^{n-j}$. How many patterns are there with j tails and $n-j$ heads? Exactly $\binom{n}{j,n-j}$.

Suppose X is the number of tails in n independent tosses. Then

$$P\{X=j\} = \binom{n}{j, n-j} p^j (1-p)^{n-j}. \tag{2.50}$$

How do we know that

$$\sum_{j=0}^{n} P\{X=j\} = 1?$$

This is true because

$$1 = (p + (1-p))^n = \sum_{j=0}^{n} \binom{n}{j, n-j} p^j (1-p)^{n-j} = \sum_{j=0}^{n} P\{X=j\},$$

using the binomial theorem.

In this case X is said to have a binomial distribution with parameters n and p, also written $X \sim B(n,p)$. The binomial distribution is the distribution of the sum of a fixed number n of independent random variables, each of which has the value 1 with some fixed probability p and is zero otherwise. The number n is often called the index of the binomial random variable.

We now extend the argument above by imagining many categories into which items might be placed, instead of just two. Suppose there are k categories, and we want to know how many ways there are of dividing n items into k categories, such that there are n_1 in category 1, n_2 in category 2, etc., subject of course to the conditions that $n_i \geq 0, i = 1, \ldots, k$ and $\sum_{i=1}^{k} n_i = n$. We already know that there are $n!$ ways of ordering the items; the first n_1 are assigned to category 1, etc. However, there are $n_1!$ ways of reordering the first n_1, which lead to the same choice of items for group 1. There are also $n_2!$ ways of reordering the second, etc. Thus the number sought must be

$$\frac{n!}{n_1! n_2! \ldots n_k!},$$

which is written $\binom{n}{n_1, n_2, \ldots n_k}$. (Now you can see why, in the case that $k=2$, I prefer to write $\binom{n}{j,n-j}$ rather than $\binom{n}{j}$ for $\frac{n!}{j!(n-j)!}$.)

Next, consider the expression $(x_1 + x_2 + \ldots + x_k)^n$, where there are n factors. Clearly this can be written in terms of the sum of products of the form $x_1^{n_1} x_2^{n_2} \ldots x_k^{n_k}$ times some coefficient. What is that coefficient? To contribute to this factor there must be n_1 x_1's, n_2 x_2's, etc., and the number of ways this can happen is exactly $\binom{n}{n_1, n_2, \ldots n_k}$. Hence we have the **multinomial theorem**: $(x_1 + x_2 + \ldots + x_k)^n = \sum \binom{n}{n_1, n_2 \ldots n_k} x_1^{n_1} x_2^{n_2} \ldots x_k^{n_k}$, where the summation extends over all $(n_1, n_2, \ldots n_k)$ satisfying $n_i \geq 0$ for $i = 1, \ldots, k$ and $\sum_{i=1}^{k} n_i = n$.

Multinomial coefficients $\binom{n}{n_1, n_2 \ldots n_k}$ satisfy the "law of motion"

$$\binom{n-1}{n_1 - 1, n_2, \ldots n_k} + \binom{n-1}{n_1, n_2 - 1, \ldots n_k} + \ldots + \binom{n-1}{n_1, n_2, \ldots n_k - 1} = \binom{n}{n_1, n_2, \ldots n_k}$$

and the "boundary conditions"

$$\binom{n}{n, 0, 0, \ldots 0} = \binom{n}{0, n, 0, \ldots 0} = \ldots = \binom{n}{0, 0, \ldots, 0, n} = 1.$$

Now consider a random process in which one and only one of k results can be obtained. Result i happens with probability p_i, where $p_i \geq 0$ and $\sum_{i=1}^{k} p_i = 1$. What is the probability, in n independent repetitions of the process, that the outcome will be that result 1 will happen n_1 times, result 2 n_2 times, ..., result k n_k times? Each such outcome has probability $p_1^{n_1} p_2^{n_2} \ldots p_k^{n_k}$, but how many ways are there of having such a result? Exactly $\binom{n}{n_1, n_2, \ldots, n_k}$ ways. Thus the probability of the specified number n_1 of result 1, n_2 of result 2, etc. is

$$\binom{n}{n_1, n_2, \ldots, n_k} p_1^{n_1} p_2^{n_2} \ldots p_k^{n_k}.$$

How do we know that these sum to 1? We use the multinomial theorem in the same way we used the binomial theorem when $k = 2$:

$$1 = (p_1 + p_2 + \ldots + p_k)^n = \sum \binom{n}{n_1, n_2, \ldots n_k} p_1^{n_1} p_2^{n_2} \ldots p_k^{n_k},$$

where the summation extends over all $(n_1, n_2 \ldots n_k)$ such that $n_i \geq 0$ for all i, and $\sum_{i=1}^{k} n_i = n$. In this case the number of results of each type is said to follow the multinomial distribution.

If $\mathbf{X} = (X_1, \ldots, X_k)$ has a multinomial distribution with parameters n and $\mathbf{p} = (p_1, \ldots, p_k)$, we write $\mathbf{X} \sim M(n, \mathbf{p})$. In this case $\mathbf{X}$ is the sum of a fixed number n independent vectors of length k, each of which has probability p_i of having a 1 in the ith position, and, if it does, it has zeros in all the other positions except the ith.

As an example of the multinomial distribution, suppose in a town there are 40% Democrats, 40% Republicans and 20% Independents. Suppose that a committee of 6 people are drawn independently at random from this town. What is the probability that the committee will consist of 3 Democrats, 2 Republicans and 1 Independent? Here there are $n = 6$ independent selections of people, who are divided into $k = 3$ categories, with probabilities $p_1 = .4$, $p_2 = .4$ and $p_3 = .2$. Consequently the probability sought is

$$\binom{6}{3, 2, 1} (.4)^3 (.4)^2 (.2)^1 = .12288.$$

If $(X_1, X_2, \ldots, X_k)$ have a multinomial distribution with parameters n and $(p_1, \ldots, p_k)$, then X_i has a binomial distribution with parameters n and p_i. This is because each of the n independent draws from the multinomial process either results in a count for X_i (which happens with probability p_i) or does not (which happens with probability $p_1 + p_2 + \ldots + p_{i-1} + p_{i+1} + \ldots + p_k = 1 - p_i$).

2.9.1 Refining and coarsening

The multinomial distribution has a special property that relates to the number of categories into which items are placed. First, I give the theorem; then I explain the vocabulary surrounding it.

Theorem 2.9.1. *If $X = (X_1, \ldots, X_k)$ has a multinomial distribution with parameters n and $(p_1, \ldots, p_k)$, then $(X_1, \ldots, X_{k-2}, X_{k-1} + X_k)$ has a multinomial distribution with parameters n and $(p_1, \ldots, p_{k-2}, p_{k-1} + p_k)$.*

Proof.

$$\begin{aligned} P\{(X_1, \ldots, X_{k-2}, X_{k-1}, +X_k) &= (n_1, n_2, \ldots, n_{k-3}, n^*)\} \\ &= \sum \binom{n}{n_1, n_2, \ldots, n_{k-2} n^*} \frac{n^*!}{n_{k-1}! n_k!} p_{k-1}^{n_{k-1}} p_k^{n_k}, \end{aligned}$$

where the sum extends over all pairs (n_{k-1}, n_k) that are non-negative and sum to n^*. But this sum is $(p_{k-1} + p_k)^{n^*}$ by the Binomial theorem. Hence

$$\begin{aligned} P\{(X_1, \ldots, X_{k-2}, X_{k-1} + X_k) &= (n_1, n_2, \ldots, n_{k-2}, n*)\} \\ &= \binom{n}{n_1, n_2, \ldots, n_{k-2}, n*} \prod_{i=1}^{k-2} p_i^{n_i} (p_{k-1} + p_k)^{n*}. \end{aligned} \tag{2.51}$$

Therefore, $(X_1, \ldots, X_2, \ldots, X_{k-2}, X_{k-1} + X_k)$ has a multinomial distribution with parameters n and $(p_1, \ldots, p_{k-2}, p_{k-1} + p_k)$. □

The distribution in (2.51) is referred to as a coarsening of the multinomial distribution with probability vector $(p_1, \ldots, p_k)$. Conversely, the latter is referred to as a refinement of (2.51).

2.9.2 *Why these distributions have these names*

The Latin word "nomen" means "name." The prefix "bi" means "two," "tri" means three and "multi" means many. Thus the binomial theorem and distribution separate objects into two categories, the trinomial into three and the multinomial into many.

2.9.3 *Summary*

$\mathbf{X} = (X_1, \ldots, X_k)$ has a multinomial distribution if $\mathbf{X}$ is the sum of n independent vectors of length k, each of which has probability p_i of having a 1 in the ith co-ordinate and 0 in all other co-ordinates where $\sum_{i=1}^{k} p_i = 1$. The special case $k = 2$ is called the binomial distribution; the special case $k = 3$ is called the trinomial distribution. The multinomial distribution permits graceful coarsening.

2.9.4 *Exercises*

1. Prove that $\binom{n}{j, n-j} + \binom{n}{j+1, n-(j+1)} = \binom{n+1}{j+1, n-j}$.
2. Prove the binomial theorem by induction on n.
3. Suppose the stronger team in the baseball World Series has probability $p = .6$ of beating the weaker team, and suppose that the outcome of each game is independent from the rest. What is the probability that the stronger team will win at least 4 of the 7 games in a World Series?
4. Prove $\binom{n-1}{n_1-1, n_2, \ldots n_k} + \binom{n-1}{n_1, n_2-1, \ldots, n_k} + \ldots + \binom{n-1}{n_1, n_2, \ldots, n_k-1} = \binom{n}{n_1, n_2, \ldots n_k}$.
5. Prove the multinomial theorem by induction on n.
6. Prove the multinomial theorem by induction on k.
7. When $k = 3$, what geometric shape generalizes Pascal's triangle?
8. Let X have a binomial distribution with parameters n and p. Find $E(X)$.
9. In section 2.5 we considered two possible opinions about the outcome of tossing a coin twice.
   (a) In the first, the probabilities offered were as follows:

$$P\{H_1H_2\} = P\{H_1T_2\} = P\{T_1H_2\} = P\{T_1T_2\} = 1/4.$$

   Does the number of heads in these two tosses have a binomial distribution? Why or why not?

(b) In the second,

$$P\{H_1H_2\} = P\{T_1T_2\} = P\{(H_1T_2 \cup T_1H_2)\} = 1/3.$$

Does the number of heads in these two tosses have a binomial distribution? Why or why not?

10. Suppose that the concessionaire at a football stadium finds that during a typical game, 20% of the attendees buy both a hot-dog and a beer, 30% buy only a beer, 20% buy only a hot-dog, and 30% buy neither. What is the probability that a random sample of 15 game attendees will have 3 who buy both, 2 who buy only a beer, 7 who buy only a hot-dog and 3 who buy neither?

11. (Continuing problem 10): What is the probability that a random sample of 15 game attendees will buy a hot-dog?

    (a) Compute this directly.

    (b) Compute this using Theorem 2.9.1.

2.10 The hypergeometric distribution: Sampling without replacement

There are many ways in which sampling can be done. Two of the most popular are sampling with replacement and sampling without replacement. In sampling with replacement the object sampled, after recording data from it, is returned to the population and might be sampled again. In sampling without replacement, the object sampled is not returned and therefore cannot be sampled again. Generally theory is easier for sampling with replacement because one continues to sample from the same population, but common sense suggests that one gets more information from sampling without replacement. As a practical matter when the population is large, the difference is negligible, because the chance of resampling the same object is vanishingly small. Nonetheless, it is worthwhile to understand the distribution that results from sampling without replacement, which is what this section is about.

Suppose that a bowl contains A apples, B bananas, C cantaloupes, D dates and E elderberries, for a total of $F = A+B+C+D+E$ fruits. Suppose that f fruits are sampled at random, with each fruit being equally likely to be chosen among those remaining at each stage, without replacement. There are exactly $\binom{F}{f,F-f}$ ways of doing this.

What proportion of those samples will contain exactly a apples, b bananas, c cantalopes, d dates and e elderberries?

The A apples have to be divided into the a that will be in the sample and the $A-a$ that will not. There are exactly $\binom{A}{a,A-a}$ distinct ways to do that. Similarly there are exactly $\binom{B}{b,B-b}$ ways to choose the bananas, etc. Thus the probability of getting exactly a apples, b bananas, etc. is

$$\frac{\binom{A}{a,A-a}\binom{B}{b,B-b}\binom{C}{c,C-c}\binom{D}{d,D-d}\binom{E}{e,E-e}}{\binom{F}{f,F-f}}$$

where $f = a+b+c+d+e$. This distribution is known as the hypergeometric distribution when there are only two kinds of fruit, and the multivariate hypergeometric distribution when there are more than two. It is denoted $HG(A, B, C, D, E)$.

As an example of the use of the hypergeometric distribution, a hand in bridge consists of 13 cards chosen at random without replacement from the 52 cards in the deck. In such a deck of cards, there are 13 of each suit: spades, hearts, diamonds and clubs. The probability that a hand of bridge has 6 spades, 4 hearts, 2 diamonds and 1 club is

$$\binom{13}{6,7}\binom{13}{4,9}\binom{13}{2,11}\binom{13}{1,12}\Big/\binom{52}{13,39} = .00196,$$

since there are $A = 13$ spades, of which $a = 6$ are chosen, $B = 13$ hearts, of which $b = 4$ are chosen, $C = 13$ diamonds, of which $c = 2$ are chosen, and $D = 13$ clubs, of which $d = 1$ is chosen. Then $F = A + B + C + D = 52$, and $f = 6 + 4 + 2 + 1 = 13$.

2.10.1 *Polya's Urn Scheme*

Polya's Urn Scheme is a generalization of sampling with and without replacement. Imagine an urn with b black balls and w white balls. A ball is chosen (equally likely!) and replaced, with k balls of the same color added to the urn. The process is repeated.

Sampling without replacement corresponds to $k = -1$, and must stop after $b + w$ stages, as the urn is empty at that point. Sampling with replacement corresponds to $k = 0$. Positive values of k indicate a "sticky" process. As k increases, strings of identical color are increasingly likely. As a model of infectious disease, k might model how infectious the disease is. Polya's Urn Scheme has many interesting properties. In this section, I'll give a little taste of them, and refer to the literature for the rest.

Let $X_i = 1$ if the ball drawn at the ith stage is black, and $X_i = 0$ otherwise.

Clearly $P\{X_1\} = b/(b + w)$. What about $P\{X_2\}$?

$$\begin{aligned} P\{X_2 = 1\} &= P\{X_2 = 1|X_1 = 1\}P\{X_1\} + P\{X_2 = 1|X_1 = 0\}P\{X_1 = 0\} \\ &= \tfrac{b+k}{b+k+w} \cdot \tfrac{b}{b+w} + \tfrac{b}{b+k+w} \cdot \tfrac{w}{b+w} \\ &= \tfrac{b[b+k+w]}{(b+w)[b+k+w]} = b/(b + w). \end{aligned}$$

In fact (not proved here) $P\{X_i = 1\} = b/(b + w)$ for all $i > 1$.

The calculation of $P\{X_2\}$ also shows

$$P\{X_2 = 1|X_1\} = \begin{cases} \frac{b+k}{b+k+w} & \text{if } X_1 = 1 \\ \frac{b}{b+k+w} & \text{if } X_1 = 0 \end{cases}.$$

What about $P\{X_1 = 1|X_2\}$?

$$P\{X_1 = 1|X_2 = 1\} = \frac{P\{X_1 = 1 \text{ and } X_2 = 1\}}{P\{X_2 = 1\}} = \frac{b + k}{b + k + w}.$$

Similarly

$$P\{X_1 = 1|X_2 = 0\} = \frac{P\{X_1 = 1 \text{ and } X_2 = 0\}}{P\{X_2 = 0\}} = \frac{w}{b + k + w}.$$

Therefore, (X_1, X_2) and (X_2, X_1) have the same distribution. If this property holds for every pair (X_i, X_j), then the X's are said to be exchangeable of order 2. The Polya Urn Scheme is exchangeable of order 2 (not proved here). If $(X_1, \ldots, X_n)$ has the same distribution as $(X_{\sigma(1)}, \ldots, X_{1,\sigma(n)})$ for any permutation σ, then $(X_1, \ldots, X_n)$ is said to be exchangeable of order n. The Polya Urn Scheme is exchangeable of order n for all finite n (also not proved here). If $(X_1, \ldots)$ has the same distribution as any finite permutation of its subscripts, then $(X_1, \ldots)$ is said to be exchangeable. The Polya Urn Scheme is exchangeable (not proved here either).

Exchangeable processes have some beautiful properties. Characterizations are given by DeFinetti (1937) for indication functions, later extended by Hewitt and Savage (1955) to arbitrary exchangeable sequences.

Is it necessary that $k \geq 0$ be an integer? The formulae still work for all real numbers k. It might stretch one's imagination to allow for non-integer numbers of balls in an urn, but stretching one's imagination is a benefit.

2.10.2 *Summary*

The hypergeometric distribution specifies the probability of each possible sample when the sampling is done at random *without* replacement. Polye's urn scheme puts sampling with and without replacement into a common framework.

2.10.3 *References*

For further reading on exchangeable sequences, see, for example, Chow and Teicher (1997); Kallenberg (2005) and Kingman (1978).

2.10.4 *Exercises*

1. Suppose there is a group of 50 people, from whom a committee of 10 is chosen at random. What is the probability that three specific members of the group, R, S and T, are on the committee?
2. How many ways are there of dividing 18 people into two baseball teams of 9 people each?
3. A deck of cards has four aces and four kings. The cards are shuffled and dealt at random to four players so that each has 13 cards. What is the probability that Joe, who is one of these four players, gets all four aces and all four kings?
4. Suppose a political discussion group consists of 30 Democrats and 20 Republicans. Suppose a committee of 8 is drawn at random without replacement. What is the probability that it consists of 3 Democrats and 5 Republicans?
5. Suppose that instead of the Polya Urn Scheme set-up, k_1 balls of the color of the ball drawn and k_2 balls of the opposite color are added to the urn.
   (a) Evaluate $P\{X_2 = 1\}$.
   (b) Show that $P\{X_1 = 1\} = P\{X_2 = 1\}$ if and only if either $k_2 = 0$ or $w = b$.
6. Suppose Polya's Urn consists of balls of several different colors, and $k \geq 0$ balls of the color drawn are added each time.
   Let $X_i^c = 1$ if color c is drawn in the ith draw.
   (a) Evaluate $P\{X_2^c = 1\}$.
   (b) Does $P\{X_2^c = 1\} = P\{X_1^c = 1\}$?

2.11 Variance and covariance

This section introduces the variance and the standard deviation, two measures of the variability of a random variable. It also introduces the covariance and the correlation, two measures of the extent to which two random variables are related.

Suppose X is a random variable, with expectation $E[X] = c$. The variance of X, written $V[X]$ is defined as follows:

$$V[X] = E\{(X - c)^2\}. \tag{2.52}$$

Because $(X - c)^2$ is non-negative, it follows that

$$V[X] \geq 0$$

for all random variables X. Furthermore $V[X] = 0$ only if $X = c$ with probability 1. $V[X]$ can be interpreted as a measure of spread or uncertainty in the random variable X. There's an alternative representation of $V[X]$ that's often useful:

$$\begin{aligned} V[X] &= E\{(X-c)^2\} = E\{X^2 - 2Xc + c^2\} = \\ &= E[X^2] - 2cE[X] + c^2 \\ &= E[X^2] - c^2 = E[X^2] - (E[X])^2; \end{aligned} \tag{2.53}$$

using (1.26) and (1.30).

Example: Letters and envelopes, once again. As an example, let's return to letters and envelopes, and compute the variance of the number of correct matchings of letters and envelopes.

Recall the notation introduced in section 1.5: Let I_i be the indicator that the ith letter is in the correct envelope. The number of letters in the correct envelope is $I = \sum_{i=1}^{n} I_i$, and we showed there that $E(I) = 1$ for all n. When $n = 1$, then a random match is sure to match the only letter with the only envelope, so I is trivial, *i.e.*, $P\{I = 1\} = 1$, and $V(I) = 0$. Thus we compute $V(I)$ when $n \geq 2$. To do so, we need $E(I^2)$.

$$E(I^2) = E(\sum_{i=1}^{n} I_i)^2 = E\left[\left(\sum_{i=1}^{n} I_i\right)\left(\sum_{j=1}^{n} I_j\right)\right].$$

This is a crucial step. The indices i and j are dummy indices (that is, any other letter could be substituted without changing the sum), but using different letters allows us to consider separately the cases when $i = j$ and when $i \neq j$.

Then we have

$$\begin{aligned} E(I^2) &= E\left[\left(\sum_{i=1}^{n} I_i\right)\left(\sum_{j=1}^{n} I_j\right)\right] \\ &= \sum_{i=1}^{n}\sum_{j=1}^{n} E(I_i I_j) \\ &= \sum_{i=j=1}^{n} E(I_i I_j) + \sum_{\substack{i,j=1 \\ i\neq j}}^{n} E(I_i I_j). \end{aligned}$$

Now when $i = j$, $I_i I_j = I_i^2 = I_i$, so $E(I_i I_j) = E(I_i) = 1/n$. However when $i \neq j$, $I_i I_j$ is the indicator of the event that both letters i and j are in their correct envelopes. This has probability $\frac{1}{n(n-1)}$. Hence, if $i \neq j$,

$$E[I_i I_j] = \frac{1}{n(n-1)}.$$

Therefore

$$\begin{aligned} E(I^2) &= \sum_{i=1}^{n} E(I_i) + \sum_{\substack{i,j=1 \\ i\neq j}}^{n} E(I_i I_j) \\ &= n\left(\frac{1}{n}\right) + n(n-1)\left(\frac{1}{n(n-1)}\right) = 2. \end{aligned}$$

Finally, using (2.53),

$$V(I) = E(I^2) - (E(I))^2 = 2 - 1 = 1$$

for all $n \geq 2$. In summary, if $n = 1, V(I) = 0$. If $n \geq 2, V(I) = 1$. □

Now consider two *independent* random variables, X and Y, with means c_1 and c_2, respectively. We know from section 1.5 that

$$E[X+Y] = c_1 + c_2,$$

then

$$\begin{aligned} V[X+Y] &= E\{[(X+Y)-(c_1+c_2)]^2\} \\ &= E\{[(X-c_1)+(Y-c_2)]^2\} \\ &= E\{(X-c_1)^2 + 2(X-c_1)(Y-c_2) + (Y-c_2)^2\} \\ &= V[X] + V[Y] + 2E[(X-c_1)(Y-c_2)]. \end{aligned}$$

Because X and Y are assumed independent, we can take $g(x) = X - c_1$ and $h(Y) = Y - c_2$ in (2.48) and conclude

$$E[(X-c_1)(Y-c_2)] = E[X-c_1]E[Y-c_2] = 0.$$

Therefore, when X and Y are independent,

$$V[X+Y] = V[X] + V[Y].$$

It is easy to forget, but important to remember, that

$$E[X+Y] = E[X] + E[Y]$$

holds without any restriction on the relationship between X and Y, but

$$V[X+Y] = V[X] + V[Y]$$

has been shown only under the restriction that X and Y are independent.

Now let's see what happens when X is transformed to $Y = kX + b$, where k and b are constants. We know, from (1.26), that $E(Y) = kE(X) + b$. Therefore the variance of Y is

$$\begin{aligned} V[Y] &= E[(Y - E(Y))^2] \\ &= E[\{(kX+b) - (kE(X)+b)\}^2] \\ &= E[k(X - E(X))]^2 \\ &= k^2 E\{[X - E(X)]^2\} \\ &= k^2 V[X]. \end{aligned}$$

Thus the variance increases as the square of k, or, as we say, scales with k^2. A transformation of the variance, namely its square root, scales with $|k|$, and is called the standard deviation. Formally,

$$SD[X] = \sqrt{V[X]}.$$

Then for any constant k,

$$SD[kX] = \sqrt{V[kX]} = \sqrt{k^2 V[X]} = |k| SD[X].$$

As an example of the computation of a variance, consider the random variable X with the following distribution:

$$X = \begin{cases} 0 & \text{with probability } 1/4 \\ 2 & \text{with probability } 1/2 \\ 4 & \text{with probability } 1/4 \end{cases}.$$

Then

$$E[X] = 0(1/4) + 2(1/2) + 4(1/4) = 0 + 1 + 1 = 2$$
$$E[X^2] = 0^2(1/4) + 2^2(1/2) + 4^2(1/4) = 0 + 2 + 4 = 6$$

so

$$V[X] = E[X^2] - (E[X])^2 = 6 - 2^2 = 2,$$

using (2.53).

Finally $SD[X] = \sqrt{V[X]} = \sqrt{2}$.

Both the variance and the standard deviation can be regarded as measures of the spread of a distribution.

We now turn to measures of the relationship between two random variables. The first concept to introduce is the covariance of X and Y, defined to be

$$\text{Cov}[X,Y] = E[(X - c_1)(Y - c_2)],$$

where $c_1 = E[X]$ and $c_2 = E[Y]$.

The covariance can be written in another form:

$$\begin{aligned} \text{Cov}[X,Y] &= E[(X - c_1)(Y - c_2)] = E[XY - c_1 Y - c_2 X + c_1 c_2] \\ &= E[XY] - c_1 c_2 - c_2 c_1 + c_1 c_2 = E[XY] - E[X]E[Y]. \end{aligned}$$

Using (2.48), if X and Y are independent, $\text{Cov}[X,Y] = 0$. However, the converse is not true.

As an example, consider the following random variables X and Y:

X	Y	probability
1	0	1/4
0	-1	1/4
0	1	1/4
-1	0	1/4

Then $E[X] = E[Y] = 0$, and $E[XY] = 0$. Therefore $\text{Cov}(X,Y) = E[XY] - (E[X])(E[Y]) = 0$. However X and Y are obviously not independent, since

$$\Pr\{Y = 0 | X = 1\} = 1, \text{but } \Pr\{Y = 0\} = 1/2.$$

The second measure of the relationship between random variables, the correlation between X and Y, is written

$$\rho(X,Y) = \frac{\text{Cov}(X,Y)}{SD(X)SD(Y)}.$$

The advantage of the correlation is that it is shift- and scale-invariant as follows: Let $W = aX + b$ and $V = kY + d$, where $a > 0$ and $k > 0$. Then $E(V) = kE(Y) + d$ and $E(W) = aE(X) + b$. Also $SD(V) = kSD(Y)$ and $SD(W) = aSD(X)$. Putting these relationships together,

$$\begin{aligned} \text{Cov }(W,V) =& E[WV] - E[W]E[V] = E[(kY + d)(aX + b)] \\ & - E(kY + d)E(aX + b) \\ =& akE(XY) + adE(X) + bkE(Y) + bd \\ & - akE(X)E(Y) - adE(X) - bkE(Y) - bd \\ =& ak\text{Cov }(X,Y). \end{aligned}$$

Hence

$$\rho(W,V) = \frac{ak\text{Cov}(X,Y)}{aSD(X)kSD(Y)} = \rho(X,Y).$$

Therefore neither the shift parameters b and d, nor the scale parameters a and k matter, which is what is meant by shift and scale invariance. This property of invariance makes the correlation especially useful, since the correlation between X and Y is the same regardless of what units X and Y are measured in.

Correlation is especially useful as a scale- and shift-invariant measure of association. A high correlation should not be confused with causation, however. Correlation is symmetric between X and Y, while causation is not. Thus, while smoking and lung cancer have a positive (and large) correlation, that in itself does not show whether smoking causes lung cancer, or lung cancer causes smoking, or both, or neither. Additional information about lung cancer and smoking (such as the mechanism by which smoking leads to lung cancer) is necessary to sort this out.

I now derive the important inequality $-1 \leq \rho \leq 1$. Suppose W and V are random variables, with means respectively $E(W)$ and $E(V)$ and standard deviations $\sigma(W)$ and $\sigma(V)$, respectively. Let $X = (W - E(W))/\sigma(W)$ and $Y = (V - E(V))/\sigma(V)$. Now X and Y have mean 0, standard deviation (and variance) 1. Furthermore, because X is a linear function of W and Y is a linear function of V, the invariance argument above shows $\rho(W,V) = \rho(X,Y)$, which I write below simply as ρ. As a consequence, $\rho = E(XY)$.

Consider the new random variable $Z = X - \rho Y$. $E(Z) = E(X) - \rho E(Y) = 0$. The variance of Z, which must be non-negative, is

$$\begin{aligned} 0 \leq V(Z) = E(X - \rho Y)^2 &= E(X^2) - 2\rho E(XY) + \rho^2 E(Y^2) \\ &= 1 - 2\rho^2 + \rho^2 = 1 - \rho^2. \end{aligned}$$

Consequently

$$-1 \leq \rho \leq 1. \tag{2.54}$$

If X is a random variable satisfying $0 < V(X) < \infty$, and $Y = aX + b$ and $a > 0$, then $\rho(X,Y) = \frac{\text{Cov(X,Y)}}{SD(X)SD(Y)} = \frac{a\text{VarX}}{aSD(X)SD(X)} = 1$. Similarly if $a < 0, \rho(X,Y) = -1$. Hence the bounds given in (2.54) are sharp, which means they cannot be improved.

The inequality (2.54) is known in mathematics as the Cauchy-Schwarz or Schwarz Inequality, and has generalizations, also known by the same name.

The next section gives a second proof of this inequality.

2.11.1 An application of the Cauchy-Schwarz Inequality

This inequality shows up often in probability theory. But here's an application that may surprise you.

There are often complaints that airplanes (and other transportation modes) are overcrowded. However, the airlines often complain that their load factors are unimpressive. Could both sides be correct?

Suppose a random flight on an airline has n seats, of which k are filled, with probability $p_{k,n}$. Then the average flight has $E(N) = \sum_{k,n} nP_{k,n}$ seats, of which $E(K) = \sum kP_{k,n}$ are filled. The airlines load factor is then

$$LF = E(K)/E(N).$$

However, the typical airline passenger on a flight with n seats, of which k are filled, experiences a plane k/n filled. There are k passengers who experience this, so the typical passenger experiences a plane that is $E(K^2/N) = \sum_{k,n} k^2/nP_{k,n}$ full, and there are $E(K)$ such passengers. So the crowdedness experienced by passengers is

$$CR = E[K^2/N]/E[K].$$

The question then is the relationship between LF and CR. To explore this, I introduce some new random variables that may look a bit strange at first. Let $X = K/\sqrt{N}$ and $Y = \sqrt{N}$. In this notation,

$$\begin{aligned} LF &= E[XY]/E[Y^2] \\ \text{and } CR &= E[X^2]/E[XY]. \end{aligned}$$

Now $CR \geq LF$ if and only if $E[X^2]/E[XY] \geq E[XY]/E[Y^2]$, *i.e.* if and only if $E[X^2]E[Y^2] \geq [EXY]^2$, which is exactly the Cauchy-Schwarz Inequality. So, yes, both the passengers and the airline can be correct. When every plane flies with exactly k/n passengers, the result is that $CR = LF$. On the other hand, if the airline flies one completely full plane and many empty ones, $CR = 1$ and the load factor LF can be virtually zero. For more on this, see Kadane (2008).

2.11.2 *Remark*

Considerations of the dimensions of the spaces involved shows why uncorrelatedness does not imply independence. Suppose that X takes on n values and Y takes on m values. Then the possible values of the set of probabilities $\{p_{ij}\}$ such that $P\{X = i, Y = j\} = p_{ij}$ are constrained only by $p_{ij} \geq 0$ and $\sum_{i=1}^{n} \sum_{j=1}^{m} p_{ij} = 1$. Consequently each such element of the $\{p_{ij}\}$ set can be expressed as vectors of length $nm - 1$, where 1 has been subtracted because of the constraint $\sum_{i=1}^{n} \sum_{j=1}^{m} p_{ij} = 1$. Uncorrelatedness, or, equivalently, covariance zero, gives a constraint on this space of the form

$$E[XY] = E[X]E[Y]$$

which is a single (quadratic) constraint in the $nm - 1$ dimensional space.

Now consider the situation in which X and Y are independent. Now the possible values that the set of probabilities $\{p_{i,+}\}$, where $P\{X_i = i\} = p_{i,+}, i = 1, \ldots, n$ may take are constrained by $p_{i,+} \geq 0$ and $\sum_{i=1}^{n} p_{i,+} = 1$. Hence the set of $\{p_{i,+}\}$ can be expressed as vectors of length $n - 1$. Similarly the set of probabilities $\{p_{+,j}\}$, where $P\{Y = j\} = p_{+,j}, j = 1, \ldots, m$ can be expressed as vectors of length $m - 1$. Under independence, we have $p_{i,j} = p_{i,+}p_{+,j}$ for $i = 1, \ldots, n$ and $j = 1, \ldots, m$, so under independence, a vector of length $(n-1) + (m-1) = n + m - 2$ suffices. The difference in the dimension of these two spaces is $(nm - 1) - (n + m - 2) = nm - n - m + 1 = (n-1)(m-1)$, which is at least 4 if n and m are both greater than one. Since independence constrains the space much more, it is the more powerful assumption.

2.11.3 *Summary*

This section introduces the variance and the standard deviation, both measures of the spread or variability of a random variable. It also introduces the covariance and the correlation, two measures of the degree of association between two random variables.

2.11.4 *Exercises*

1. Vocabulary. State in your own words the meaning of:
   (a) variance
   (b) standard deviation

   (c) covariance
   (d) correlation
2. Find your own example of two random variables that have correlation 0 but are not independent.
3. Let X and Y have the following values with the stated probabilities:

X	Y	Probability
1	1	1/9
1	0	1/9
-1	1	1/9
-1	0	1/9
0	1	1/9
0	0	1/9
1	-1	1/9
0	-1	1/9
-1	-1	1/9

   (a) Find $E(X)$ and $E(Y)$.
   (b) Find $V(X)$ and $SD(X)$.
   (c) Find $V(Y)$ and $SD(Y)$.
   (d) Find $E(XY)$ and $\text{Cov}(X, Y)$.
   (e) Find $\rho(X, Y)$.
4. Find the variance of a binomial random variable with parameters n and p.
5. Prove $\text{Cov}(X, X) = V[X]$.
6. Prove $\text{Cov}(X, Y) = \text{Cov}(Y, X)$.
7. Let a and b be integers such that $a < b$. Let X be a uniformly distributed random interval on the integers from a to b, so X has the following distribution:

$$X = \begin{cases} i & \text{if } a \leq i \leq b \text{ with probability } \frac{1}{b-a+1} \\ 0 & \text{otherwise .} \end{cases}$$

   (a) Find $E(X)$.
   (b) Find $V(X)$.

2.12 A short introduction to multivariate thinking

This section introduces an essential tool for thinking about random variables. Here, instead of thinking about a single random variable X, or a pair of them (X, Y), we think about a whole vector of them $\mathbf{X} = (X_1, \ldots, X_n)$. To manage this mathematically, we need to introduce notation for vectors and matrices, and to review some of their properties. We then move on to use these results to prove again that the correlation between two random variables is bounded between -1 and 1. Finally we prove a result about conditional covariances and variances.

2.12.1 A supplement on vectors and matrices

A rectangular array, written $A = (a_{i,j})$ is displayed as

$$A = \begin{bmatrix} a_{1,1} & a_{1,2} & \ldots & a_{1,n} \\ a_{2,1} & & & \\ \vdots & & & \\ a_{m,1} & a_{m,2} & \ldots & a_{m,n} \end{bmatrix}$$

and is called a matrix of size $m \times n$, or, more simply, an $m \times n$ matrix. Such a matrix has m rows and n columns. An $m \times 1$ matrix is called a column vector; a $1 \times n$ matrix is called a row vector. Matrices have certain rules of combination, to be explained.

If $A = (a_{i,j})$ and $B = (b_{i,j})$ are matrices of order $m \times n$ and $n \times p$ respectively, then the product AB is an $m \times p$ matrix $C = (c_{i,j})$ with elements given by $c_{i,j} = \sum_{k=1}^{n} a_{i,k} b_{k,j}$. Such a product is defined only when the number of columns of A is the same as the number of rows of B.

It is easy to see that $(AB)C = A(BC)$, since the i, ℓ^{th} element of the matrix $(AB)C$ is

$$\sum_j \left(\sum_k a_{i,k} b_{k,j} \right) c_{j,\ell} = \sum_k a_{i,k} \sum_j b_{k,j} c_{j,\ell},$$

which is the i, ℓ^{th} element of $A(BC)$. Then $(AB)C$ and $A(BC)$ can be written as ABC without confusion.

If $A = (a_{i,j})$ is an $m \times n$ matrix, then $A' = (a'_{i,j})$, pronounced "A-transpose," is the $n \times m$ matrix whose i, j^{th} element is $a'_{i,j} = a_{j,i}$, for $i = 1, \ldots, n$ and $j = 1, \ldots, m$. $(AB)'$ is a $p \times m$ matrix whose j, i^{th} element is given by $c_{j,i} = \sum_{k=1}^{n} b_{k,j} a_{i,k}$, which is what would be obtained by multiplying B' by A'. Hence $(AB)' = B'A'$. The transpose operator permits writing $(a_{1,1}, \ldots, a_{1,n})'$, a convenient way to write a column vector in horizontal format to save space.

A matrix is said to be square if the number of rows m is the same as the number of columns n. A square matrix is said to be symmetric if $A = A'$, or, equivalently, if $a_{i,j} = a_{j,i}$ for all i and j.

The identity matrix, denoted by I, is a symmetric matrix, whose i, j^{th} element is 1 if $i = j$ and 0 otherwise. It is easy to check that

$$AI = IA = A$$

for all square matrices A.

For some (but not all) square matrices A, there is an inverse matrix A^{-1} having the property that

$$A^{-1}A = AA^{-1} = I.$$

Chapter 5 displays a characterization of which matrices A have inverses and which do not. If A has an inverse, it is unique. To see this, suppose A had two inverses, A_1^{-1} and A_2^{-1}. Then

$$A_1^{-1} = A_1^{-1}I = A_1^{-1}(AA_2^{-1}) = (A_1^{-1}A)A_2^{-1} = IA_2^{-1} = A_2^{-1}.$$

There is one class of matrices for which it is easy to see that inverses exist. A diagonal matrix $D_{\boldsymbol{\lambda}}$ has the vector $\boldsymbol{\lambda}$ down the diagonal, and is zero elsewhere. It is easy to see that $D_{\boldsymbol{\lambda}} D_{\boldsymbol{\mu}} = D_{\boldsymbol{\lambda\mu}}$, where the ith element of $\boldsymbol{\lambda\mu}$ is given by $\lambda_i \mu_i$. Then, provided $\lambda_i \neq 0$ for all i, the vector with elements $\mu_i = 1/\lambda_i$ has the property that $D_{\boldsymbol{\lambda}} D_{\boldsymbol{\mu}} = D_{\boldsymbol{\mu}} D_{\boldsymbol{\lambda}} = I$, so $(D_{\boldsymbol{\lambda}})^{-1} = D_{\boldsymbol{\mu}}$.

2.12.2 Least squares

As a statistical application of matrix algebra, we'll use the algebra of matrices to prove a familiar fact about least squares.

Let y be a $n \times 1$ vector (think of it as a vector of observations). Let X be a known $n \times k$ matrix (think of them as variables used to predict y), and let β be a $k \times 1$ vector of coefficients. The question is what choice of β to make to best predict y The particular sense of "best choice" to be used in this calculation is that choice that minimizes S, the sum of squared discrepancies between y and $X\beta$. It will turn out that the choice of β that minimizes the sum of squares is

$$\hat{\beta} = (X'X)^{-1}X'y, \tag{2.55}$$

assuming that $X'X$ has an inverse.

To start, suppose that $Z = \begin{pmatrix} Z_1 \\ \vdots \\ Z_n \end{pmatrix}$ is a column vector. Then $Z'Z = (Z_1, Z_2, \ldots, Z_n)\begin{pmatrix} Z_1 \\ \vdots \\ Z_n \end{pmatrix} = \sum_{i=1}^n Z_i^2$. This fact will be used twice in the argument to come. The first use comes immediately: S is the sum of squared discrepancies between y and $X\beta$, i.e.

$$S = (y - X\beta)'(y - X\beta). \tag{2.56}$$

A convenient way to think about S is to allow S to be expressed in terms of $\hat{\beta}$ and a discrepancy β^* from $\hat{\beta}$. That is, I can write

$$S = (y - X\hat{\beta} + X\beta^*)'(y - X\hat{\beta} + X\beta^*) \tag{2.57}$$

and seek to show that S is minimized by the choice $\beta^* = 0$.

To proceed, I first study $y - X\hat{\beta}$. By substitution, $y - X\hat{\beta} = y - X(X'X)^{-1}X'y = [I - X(X'X)^{-1}X']y$. The matrix in brackets is a special matrix, called a projection, and written $\bar{P}_X$. (We return to projection matrices in Chapter 5, after more linear algebra has been developed). For our purposes, there are two facts about $\bar{P}_X$ that are important.

(i) $\bar{P}_X$ is symmetric, i.e. $\bar{P}_X = \bar{P}'_X$.

$$\bar{P}'_X = [I - X(X'X)^{-1}X']' = I - (X(X'X)^{-1}X')' = I - X(X'X)^{-1}X' = \bar{P}_X.$$

(ii) $X'\bar{P}_X = X'[I - X(X'X)^{-1}X'] = X' - (X'X)(X'X)^{-1}X' = X' - X' = 0.$

Consequently, $\bar{P}_X X = 0$.

Then (2.57) can be rewritten as follows:

$$\begin{aligned} S &= (y - X\hat{\beta} + X\beta^*)'(y - X\hat{\beta} + X\beta^*) \\ &= (\bar{P}_X y + X\beta^*)'(\bar{P}_X y + X\beta^*) \\ &= (y'\bar{P}_X + \beta^* X')(\bar{P}_X y + X\beta^*) \\ &= y'\bar{P}_X\bar{P}_X y + \beta^* X'X\beta^* \end{aligned}$$

Now $\beta^* X'$ is a vector of length n, so $\beta^* X'X\beta^*$ is a sum of squares (here's the second use of this convenient fact).

A sum of squares is non-negative. Therefore, $\beta^* X'X\beta^*$ is minimized at $\beta^* = 0$, which completes the proof.

There's some interesting geometry going on here, but a full explanation will have to wait until Chapter 5.

2.12.3 A limitation of correlation in expressing negative association between non-independent random variables

Suppose $X_1, X_2, \ldots, X_m$ are m random variables satisfying $E(X_i) = \mu_i$ and $V(X_i) = \sigma_i^2$ for $i = 1, \ldots, m$. Suppose they also satisfy $\rho(X_i, X_j) = \rho$ for $i \neq j$. Then $\rho \geq -1/(m-1)$.

Proof. Let $Y_i = (X_i - \mu_i)/\sigma_i, i = 1, \ldots, m$. Then $E(Y_i) = 0$ and $V(Y_i) = 1$ for $i = 1, \ldots, m$. Also, $\rho(Y_i, Y_1) = \rho$ for $i \neq j$. Now $E(\sum_{i=1}^m Y_i) = 0$ and

$$\begin{aligned} 0 &\leq V(\textstyle\sum_{i=1}^m Y_i) = E(\sum_{i=1}^m Y_i)^2 = E[(\sum_{i=1}^m Y_i)(\sum_{j=1}^m Y_j)] \\ &= E(\textstyle\sum_{i,j=1}^m Y_i Y_j) = \sum_{i,j=1}^m E(Y_i Y_j) \\ &= \textstyle\sum_{i=1}^m E(Y_i)^2 + \displaystyle\sum_{\substack{i,j=1 \\ i \neq j}}^{m} E(Y_i Y_j) \\ &= m + m(m-1)\rho \end{aligned}$$

Hence

$$\rho \geq -1/(m-1)$$

□

When $m = 2$, this result gives the same bound as before, namely $\rho \geq -1$. However, for large m, it severely restricts the space of possible values of ρ.

2.12.4 Covariance matrices

Suppose that Y_{ij} is a random variable for each $i = 1, \ldots, m$ and each $j = 1, \ldots, n$. These random variables can be assembled into a matrix Y whose $(i, j)^{th}$ element is $Y_{i,j}$. The numbers $E[Y_{ij}]$, which, for each i and j are the expectations of the random variables Y_{ij}, can also be assembled into an $m \times n$ matrix, with the obvious notation $E[Y]$. In particular, if $\mathbf{X} = (X_1, \ldots, X_n)'$ is a length n column vector of random variables, then $E(\mathbf{X})$ is also a column vector of length n, with i^{th} element $E(X_i)$. That is, $E(\mathbf{X}) = E[(X_1, \ldots, X_n)'] = (E(X_1), E(X_2), \ldots, E(X_n))'$.

If $\mathbf{X}$ is such a column vector of random variables, it is natural to assemble the covariances of X_i and X_j, $\mathrm{Cov(X_i, X_j)}$ into an $n \times n$ square matrix, called a covariance matrix. Let Ω be an $n \times n$ matrix whose $(i, j)^{th}$ element is $\omega_{i,j} = \mathrm{Cov(X_i, X_j)}$. Such a matrix is symmetric, because $\mathrm{Cov(X_i, X_j)} = \mathrm{Cov(X_j, X_i)}$.

Now suppose that $E(\mathbf{X}) = \mathbf{0} = (0, 0, \ldots, 0)'$. Then $\mathrm{Cov(X_i, X_j)} = \mathrm{E(X_i X_j)}$, and $\mathbf{\Omega} = E(\mathbf{XX}')$. Let $f' = (f_1, \ldots, f_n)$ be a row vector of n constants. Then $Y = \sum_{i=1}^n f_i X_i = f'X$ is a new random variable, a linear combination of $X_1, \ldots, X_n$ with coefficients $f_1, \ldots, f_n$, respectively. Y has mean $E(Y) = E(f'x) = f'E\mathbf{X} = 0$ and variance $V(Y) = E[f'XX'f] = f'E(XX')f = f'\Omega f$.

Since $\mathrm{Var(Y)} \geq 0$, we have $0 \leq f'\mathbf{\Omega} f$ for all vectors f. Such a matrix Ω is called positive-semi-definite. A matrix Ω such that $f'\Omega f > 0$ for all vectors $f \neq 0$ is called positive definite.

This result can be used to prove again the bounds on the correlation ρ. A general 2×2 covariance matrix can be written as

$$\Omega = \begin{pmatrix} \sigma_1^2 & \rho\sigma_1\sigma_2 \\ \rho\sigma_1\sigma_2 & \sigma_2^2 \end{pmatrix}$$

where $\sigma_i^2 = V(X_i)$, and $\rho = \rho(X_1, X_2)$. Suppose $f = \left(\frac{1}{\sigma_1}, \pm\frac{1}{\sigma_2}\right)$. [I'll do the calculation for both + and – together, to avoid repetition.]

Then

$$\begin{aligned}
0 \leq V(Y) &= f\Omega f' \\
&= \left(\frac{1}{\sigma_1}, \pm\frac{1}{\sigma_2}\right)\begin{pmatrix}\sigma_1^2 & \rho\sigma_1\sigma_2 \\ \rho\sigma_1\sigma_2 & \sigma_2^2\end{pmatrix}\begin{pmatrix}\frac{1}{\sigma_1} \\ \pm\frac{1}{\sigma_2}\end{pmatrix} \\
&= (\sigma_1 \pm \rho\sigma_1, \rho\sigma_2 \pm \sigma_2)\begin{pmatrix}\frac{1}{\sigma_1} \\ \pm\frac{1}{\sigma_2}\end{pmatrix} \\
&= (1 \pm \rho) \pm \rho + 1 \\
&= 2 \pm 2\rho.
\end{aligned}$$

Therefore

$$-1 \leq \rho \leq 1. \tag{2.58}$$

2.12.5 *Conditional variances and covariances*

The purpose of this section is to demonstrate the following result:

$$\text{Cov}(X, Y) = E[\text{Cov}(X, Y)|Z] + \text{Cov}[E(X|Z), E(Y|Z)], \tag{2.59}$$

which will be useful later.

Proof. First we will expand out each of the summands using the computational form for the covariance.

$$\text{Cov}\,[(X, Y)|Z] = E[XY|Z] - E[X|Z]E[Y|Z]$$

Then, taking the expectation of both sides,

$$E\,[\text{Cov}[X, Y|Z]] = E[E[XY|Z]] - E[E[X|Z]E[Y|Z]].$$

Also,

$$\text{Cov}[E[X|Z], E[Y|Z]] = E[E[X|Z]E[Y|Z]] - EE[X|Z]EE[Y|Z].$$

Now we can use the formula for iterated expectation (see section 2.8) to simplify the two expressions.

$$\begin{aligned}
EE[X|Z] &= E[X], \\
EE[Y|Z] &= E[Y]
\end{aligned}$$

and

$$EE[XY|Z] = E[XY].$$

Hence

$$\begin{aligned}
&E[\text{Cov}(X, Y)|Z] + \text{Cov}[E[X|Z], E[Y|Z]] = \\
&E[XY] - E[X]E[Y] = \text{Cov}(X, Y).
\end{aligned}$$

□

2.12.6 Summary

This section develops vectors and matrices of random variables, and introduces covariance matrices. From a property of covariance matrices, the important bound for the correlation ρ, $-1 \leq \rho \leq 1$ is derived. Finally a result about conditional covariances is derived.

2.12.7 Exercises

1. Vocabulary. State in your own words the meaning of:
   (a) matrix
   (b) matrix muliplication
   (c) square matrix
   (d) symmetric matrix
   (e) inverse of a matrix
   (f) diagonal matrix
   (g) covariance matrix
   (h) positive semi-definite matrix
   (i) positive definite matrix
2. Show that
$$AI = IA = A$$
   for all square matrices A.
3. Prove $V[X] = E[V[X|Z]] + V[E(X|Z)]$.
4. Recall the distribution given in problem 3 of section 2.11.4:

X	Y	Probability
1	1	1/9
1	0	1/9
-1	1	1/9
-1	0	1/9
0	1	1/9
0	0	1/9
1	-1	1/9
0	-1	1/9
-1	-1	1/9

   Let $W = (X, Y)$.
   (a) What is $E[W]$?
   (b) Find the covariance matrix of W.
5. Suppose A and B are 2×2 square matrices.
   (a) Find A and B such that $AB = BA$.
   (b) Find A and B such that $AB \neq BA$.
6. Suppose $X_1, X_2, \ldots, X_m$ are m random variables satisfying $E(X_i) = \mu_i$, and $V(X_i) = \sigma_i^2, i = 1, \ldots, m$. Also, $\rho(X_i, X_j) = \rho_{i,j}$ for $i \neq j$. Let $\rho^* = \sum_{\substack{i \neq j \\ i,j=1}}^{m} \rho_{i,j}/m(m-1)$, so ρ^* is the average of $\rho_{i,j}(i \neq j)$.
   Prove that $\rho^* \geq -1/(m-1)$.

2.13 Tchebychev's Inequality and the (weak) law of large numbers

The weak law of large numbers (WLLN) is an important and famous result in probability. It says that, under certain conditions, averages of independent and identically distributed random variables approach their expectation as the number of averages grows. Tchebychev's Inequality is introduced and used in this section to prove a form of the WLLN.

To start, suppose that $X_1, X_2, \ldots, X_n$ are independent random variables that have the same distribution. The expectation of each of them is $E(X_1) = m$, and their variance is σ^2. Now consider the average of these random variables, denoted, $\overline{X}_n$:

$$\overline{X}_n = \sum_{i=1}^{n} X_i/n.$$

Clearly $\overline{X}_n$ is a new random variable. Its mean is $E(\overline{X}_n) = E(\sum_{i=1}^{n} X_i/n) = \frac{1}{n}\sum_{i=1}^{n} E(X_i) = \frac{nm}{n} = m$, and its variance is $V(\overline{X}_n) = V(\sum_{i=1}^{n} X_i/n) = \frac{1}{n^2}V(\sum_{i=1}^{n} X_i) = \frac{1}{n^2}\sum_{i=1}^{n} V(X_i) = \frac{n\sigma^2}{n^2} = \sigma^2/n$. The fact that the variance of $\overline{X}_n$ decreases as n increases is critical to the argument that follows. Consider the random variable $\overline{X}_n - m$, which has mean 0 and variance σ^2/n.

Now we switch our attention to Tchebychev's Inequality. Suppose Y has mean 0 and variance τ^2. (When we apply this inequality, we're going to think of $Y = \overline{X}_n - m$ and $\tau^2 = \sigma^2/n$.) Let $P\{Y = y_i\} = p_i$ for $i = 1, \ldots, n$. Then, for any $k > 0$ we choose,

$$\tau^2 = E(Y^2) = \sum_{i=1}^{n} y_i^2 p_i = \sum_{\substack{i=1 \\ |y_i| \leq k}}^{n} y_i^2 p_i + \sum_{\substack{i=1 \\ |y_i| > k}}^{n} y_i^2 p_i$$

$$\geq \sum_{\substack{i=1 \\ |y_i| > k}}^{n} y_i^2 p_i \geq k^2 \sum_{\substack{i=1 \\ |y_i| > k}}^{n} p_i = k^2 P\{|Y| > k\}.$$

Here the first inequality holds because we dropped the entire sum for $y_i < k$. The second inequality holds since the sum is over only those indices i for which $|y_i| > k$ and for each of them, $y_i^2 > k^2$. Finally the equality holds by substititution.

Rearranged,

$$\tau^2/k^2 \geq P\{|Y| > k\},$$

which is Tchebychev's Inequality.

Now let $Y = \overline{X}_n - m$. Making this substitution, we have

$$\sigma^2/nk^2 \geq P\{|\overline{X}_n - m| > k\}.$$

This inequality says that for each k, there is an n large enough so that $P\{|\overline{X}_n - m| > k\}$ can be made as small as we like, so almost all of the probability distribution is piled up at m. In this sense, $\overline{X}_n$ approaches m as n gets large.

Phrased formally, the weak law of large numbers says that for every $\epsilon > 0$ and every $\eta > 0$ there is an N such that for every $n \geq N$, $P\{|\overline{X}_n - m| > \eta\} < \epsilon$. [If this is too formal for your taste, don't let it bother you.]

2.13.1 Interpretations

Since the weak law of large numbers is sometimes used to interpret probability, it is useful to visit that subject at this point. As mentioned in section 1.1.2, some authors propose that probability of an event A should be defined as the limiting relative frequency with which A occurs in an infinite sequence of independent trials. Let I_{A_i} be the indicator function for the ith trial. Then $I_{A_1}, I_{A_2}, \ldots$ is an infinite sequence of independent and identical trials, whose average, $\overline{X}_n = \sum_{i=1}^{n} I_{A_i}/n$ is the relative frequency with which the event A occurs. Also $E(I_{A_i}) = P\{A_i = 1\} = p$, say. Also $\text{Var}(I_{A_i}) = p(1-p)$. Then the WLLN applies, and says that the limiting relative frequency of the occurrence of A approaches p.

Let A be an event, and consider an infinite sequence of indicator random variables $\mathbf{t}_A$ indicating whether A has occurred in each of infinitely many repetitions. Suppose A has a limiting relative frequency which I write as p_A. Then it is immediate that $0 \leq p_A \leq 1$. Also if S is the sure event, then $\mathbf{t}_s$ has a limiting relative frequency, and $p_s = 1$. If A and B are two disjoint events with sequences $\mathbf{t}_A$ and $\mathbf{t}_B$, respectively, we may associate with $A \cup B$ the sequence which is the component-by-component maximum of $\mathbf{t}_A$ and $\mathbf{t}_B$. This corresponds to the union of A and B because it is zero if and only if neither A, nor B, nor both, are one. But the maximum of two binary numbers that are not simultaneously 1 is the same as the sum. If $\mathbf{t}_A$ and $\mathbf{t}_B$ both have limiting relative frequencies, then so does $\mathbf{t}_{A\cup B}$, and $p_{A\cup B} = p_A + p_B$. Thus limiting relative frequency, in this sense, satisfies the requirements for coherence, (1.1), (1.2) and (1.3).

There are difficulties, however, with looking to this argument to support a view of probability that is independent of the observer.

The first difficulty is that, conceived in this way, probability is a function not of events, but of infinite (or long) sequences of them. Consider two infinite sequences of indicators of events, s_1 and s_2, with respective relative frequencies ℓ_1 and ℓ_2, where $\ell_1 \neq \ell_2$. Let A be the indicator of an event not an element of s_1 or s_2. Consider new sequences $s_1' = (A, s_1)$ and $s_2' = (A, s_2)$. These sequences have relative frequencies ℓ_1 and ℓ_2, respectively. Hence within this theory the event A does not have a well-defined probability. While this may or may not be a defect of limiting relative frequency, its use would require a substantial shift in how probability is discussed and used.

A second issue is that some limitation must be found to the sequences to which limiting relative frequency is applied. It is necessary to avoid circularity (independence defined in terms of probability, as in this chapter, but probability defined in terms of sequences of independent events to get the weak law of large numbers). Consequently there has grown up a study of "randomness" of a sequence (see von Mises (1939), Richenbach (1948), Church (1940), Ville (1936, 1939), Martin-Lof (1970) and Li and Vitanyi (1993)). This literature has not yet, I think, been successful in finding a satisfactory way to think about randomness as it might apply to a single event.

The frequency view of probability is not operational. There is no experiment I can conduct to determine the probability of an event A that comports with frequencies. This contrasts with the subjective view used in this book, which is based on the experiment of asking at what price you would buy or sell a ticket that pays \$1 if A occurs and nothing if A does not occur.

There is a fourth issue with limiting relative frequency that is examined in section 3.5.3.

As a person who applies probability theory with the intention of making inferences, I note that many of my colleague statisticians claim to base their viewpoint on relative frequency without taking into account its limitations and unsettled nature.

From the perspective of this book, the meaning of the weak law of large numbers is as follows: If you believe that $X_1, X_2, \ldots,$ are independent and identically distributed, with mean m and variance σ^2 , then, in order to avoid sure loss, you also must bet that $\overline{X}_n$ will diverge from m only by an arbitrarily small amount as n gets large.

2.13.2 Summary

Tchebychev's Inequality is used to prove the weak law of large numbers. The weak law of large numbers says that for a sequence of independent and identically distributed random variables, the sample average $\overline{X}_n$ approaches the expectation m as the number, n, of random variables in the sequence grows.

2.13.3 *Exercises*

1. Vocabulary. Explain in your own words
   (a) Tchebychev Inequality
   (b) Weak Law of Large Numbers
2. Recall the rules of "Pick Three" from the Pennsylvania Lottery (see problem 3 in section 1.5.2): A contestant chooses a three-digit number, between 000 and 999. A number is drawn, where each possibility is intended to be equally likely. Each ticket costs \$1, and you win \$600 if your chosen number matches the number drawn. Your winnings in a particular lottery i can be described by the following random variable:

$$X_i = \begin{cases} -\$1 \text{ with probability } .999 \\ \$599 \text{ with probability } .001 \end{cases}$$

   (Check this to be sure you agree.)
   (a) Find the mean and variance of X_i.
   (b) Suppose that you play the lottery for n days, where n is large. Toward what number will your average winning tend? Does the WLLN apply? Why or why not?
   (c) The advertising slogan of the Pennsylvania Lottery is "you have to play to win." Discuss this slogan, together with its counterpart, "you have to play to lose." Which is more likely?
3. A multivariate Tchebychev Inequality.
   Let $X_1, \ldots, X_n$ be random variables with $E(x_i) = \mu_i$ and $V(x_i) = \sigma_i^2$, for $i = 1, \ldots, n$. Let $A_i = \{x_i || X_i - \mu_i| \leq \sqrt{n}\sigma_n\delta\}$, where $\delta > 0$. Prove

$$P(A_1, \ldots, A_n) \geq 1 - \delta^{-2}.$$

   Hint: Use Boole's Inequality from section 1.2.
4. Consider the following random variable, X.

$$\text{Let } X = \begin{cases} \text{Value} & \text{Probability} \\ -2 & 1/10 \\ -1 & 1/5 \\ 0 & 2/5 \\ 1 & 1/5 \\ 2 & 1/10 \end{cases}$$

   (a) Compute $E(X)$ and $\text{Var}(X)$.
   (b) For each $k = .5, 1, 1.5, 2, 2.5$, compute $P\{|X| > k\}$.
   (c) For each such k, compute σ^2/k.
   (d) Compare the answers to (b) and (c). Does the relationship given by the Tchebychev Inequality hold?
5. Consider the random variable X defined in problem 4.
   (a) Write a program in `R` to draw a random variable with the same distribution as X.
   (b) Use that program to draw a sample of size n with that distribution, where $n = 10$.

(c) Compute the average $\overline{X}_{10}$ of these 10 observations.

(d) Do the computation in (c) $m = 100$ times. Use `R` to draw a plot of these 100 values of $\overline{X}_{10}$.

(e) Do the same for $n = 100$, drawing $m = 100$ times. Again draw a plot of the resulting 100 values of $\overline{X}_{100}$.

(f) Compare the plots of (d) and (e). Does the comparison comport with the WLLN's?

Chapter 3

Discrete Random Variables

"Don't stop thinking about tomorrow"

—Fleetwood Mac

"Great fleas have little fleas upon their backs to bite 'em
And little fleas have lesser fleas
and so on, ad infinitum"

—Augustus DeMorgan

3.1 Countably many possible values

Since section 1.5 of Chapter 1, the random variables considered have been limited to those taking at most a finite number of possible values. This chapter deals with random variables taking at most countably many values; the next chapter deals with the continuous case, in which random variables take uncountably many values. However, since the material of sections 1.1 to 1.4 was developed without reference to the limitation imposed in section 1.5, those results still apply.

There is a superficially attractive position that holds that everything we do in probability and statistics occurs in a space of a finite number of possibilities. After all, computers express numbers to only a finite number of significant digits, all measurements are taken from a discrete, finite set of possibilities, and so on. Why then do we need to bother with random variables taking infinite numbers of possible values?

One answer is that much of what we want to do is more conveniently expressed in a space of infinite possibilities. But, as will soon be apparent, random variables taking infinitely many values have costs in terms of additional assumptions. Additionally, it seems to me that it is the "job" of mathematics to accept the assumptions that are most reasonable for the application, and not the "job" of the application to accept mathematically convenient, but inappropriate assumptions.

I think a better answer to this question is that even in a discrete world, infinite spaces of possibilities come up very naturally. For example, consider independent flips of a coin, each with probability $p > 0$ of success. Define the random variable F, which is the number of failures before the first success. The event that $F = k$ is the event that the first k flips were failures and that the $(k+1)^{st}$ was a success. Therefore

$$P\{F = k\} = (1-p)^k p \text{ , for } k \text{ in some set.}$$

Over what set is it natural to think of this random variable as ranging? I think it natural to think of k as having any finite value, and therefore as having no upper bound. Thus I would write

$$P\{F = k\} = (1-p)^k p \quad k = 0, 1, 2, \ldots \quad .$$

What would be the consequence of imposing a finite world on F? After all, the argument might be that the probability that F is very large goes to zero exponentially fast and hence truncating it at some large number would do no harm. And of course if F were simulated on a computer, there is some upper bound for the simulated F beyond which the computer would report an overflow or other numerical problem.

While all of this is true, it is also true that if we decided to truncate F by eliminating the upper tail, choosing to omit only $\epsilon > 0$ of the probability in the tail, the truncation point would depend on p (and we might not know p). Thus, considering every sample space to be finite gets awkward and inconvenient, even in this simple example.

Allowing F to take any integer value leads to a random variable that we will study later in more detail, the geometric random variable F. Because

$$\begin{aligned}\sum_{k=0}^{\infty} P\{F=k\} &= \sum_{k=0}^{\infty}(1-p)^k p = \lim_{j\to\infty}\sum_{k=0}^{j}(1-p)^k p \\ &= p \lim_{j\to\infty}\frac{1-(1-p)^{j+1}}{p} \text{(using 2.24)} \\ &= \lim_{j\to\infty}\left[1-(1-p)^{j+1}\right] \\ &= 1,\end{aligned}$$

if you flip long enough, you're sure of getting a success. In this chapter we study random variables that take a countably infinite number of values, as does F.

Intuitively one might think that the extension to an infinite number of possibilities should be trivial, and some results do extend easily. But others do not. Accordingly this chapter is organized to help you see which results extend, which don't and why. First (in section 3.1.1) I explain that there are different "sizes" of infinite sets. This leads to the surprise that there are as many positive even integers as there are positive integers, and that there are "many" more real numbers than positive numbers. Section 3.2 examines some peculiar behavior of probabilities on infinite sets. Section 3.3 introduces an additional assumption (countable additivity) that resolves the peculiarities. The discussion of the properties of expectations leads to a discussion of why an expectation is defined only when the expectation of the absolute value is finite. Finally generating functions, cumulative distribution functions and some standard discrete probability distributions are discussed.

3.1.1 *A supplement on infinity*

Above I have used the word "countable" without explaining what is meant. So this supplement describes the mathematical theory of infinity, and, along the way, which sets are countable (also called "denumerable") and which are not.

We all understand how many elements there are in a finite set. For example, the set $A = \{1, 3, 8\}$ consists of three elements. Suppose $B = \{2, 4, 13\}$. Then the elements of A and B can be put in one-to-one correspondence with each other, for instance with the mapping $1 \leftrightarrow 4$, $8 \leftrightarrow 2$, $3 \leftrightarrow 13$. The existence of such a mapping formally assures us that A and B have the same number of elements, namely three.

Now we apply the same idea to infinite sets. The simplest, and most familiar such set is the set of positive natural numbers:

$$1, 2, 3, \ldots \quad .$$

In this section, I'll refer to this set as the natural numbers. (Sometimes 0 is included as well.) The set of natural numbers is defined to be countable. Every set that can be mapped in one-to-one correspondence with the natural numbers is also countable. For example, consider the set of even natural numbers. Consider the following mapping:

$$\begin{array}{cccccccc} 1 & 2 & 3 & 4 & 5 & & & n \\ \updownarrow & \updownarrow & \updownarrow & \updownarrow & \updownarrow & \cdots & & \updownarrow \\ 2 & 4 & 6 & 8 & 10 & & & 2n \end{array}$$

Clearly each natural number is mapped into an even natural number, and each even natural number is mapped into a natural number. Thus there is a one-to-one map between the natural numbers and the even natural numbers, so the set of even natural numbers is countable.

Another important set in mathematics is the set of positive rational numbers, that is, the set of numbers that can be expressed as the ratio of two natural numbers. Surprisingly, the set of rational numbers is also countable, as the following construction shows: Consider displaying the rational numbers in the following matrix:

$$\begin{array}{cccccc} 1 & 2 & 3 & 4 & 5 & \ldots \\ \frac{1}{2} & \frac{2}{2} & \frac{3}{2} & \frac{4}{2} & \frac{5}{2} & \\ \frac{1}{3} & \frac{2}{3} & \frac{3}{3} & \frac{4}{3} & \frac{5}{3} & \\ \frac{1}{4} & \frac{2}{4} & \frac{3}{4} & \frac{4}{4} & \frac{5}{4} & \ldots \\ \vdots & & & & & \end{array}$$

where the rational number p/q is placed in the pth column and q^{th} row. Now we traverse this matrix on adjacent upward sloping diagonals, eliminating those rational numbers that have already appeared. Hence the first few elements of this ordering of the positive rational numbers would be

$$1, \frac{1}{2}, 2, \frac{1}{3}, [2/2 = 1 \text{ is omitted}], 3, \frac{1}{4}, \frac{2}{3}, \frac{3}{2}, 4, \text{etc.}$$

In this way, every positive rational number appears once in the countable sequence, so the positive rational numbers are countable. So are all the rational numbers.

The final set we will discuss is the set of real numbers. It turns out that the set of real numbers is not countable. While this may seem like a nearly impossibly difficult fact to prove, the proof is remarkably simple. It proceeds by contradiction. Thus we suppose that the real numbers are countable, and show that the assumption leads to an impossibility. So let's suppose that the real numbers can be put in one-to-one correspondence with the natural numbers. Then every real number must appear somewhere in the resulting sequence. I'll now produce a real number that I can show is not in the sequence, which will establish the contradiction.

Suppose the first real number in the sequence has a decimal expansion that looks like

$$N_1.a_1a_2\ldots,$$

where the dot is the decimal point. So a_1 is some natural number between 0 to 9. Let a be a natural number that is not 0, 9 nor a_1. (There are at least seven such choices. Choose your favorite.) Now consider the second number in the sequence. It has a decimal expansion of the form

$$N_2.b_1b_2b_3\ldots$$

Choose a number b that is not 0, 9 nor b_2. (Again, you have at least seven choices.) Keep doing this process indefinitely.

Now consider the number with the decimal expansion.

$$x = .abc \ldots \quad .$$

If this number were in the sequence, it would have to be the nth in the sequence for some n. The nth element in its decimal expansion is chosen to be different from the nth element in the decimal expansion of the nth number in the sequence. Therefore x does not appear in the sequence. Therefore this proposed sequence does not have the number x in it. Since the proposed mapping from the natural numbers to the reals is arbitrary, there is no such mapping. Hence the real numbers are not countable. In this argument, we avoided 0 and 9, so there would be no ambiguity arising from equalities like

$$2.4999 \ldots = 2.5000 \ldots \quad .$$

Thus the real numbers are not countable.

3.1.2 Notes

This way of thinking about infinite sets is due to Cantor. A friendly introduction is found in Courant and Robbins (1958, pp. 77-88).

3.1.3 Summary

Two sets (finite or infinite) have the same number of elements if there is a one-to-one mapping between them. The sets that have a one-to-one mapping with the natural numbers are called countable or denumerable. The even natural numbers and the rational numbers are countable, but the real numbers are not.

3.1.4 Exercises

1. Vocabulary: Explain in your own words the meaning of:
   (a) natural number
   (b) rational number
   (c) real number
   (d) countable set
2. Find mappings that show that each of the following is countable:
   (a) the positive and negative natural numbers $\ldots -3, -2, -1, 0, 1, 2, 3, \ldots$
   (b) all rational numbers, both positive and negative
3. Show that the set of positive real numbers can be put in one-to-one correspondence with the set of real numbers $x, 0 < x < 1$. Hint: think about the function $g(x) = \frac{1}{x} - 1$.
4. Show that the set of real numbers satisfying $a < x < b$ for some a and b, can be put in one-to-one correspondence with the set of real numbers y satisfying $c < y < d$ for every c and d.

3.2 Finite additivity and countably infinite random variables

This section reviews the axioms of Chapter 1 in the context of random variables that take on more than a finite number of values. It turns out that some rather strange consequences ensue, in particular a dynamic sure loss. The next sections show what additional assumption removes the possibility of dynamic sure loss, and the other bizarre behavior uncovered in this section.

Coherence is defined in Chapter 1 by prices (probabilities) satisfying the following equations:

$$P\{A\} \geq 0 \text{ for all } A \subseteq S, \tag{1.1}$$

$$P\{S\} = 1, \text{ where S is the sure event} \tag{1.2}$$

and

$$P\{A \cup B\} = P\{A\} + P\{B\}, \text{ where } A \text{ and } B \text{ are disjoint.} \tag{1.3}$$

When S has only a finite number of elements, one can specify the whole distribution by specifying the probability of each element, where these probabilities are non-negative and sum to 1. Then the probability of any set A can be found by adding the probabilities of the elements of A, using (1.3) a finite number of times.

However, when S has a countable number of elements, like the integers, that strategy no longer gives the probability of every set A. Since the strategy works on every finite subset, it can be extended to complements of finite sets. A *cofinite* set is a set that contains all but a finite number of elements. The probability of every cofinite set is also determined by the strategy of adding up the probabilities of a finite number of disjoint sets and subtracting the result from 1. But there are many subsets of the integers whose probabilities cannot be determined this way. These are infinite sets whose complement is also infinite. Examples of such sets include the even integers (whose complement is the odd numbers), and the prime numbers (whose complement is every number that is the product of two or more prime numbers larger than one).

A uniform distribution on the set of all integers is key to the example that is discussed below. Recall the definition of a uniform distribution on the finite set $\{1, 2, \ldots, n\}$, as given in problem 7 in section 2.11.4: each point has probability $1/n$. What happens when we look for a uniform distribution on the infinite set $\{1, 2, \ldots\}$? The only way each point can have the same probability is for each point to have probability zero. Then, using (1.3), each finite set has probability zero. By (1.2), the set $S = \{1, 2, \ldots\}$ has probability one, as does every cofinite set.

There are many ways in which the probability of infinite sets whose complement is infinite might be specified. For example, one could extend the specification of probability by considering, for fixed k, each set of the type $\{kn + i, n \in \{0, 1, 2, \ldots\}, 0 \leq i \leq k - 1\}$. These sets are called residue classes mod k, or cylinder sets mod k. (Thus if $k = 2$ and $i = 0$, the even numbers result; if $k = 2$ and $i = 1$, the odd numbers result.) It is consistent with coherence to give each residue class mod k the probability $1/k$. Indeed, using advanced methods it is possible to show that there is a whole class of coherent probabilities satisfying these constraints. (See Kadane and O'Hagan (1995); Schirokauer and Kadane (2007).) Since the example that follows is true of each member of that class, it won't matter which of them one has in mind.

Now that you know about uniform distributions on the integers, I can introduce you to the central example of this section. It illustrates the anomalies that can occur when one attempts to apply coherence to random variables with countably many possible values.

Suppose there are two exclusive and exhaustive states of the world, each of which currently has probability 1/2 to you, A and $\overline{A}$. Let X be a random variable taking the value 1 if A occurs and 0 if $\overline{A}$ occurs. Then $E\{X\} = (1/2)(1) + (1/2)(0) = 1/2$. Now let Y be a random variable taking values on the natural numbers as follows: $\{Y|X = 0\}$ has a uniform distribution on the integers, while $P\{Y = j|X = 1\} = 1/2^{(j+1)}$, for $j = 0, 1, 2, \ldots$. These numbers, $1/2, 1/4, 1/8$, etc. sum to 1. To verify that these choices are coherent, notice that all probabilities are non-negative, so (1.1) is satisfied. Also

$P\{S\} = P\{X = 1\} + P\{X = 0\} = 1/2 + 1/2 = 1$ so (1.2) is also satisfied. Only finite and cofinite sets have probabilities determined by (1.3), but this is all we need for the example.

Now we're in a position to do the critical calculation. Let's see what happens if Y is known to take a specific value, say 3:

$$\begin{aligned} P\{X = 1|Y = 3\} &= \frac{P\{X=1 \text{ and } Y=3\}}{P\{Y=3\}} \\ &= \frac{P\{Y=3|X=1\}P\{X=1\}}{P\{Y=3|X=1\}P\{X=1\}+P\{Y=3|X=0\}P\{X=0\}} \\ &= \frac{(1/2)^4 1/2}{(1/2)^4(1/2)+0(1/2)} = 1. \end{aligned}$$

Furthermore, the same result applies if any other value for Y is substituted instead of 3. This leads to a very odd circumstance: The value of a dollar ticket on $\overline{A}$ is \$0.50 to you at this time. But if tonight you observe the random variable Y, tomorrow you would, in order not to be a sure loser, value the same ticket at \$0. It seems that you should anticipate being a dynamic sure loser, as you would buy from me for 50 cents the ticket you know you expect to be valueless to you tomorrow, regardless of the value of Y observed.

Dynamic sure loss (defined formally in section 3.5) faces you with the following difficult question: Would you pay \$0.25 never to see Y? It would seem that this would be a good move on your part. But it is a very odd world in which you would pay not to see data. To be sure of not seeing Y, you would have to make a deal with every other person in the world, or at least those who know Y, not to tell you. This would lead to a thriving market for non-information! Dynamic sure loss is uncomfortable.

Our example involves dynamic sure loss, but it is coherent. Therefore, avoidance of dynamic sure loss involves an additional constraint, beyond coherence, on the prices offered for tickets. It must apply when random variables, such as Y, take infinitely many possible values. The needed constraint is developed in the next section. First, however, we need to understand what went wrong. The heart of the example above is the random variable X, which is an indicator random variable, with expectation $E(X) = \left(\frac{1}{2}\right) 1 + \left(\frac{1}{2}\right)\ 0 = 1/2$. However, the conditional expectation $E(X|Y = k) = 1$ for all k. Hence this example violates the theorem given in section 2.8 that

$$E[X] = E\{E[X|Y]\}. \tag{3.1}$$

Let's see where the proof of the iterated expectation law breaks down when Y can take an infinite number of possible values.

Recall the notation

$$P\{X = x_i, Y = y_j\} = p_{i,j}.$$

In this example,

$$\begin{aligned} P\{X = 1, Y = j\} &= (1/2)^{j+2} \quad j = 0, 1, 2, \ldots \\ P\{X = 0, Y = j\} &= 0 \quad\quad\quad\quad j = 0, 1, 2, \ldots \end{aligned}$$

Then the $p_{+,j}$'s and $p_{i,+}$'s are the marginal totals of these probabilities, and take the values

$$\begin{aligned} p_{+,j} &= p_{1,j} + p_{0,j} = (1/2)^{j+2}, \\ p_{1,+} &= \textstyle\sum_{j=0}^{\infty} p_{1,j} = 1/2, \text{ as the sum of a geometric series} \\ p_{0,+} &= \textstyle\sum_{j=0}^{\infty} p_{0,j} = 0. \end{aligned}$$

Hence $\sum_{j=0}^{\infty} p_{+,j} = 1/2$ and $p_{1,+} + p_{0,+} = 1/2$. Thus the constraint, imposed in section 2.8, that $\sum p_{+,j} = \sum p_{i,+} = 1$ is violated. Why does this matter? We can compute the conditional probability that $X = x_i$, given $Y = y_j$; indeed, that is done above. The answers are

$$P\{X = 1|Y = j\} = 1 \text{ and } P\{X = 0|Y = j\} = 0 \text{ for all } j.$$

Thus for each value of k, $X|Y=k$ is a random variable taking the value 1 with probability 1, and hence has expectation 1. However, the next line in section 2.8 creates a problem: "Now, for various values of y_j, the conditional expectation itself can be regarded as a random variable, taking the value $\sum_i x_i p_{i,j}/p_{+,j}$ with probability $p_{+,j}$." In our example $\sum_i x_i p_{i,j}/p_{+,j} = \frac{(1)(1/2)^{j+2}}{(1/2)^{j+2}} = 1$ for all j. The issue is that because in the example $\sum p_{+,j} = 1/2$, not 1, this conditional expectation is not a random variable. Half the probability has escaped, and has vanished from the calculation. Hence our assumptions are not strong enough to conclude that $E(X) = E\{E(X|Y)\}$.

There is another way to understand this example. A **partition** of the sure event S is a class of non-empty subsets H_j of S that are disjoint and whose union is S. Thus every element of S is a member of one and only one H_j.

There's one other mathematical concept we'll need. The supremum of a set of real numbers B is the smallest number y such that $x \leq y$ for all $x \in B$, and is written $\sup B$. Similarly the infinum of B (written $\inf B$) is the largest number z such that $z \leq x$ for all $x \in B$. Hence $\inf[0,1] = \inf(0,1] = 0$ and $\sup[0,1) = \sup[0,1] = 1$. For unbounded sets B, it is possible that $\inf B = -\infty$ and/or $\sup B = \infty$. If $B = \emptyset$, $\inf B = \infty$ and $\sup B = -\infty$.

In our example, we have $P\{X=1\} = 1/2$, but $P\{X=1|Y=j\} = 1$ for all j. Further, the events $H_j = \{Y=j\}$ are a partition of the sure event S. In general, the **conglomerative property** is said to be satisfied by an event A and a partition $H = \{H_j\}$ if

$$\inf_j P\{A|H_j\} \leq P\{A\} \leq \sup_j P\{A|H_j\}.$$

In the example, $P\{X=1\} = 1/2 < \inf_j P\{X=1|Y=j\} = 1$, so the conglomerative property fails in the example.

Every event A and every partition $\{H_j\}$ of S satisfies the conglomerative property if S has only finitely many elements and P is coherent. To show this, note that the index j can take only finitely many values because S has only finitely many elements. Suppose there are J members of the partition $\{H_j\}$. Then

$$\begin{aligned} P\{A\} =& EI_A = EE(I_A|H_j) = \sum_{j=1}^{J} E(I_A|H_j)P\{H_j\} \\ =& \sum_{j=1}^{J} P\{A|H_j\}P\{H_j\}. \end{aligned}$$

This displays $P\{A\}$ as a weighted sum of the numbers $P\{A|H_j\}$, $j=1,\ldots,J$. The weight on $P\{A|H_j\}$ is $P\{H_j\}$, where $P\{H_j\}$'s are non-negative and sum to one. Therefore

$$\min_{j=1,\ldots,J} P\{A|H_j\} \leq P\{A\} \leq \max_{j=1,\ldots,J} P\{A|H_j\}$$

and the conglomerative property is satisfied, since inf = min and sup = max for a finite set.

The next section introduces an additional assumption, countable additivity, and shows that the conglomerative property holds if countable additivity is assumed.

3.2.1 *Summary*

This section discusses an example in which your prices are coherent, but you are a dynamic sure loser, in the sense that you will accept bets at prices that ensure loss after new information becomes available. In the same example, you would rationally pay not to see data. It also violates the conglomerative property. The next three sections discuss what to do about this.

3.2.2 References

For more information about finitely additive probabilities on the integers, see Kadane and O'Hagan (1995) and Schirokauer and Kadane (2007). For the peculiar consequences of mere finite additivity in the example discussed in this section see Kadane et al. (1996) and Kadane et al. (2008). DeFinetti (1974) pointed to conglomerability as an important property. Stern and Kadane (2015) explore when countably many bets imply a countably additive probability. The answer depends on the conditions on convergence required on the bets.

3.2.3 Exercises

1. Vocabulary: Explain in your own words:
   (a) dynamic sure loss
   (b) conglomerability
   (c) cofinite set
   (d) residue class, mod k
   (e) partition
   (f) uniform distribution

   Why are these important?
2. Make your own example in which the conglomerative property fails.
3. Calculate $P\{X = 1, Y = 8\}$ and $P\{X = 0, Y = 8\}$ in the example discussed in this section.
4. Verify $p_{1,+} = 1/2$ from the example.

3.3 Countable additivity and the existence of expectations

To avoid an uncomfortable example like that shown in section 3.2 requires accepting an additional constraint on the prices you would pay for certain tickets. Reasonable opinions can differ about whether the constraint is worthwhile: more "regular" behavior, but only for assignments of probability that satisfy the additional constraints.

The additional constraint on P that prevents the "pathological" behavior shown in 3.2 is called countable additivity and is defined as follows: Let $A_1, A_2, \ldots$ be a (infinite, but countable) collection of disjoint sets.

Then

$$P(\cup_{i=1}^{\infty} A_i) = \sum_{i=1}^{\infty} P(A_i). \tag{3.2}$$

First, I must show that if your probabilities are countably additive, then they also are finitely additive. That is, I propose:

Theorem 3.3.1. *If $P(\cdot)$ satisfies (1.1), (1.2) and (3.2), then it is coherent, i.e., it satisfies (1.1), (1.2) and (1.3).*

Before proving the theorem, I will first prove the following lemma (A lemma is a path. Remember a dilemma? That's two paths, presumably hard to choose between.): If (1.1), (1.2) and (3.2) hold, then $P(\varnothing) = 0$.

Proof. Let $A_1, A_2, \ldots$ be such that $A_i = \varnothing$ for all $i = 1, 2, \ldots$ The A_i's are disjoint and their union is $\varnothing$. Therefore, using (3.2),

$$P(\varnothing) = \sum_{i=1}^{\infty} P(\varnothing).$$

The only value for $P(\emptyset)$ that can satisfy this equation is $P(\emptyset) = 0$. This concludes the proof of the lemma. □

Now I will prove the theorem. Since (1.1) and (1.2) are assumed, they need not be proved. I can assume that (3.2) holds for every countable infinite sequence of disjoint events, and must prove (1.3) for every finite sequence of disjoint events. Suppose $A_1, A_2, \ldots, A_n$ is a given finite collection of disjoint sets. I choose to let $A_{n+1} = A_{n+2} = \ldots = \emptyset$. Then $A_1, A_2, \ldots$ is an infinite collection of disjoint sets, and $\cup_{i=1}^{\infty} A_i = \cup_{i=1}^{n} A_i$. Therefore

$$P(\cup_{i=1}^{n} A_i) = P(\cup_{i=1}^{\infty} A_i) = \sum_{i=1}^{\infty} P(A_i) = \sum_{i=1}^{n} P(A_i) + \sum_{i=n+1}^{\infty} P(A_i) = \sum_{i=1}^{n} P(A_i),$$

so (1.3) holds. □

Not every finitely additive probability is countably additive, however. For example, uniform distributions on the integers cannot be countably additive. Therefore the converse of this theorem is false.

Since every countably additive probability is finitely additive, any result or theorem proved for finitely additive probabilities holds for countably additive probabilities as well. By the same token, every countably additive example is an example for finite additivity as well.

The final sentence of the previous section promised a result showing that countably additive probabilities satisfy the conglomerative property. Here is that result.

Theorem 3.3.2. *If P satisfies countable additivity* (3.2), *it satisfies the conglomerative property with respect to every set A and every countable partition $\{H_j\}$.*

Proof. Let $\{H_j\}$ be a partition and A a set. Then

$$A = \cup_{j=1}^{\infty} AH_j.$$

The sets $\{AH_j\}$ are disjoint. Then, using countable additivity,

$$P(A) = \sum_{j=1}^{\infty} P\{AH_j\} = \sum_{j=1}^{\infty} P\{A|H_j\}P\{H_j\}.$$

This displays $P(A)$ as a weighted sum of the numbers $P\{A|H_j\}$. The weights are $P\{H_j\}$, which are non-negative and sum to 1. Hence

$$\inf_j P\{A|H_j\} \leq P(A) \leq \sup_j P\{A|H_j\}$$

and the conglomerative property is satisfied. □

There is a simple argument that connects countable additivity with avoidance of sure loss. Since every countably additive probability is finitely additive, sure loss is impossible using countable additivity. The relevant question, then, is whether avoiding sure loss requires countable additivity. Let $A_1, A_2 \ldots$ be a countable sequence of disjoint events, and let $A_0 = \cup_{i=1}^{\infty} A_i$.

Let p_i be your probability for the event A_i, $i = 0, 1, 2, \ldots$. Now suppose your opponent buys from you α tickets on A_i, for $i = 1, 2, \ldots$ and sells to you α tickets on A_0. Then your winnings are

$$W = \sum_{i=1}^{\infty} \alpha(I_{A_i} - p_i) - \alpha(I_{A_0} - p_0) \tag{3.3}$$

$$= \alpha[\sum_{i=1}^{\infty} I_{A_i} - I_{A_0}] + \alpha[p_0 - \sum_{i=1}^{\infty} p_i].$$

Now

$$\sum_{i=1}^{\infty} I_{A_i} = I_{A_0}, \tag{3.4}$$

so $W = \alpha[p_0 - \sum_{i=1}^{\infty} p_i]$, and is non-stochastic.

A negative value of W would indicate sure loss. The only way to avoid it, for both positive and negative α, is

$$p_0 = \sum_{i=1}^{\infty} p_i, \tag{3.5}$$

which is the formula for countable additivity.

DeFinetti (1974, Volume 1, p. 75) would object to (3.3) because it involves being ready to bet on infinitely many events at once. If one is willing to bet on countably many events at once, why not uncountably many? This leads to "perfect" additivity, which in turn bans continuous uncertain quantities. I don't find this argument particularly appealing, in that I could imagine stopping at countable infinities.

There is a second objection to (3.3), that I find more persuasive. It is discussed in section 3.3.3.

Viewed in this light, what I have called dynamic sure loss is not so surprising. It involves an infinite partition, and displays a sure loss that develops if countable additivity is not assumed.

In order to progress, it is now necessary to review the theorems of Chapters 1 and 2, to see what modifications are required by the extension of random variables to countable many values under the assumption of countable additivity. For some of the proofs serious rethinking is needed, while in others allowing $n = \infty$ suffices.

As an example of the latter, let's look at the third form of Bayes Theorem given in section 2.4. Recall that the first form yields the result

$$P\{A|B\} = P\{B|A\}P\{A\}/P\{B\}, \tag{3.6}$$

when $P\{B\} > 0$.

Now suppose that $A_1, A_2, \ldots$ are mutually exclusive and exhaustive sets. Then B can be written as

$$B = \cup_{i=1}^{\infty} BA_i. \tag{3.7}$$

The BA_i's are disjoint. Then

$$P\{B\} = \sum_{i=1}^{\infty} P\{BA_i\} = \sum_{i=1}^{\infty} P\{B|A_i\}P\{A_i\}, \tag{3.8}$$

where the first equality uses (3.2). Now substituting (3.8) into (3.6) and replacing A by A_1 yields

$$P\{A_1|B\} = \frac{P\{B|A_1\}P\{A_1\}}{\sum_{i=1}^{\infty} P\{B|A_i\}P\{A_i\}}. \tag{3.9}$$

This result generalizes (2.10).

However, things are not so easy for the properties of random variables, and especially for expectations of random variables. The first issue occurs in formula (1.24), the definition of the expectation of a random variable, which reads as follows:

$$E(W) = \sum_{i=1}^{n} w_i p_i, \tag{1.24}$$

where $P\{W = w_i\} = p_i$ for $i = 1, \dots n$ and $\sum_{i=1}^{n} p_i = 1$. Now why can't we substitute $n = \infty$ in (1.24) and go about our business?

First, let's be careful about what such a sum would mean. If $u_1, u_2, u_3, \dots$ are the terms of such a series, and $S_1 = u_1, S_2 = u_1 + u_2,\ S_3 = u_1 + u_2 + u_3$, etc. are the partial sums, then what is meant by $S = \sum_{i=1}^{\infty} u_i$ is $S = \lim_{n\to\infty} S_n$? Such a limit does not always exist. This is unlike the case of (1.24), which always exists, because it is a finite sum.

We begin with an example to show that S does not always exist.. Consider the sum

$$T = \sum_{i=1}^{\infty} 1/i^2.$$

It is not immediately obvious whether T is finite or infinite. It is finite, as the following argument shows:

Let $u_1 = 1 - 1/2\ ;\ u_2 = 1/2 - 1/3; \dots u_n = \frac{1}{n} - \frac{1}{n+1}$.

Each of these is positive, and its partial sums are

$$S_n = \sum_{i=1}^{n} u_i = 1 - 1/(n+1)$$

which converges to 1 as $n \to \infty$. However,

$$u_n = \frac{1}{n} - \frac{1}{n+1} = \frac{1}{n(n+1)} > \frac{1}{(n+1)^2}, \text{ so}$$

$$1 + S_n = 1 + \sum_{i=1}^{n} u_i > 1 + \sum_{i=2}^{n+1} 1/i^2 = \sum_{i=1}^{n+1} 1/i^2.$$

Taking the limit as $n \to \infty$ of both sides, we have

$$2 = \lim_{n\to\infty} (1 + S_n) \geq \lim_{n\to\infty} \sum_{i=1}^{n+1} 1/i^2 = T.$$

So T is no greater than 2. For our purposes, it doesn't matter exactly what the value of T is; we care only that it is finite for this argument.

This means that the following is a probability distribution on the natural numbers $\mathcal{N}$:

$$P\{W = i\} = p_i = \frac{1}{Ti^2} \quad i = 1, 2, \dots \ .$$

Now let's investigate the expectation of W:

$$E(W) = \sum_{i=1}^{\infty} w_i p_i = \sum_{i=1}^{\infty} i \cdot \frac{1}{Ti^2} = \frac{1}{T} \sum_{i=1}^{\infty} \frac{1}{i}.$$

What can be said about the sum $\sum_{i=1}^{\infty} \frac{1}{i}$?

We group terms together in this sum, taking together the 2^{n-1} terms that end in $\frac{1}{2^n}$. We then bound these terms below by $\frac{1}{2^n}$, and add. This yields an infinite number of summands of $\frac{1}{2}$, as follows:

$$\begin{array}{ccccccccc}
1 + \frac{1}{2} & + & \frac{1}{3} + \frac{1}{4} & + & \frac{1}{5} + \frac{1}{6} + \frac{1}{7} + \frac{1}{8} & + & \dots \geq \\
1 + \frac{1}{2} & + & (\frac{1}{4} + \frac{1}{4}) & + & (\frac{1}{8} + \frac{1}{8} + \frac{1}{8} + \frac{1}{8}) & + & \dots = \\
1 + \frac{1}{2} & + & \frac{1}{2} & + & \frac{1}{2} & + & \dots = \infty.
\end{array}$$

Hence $\sum_{i=1}^{\infty} \frac{1}{i}$ diverges to infinity. In this case, $E(W)$ is said not to exist. Therefore we have an example of a random variable whose expectation does not exist. The important lesson to learn is that when random variables have countably many values, the expectation may or may not exist.

More mischief can happen when W can take both positive and negative values. To give a flavor of the difficulties, consider the sum

$$1 - 1 + 1 - 1 + 1 - 1 \ldots \quad .$$

The terms in this sum can be grouped in two natural ways. The grouping

$$(1 - 1) + (1 - 1) + (1 - 1) \ldots$$

makes it seem that the sum must be 0. However the grouping

$$1 + (-1 + 1) + (-1 + 1) + (-1 + 1)$$

makes it seem that the sum must be 1. Indeed, the even partial sums are 0, while the odd partial sums are 1, so $\lim_{n \to \infty} S_n$ does not exist.

I now begin an examination of the convergence of series to determine for which random variables one may usefully write an expectation.

Theorem 3.3.3. *Let $u_1, u_2, \ldots$ be terms in an infinite series. Denote by $a_1, a_2, \ldots$ the positive terms among $u_1, u_2, \ldots$ taken in the order of their occurrence. Similarly let $-b_1, -b_2, \ldots$ be the negative terms, again in order of occurrence.*

If $\sum |u_n| < \infty$, then $\sum a_n$ and $\sum b_n$ are both convergent, and

$$\sum u_n = \sum a_n - \sum b_n.$$

Proof. If $\sum |u_n| = M < \infty$, then for all N,

$$\sum_{n=1}^{N} |u_n| \leq M.$$

Now consider a partial sum of the a's, $\sum_{i=1}^{m} a_i$. Since each of the a's is a u_n for some n, there is an N large enough so that each of the terms $a_1 \ldots, a_m$ occur in $u_1 \ldots, u_N$. But $\sum_{i=1}^{m} a_i \leq \sum_{j=1}^{N} |u_j| \leq M$. It then follows that $\sum_{i=1}^{\infty} a_i$ converges. Similarly each b_i occurs somewhere in the sequence $\sum_{n=1}^{\infty} u_n$, so by the same argument $\sum b_n$ converges.

Finally suppose that in the sum $u_1 + \ldots + u_n$ there are r_n positive terms and s_n negative ones. Then

$$u_1 + u_2 + \ldots + u_n = a_1 + \ldots + a_{r_n} - (b_1 + \ldots + b_{s_n}).$$

Letting $n \to \infty$, we have

$$\sum_{n=1}^{\infty} u_n = \sum_{n=1}^{\infty} a_n - \sum_{n=1}^{\infty} b_n \text{ as claimed.}$$

□

There is one more property of series critical for the application to the expectation of random variables. The expectation of a random variable, which we are thinking of as

$$E(W) = \sum_{i=1}^{\infty} w_i p_i$$

cannot depend on the order of the terms in the summation. The next theorem shows that if a series is absolutely convergent, or, equivalently, if $\sum |u_n| < \infty$, then the order of terms doesn't matter.

Theorem 3.3.4. *Let the series $\sum u_n$ be convergent, with sum s and suppose $\sum |u_n| < \infty$. Let $\sum v_n$ be a series obtained by rearranging the order of terms in $\sum u_n$ (i.e., every v_i is some u_j and every u_j is some v_i). Then $\sum v_n$ is convergent, with sum s.*

Proof. Consider first the situation in which all the u's (and hence all the v's) are non-negative. Since $s = \sum u_n$ and each v_i is some u_j, the partial sums of the series $\sum v_n$ must each be less than s. Therefore the series $\sum v_n$ converges, and its sum s' must satisfy $s' \leq s$. But this argument can be reversed, yielding $s \leq s'$. Hence $s = s'$.

Now consider the case in which the u's can be negative. By Theorem 3.3.3, we can write

$$\sum u_n = \sum a_n - \sum b_n.$$

Similarly for the rearranged series, we can write

$$\sum v_n = \sum a'_n - \sum b'_n.$$

But a'_n is a rearrangement of a_n and b'_n is a rearrangement of b_n. Hence $\sum a_n = \sum a'_n$ and $\sum b_n = \sum b'_n$. Therefore $\sum v_n$ converges, and to the same sum as $\sum u_n$. □

The case not yet considered is when $\sum |u_n|$ is not convergent, but $\sum u_n$ converges. The following (classic) example shows that odd things happen to rearrangements under these conditions.

We already know, from section 2.7, the sum of a finite geometric series:

$$\sum_{i=0}^{k} r^i = \frac{1 - r^{k+1}}{1 - r} = \frac{1}{1 - r} - \frac{r^{k+1}}{1 - r}.$$

From elementary calculus, we know

$$\int_0^x \frac{dt}{1 + t} = \log(1 + t)|_0^x = \log(1 + x) - \log 1 = \log(1 + x).$$

Then

$$\begin{aligned}
\log(1 + x) &= \int_0^x \frac{1}{1 + t} dt = \int_0^x \frac{1}{1 - (-t)} dt \\
&= \int_0^x \left[\sum_{i=0}^{n} (-t)^i + \frac{(-t)^{n+1}}{1 + t} \right] dt \\
&= \sum_{i=0}^{n} \int_0^x (-t)^i dt + \int_0^x \frac{(-t)^{n+1}}{1 + t} dt \\
&= \sum_{i=0}^{n} - \frac{(-x)^{i+1}}{i + 1} + \int_0^x \frac{(-t)^{n+1}}{1 + t} dt \\
&= x - \frac{x^2}{2} + \frac{x^3}{3} + \ldots + \frac{(-1)^n(-x)^{n+1}}{n + 1} + R_n
\end{aligned}$$

where $R_n = \int_0^x \frac{(-1)^{n+1} t^{n+1}}{(1+t)} dt$. Now

$$|R_n| \leq \int_0^x t^{n+1} dt = \frac{x^{n+2}}{n + 2},$$

which goes to zero for $0 \leq x \leq 1$, as $n \to \infty$. Therefore, taking the limit as $n \to \infty$, we may write

$$\log(1 + x) = x - \frac{x^2}{2} + \frac{x^3}{3} - \frac{x^4}{4} + \ldots \quad , 0 \leq x \leq 1$$

and in particular

$$\log 2 = 1 - \frac{1}{2} + \frac{1}{3} - \frac{1}{4} + \frac{1}{5} - \frac{1}{6} + \ldots. \tag{3.10}$$

This series (called the alternating harmonic series) converges but is not absolutely convergent since $\sum_{i=1}^{\infty} \frac{1}{i} = \infty$. Therefore it may serve as an example not covered by Theorems 3.3.3 and 3.3.4. It is convenient to re-express (3.10) as follows:

$$\log 2 = \sum_{i=1}^{\infty} \frac{(-1)^{i+1}}{i} = \sum_{k=0}^{\infty} \left[\frac{1}{4k+1} - \frac{1}{4k+2} + \frac{1}{4k+3} - \frac{1}{4k+4} \right]. \tag{3.11}$$

There are three operations we may perform on a convergent series. We may multiply each term by a constant, and get the constant times the sum. Thus we may write

$$\frac{1}{2} \log 2 = \frac{1}{2} - \frac{1}{4} + \frac{1}{6} - \frac{1}{8} + \frac{1}{10} - \frac{1}{12} + \ldots. \tag{3.12}$$

Another thing we may do, without changing the sum, is to add zeros where we wish. Hence I can write

$$\frac{1}{2} \log 2 = 0 + \frac{1}{2} + 0 - \frac{1}{4} + 0 + \frac{1}{6} + 0 - \frac{1}{8} + 0 + \frac{1}{10} + \ldots. \tag{3.13}$$

Again it is convenient to re-express (3.13) as follows:

$$\frac{1}{2} \log 2 = \sum_{k=0}^{\infty} \left[\frac{0}{4k+1} + \frac{1}{4k+2} + \frac{0}{4k+3} - \frac{1}{4k+4} \right]. \tag{3.14}$$

Term-by-term addition of two convergent series yields a series that converges to the sum of the two series. To see this, suppose $\{a_n\}$ is a series converging to A, and $\{b_n\}$ is a series converging to B. Then I claim that $\{c_n\}$, where $c_n = a_n + b_n$, is a series converging to $C = A + B$. Then we have

$$\lim_{n\to\infty} \sum_{i=1}^{n} a_i = A \text{ and } \lim_{n\to\infty} \sum_{i=1}^{n} b_i = B.$$

Therefore

$$\begin{aligned} \lim_{n\to\infty} \sum_{i=1}^{n} c_i &= \lim_{n\to\infty} \sum_{i=1}^{n} (a_i + b_i) = \lim_{n\to\infty} \left(\sum_{i=1}^{n} a_i + \sum_{i=1}^{n} b_i \right) \\ &= \lim_{n\to\infty} \sum_{i=1}^{n} a_i + \lim_{n\to\infty} \sum_{i=1}^{n} b_i = A + B = C. \end{aligned}$$

Returning to our example, I now add (3.11) and (3.14) term-by-term, and obtain

$$\begin{aligned} \frac{3}{2} \log 2 &= \sum_{k=0}^{\infty} \left[\frac{1}{4k+1} + \frac{0}{4k+2} + \frac{1}{4k+3} - \frac{2}{4k+4} \right] \\ &= \sum_{k=0}^{\infty} \left[\frac{1}{4k+1} + \frac{1}{4k+3} - \frac{1}{2k+2} \right]. \end{aligned} \tag{3.15}$$

The terms $\frac{1}{4k+1} + \frac{1}{4k+3}$, for $k = 0, 1, \ldots$, give the sum of the reciprocals of the odd integers, once each, and each with a coefficient of $+1$. Similarly, $\frac{-1}{2k+2}$, for $k = 0, 1, 2, \ldots$, gives the reciprocals of the even integers, once each, and each with a coefficient of -1. Thus (3.15) is a rearrangement of (3.10). Hence a rearrangement of the series for $\log 2$ yields a series for $\frac{3}{2} \log 2$.

The next theorem shows that the situation is in fact much worse than is hinted by this example: if $\sum |u_n|$ is not convergent, but $\sum u_n$ converges, a rearrangement of the terms in u_n can be found to yield any desired sum R. More formally,

Theorem 3.3.5. *(Riemann's Rainbow Theorem) Suppose that $\sum u_n$ converges, but $\sum |u_n|$ does not. Let R be any real number. Then there is a rearrangement of the terms in u_n such that the partial sums approach R.*

Proof. First, it is clear that $u_n \to 0$, because $u_n = s_n - s_{n-1}$. Since $\sum u_n$ converges to some number s, $s = \lim_{n\to\infty} s_n = \lim_{n\to\infty} s_{n-1}$. Hence $\lim_{n\to\infty} u_n = \lim_{n\to\infty}(s_n - s_{n-1}) = \lim s_n - \lim s_{n-1} = s - s = 0$.

We may separate the positive and negative terms in u_n as in the proof of Theorem 3.3.3 into a_n and $-b_n$, respectively. Now because $\sum |u_n|$ does not converge, at least one of $\sum a_n$ and $\sum b_n$ must not converge. Indeed, neither can converge, since if only one converged, $\sum u_n$ would be either ∞ or $-\infty$, and hence would not converge.

The idea of the construction is as follows: If $R \geq 0$, we start with a_n's (in order) and append to the series as many a's as required to lift the partial sum s_n to be above R for the first time. Then we append just as many b's as necessary to reduce the partial sum to be below R. This process is repeated indefinitely. Since $u_n \to 0$, it follows that $a_n \to 0$ and $b_n \to 0$. I show below that the consequence is that the partial sums from this construction approach R. Similarly, if $R < 0$, begin with the b's.

Because $\sum u_n$ converges, we have $u_n \to 0$. Then for every $\epsilon > 0$ there is some N such that for all $n \geq N, |u_n| < \epsilon$.

Suppose $R \geq 0$. The construction above specifies a particular order of the terms in the rearrangement. Let w_n be the nth term in the rearranged series, and let v_n be the nth partial sum, so that

$$v_n = \sum_{i=1}^{n} w_i.$$

Because we have assumed $R \geq 0$, we have $w_1 = v_1 = a_1$. Because $\sum a_i = \infty$ and $a_i > 0$ for all i, after some positive, finite number t_1 of terms, we have

$$v_{t_1} > R$$

for the first time. By construction,

$$|v_{t_1} - R| = v_{t_1} - R < a_{t_1}.$$

Then we subtract some positive, finite number s_1 of b-terms from the sum, until, for the first time

$$v_{t_1+s_1} = \sum_{i=1}^{t_1} a_i - \sum_{J=1}^{s_1} b_j \leq R.$$

Again, note that

$$|R - v_{t_1+s_1}| = R - v_{t_1+s_1} < b_{s_1}.$$

Now the process proceeds by adding some finite, positive number of a-terms, so, that, for the first time, for $t_2 > t_1$,

$$u_{t_2+s_1} = \sum_{i=1}^{t_2} a_i - \sum_{j=1}^{s_1} b_j \leq R,$$

and, once again,

$$|u_{t_2+s_1} - R| = R - u_{t_2+s_1} \leq a_{t_2}.$$

After at most $2N$ switches of sign, suppose $\sum_{i=1}^{N}(t_i + s_i) = M$. Then for all $n \geq M$, we have $|w_n| < \epsilon$. [This is where we use that $t_i > 0$ and $s_i > 0$. Because of that, after $2N$

switches of sign, each of $u_1, \ldots, u_N$ has already appeared among $w_1, \ldots, w_M$.] At the first switch of sign after the $2N^{th}$, we also have

$$|R - v_{m'}| < |w_{m'}| < \epsilon, \tag{3.16}$$

where m' is the index of the next sign switch. I now proceed by induction on n to show

$$|R - v_n| < \epsilon$$

for all $n \geq m'$.

Equation (3.16) gives the result for $n = m'$. Now suppose the result is true for n. We have the following facts:

(i) $|v_n - R| < \epsilon$ [inductive hypothesis]

(ii) $|w_{n+1}| < \epsilon$

(iii) $v_n - R$ and w_n have opposite signs.

If x and y are two numbers with opposite signs, then

$$|x + y| \leq \max\{|x|, |y|\}.$$

Let $x = v_n - R$ and $y = w_{n+1}$. Then

$$|v_n - R + w_{n+1}| \leq \max\{|v_n - R, |w_{n+1}|\} < \epsilon.$$

But $v_n - R + w_{n+1} = v_{n+1} - R$. Therefore

$$|v_{n+1} - R| < \epsilon.$$

This completes the induction.

Consequently, for all $n \geq m'$,

$$|v_n - R| < \epsilon.$$

But this shows

$$\lim_{n \to \infty} v_n = R.$$

If $R < 0$, the only change is to start with the b's first. Again, we have

$$\lim_{n \to \infty} v_n = R.$$

Since R is arbitrary, this completes the proof of the theorem. □

We now return to the topic that necessitated this investigation into the convergence of series, namely the expectation of random variables taking a countable number of possible values. The results of Theorems 3.3.4 and 3.3.5 are as follows:

(i) Consider a series $u_1, u_2, \ldots$ that is absolutely convergent, which means $\sum |u_n| < \infty$. Suppose that $\sum u_i$ converges to s. Let $v_1, v_2, \ldots$ be a rearrangement of $u_1, u_2, \ldots$. Then $\sum v_n$ also converges to s.

(ii) If the series $u_1, u_2, \ldots$ converges but is not absolutely convergent, then there is a rearrangement of $u_1, u_2, \ldots$ that converges to any number R you choose.

Hence if we allow convergent but not absolutely convergent series for the expectation of a random variable, it would also be necessary to specify the order in which the terms are to enter the series. Since the order of the summands $w_i p_i$ has no probabilistic meaning, we choose the second possibility, and require absolute convergence before we regard the expectation as being defined. It is for this reason that we must have $E|W| < \infty$ as a condition before $E(W)$ is regarded as defined.

3.3.1 Summary

Theorems 3.3.4 and 3.3.5 can be summarized in the following statement. Suppose $\sum u_n$ converges. Then every rearrangement of the terms in u_n converges to the same limit if and only if $\sum |u_n|$ converges.

Therefore we say that the random variable W, which satisfies $P\{W = w_i\} = p_i$, $i = 1, 2, \ldots$ and $\sum p_i = 1$ has an expectation provided

$$E(|W|) = \sum_{i=1}^{\infty} |w_i| p_i < \infty. \tag{3.17}$$

3.3.2 References

Much of this discussion comes from Taylor (1955, chapter 17), and from Courant (1937) to which the reader is referred for further details. Both Courant (1937) and Hardy (1955) use the term "conditional convergence" to refer to what I have called convergence.

3.3.3 Can we use countable additivity to handle countably many bets simultaneously?

Sections 1.1 and 1.7 show that you cannot be made a sure loser for any finite set of gambles if and only if your probabilities are coherent, or, equivalently, if and only if they are finitely additive. Since countably additive probabilities are a subset of finitely additive probabilities, it follows that if your probabilities are countably additive, you cannot be made a sure loser for any finite set of gambles. It is natural to hope that, if your probabilities are countably additive, you might avoid being a sure loser against a countable set of gambles. This is not the case, however, as the following example, due to Beam (2007), shows.

Let c be a real number (whose selection will be discussed later).

Recalling (3.7), the series

$$1 - 1/2 + 1/3 - 1/4 + 1/5 \ldots$$

converges (and in fact, converges to log 2). However, this series does not converge absolutely. Using Theorem 3.5, there is a rearrangement of these terms, which can be expressed as a permutation i_n of the integers n, such that

$$\sum_{n=1}^{\infty} (-1)^{i_n} \cdot \frac{1}{i_n} = c.$$

Let $a_n = (-1)^{i_n+1}$ and $A_n = (0, 1/i_n)$, and let w have a uniform distribution on $(0, 1)$. (Continuous random variables, such as a uniform distribution on $(0, 1)$ used here, are introduced in Chapter 4. For our purposes here, the only property needed is that $P\{0 < w\} = w$ for $0 < w < 1$.) Thus $P(A_n) = 1/i_n$. We now study the payoff from these bets, which is

$$\sum_{n \geq 1} a_n (I_{A_n} - P(A_n)). \tag{3.18}$$

Suppose $w \epsilon (0, 1)$ is the random outcome. Then only finitely many of the terms of the form $a_n I_{A_n}(w)$ are non-zero, so the contribution

$$\sum_{n>1} a_n I_{A_n}(w)$$

is independent of the ordering i_n.

In particular, there is a value of $k \geq 1$ such that $\frac{1}{k} > w \geq 1/(k+1)$, so

$$\sum_{n \geq 1} a_n I_{A_n}(w) = \sum_{i=1}^{k} (-1)^{i+1} I_{(0,1/i)}(w) = \sum_{i=1}^{k} (-1)^{i+1} = I_{\{k \text{ is odd}\}}(w).$$

Then

$$\sum_{n \geq 1} a_n (I_{A_n} - P(A_n))(w) = c + I_{\{k \text{ is odd}\}}(w). \tag{3.19}$$

Thus, choosing an order of the terms i_n so that $c > 0$ leads to a sure gain, while $c < -1$ leads to a sure loss. Since the permutation i_n corresponds to the order in which the bets are settled, this means that whether this countably infinite set of bets is favorable or not depends on the order of settlement, which is unsatisfactory.

3.3.4 Exercises

1. Vocabulary: Explain in your own words:
   (a) convergent series
   (b) absolutely convergent series
2. Let $T = \sum_{i=1}^{\infty} 1/i^2$. Recall $0 < T \leq 2$.
   Let Y be defined as follows:

$$y = \begin{cases} Y = & i, \text{if } i \text{ is even, with probability } 1/Ti^2 \\ = & -i, \text{if } i \text{ is odd, with probability } 1/Ti^2 \\ = & 0 \text{ otherwise} \end{cases}$$

   (a) Show that Y is a random variable, that is, show

$$\sum_{i=-\infty}^{\infty} P\{Y = i\} = 1.$$

   (b) Does $E(Y)$ exist? Explain why or why not.
3. Let $T^* = \sum_{i=1}^{\infty} \frac{1}{i^3}$.
   (a) Show that $T^* < \infty$.
   (b) Define W as follows:

$$P\{W = i\} = \frac{1}{T^* i^3} \quad i = 1, 2, \ldots$$

   Show that W is a random variable.
   (c) Show that $E(W)$ exists.
   (d) Show that $E(W^2)$ does not exist.
4. (Rudin (1976, p. 196)) Consider the following two-dimensional array of numbers:

$$\begin{array}{ccccc} -1 & 0 & 0 & 0 & \ldots \\ 1/2 & -1 & 0 & 0 & \ldots \\ 1/4 & 1/2 & -1 & 0 & \ldots \\ 1/8 & 1/4 & 1/2 & -1 & \ldots \\ \vdots & \vdots & \vdots & \vdots & \vdots \end{array}$$

   (a) Show that the row sums are respectively $-1, -1/2, -1/4, -1/8, \ldots$.

(b) Show that the sum of the row sums is -2.

(c) Show that each column sum is 0, and therefore that the sum of the column sums is 0.

(d) Explain why the sum in (b) is different from the sum in (c), using the theorems of this section.

5. In Chapter 1, at equation 1.8, a proof is given that $P\{B\} = 0$. Is it necessary to prove the Lemma that is part of the proof of Theorem 3.3.1? Why or why not?

3.3.5 A supplement on calculus-based methods of demonstrating the convergence of series

There is a reason that will be immediate from calculus as to why $\sum 1/j^2$ converges but $\sum 1/j$ diverges. Both sums can be thought of as bounded by the area under a step function whose height at some positive number x is respectively $\frac{1}{\lfloor x \rfloor^2}$ or $\frac{1}{\lfloor x \rfloor}$ where $\lfloor x \rfloor$ is the largest integer smaller than or equal to x. Since the function

$$g(x) = \begin{cases} 1 & 0 < x < 1 \\ \frac{1}{x^2} & x > 1 \end{cases}$$

is everywhere greater than or equal to $\frac{1}{\lfloor x \rfloor^2}$, the area under it is at least as large as $\sum \frac{1}{i^2}$. But this is simply $1 + \int_1^\infty \frac{1}{x^2} dx = 1 - (\frac{1}{x})|_1^\infty = 1 + 1 = 2$, so $\sum 1/i^2$ is bounded above by 2 and hence converges.

Similarly, the function

$$f(x) = \begin{cases} 1/2 & 0 < x < 2 \\ \frac{1}{x+1} & x > 2 \end{cases}$$

is always less than $1/x$, so its integral is a lower bound to the sum. But

$$\int_0^\infty f(x)dx = 1 + \int_2^\infty \frac{1}{x+1} dx = 1 + \int_1^\infty \frac{1}{x} dx = 1 + \log(x)|_1^\infty = \infty.$$

Hence $\sum 1/j$ diverges.

These arguments can be generalized to show that $\sum_{i=1}^\infty \frac{1}{j^p}$ converges for $p > 1$ and diverges if $p \leq 1$.

3.4 Properties of expectations of random variables taking at most countably many values, assuming countable additivity

This section explores the properties of expectation stated in sections 1.5 and 1.6 to see which of them extend to random variables taking a countable number of possible values, assuming that the underlying probability is countably additive.

1. Suppose X is a random variable having an expectation. Let k and b be constants, and let $Y = kX + b$. Then Y has an expectation, and its value is

$$E(Y) = kE(X) + b.$$

Proof. Suppose $P\{X = x_i\} = p_i$, $i = 1, 2, \ldots$, with $\sum p_i = 1$. Then

$P\{Y = kx_i + b\} = p_i$, $i = 1, 2, \ldots$ From this,

$$\begin{aligned} E|Y| &= \sum_{i=1}^{\infty} |kx_i + b|p_i \\ &\leq \sum_{i=1}^{\infty} (|kx_i| + |b|)p_i \\ &= |k| \sum_{i=1}^{\infty} |x_i|p_i + \sum_{i=1}^{\infty} |b|p_i \\ &= |k|E|X| + |b| < \infty. \end{aligned}$$

Therefore the expectation of Y exists. Its value is

$$\begin{aligned} E(Y) &= \sum_{i=1}^{\infty} (kx_i + b)p_i \\ &= k \sum_{i=1}^{\infty} x_i p_i + b \sum_{i=1}^{\infty} p_i \\ &= kE(X) + b. \end{aligned}$$

□

2. Suppose X and Y are random variables whose expectations exist. Then $X + Y$ is a random variable whose expectation exists, and

$$E(X + Y) = E(X) + E(Y).$$

Proof. The argument is parallel to that in section 1.5.
Let

$$p_{i,j} = P\{X = x_1, Y = y_j\} \text{ for } i = 1, \ldots \text{ and } j = 1, \ldots.$$

The events $\{X = x_i, Y = y_j\}$, for $j = 1, \ldots$ are disjoint and

$$\{X = x_i\} = \cup_{j=1}^{\infty} \{X = x_i, Y = y_j\}.$$

Consequently, using countable additivity, it follows that

$$\begin{aligned} P\{X = x_i\} &= \sum_{j=1}^{\infty} P\{X = x_i, Y = y_i\} \\ &= \sum_{j=1}^{\infty} p_{i,j} = p_{i,+} \;, i = 1, 2, \ldots \end{aligned}$$

Similarly, reversing the roles of X and Y, we have

$$p_{+,j} = \sum_{i=1}^{\infty} p_{i,j} \;, \quad j = 1, 2, \ldots$$

Now

$$\begin{aligned}
E|X+Y| &= \sum_{i=1}^{\infty}\sum_{j=1}^{\infty} |x_i + y_j| P\{X = x_i, Y = y_j\} \\
&= \sum_{i=1}^{\infty}\sum_{j=1}^{\infty} |x_i + y_j| p_{i,j} \\
&\leq \sum_{i=1}^{\infty}\sum_{j=1}^{\infty} (|x_i| + |y_j|) p_{i,j} \\
&= \sum_{i=1}^{\infty}\sum_{j=1}^{\infty} |x_i| p_{i,j} + \sum_{i=1}^{\infty}\sum_{j=1}^{\infty} |y_j| p_{i,j} \\
&= \sum_{i=1}^{\infty} |x_i| \sum_{j=1}^{\infty} p_{i,j} + \sum_{j=1}^{\infty} |y_j| \sum_{i=1}^{\infty} p_{i,j} \\
&= \sum_{i=1}^{\infty} |x_i| p_{i,+} + \sum_{j=1}^{\infty} |y_j| p_{+,j} \\
&= E|X| + E|Y| < \infty.
\end{aligned}$$

Therefore $X + Y$ has an expectation. Its value is

$$\begin{aligned}
E(X+Y) &= \sum_{i=1}^{\infty}\sum_{j=1}^{\infty} (x_i + y_j) P\{X = x_i, Y = y_j\} \\
&= \sum_{i=1}^{\infty}\sum_{j=1}^{\infty} p_{i,j}(x_i + y_j) \\
&= \sum_{i=1}^{\infty} x_i p_{i,+} + \sum_{j=1}^{\infty} y_j p_{+,j} = E(X) + E(Y)
\end{aligned}$$

□

Again, by induction, if $X_1 \ldots X_k$ are random variables having expectations, then $X_1 + \ldots + X_k$ has an expectation, whose value is

$$E(X_1 + \ldots + X_k) = \sum_{i=1}^{k} E(X_i).$$

This result holds regardless of any dependencies between the X_i.

3. Suppose X is non-trivial and has an expectation. Then

$$\min X < E(X) < \max X.$$

(We must extend the possible values of $\min X$ to include $-\infty$, and of $\max X$ to include ∞.)

Proof. Since X is non-trivial, it takes at least two distinct values each with positive probability. Then

$$-\infty \leq \min X = \sum_{i=1}^{\infty} p_i(\min X) < \sum_{i=1}^{\infty} p_i x_i = E(X)$$

$$< \sum_{i=1}^{\infty} p_i(\max X) = \max X \leq \infty.$$

□

4. If X is non-trivial and has an expectation c, then there is some positive probability $\epsilon > 0$ and some $\eta > 0$ such that X exceeds c by η, and such that c exceeds X by the fixed amount η, that is $P\{X - c > \eta\} > \epsilon$ and $P\{c - X > \eta\} > \epsilon$.

Proof. Let $A_i = \{\frac{1}{i} > X - c \geq \frac{1}{i+1}\}$ $i = 0, 1 \ldots \infty$, where $\frac{1}{0}$ is taken to be ∞. The A_i's are disjoint, and

$$\cup_{i=1}^{\infty} A_i = \{X - c > 0\}.$$

Similarly let $B_j = \{\frac{1}{j} > c - X \geq \frac{1}{j+1}\}$, $j = 0, 1, \ldots, \infty$. The B_j's are disjoint and

$$\cup_{j=1}^{\infty} B_j = \{c - X > 0\}.$$

Since X is non-trivial, $P\{X \neq c\} > 0$. But

$$\begin{aligned} 0 < P\{X \neq c\} &= P\{X > c\} + P\{X < c\} \\ &= \sum_{i=0}^{\infty} P\{A_i\} + \sum_{j=0}^{\infty} P\{B_j\}, \end{aligned}$$

using countable additivity.
Hence there must be some i or j such that $P\{A_i\} > 0$ or $P\{B_j\} > 0$. Suppose first that $P\{A_i\} > 0$. If it were the case that $P\{B_j\} = 0$ for all j, then

$$0 = E(X - c) \geq (1/(i+1))P\{A_i\} > 0$$

a contradiction.
Therefore if $P\{A_i\} > 0$ for some i, there is a j such that $P\{B_j\} > 0$.
Conversely, if $P\{B_j\} > 0$, but $P\{A_i\} = 0$ for all i, then

$$0 = E(c - X) \geq (1/(j+1))\, P\{B_j\} > 0$$

a contradiction. Therefore there is both an i such that $P\{A_i\} > 0$ and a j such that $P\{B_j\} > 0$.
Now taking $\epsilon = \min(P\{A_i\}, P\{B_j\}) > 0$ and $\eta = \min\{\frac{1}{i+1}, \frac{1}{j+1}\} > 0$ suffices. □

5. If g is a real valued function, then $Y = g(X)$ has expectation

$$E(Y) = \sum g(x_k) P\{X = x_k\}$$

where $x_1, x_2, \ldots$ are the possible values of X, provided $E(|Y|) < \infty$.

Proof. This proof is very similar to the proof of theorem 1.6.3 in section 1.6.
The values of $Y = g(X)$ with positive probability are countable, since the values of X with positive probability are countable. Let those values be y_j, $j = 1, 2, \ldots$. Let I_{kj} be an indicator for the event $X = x_k$ and $Y = y_j$ for $j = 1, 2, \ldots$ and $k = 1, 2, \ldots$. Note

that $y_j I_{kj} = g(x_k)I_{kj}$. Then

$$\begin{aligned}
E(Y) =& \sum_{j=1}^{\infty} y_j P\{Y = y_j\} \\
=& \sum_{j=1}^{\infty} y_j E(\sum_{k=1}^{\infty} I_{kj}) \\
=& E \sum_{j=1}^{\infty} \sum_{k=1}^{\infty} y_j I_{kj} \\
=& E \sum_{j=1}^{\infty} \sum_{k=1}^{\infty} g(x_k) I_{kj} \\
=& \sum_{k=1}^{\infty} g(x_k) E(\sum_{j=1}^{\infty} I_{kj}) \\
=& \sum_{k=1}^{\infty} g(x_k) P\{X = x_k\}.
\end{aligned}$$

The reordering of the terms does not affect the sum because we have $E|Y| < \infty$. □

6. Let X and Y be random variables taking at most countably many values. Suppose that $E[X]$ and $E[X|Y = y_j]$ exist for all possible values of Y. Then

$$E[X] = E\{E[X|Y]\}.$$

Proof. Let $P\{X = x_i, Y = y_j\} = p_{i,j} \quad i, j = 1, 2, \ldots$ where $\sum_{i=1}^{\infty} p_{i,j} = p_{+,j}$ and $\sum_{j=1}^{\infty} p_{i,j} = p_{i,+}$ for all i and j, and $\sum_{i=1}^{\infty} p_{i,+} = \sum_{j=1}^{\infty} p_{+,j} = 1$.
Without loss of generality, we may eliminate any values of X that have zero probability. Hence we may assume $p_{i,+} > 0$ for $i = 1, 2, \ldots$. Similarly, we may eliminate any values of Y with zero probability, and thus assume $p_{+,j} > 0$.
Now the conditional probability that $X = x_i$, given $Y = y_j$, is

$$P\{X = x_i|Y = y_j\} = \frac{P\{X = x_i, Y = y_j\}}{P\{Y = y_j\}} = p_{i,j}/p_{+,j}.$$

For each fixed value of y_j, $X|Y = y_j$ is a random variable, taking the value x_i with probability $p_{i,j}/p_{+,j}$.
Now $E[X]$ exists by assumption, and satisfies $E[X] = \sum_{i=1}^{\infty} x_i p_{i,+} = \sum_{i=1}^{\infty} x_i \sum_{j=1}^{\infty} p_{i,j}$.
Because $E|X| < \infty$, we may interchange the order of summation, so

$$\begin{aligned}
E[X] =& \sum_{j=1}^{\infty} \sum_{i=1}^{\infty} x_i p_{i,j} \\
=& \sum_{j=1}^{\infty} p_{+,j} \sum_{i=1}^{\infty} x_i p_{i,j}/p_{+,j} \\
=& E\{E[X|Y]\}
\end{aligned}$$

This result is sometimes called the law of iterated expectation. □

7. Let X and Y be independent random variables, and $g(X)$ and $h(Y)$ are functions of X and Y, respectively. Suppose both $g(X)$ and $h(Y)$ have expectations. Then the expectation of $g(X)h(Y)$ exists, and

$$E[g(X)h(Y)] = E[g(X)]E[h(Y)].$$

Proof. If X and Y are independent,

$$P\{X = x_i, Y = y_j\} = P\{X = x_i\}P\{Y = y_j\} = s_i t_j.$$

Then

$$\begin{aligned}
E|g(X)h(Y)| &= \sum_{i=1}^{\infty}\sum_{j=1}^{\infty} |g(x_i)||h(y_j)|s_i t_j \\
&= \sum_{i=1}^{\infty} |g(x_i)|s_i \sum_{j=1}^{\infty} |h(y_j)|t_j \\
&= E|g(X)|E|h(Y)| < \infty.
\end{aligned}$$

Therefore the expectation of $g(X)h(Y)$ exists.
Then

$$\begin{aligned}
E[g(X)h(Y)] &= \sum_{i=1}^{\infty}\sum_{j=1}^{\infty} g(x_i)h(y_j)s_i t_j = \sum_{i=1}^{\infty} g(x_i)s_i \sum_{j=1}^{\infty} h(y_j)t_j \\
&= E[g(X)]E[h(Y)]
\end{aligned}$$

□

8. Suppose $E|X|^k < \infty$ for some k. Let $j < k$. Then $E|X|^j < \infty$.

   Proof. Let $P\{X = i\} = p_i$. Then

$$\begin{aligned}
E|X|^j &= \sum |X_i|^j p_i = \sum |X_i|^j p_i I(|X_i| \leq 1) + \sum |X_i|^j p_i \; I(|X_i| \geq 1) \\
&\leq 1 + \sum |X_i|^k p_i \; I(|X_i \geq 1) \\
&\leq 1 + E|X|^k < \infty
\end{aligned}$$

□

   In particular, if $E(X^2) < \infty$ then $E|X| < \infty$.

9. All the properties of covariances and correlations given in section 2.11 hold for all discrete random variables provided that each of the sums is absolutely convergent, that is, provided $E(X^2) < \infty$, $E(Y^2) < \infty$ and $E(|XY|) < \infty$.
   Thus, once the question of the existence of expectations is clarified, the properties of expectations of random variables taking countably many values, under countable additivity, are the same as those of random variables taking only finitely many values under finite additivity.

3.4.1 Summary

This section proves the properties of expectations of discrete random variables that may have countably many values, under the assumption of countable additivity.

3.5 Dynamic sure loss

Having found the correct condition for the existence of expectations of discrete random variables and having checked their properties, it is now possible to return to the subject of dynamic sure loss and countable additivity.

Before we do so, though, I must address a subtle point about what is to be considered a sure loss or a sure gain. Suppose that U has a finitely-additive uniform distribution on the positive numbers, and suppose $X = 1/U$. What shall we think of a gain or loss in the amount X?

While it is certainly true that $\Pr\{X > 0\} = 1$, so we can be sure that X is positive, it is also true that for any positive amount $\eta > 0$, $P\{X > \eta\} = 0$. Thus if X is a gain you will experience, you are sure to gain something, but you are also sure that gain will be less than $\eta = 1$ millionth of a penny. Is such a gain (or loss, for that matter) worth noticing?

This comes up in thinking about what it means to avoid sure loss when betting on random variables that can take a countable number of values. I prefer to count as a gain that there is positive probability $\epsilon > 0$ that you will gain at least some amount $\eta > 0$. This distinction makes no difference in the context of Chapter 1, where random variables take only a finite number of values. Consequently the Fundamental Theorem of that chapter uses this concept without comment. However in the context of this chapter, it does matter, and I believe that insisting on positive probability $\epsilon > 0$ of gaining some positive amount $\eta > 0$ is the best choice.

Dynamic sure loss is said to exist if (1) there is an event A and a partition $\{B_i\}$ such that $P(A) > P(A|B_i) + \eta$ for all i and some $\eta > 0$ or if (2) there is an event A and a partition B_i such that $P(A) < P(A|B_i) - \eta$ for all i, and some $\eta > 0$. If (1) holds, then $\overline{A}$ suffices for (2), and conversely.

If (1) holds, then I can sell you a ticket on A, and buy from you tickets on $A|B_i$ for each i. Whatever i ensues, I am sure to come out at least $\eta > 0$ ahead. Conversely, if (2) holds, I can buy a ticket on A from you, and sell you tickets on $A|B_i$. Again, whatever i ensues, I am sure to come out at least $\eta > 0$ ahead.

Next, I show that dynamic sure loss is incompatible with countable additivity. Let I_A be an indicator random variable for A. Suppose your price for a ticket on A is p, which means

$$E(I_A) = p.$$

Let Y be a random variable that takes the value i when B_i occurs. Then a ticket on A if B_i occurs has price

$$E(I_A|B_i) = E(I_A|Y = i).$$

Using property 6 of section 3.4, we can put these together as

$$p = E(I_A) = E\Big[E[I_A|Y]\Big].$$

The random variable

$$E[I_A|Y]$$

might be trivial (meaning that $E[I_A|Y = i] = p$ for all i), in which case dynamic sure loss cannot ensue. However, if $E[I_A|Y]$ is not trivial, then property 4 of section 3.4 applies. Property 4 applied in this notation says that there is a set B_i and a positive probability $\epsilon > 0$ such that $P\{B_i\} > \epsilon$ and

$$p = P\{A\} < P\{A|B_i\} - \eta,$$

for some $\eta > 0$. Therefore there is at least one i for which

$$p = P\{A\} < P\{A|B_i\} - \eta,$$

so (1) in the definition of dynamic sure loss does not occur. The argument for why (2) cannot occur is similar.

Hence there is no sure loss. This argument proves the following result:

Theorem 3.5.1. *If P is countably additive, then no dynamic sure loss is possible.*

Now what about the converse? Can there be assurance against dynamic loss if P is finitely but not countably additive?

I cite a theorem that shows that nonconglomerability is characteristic of those probabilities that are finitely but not countably additive.

Theorem 3.5.2. *(Schervish et al. (1984)) Suppose $P(\cdot)$ is a coherent probability that is finitely but not countably additive. Then there is a set A and a countable partition $B_1, B_2, \ldots$ of disjoint sets whose union is S, on which conglomerability fails.*

Since conglomerability fails, it is not the case that

$$\inf_j P(A|B_j) \leq P(A) \leq \sup_j P(A|B_j).$$

Therefore either

$$\inf_j E(A|B_j) > P(A) \tag{3.20}$$

or

$$\sup_j E(A|B_j) < P(A). \tag{3.21}$$

If (3.20) is the case, then (2) in the definition of dynamic sure loss holds, with

$$\eta = (\inf_j E(A|B_j) - P(A)).$$

Similarly, if (3.21) holds, then (1) in the definition of dynamic sure loss holds, with $\eta = P(A) - \sup_j E(A|B_j)$. Hence dynamic sure loss exists.

3.5.1 Summary

Consider a coherent probability $P(\cdot)$. P avoids dynamic sure loss if and only if P is countably additive.

3.5.2 Discussion

Given these results, how is it reasonable to view countable additivity? The strategy of extending the results of Chapter 1 to a countable number of bets does not work, as shown both by the example in section 3.3.3, and by the consideration that covering a countable number of bets could require infinite resources.

I think from a foundational point of view that both finite and countable additivity are worth exploring. Perhaps dynamic sure loss, non-conglomerability, etc. will come to be regarded as so damaging as to preclude the use of probabilities that are finitely but not countably additive. Perhaps not. Meanwhile the vast preponderance of work on probability is done in the context of countable additivity. It would be useful to have a corresponding effort into the more general case of finite additivity.

While the remainder of this book concentrates on countable additivity, it does so mostly out of ignorance about which results might extend to the full finitely additive case, and which do not.

3.5.3 Other views

The dominant view in probability and statistics at this time comes from Kolmogorov (1933), who takes countable additivity as an axiom. Similarly DeGroot (1970) regards it as an assumption of continuity.

There is, however, a vociferous minority centered around DeFinetti (1974) and carried on by Coletti and Scozzafava (2002). These authors take finite additivity as basic, and regard countable additivity as an (unwarranted) restriction on the opinions you are permitted to express. Perhaps the most eloquent expression of this view is given by DeFinetti in 1970, section 3.11.

Goldstein (1983) advocates finitely additive probability together with property 4. However, Kadane et al. (2001) show that property 4 implies countable additivity. Heath and Suddereth (1978) propose using finitely additive probability for those events and partitions in which dynamic sure loss does not occur. (And see Kadane et al. (1986) for further comment.)

It is notable that limiting relative frequency does not support a limitation of probability to countable additivity. To see this, let $\mathbf{t}_i$ be a sequence with a 1 in position i and 0 elsewhere, for $i = 1, 2, \ldots$. These sequences are mutually exclusive, since a 1 never occurs in the same position for two of them. Each such $\mathbf{t}_i$ has limiting relative frequency 0. However the sum of the $\mathbf{t}_i$'s has a 1 in each position, and hence limiting relative frequency 1. Thus countable additivity is contradicted. Provided the issues mentioned in 2.13.1 can be overcome, a principled frequentist treatment would either accept finite additivity and give up conglomerability, or would explain how only countable additivity is consistent with limiting relative frequency. However, this is impossible since the class of sets with limited relative frequency does not form a field (*i.e.*, the intersection of two sets with limiting relative frequency may not have a limiting relative frequency, (see van Fraassen (1977)).

3.6 Probability generating functions

For the remainder of this chapter, I restrict attention to distributions on $\mathcal{N}$, the set of natural numbers. On this set, I introduce a function that can be used to summarize a distribution, called the probability generating function.

Suppose X is a random variable taking values on the non-negative integers, so that

$$P\{X = j\} = p_j \quad j = 0, 1, \ldots, \infty \tag{3.22}$$

and

$$\sum_{j=0}^{\infty} p_j = 1. \tag{3.23}$$

Consider the function

$$\alpha_X(t) = Et^X = p_0 + p_1 t + p_2 t^2 + \ldots \quad . \tag{3.24}$$

This function is called the **probability generating function** for X.

Some immediate properties of α are as follows:

1. $\alpha_X(1) = 1$. This follows from (3.23).
2. If X and Y are independent random variables, then

$$\alpha_{X+Y}(t) = \alpha_X(t)\alpha_Y(t).$$

Proof.

$$\begin{aligned}\alpha_{X+Y}(t) =& Et^{(X+Y)} = Et^X t^Y = Et^X Et^Y \\ =& \alpha_X(t)\alpha_Y(t)\end{aligned}$$

using property 7 of 3.5. □

3. If $\alpha_X(t) = \alpha_Y(t)$, then X and Y have the same distribution. This relies on the uniqueness of power series.

4. The next property relies on differentiation of α. First I will show a formal calculation, which demonstrates why a statistician would want to do this. Then I will cite a theorem showing when the differentiation is valid.

$$\frac{d}{dt}\alpha_X(t) = \frac{d}{dt}\left\{\sum_{j=0}^{\infty} p_j t^j\right\} = \sum_{j=1}^{\infty} j p_j t^{j-1}$$

(differentiating through an infinite sum, which is not yet justified).

$$\alpha_X'(1) = \frac{d}{dt}\alpha_X(t)|_{t=1} = \sum_{j=1}^{\infty} j p_j = E(X).$$

Taking a second derivative, (again, only formally)

$$\frac{d^2}{dt^2}\alpha_X(t) = \sum_{j=2}^{\infty} j(j-1)p_j t^{j-2} \text{ , so}$$
$$\alpha_X''(1) = E[X(X-1)] = EX^2 - EX.$$

Hence (again formally)

$$V[X] = E(X^2) - [E(X)]^2 = \alpha_X''(1) + \alpha_X'(1) - [\alpha_X'(1)]^2.$$

Thus if the formal calculation can be justified, both the mean and variance of X can be found easily from the probability generating function. The justification of the formal calculation is discussed next.

A power series $\sum_{n=0}^{\infty} a_n t^n$ is said to have radius of convergence ρ if it is convergent for all $|t| < \rho$ and divergent for all $|t| > \rho$. Then the following theorem applies:
If $\sum_{n=0}^{\infty} a_n t^n$ has radius of convergence ρ, then it has derivatives of all orders on $[-r, r]$, where $r < \rho$, and $\frac{d^k}{dt^k}\left[\sum_{n=0}^{\infty} a_n t^n\right] = \sum_{n=k}^{\infty} \frac{a_n n!}{(n-k)!} t^{n-k}, k = 1, 2, \ldots, -r \leq t \leq r$. (See Khuri (2003, Theorem 5.4.4, pp. 176-177).)

Hence we can conclude: If $\alpha_X(t)$ has radius of convergence $\rho > 1$, then

$$E[X] = \alpha_X'(1)$$
$$V[X] = \alpha_X''(1) + \alpha_X'(1) - [\alpha_X'(1)]^2.$$

What happens when the mean of a random variable does not exist? Section 3.3 discusses such an example, namely $P\{W = i\} = \frac{1}{Ti^2}$ $i = 1, 2, \ldots$. Then in this example the probability generating function of W is

$$\alpha_W(t) = \sum_{i=1}^{\infty} \frac{1}{T}\frac{t^i}{i^2}.$$

By the ratio test, Khuri (2003, p. 174, 175) we have

$$\lim_{i\to\infty} \left|\frac{a_{i+1}}{a_i}\right| = \lim_{i\to\infty} \frac{(i+1)^2}{i^2} = \lim_{i\to\infty} \left(1 + 2/i + 1/i^2\right) = 1.$$

Therefore $\rho = 1$. Consequently the Theorem does not apply. $\alpha_W(t)$ cannot be differentiated at 1. This example shows why the condition that $\alpha_X(t)$ should have a radius of convergence $\rho > 1$ is critical for finding the moments of X.

3.6.1 Summary

The probability generating function $\alpha_X(t) = E[t^X]$ has the following properties:

1. $\alpha_X(1) = 1$.
2. If X and Y are independent
$$\alpha_{X+Y}(t) = \alpha_X(t)\alpha_Y(t).$$
3. If $\alpha_X(t) = \alpha_Y(t)$ then X and Y have the same distribution.
4. When $\alpha_X(t)$ has radius of convergence $\rho > 1$,
$$\begin{aligned} E(X) =& \alpha'_X(1) \\ V(X) =& \alpha''_X(1) + \alpha'_X(1) - (\alpha'_X(1))^2. \end{aligned}$$

3.6.2 Exercises

1. Vocabulary. Define in your own words:
   (a) radius of convergence
   (b) probability generating function
2. If you knew the distribution of X, can you always find $\alpha_X(t)$? If you knew $\alpha_X(t)$, can you always find the distribution of X?
3. Consider the random variable X, which takes the value 1 with probability p, and the value 0 with probability $q = 1 - p$. Find the probability generating function of X. [Such a random variable is called a Bernoulli random variable with parameter p.]
4. Let $S = X_1 + X_2 + \ldots + X_n$ be the sum of n independent random variables each of which is a Bernoulli random variable with parameter p. Find the probability generating function of S, using property 2.
5. Find the probability generating function of S directly, using
$$P\{S = j\} = \binom{n}{j, n-j} p^j q^{n-j} \quad j = 0, 1, \ldots, n, \text{ where } q = 1 - p.$$
6. Using the answer to (4) and/or (5), find $E[S]$ and $V[S]$.

3.7 Geometric random variables

Section 3.1 has already introduced the geometric distribution without naming it, namely the number F of failures before the first success in a sequence of independent Bernoulli trials each with known probability $p > 0$ of success. The probability distribution of F is given by
$$P\{F = k\} = (1 - p)^k p \quad k = 0, 1, 2, \ldots \tag{3.25}$$
Then the probability generating function of F is
$$\begin{aligned} \alpha_F(t) = E(t^X) =& \sum_{k=0}^{\infty} t^k (1-p)^k p \\ =& p \sum_{k=0}^{\infty} [t(1-p)]^k = \frac{p}{1 - t(1-p)}. \end{aligned} \tag{3.26}$$

(The reason (3.25) is called a Geometric random variable is that the sum involved in showing that (3.25) sums to one is a geometric sum, as is the sum involved in its probability generating function.)

The latter geometric sum converges if $t(1-p) < 1$ and diverges if $t(1-p) > 1$. Hence the radius of convergence is $\rho = 1/(1-p) > 1$. Therefore we can differentiate (3.26) as many times as we please at $t = 1$. In particular, we can now apply property 4 of section 3.6 to find the mean and variance of F, as follows.

$$\begin{aligned} E(F) =& \frac{d}{dt}\alpha_F(t)|_{t=1} = \frac{d}{dt}\left[\frac{p}{1-t(1-p)}\right]|_{t=1} \\ =& p\left[\frac{1-p}{(1-t(1-p))^2}\right]|_{t=1} = \frac{p(1-p)}{p^2} = \frac{1-p}{p}. \end{aligned} \tag{3.27}$$

This is a reasonable result. It says that the smaller p is (*i.e.*, the harder it is to get a success), the longer one should expect to wait for a success.

$$\begin{aligned} \frac{d^2\alpha_F(t)}{dt^2}|_{t=1} =& \frac{d}{dt}\left[\frac{p(1-p)}{(1-t(1-p))^2}\right]|_{t=1} \\ =& p(1-p)\left[\frac{2(1-t(1-p))(1-p)}{(1-t(1-p))^4}\right]|_{t=1} \\ =& \frac{2p(1-p)^2p}{p^4} = \frac{2(1-p)^2}{p^2}. \end{aligned} \tag{3.28}$$

Then

$$\begin{aligned} V(F) =& \frac{d^2\alpha_F(t)}{dt^2}|_{t=1} + E(F) - (E(F))^2 \\ =& \frac{2(1-p)^2}{p^2} + \frac{1-p}{p} - (\frac{1-p}{p})^2 \\ =& \frac{(1-p)^2}{p^2} + \frac{(1-p)}{p} = \frac{(1-p)^2 + (1-p)p}{p^2} = \frac{(1-p)}{p^2}. \end{aligned} \tag{3.29}$$

The geometric distribution has an important memoryless property. Suppose a sequence of independent Bernoulli random variables has been observed, the first k of which have resulted in failures. Then the probability that the first success will occur after exactly t more failures is the same as if one had started over at that point. (Such a time is called a recurrence time. These are an important tool in probability theory.) The memoryless property for the geometric distribution can be expressed formally as follows:

If F has a geometric distribution with parameter p, then for any integers k and t:

$$P\{F = k + t | F \geq k\} = P\{F = t\}. \tag{3.30}$$

3.7.1 Summary

A geometric distribution is given by (3.25). Its probability generating function is $p/[1 - t(1-p)]$.

Its mean and variance are $(1-p)/p$ and $(1-p)/p^2$, respectively. It has the memoryless property (3.30).

3.7.2 Exercises

1. Suppose a person plays the Pennsylvania Lottery every day, waiting for a win. (See problem 3 in section 1.5.2 for the rules.) Suppose that the person has played the lottery for k days, with no success so far. The person feels "due" for a win, that is, the person thinks incorrectly that the probability of a win is increased by the fact of having lost k days in a row. Is this line of reasoning consistent with the assumption that each day's drawing is independent of all the others? Explain why or why not.

2. Prove (3.30).
3. Suppose a person is waiting at a bus stop. The person believes that the event of a bus coming in each five-minute period is independent of each other five-minute period, and that the probability that a bus will come in any given five-minute period is p, where p is assumed to be known. This belief is different from believing that the buses operate on a fixed schedule. Having waited 20 minutes already, is the bus more likely, less likely or equally likely to come in the next five-minute period. Why?
   Now suppose that the person is unsure what the value of p is, so that the person is gaining information about p during the wait. Argue intuitively why the person might reasonably believe that the bus is more, equally or less likely to come in the next five-minute period.
4. Prove that if F has a geometric distribution with known probability p of success on each trial, then
$$P\{F \geq k\} = (1-p)^k.$$

3.8 The negative binomial random variable

The geometric distribution has the following generalization: Let r be a fixed positive integer. Let F be the number of failures until the rth success among a sequence of independent Bernoulli trials, each of which has probability $p > 0$ of success. Clearly the geometric distribution is the special case $r = 1$.

How can it happen that the rth success is preceded by exactly n failures? It must be that among the first $n + r - 1$ trials there are exactly n failures and $r - 1$ successes, and the last trial is a success. The probability of this event is

$$\begin{aligned} P\{F = n\} &= \binom{n+r-1}{n, r-1} p^{r-1}(1-p)^n \cdot p \qquad n = 0, 1, 2, \ldots \\ &= \binom{n+r-1}{n, r-1} p^r (1-p)^n \qquad n = 0, 1, 2, \ldots \end{aligned} \tag{3.31}$$

Now suppose $X_1, X_2, \ldots, X_r$ are r independent geometric random variables each with parameter p. Then it is immediately obvious that F has the same distribution as $X_1 + X_2 + \ldots + X_r$.

This convenient fact has the following consequences: If F has a negative binomial distribution with parameters p and r, then

$$\alpha_F(t) = \left[\frac{p}{1 - t(1-p)}\right]^r \tag{3.32}$$

$$E(F) = r(1-p)/p \tag{3.33}$$

and

$$V(F) = r(1-p)/p^2. \tag{3.34}$$

Since F has a finite mean, F is finite with probability 1. Therefore the following infinite sum can be derived:

$$\sum_{n=0}^{\infty} \binom{n+r-1}{n, r-1} p^r (1-p)^n = 1.$$

This is an example of using probabilistic reasoning to prove a mathematical fact in an intuitive way.

3.8.1 Summary

A negative binomial random variable has distribution given by (3.31), probability mass function by (3.32), mean (3.33) and variance by (3.34).

3.8.2 Exercises

1. Is it reasonable to suppose that the negative binomial distribution has the memoryless property? Why or why not?
2. Prove your answer to problem 1.
3. Suppose X has a negative binomial distribution with parameters r and p, and Y has a negative binomial distribution with parameters s and p. Show that $X+Y$ has a negative binomial distribution with parameters $r+s$ and p.
4. Hypergeometric Waiting Time. Suppose a bowl of fruit contains five apples, three bananas and four cantaloupes. Suppose these fruits are sampled without replacement, choosing each fruit equally likely from those that remain. Find the distribution for the number of fruit selected before the third apple is chosen.
5. Do the same problem as 4 when there are a apples, b bananas and c cantaloupes in the bowl. Find the distribution of the number of fruit selected before the a^*th apple is selected, for each a^*, $0 \leq a^* \leq a$.

3.9 The Poisson random variable

Every sequence of non-negative numbers that has a finite sum can be made into a probability distribution by dividing by that sum. In formula (1.34), we encountered the sum

$$e^{\lambda} = 1 + \lambda + \frac{\lambda^2}{2!} + \frac{\lambda^3}{3!} + \ldots \tag{1.33}$$

Therefore there is a random variable X having distribution given by

$$P\{X = k\} = \frac{e^{-\lambda}\lambda^k}{k!} \quad k = 0, 1, 2, \ldots. \tag{3.35}$$

Such a random variable is said to have the Poisson distribution with parameter $\lambda > 0$.

The probability generating function of X is

$$\begin{aligned} \alpha_X(t) =& Et^X = \sum_{j=0}^{\infty} \frac{e^{-\lambda}\lambda^j}{j!} t^j = e^{-\lambda} \sum_{j=0}^{\infty} \frac{(\lambda t)^j}{j!} \\ =& e^{-\lambda} e^{\lambda t} = e^{\lambda(t-1)}. \end{aligned} \tag{3.36}$$

The radius of convergence for this sum is $\rho = \infty$, so differentiation at $t = 1$ is justified. Then

$$E(X) = \frac{d\alpha_X(t)}{dt}|_{t=1} = \lambda e^{\lambda(t-1)}|_{t=1} = \lambda. \tag{3.37}$$

Also

$$\frac{d^2\alpha_X(t)}{dt^2}|_{t=1} = \frac{d}{dt}\lambda e^{\lambda(t-1)}|_{t-1} = \lambda^2 e^{\lambda(t-1)}|_{t=1} = \lambda^2.$$

Consequently

$$\begin{aligned} V(X) =& \frac{d^2\alpha_X(t)}{dt^2}|_{t=1} + E(X) - (E(X))^2 \\ =& \lambda^2 + \lambda - \lambda^2 = \lambda. \end{aligned} \tag{3.38}$$

Hence both the mean and variance of the Poisson distribution are λ.

The Poisson distribution is often used as a model for the distribution of rare events. Consider the number of successes in a sequence of n independent Bernoulli trials in which the probability of success, p, is decreasing to 0 but the number of trials in the sequence, n, is increasing without limit in such a way that np approaches some number λ.

From equation 2.50 for the binomial distribution for the sum of n independent Bernoulli trials,

$$\begin{aligned} P\{X_n = j\} &= \binom{n}{j, n-j} p^j (1-p)^{n-j} \qquad j = 0, 1, \ldots, n \\ &= \frac{n!}{j!(n-j)!} (\frac{\lambda}{n})^j (1-\lambda/n)^n (1-\lambda/n)^{-j} \\ &= \frac{\lambda^j}{j!} \left\{ \left[\frac{n(n-1)\ldots(n-j+1)}{n \quad n \qquad\quad n} \right] (1-\lambda/n)^{-j} \right\} (1-\lambda/n)^n. \end{aligned} \tag{3.39}$$

Now as $n \to \infty$, the factor in curly brackets approaches 1. In addition, $\lim_{n\to\infty}(1-\lambda/n)^n \to e^{-\lambda}$, a fact which follows from the Taylor expansion of $\log(1+x)$ (see section 2.2) as follows:

$$\begin{aligned} \log \lim_{n\to\infty} (1-\lambda/n)^n &= \lim_{n\to\infty} \log(1-\lambda/n)^n \\ &= \lim_{n\to\infty} n \log(1-\lambda/n) \\ &= \lim_{n\to\infty} n[-\lambda/n + HOT] = -\lambda. \end{aligned}$$

Again, remember that "HOT" stands for Higher Order Terms.

Returning to (3.39),

$$\lim_{n\to\infty} P\{X_n = j\} = \frac{\lambda^j}{j!} e^{-\lambda}, \tag{3.40}$$

which is the Poisson probability.

Example: Letters and envelopes again. Finally, we return to the problem of the letters and envelopes. To review, n letters are matched to n envelopes randomly, and our interest is in the number of correctly matched letters and envelopes. Recall that in sections 1.5, 1.6 and 2.11, respectively, gave three results about this problem:

1. $P_{o,n}$ the probability that no letter gets matched to its correct envelope, satisfies $\lim_{n\to\infty} P_{o,n} = e^{-1}$.
2. The expected number of correct matchings is 1 for all n.
3. The variance of the number of current matchings is 1 for all $n \geq 2$.

Now we seek a general formula for $\lim_{n\to\infty} P_{k,n}$, the probability that exactly k envelopes and letters match, as the number of them, n, goes to infinity.

We have

$$P_{k,n} = \binom{n}{k, n-k} \cdot \left(\frac{1}{n}\right)^k P_{0,n-k},$$

because there are $\binom{n}{k,n-k}$ ways of choosing which k letters and envelopes will match, each of those that match have probability $1/n$ of doing so, and the other $n-k$ letters and envelopes have probability $P_{0,n-k}$ of not matching.

Then

$$P_{k,n} = \frac{1}{k!} \left[\frac{n(n-1)\ldots(n-k+1)}{n \quad n \qquad\quad n} \right] P_{0,n-k}.$$

Again the expression in square brackets approaches 1, so

$$\lim_{n\to\infty} P_{k,n} = \frac{1}{k!}e^{-1}, \quad k = 0, 1, \ldots$$

which is a Poisson distribution with parameter 1. □

An incautious reader might conclude that the limiting distribution had to have mean 1 because each of the random variables in the sequence has expectation 1. However this inference is not valid, as the following example shows: Let X_n take the value n with probability $1/n$, and otherwise take the value 0. Then

$$E(X_n) = n(1/n) + 0(1 - 1/n) = 1 \text{ for all } n.$$

However $\lim_{n\to\infty} P\{X_n = 0\} = 1$, so the limiting random variable is trivial, putting all its mass at 0, and has mean 0.

3.9.1 Summary

The Poisson distribution with parameter λ has mean and variance λ, and probability generating function $e^{\lambda(t-1)}$. It is used as a distribution for rare events, and is the limiting distribution for binomial random variables as $p \to 0$ and $n \to \infty$ in such a way that $np \to \lambda$ for some λ. Additionally a Poisson distribution with parameter 1 is the limiting distribution for the number of randomly matched letters and envelopes.

3.9.2 Exercises

1. Suppose that X_1 and X_2 are independent Poisson random variables with parameters λ_1 and λ_2, respectively.
   (a) Find the probability generating function of $X_1 + X_2$.
   (b) What is the distribution of $X_1 + X_2$?
   (c) What is the conditional distribution of X_1 given that $X_1 + X_2 = k$?
2. Suppose that a certain disease strikes .01% of the population in a year, and suppose that occurrences of it are believed to be independent from person to person. Find the probability of three or more cases in a given year in a town of 20,000 people.
   Note: This problem gives a slight flavor of the problems faced in the field of epidemiology. They are often confronted with the difficult problem of determining whether an apparent cluster of persons with a specific disease is due to natural variation or to some unusual underlying common cause, and, if so, what that common cause is.
3. Recall the rules for "Pick Three" from the Pennsylvania Lottery (see problem 3 in section 1.5.2). Suppose that 2000 players choose their three digit numbers independently of the others on a particular day.
   (a) Find the mean and variance of the number of winners.
   (b) Find an approximation to the probability of at least three winners on that day.

3.10 Cumulative distribution function

3.10.1 Introduction

This section introduces a useful analytic tool and an alternative way of specifying the distribution of a random variable, the cumulative distribution function, abbreviated cdf.

Suppose X is a discrete random variable. Then the cdf of X, written $F(x)$, is a function defined by

$$F_X(x) = P\{X \leq x\} \tag{3.41}$$

Thus the cdf has the following properties:

(i) $\lim_{x \to -\infty} F_X(x) = 0$.

(ii) $\lim_{x \to \infty} F_X(x) = 1$.

(iii) $F_X(x)$ is non-decreasing in x.

(iv) If $P\{X = x_i\} = p_i > 0$, then $F_X(x)$ has a jump of size p_i at x_i, so $F_X(x + \epsilon) - F_X(x - \epsilon) = p_i$ for all sufficiently small, positive ϵ.

Suppose X and Y are two random variables. It is important to understand the distinction between (i) X and Y are the same random variable and (ii) X and Y have the same distribution. For example, suppose X and Y are both 1 if a flipped coin comes up heads, and are 0 otherwise. If they refer to the same flip of the same coin, then $X = Y$. If they refer to different flips of the same coin, it is reasonable to suppose that they have the same distribution, but of course it is possible that one coin would show heads and the other tails. When X and Y have the same distribution, this is equivalent to the condition $F_X(t) = F_Y(t)$ for all t.

3.10.2 An interesting relationship between cdf's and expectations

Suppose X is a random variable taking values on the non-negative integers. Then

$$E(X) = \sum_{j=0}^{\infty}(1 - F_X(j))$$

provided the expectation of X exists.

Proof. Suppose $P\{X = i\} = p_i \quad i = 0, 1, \ldots,$ where $\sum_{i=0}^{\infty} p_i = 1$.

$$\begin{aligned} E(X) &= \textstyle\sum_{i=0}^{\infty} ip_i = \sum_{i=0}^{\infty}\sum_{j=0}^{i-1} p_i = \sum_{0 \le j < i < \infty} p_i \\ &= \textstyle\sum_{j=0}^{\infty}\sum_{i=j+1}^{\infty} p_i = \sum_{j=0}^{\infty}(1 - F_X(j)). \end{aligned}$$

The first equality is just the definition of expectation. The second equality makes use of the fact that the number of integers j starting at 0 and ending at $i-1$ is exactly i. The third equality reorganizes the double sum into a single sum over both indices, in preparation for the fourth equality, which reverses the order of summation (justified by Theorem 3.3.4). Finally, the fifth equality makes use of the definition of the cdf. □

3.10.3 Summary

The cdf is defined by equation 3.41, and has properties (i), (ii), (iii) and (iv). Additionally random variables taking values on the non-negative integers have an expectation satisfying $E(X) = \sum_{j=0}^{\infty}(1 - F_X(j))$ provided the expectation exists.

3.10.4 Exercises

1. Suppose X has a binomial distribution with $n = 2$, for some p, $0 < p < 1$. Find the cdf of X.
2. Use the cdf of X found in exercise 1 to find the expectation of X. Check your answer against the expectation of X where X has a binomial distribution found in exercise 8 of section 2.1.2.

3.11 Dominated and bounded convergence for discrete random variables

I now examine the circumstances under which I may write

$$\lim_{n\to\infty} E(X_n) = E \lim_{n\to\infty} X_n. \tag{3.42}$$

Section 1.8 shows that 1.37 holds if X_n and hence $X = \lim_{n\to\infty} X_n$ have domain limited to a finite set. This section considers what happens when this constraint is relaxed, to allow a countable but infinite domain. Since the domain is countable, without loss of generality I may consider it to be the natural numbers $1, 2, \ldots$.

The example given at the end of section 3.9, already hints that restrictions need to be imposed on the random variables X_n for (3.42) to hold. Suppose that integer i occurs with probability p_i, for $i = 1, 2, \ldots$. If only finitely many p_i's are positive, then the theorem proved in section 1.8 applies, and assures that (3.42) holds. Suppose then, that $p_i > 0$ for infinitely many values of i. Without loss of generality, renumbering if necessary, we may assume that $p_i > 0$ for all $i = 1, 2, \ldots$.

Suppose $X_n = \begin{cases} c/p_n & \text{with probability } p_n, \\ 0 & \text{otherwise} \end{cases}$ for some number c to be discussed later.

Then the limiting random variable X takes the value 0 with probability 1, and hence

$$E(X) = E(\lim_{n\to\infty} X_n) = 0.$$

However

$$E(X_n) = p_n(c/p_n) + (1 - p_n)(0) = c.$$

Hence, if $c \neq 0$, (3.42) fails to hold. This example shows that restrictions on X_n are necessary if (3.42) is to hold in the countable case. A hint about the issue lies in the fact that $p_n \to 0$ as $n \to \infty$, so $c/p_n \to \infty$. Considerations of this type lead to the following.

Theorem 3.11.1. *(Dominated convergence for random variables with a countable domain.)*

Let a_{in} be a double sequence of numbers, and let X_n be a random variable such that

(i) $P\{X_n = a_{in}\} = p_i$ for all $n = 1, 2, \ldots,$ and all $i = 1, \ldots$.
Also suppose that for each $i, i = 1, \ldots$

(ii) $\lim_{n\to\infty} a_{in} = a_i$.
Let X be the random variable such that $P\{X = a_i\} = p_i$.
Finally suppose that there is a random variable Y such that

(iii) $P\{Y = b_i\} = p_i, |a_{in}| \leq b_i$ and $E(Y)$ exists.

Then (a) $E(X)$ exists
and
(b) (3.42) holds.

Proof. (a) $\sum_{i=1}^{\infty} p_i|a_i| \leq \sum_{i=1}^{\infty} p_i b_i < \infty$ since $E(Y)$ exists. Hence $E(X)$ exists, and (a) is proved.

(b) Let $\epsilon > 0$ be given. Then let M be large enough so that $\sum_{i=M+1}^{\infty} p_i b_i < \epsilon/4$. $|\sum_{i=1}^{\infty} p_i a_{in} - \sum_{i=1}^{\infty} p_i a_i| \leq \sum_{i=1}^{M} p_i|a_{in} - a_i| + \sum_{i=M+1}^{\infty} p_i|a_{in} - a_i|$.

Then using $a_{in} \to a_i$ for each $i = 1, \ldots, M$ there exists an N_i such that for all $n \geq N_i$, $|a_{in} - a_i| < \epsilon/2M$. Let $N = \max_{i=1,\ldots,M} N_i$. Then for $i = 1, \ldots, M$ and for all $n \geq N$, $|a_{in} - a_i| < \epsilon/2M$. Then for all $n \geq N$,

$$\begin{aligned}
\left|\sum_{i=1}^{\infty} p_i a_{in} - \sum_{i=1}^{\infty} p_i a_i\right| &\leq \sum_{i=1}^{M} p_i(\epsilon/2M) + \sum_{i=M+1}^{\infty} 2p_i b_i \\
&\leq M(\epsilon/2M) + 2(\epsilon/4) \\
&= \epsilon.
\end{aligned}$$

This proves (b). □

Corollary 3.11.2. *(Bounded convergence for random variables with a countable domain.)*

Using the same notation as in the theorem, and assumptions (i) and (ii), replace assumption (iii) with (iii'):

$$|a_{in}| \leq d, \text{ for some number } d.$$

Then (a) and (b) hold.

Proof. Let $Y = d$ with probability 1. Then $E(Y) = d < \infty$, so condition (iii) holds. □

3.11.1 Summary

When the domain of the random variables is countably infinite, $P\{X_n = a_{in}\} = p_i$, and $\lim a_{in} = a_i$ for $i = 1, \ldots$

$$\lim_{n \to \infty} E[X_n] = E[X]$$

where $P\{X = a_i\} = p_i$, for $i = 1, \ldots,$ provided either

(I) there exists a random variable Y such that

$$P\{Y = b_i\} = p_i, |a_{in}| \leq b_i \text{ for all } n = 1, 2, \ldots, \text{ and all } i = 1, \ldots$$

and
$E(Y)$ exists (dominated convergence), or

(II) $|a_{in}| \leq d$ for all $n = 1, 2, \ldots,$ and all $i = 1, \ldots$ and some constant d (bounded convergence).

3.11.2 Exercises

1. Vocabulary. Explain in your own words:
   (a) bounded convergence
   (b) dominated convergence
2. Let $p_i > 0$ for $i = 1, 2, \ldots$ and $\sum_{i=1}^{\infty} p_i = 1$.

   Let $a_{i,n} = \begin{cases} 0 & n < i \\ 1/p_i & i \geq n \end{cases}$

   (i) Find $a_i = \lim_{n \to \infty} a_{i,n}$.
   (ii) What is the distribution of $X = \lim_{n \to \infty} X_n$?
   (iii) Find a positive lower bound for $E(X_n)$.
   (iv) Is it possible that (3.42) holds? Explain why or why not? How does this example relate to bounded convergence? to dominated convergence?

Chapter 4

Continuous Random Variables

"Take it to the limit, one more time"

—The Eagles

"Does anyone believe that the difference between the Lebesgue and Riemann integrals can have physical significance, and that whether say, an airplane would or would not fly could depend on this difference? If such were claimed, I should not care to fly in that plane."

—Richard Wesley Hamming

4.1 Introduction

Suppose we want to model the idea of an instant in time uniformly distributed within a particular hour. We could use the idea of a randomly chosen minute by letting each minute have probability $1/60$. If we wanted to measure time in tenths of seconds, we could let each tenth of a second have probability $1/600$, etc. But it is often convenient to think of time in a continuous way, even though as a practical matter time can be measured only up to some degree of precision, and whatever that degree of precision, the result is some finite set of possibilities. Looking to continuous random variables means seeking a treatment that does not depend on the precision of measurement, pretending that time can be measured arbitrarily accurately.

A natural way of making this intuition precise is to think of the probability of this random time T falling between time a and b (expressed in minutes but thought of as real numbers) as

$$P\{a \leq T \leq b\} = \frac{b-a}{60} = \int_a^b \frac{1}{60} dx. \tag{4.1}$$

Here the integral is taken in the ordinary, Riemann sense, which is discussed more precisely below.

More generally, if $f_X(x)$ is any (Riemann-integrable) function, then the goal is to define

$$P\{X \in A\} = \int_A f_X(x) dx \tag{4.2}$$

to be the probability that the random variable X lies in the set A for all sets A for which the integral is defined (typically, in the one-dimensional case, intervals and unions of intervals). Probabilities defined this way are called Riemann probabilities. Now (4.1) is a special case of (4.2), where

$$f_T(x) = \begin{cases} 1/60 & 0 < x < 60 \\ 0 & \text{otherwise} \end{cases}. \tag{4.3}$$

There are conditions that must be imposed on $f_X(x)$ in order to have a hope that (4.2) might represent probabilities. A function $f_X(x)$ is called a probability density function (pdf), or more simply a density function, if it has the following properties:

1. $f(x) \geq 0$ for all x. (Otherwise it would be possible to find a set with negative probability.)
2. $\int_{\Re} f(x)dx = 1$, so that the probability over the whole space is 1, as it must be.

There is a consequence of equation (4.2) that deserves some discussion. Consider some real number a, and ask what the meaning is of $f_X(a)$. It is NOT the probability that $X = a$. Indeed, for every continuous random variable, the probability that $X = a$ is zero for all a, since

$$P\{X \epsilon \{a\}\} = \int_a^a f_X(x)dx = 0. \tag{4.4}$$

However, we can say that if we imagine a small interval around a of length Δx, say the interval $(a - (\Delta x)/2,\ a + (\Delta x)/2)$, we have

$$P\{X \epsilon (a - (\Delta x)/2,\ a + (\Delta x)/2)\} = \int_{a-(\Delta x)/2}^{a+(\Delta x)/2} f_X(x)dx \doteq (f_X(a))(\Delta x),$$

for continuous functions $f_X(x)$.

The statement (4.2) is simultaneously a statement of an uncountable number of probabilities, and hence, in the terms of this book, of an offer to buy or sell an uncountable number of dollar tickets at prices specified by (4.2). Therefore, we must be somewhat careful in relating these statements to our previous theory. Conditions 1 and 2 on $f_X(x)$ obviously imply (1.1) (non-negative probabilities) and (1.2) (the sure event has probability one). However the additivity properties of (4.2) are less obvious.

One of the elementary properties of the Riemann integral is finite additivity. Thus if $g(x)$ and $h(x)$ are Riemann integrable functions, then so is $g(x) + h(x)$, and

$$\int \left[g(x) + h(x)dx\right] = \int g(x)dx + \int h(x)dx. \tag{4.5}$$

(For completeness, a formal proof of (4.5) is provided in section 4.7.1.)

Riemann probabilities are probabilities defined with respect to a density $f(x)$ integrated with a Riemann integral.

Formula (4.5) has the following consequence for Riemann probabilities:

Theorem 4.1.1. *Let $f(x)$ be a density function, and let A and B be disjoint sets with defined Riemann probabilities with respect to the density $f(x)$. Then*

$$P\{A \cup B\} = P\{A\} + P\{B\}.$$

Proof. Let $g(x) = I_A(x)f(x)$ and $h(x) = I_B(x)f(x)$. Then

$$\begin{aligned} P\{A \cup B\} &= \int \Big(I_A(x) + I_B(x)\Big) f(x)dx & (4.6)\\ &= \int \Big(g(x) + h(x)\Big)dx = \int g(x)dx + \int h(x)dx & (4.7)\\ &= P\{A\} + P\{B\} & (4.8) \end{aligned}$$

□

Corollary 4.1.2. *Let $f(x)$ be a density function, and let $A_1, \ldots, A_n$ be a finite collection of disjoint sets with defined Riemann probabilities with respect to the density $f(x)$. Then*

$$P\{\cup_{i=1}^n A_i\} = \sum_{i=1}^n P\{A_i\}.$$

Proof. By induction on n using Theorem 4.1.1. □

There is a sense in which Riemann probabilities are countably additive, and a sense in which they are not. This subject is postponed until section 4.7, at which point more tools will have been developed.

Later this chapter goes into the theory of Riemann integration much more carefully.

4.1.1 The cumulative distribution function

Recall from section 3.10 that the cumulative distribution function (cdf) $F_X(x)$ of a random variable X is defined as follows:

$$F_X(x) = P\{X \leq x\}. \tag{4.9}$$

Cumulative distribution functions have the following properties:

(i) $\lim_{x \to -\infty} F_X(x) = 0$.

(ii) $\lim_{x \to \infty} F_X(x) = 1$.

(iii) If $x_1 \leq x_2$, then $F_X(x_1) \leq F_X(x_2)$. (non-decreasing)

(iv) $F_X(x) = \lim_{y > x, y \to x} F_X(y)$. (continuous from above)

Property (iv) follows from the fact the cumulative density function $F_X(x)$ defined in (4.9) is the probability of the event $\{X \leq x\}$, not the event $\{X < x\}$. When there is a lump of probability at x, the distinction matters. Thus when $y > x$, the lump of probability at x is included in $F_X(y)$ for each y, as well as being included in $F_X(x)$. However, if $z < x$, $F_X(z)$ does not include the lump of probability at x. With the cumulative distribution function as defined in (4.9) (which is the traditional choice), $F_X(x)$ is said to be continuous from above (as in property (iv)), but not necessarily from below.

The cumulative distribution function of the random variables studied in Chapters 1 to 3 rise only at the discrete points x at which $P\{X = x\} > 0$. The cumulative distribution functions of the random variables X considered in this chapter arise from density functions $f_X(x)$. There is a third case, cdf's that are continuous but do not have associated densities. These are called singular, and their study is postponed.

In addition, there are random variables that mix types. For example, consider a random variable that with probability $\frac{1}{2}$ takes the value 0, and with probability $\frac{1}{2}$ is uniform on the interval $(0, 1)$. This random variable has cdf

$$F(x) = \begin{cases} 0 & x < 0 \\ \frac{1}{2} & x = 0 \\ \frac{1}{2} + \frac{x}{2} & 0 < x < 1 \\ 1 & x \geq 1 \end{cases}$$

It is easy to check that the cdf satisfies the four conditions stated above.

4.1.2 Summary and reference

The sense of integral being used here is the usual Riemann integral, defined when the limit of the lower sum of rectangular areas below the curve equals that of the upper sum. This is explained in many calculus books, of which my favorite is Courant (1937), see Chapter II.

The kinds of continuous random variable X addressed here, known more precisely as absolutely continuous random variables, are characterized by their densities $f_X(x)$, which gives the probability of a set A to be $\int_A f_X(x)dx$.

Densities satisfy the following conditions:

1. $f(x) \geq 0$ for all x.
2. $\int_{\Re} f(x)dx = 1$.

Cumulative distribution functions are defined in (4.9), and have the four properties stated above.

4.1.3 *Exercises*

1. Vocabulary. Explain in your own words:
   (a) probability density function
   (b) absolutely continuous random variable
2. Show whether each of the following satisfy the conditions to be a probability density:

   (a) $f(x) = \begin{cases} 0 & x < 0 \\ x & 0 \leq x \leq 2 \\ 0 & x > 2 \end{cases}$

   (b) $f(x) = \begin{cases} 1/2 & -1 < x < 1 \\ 0 & \text{otherwise} \end{cases}$

   (c) $f(x) = \begin{cases} \frac{2x}{3} & -1 < x < 2 \\ 0 & \text{otherwise} \end{cases}$

   (d) $f(x) = \begin{cases} e^{-x} & x > 0 \\ 0 & \text{otherwise} \end{cases}$
3. For each of the functions $f(x)$ in problem 2 that satisfies the conditions to be a probability density, find the cumulative density function.

4.2 Joint distributions

Suppose X and Y are two random variables defined over the same probability space. Then we can consider their joint cumulative distribution function

$$F_{X,Y}(x,y) = P\{X \leq x, Y \leq y\}. \tag{4.10}$$

What properties must such a cumulative distribution function have?

First, it must have the appropriate relationship to the univariate cumulative distribution functions:

$$F_{X,Y}(x,\infty) = P\{X \leq x, Y \leq \infty\} = P\{X \leq x\} = F_X(x) \tag{4.11}$$

$$F_{X,Y}(\infty,y) = P\{X \leq \infty, Y \leq y\} = P\{Y \leq y\} = F_Y(y). \tag{4.12}$$

The distribution functions F_X and F_Y are called marginal cumulative distribution functions.

Similarly,

$$F_{X,Y}(-\infty,y) = F_{X,Y}(x,-\infty) = 0 \text{ for all } x \text{ and } y.$$

Now suppose we wish to find the probability content of a rectangle of the form $a < X \leq b, c < Y \leq d$. Then

$$\begin{aligned} P\{a &< X \leq b, c \leq Y \leq d\} \\ &= P\{a < X \leq b, Y \leq d\} - P\{a < X \leq b, Y \leq c\} \\ &= P\{X \leq b, Y \leq d\} - P\{X \leq a, Y \leq d\} - P\{X \leq b, Y \leq c\} \\ &+ P\{X \leq a, Y \leq c\} \\ &= F_{X,Y}(b,d) - F_{X,Y}(a,d) - F_{X,Y}(b,c) + F_{X,Y}(a,c). \end{aligned} \tag{4.13}$$

Since the probability of every event is non-negative, (4.13) must be non-negative for every choice of a, b, c and d. This is the bivariate generalization of condition (iii) of section 4.1 that every univariate cumulative distribution function must be non-decreasing. Of course $F_{X,Y}(x,y)$ is symmetric, in the sense that $F_{X,Y}(x,y) = F_{Y,X}(y,x)$.

Now suppose that X and Y have a probability density function f defined over the xy plane and satisfying

$$P\{(X,Y)\epsilon A\} = \int\int_A f_{X,Y}(s,t)dsdt. \tag{4.14}$$

Here s and t are simply dummy variables of integration: any other symbols would do as well.

Such a probability density function satisfies

$$f_{X,Y}(x,y) \geq 0 \text{ for } -\infty < x < \infty \text{ and } -\infty < y < \infty \tag{4.15}$$

and

$$\int_{-\infty}^{\infty}\int_{-\infty}^{\infty} f_{X,Y}(x,y)dxdy = 1. \tag{4.16}$$

In this case, every single point, every finite collection of points and every one-dimensional curve in the xy plane has probability zero.

A marginal probability density can be found from a joint probability density as follows:

$$f_X(x) = \int_{\{(x,y)|-\infty<y<\infty\}} f_{X,Y}(x,y)dy = \int_{-\infty}^{\infty} f_{X,Y}(x,t)dt. \tag{4.17}$$

By symmetry of course

$$f_Y(y) = \int_{-\infty}^{\infty} f_{X,Y}(s,y)ds. \tag{4.18}$$

As an example, suppose

$$f_{X,Y}(x,y) = \begin{cases} cx(y+y^2) & \text{if } 0 < x < 2,\ 0 < y < 1 \\ 0 & \text{otherwise} \end{cases}$$

We'll find:

(a) the value of c that makes $f_{X,Y}$ a joint density function

(b) the cumulative distribution function $F_{X,Y}$

(c) $P\{X < Y\}$.

To do (a), we start with

$$\begin{aligned} 1 &= \int_{-\infty}^{\infty}\int_{-\infty}^{\infty} f_{X,Y}(x,y)dxdy = \int_0^2\int_0^1 cx(y+y^2)dydx \\ &= \int_0^2 cx\left(\frac{y^2}{2}+\frac{y^3}{3}\right)\Big|_0^1 dx = \int_0^2 cx\left(\frac{1}{2}+\frac{1}{3}\right)dx \\ &= \frac{5c}{6}\int_0^2 xdx = \frac{5c}{6}\frac{x^2}{2}\Big|_0^2 = \frac{5c}{6}\cdot 2 = \frac{5c}{3}. \end{aligned}$$

Therefore $c = \frac{3}{5}$.

Addressing (b), we have, for $0 < x < 2$ and $0 < y < 1$,

$$\begin{aligned}
F_{X,Y}(x,y) &= \int_{-\infty}^{x}\int_{-\infty}^{y} f_{X,Y}(s,t)dtds \\
&= \int_{0}^{x}\int_{0}^{y} \frac{3}{5}s(t+t^2)dtds \\
&= \int_{0}^{x} \frac{3}{5}s \int_{0}^{y} (t+t^2)dtds \\
&= \int_{0}^{x} \frac{3}{5}s\left(\frac{t^2}{2}+\frac{t^3}{3}\mid_0^y\right)ds \\
&= \int_{0}^{x} \frac{3}{5}s\left(\frac{y^2}{2}+\frac{y^3}{3}\right)ds \\
&= \frac{3}{5}\left(\frac{y^2}{2}+\frac{y^3}{3}\right)\int_{0}^{x} sds \\
&= \frac{3}{5}\left(\frac{y^2}{2}+\frac{y^3}{3}\right)\frac{x^2}{2}.
\end{aligned}$$

Additionally,

$$\begin{aligned}
F_{X,Y}(x,y) &= 0 \text{ if } x<0 \text{ or } y<0 \\
F_{X,Y}(x,y) &= 1 \text{ if } x>2 \text{ and } y>1 \\
F_{X,Y}(x,y) &= \frac{x^2}{4} \text{ if } 0<x<2 \text{ and } y>1 \\
F_{X,Y}(x,y) &= \frac{6}{5}\left(\frac{y^2}{2}+\frac{y^3}{3}\right) \text{ if } x>2 \text{ and } 0<y<1.
\end{aligned}$$

Together, these equations define F over the whole xy plane.

To do (c),

$$\begin{aligned}
P\{X<Y\} &= \int\int_{0<x<y<1} f_{X,Y}(x,y)dydx \\
&= \int_{0}^{1}\int_{x}^{1} \frac{3x}{5}(y+y^2)dydx \\
&= \int_{0}^{1} \frac{3x}{5}\left(\frac{y^2}{2}+\frac{y^3}{3}\right)\mid_x^1 dx \\
&= \int_{0}^{1} \frac{3x}{5}\left(\frac{1}{2}+\frac{1}{3}-\frac{x^2}{2}-\frac{x^3}{3}\right)dx \\
&= \frac{3}{5}\int_{0}^{1}\left(\frac{5x}{6}-\frac{x^3}{2}-\frac{x^4}{3}\right)dx \\
&= \frac{3}{5}\left(\frac{5x^2}{12}-\frac{x^4}{8}-\frac{x^5}{15}\right)\mid_0^1 \\
&= \frac{3}{5}\left(\frac{5}{12}-\frac{1}{8}-\frac{1}{15}\right) = \frac{3}{5}\frac{(50-15-8)}{120} \\
&= \left(\frac{3}{5}\right)\left(\frac{27}{120}\right) = \frac{27}{200}.
\end{aligned}$$

This completes the example.

4.2.1 Summary

The joint cumulative distribution function of two random variables X and Y is defined by

$$F_{X,Y}(x,y) = P\{X \leq x, Y \leq y\}.$$

It satisfies equations (4.11), (4.12) and (4.13). The marginal cdf can be calculated from the joint cdf as follows:

$$F_{X,Y}(x,\infty) = F_X(x)\ ;\ F_{X,Y}(\infty,y) = F_Y(y).$$

When (X,Y) are jointly continuous and have probability density $f_{X,Y}(x,y)$, then

$$F_{X,Y}(x,y) = \int_{-\infty}^{x}\int_{-\infty}^{y} f_{X,Y}(x,y)dydx.$$

Marginal densities can be calculated from the joint density as follows:

$$f_X(x) = \int_{-\infty}^{\infty} f_{X,Y}(x,y)dy$$

and

$$f_Y(y) = \int_{-\infty}^{\infty} f_{X,Y}(x,y)dx.$$

4.2.2 Exercises

1. Vocabulary. Define in your own words:
   (a) Riemann probability
   (b) joint cumulative distribution function
   (c) marginal cumulative distribution function
   (d) joint probability density function
   (e) marginal probability density function
2. Suppose X and Y are continuous random variables that are uniform within the unit circle, that is

$$f_{X,Y}(x,y) = \begin{cases} c & \text{if } x^2+y^2 \leq 1 \\ 0 & \text{otherwise} \end{cases}$$

   (a) Find c.
   (b) Find the marginal probability density function of X.
   (c) Find the cumulative distribution function of X.
   (d) Find $P\{X < Y\}$.
3. Suppose X and Y are continuous random variables having the probability density

$$f_{X,Y}(x,y) = \begin{cases} k \mid x+y \mid & -1 < x < 1, -2 < y < 1 \\ 0 & \text{otherwise} \end{cases}$$

   (a) Find k.
   (b) Find the marginal probability density function of Y.
   (c) Find $P\{Y > X + \frac{1}{2}\}$.

4.3 Conditional distributions and independence

We have already studied discrete conditional distributions in sections 1.6, 2.8 and 3.4. We now wish to find an analog for continuous distributions. In particular, we seek a conditional density $f_{Y|X}(y \mid x)$.

The principal issue here is that the event $\{X = x\}$ has probability zero. Therefore we'll consider $X \epsilon N_\Delta(x)$ where $N_\Delta(x) = (x - \frac{\Delta}{2}, x + \frac{\Delta}{2})$ is a neighborhood of size $\Delta > 0$ around x. We assume that the density $f_X(x)$ of X at the point x, is positive and continuous there. Considering the limit as $\Delta \to 0$ gives us the concept we want. Therefore we have

$$f_{Y|X}(y \mid x) = \lim_{\Delta \to 0} \frac{d}{dy} P\{Y \leq y \mid X \epsilon N_\Delta(x)\}. \tag{4.19}$$

This relationship can be simplified as follows:

$$\begin{aligned}
f_{Y|X}(y \mid x) &= \lim_{\Delta \to 0} \frac{d}{dy} P\{Y \leq y \mid X \epsilon N_\Delta(x)\} \\
&= \lim_{\Delta \to 0} \frac{d}{dy} \frac{P\{Y \leq y, X \epsilon N_\Delta(x)\}}{P\{X \epsilon N_\Delta(x)\}} \\
&= \lim_{\Delta \to 0} \frac{d}{dy} \frac{F_{X,Y}(x + \frac{\Delta}{2}, y) - F_{X,Y}(x - \frac{\Delta}{2}, y)}{F_X(x + \frac{\Delta}{2}) - F_X(x - \frac{\Delta}{2})} \\
&= \frac{\lim_{\Delta \to 0} \frac{d}{dy} \left[F_{X,Y}(x + \frac{\Delta}{2}, y) - F_{X,Y}(x - \frac{\Delta}{2}, y)\right] / \Delta}{\lim_{\Delta \to 0} \left[F_X(x + \frac{\Delta}{2}) - F_X(x - \frac{\Delta}{2})\right] / \Delta} \\
&= \frac{f_{X,Y}(x, y)}{f_X(x)},
\end{aligned}$$

using the limit definition of derivative. Hence we have

$$f_{Y|X}(y \mid x) = \frac{f_{X,Y}(x, y)}{f_X(x)}. \tag{4.20}$$

It's important to check that $f_{Y|X}$ is a probability density. It certainly is non-negative.

In addition,

$$\int_{-\infty}^{\infty} f_{Y|X}(y \mid x) dy = \int_{-\infty}^{\infty} \frac{f_{X,Y}(x, y)}{f_X(x)} dy = \frac{f_X(x)}{f_X(x)} = 1. \tag{4.21}$$

Therefore $f_{Y|X}(y \mid x)$ satisfies the conditions for a density, for each value of x.

Of course, having found the conditional density, there is a related cdf:

$$F_{Y|X}(y \mid x) = P\{Y \leq y \mid x\} = \int_{-\infty}^{y} f_{Y|X}(y \mid x) dy. \tag{4.22}$$

Discrete random variables X and Y are defined to be independent in section 2.8 if any event defined on X is independent of any event defined on Y, or, equivalently, if

$$P\{X \epsilon A, Y \epsilon B\} = P\{X \epsilon A\} P\{Y \epsilon B\} \tag{4.23}$$

for any events A and B. This definition is also used for continuous random variables.

Suppose that X and Y are independent random variables. Then

$$\begin{aligned}
F_{X,Y}(x, y) &= P\{X \leq x, Y \leq y\} = P\{X \leq x\} P\{Y \leq y\} \\
&= F_X(x) F_Y(y).
\end{aligned} \tag{4.24}$$

If X and Y have a joint probability density function $f_{X,Y}(x,y)$, then X and Y are independent if and only if $f_{X,Y}(x,y) = f_X(x)f_Y(y)$. In this case

$$\begin{aligned}
P\{X \epsilon A, Y \epsilon B\} &= \int I_{AB}(x,y) f_{X,Y}(x,y) dxdy \\
&= \int I_A(x) I_B(y) f_X(x) f_Y(y) dxdy \\
&= \int I_A(x) f(x) dx \int I_B(y) f_Y(y) dy \\
&= P\{X \epsilon A\} P\{Y \epsilon B\}
\end{aligned} \tag{4.25}$$

for all sets A and B for which the integrals are defined.

A consequence of (4.25) is that when X and Y have a joint density and are independent,

$$f_{Y|X}(y \mid x) = \frac{f_{X,Y}(x,y)}{f_X(x)} = \frac{f_X(x)f_Y(y)}{f_X(x)} = f_Y(y), \tag{4.26}$$

so the conditional density of Y given X does not depend on x. This is the analog of 2.47 in section 2.8.

As an example, consider again X and Y with the joint density defined in section 4.2.2, exercise 2. The density is uniform in the shaded area of Figure 4.1 The square box is the region $\{1 > X > 1/\sqrt{2},\ 1 > Y > 1/\sqrt{2}\}$, and has zero probability. However $P\{1 > X > 1/\sqrt{2}\}$ and $P\{1 > Y > 1/\sqrt{2}\}$ are both positive. Therefore X and Y are not independent. (To understand the commands given for Figure 4.1, you should know that `R` ignores the rest of a line after the "#" symbol.)

This can also be checked by observing that the conditional distribution of X and Y depends on Y (see section 4.2, problem 2). This phenomenon is deceptive, because the density appears to factor (if you forget about the range $x^2 + y^2 \leq 1$ of positive probability). Thus it is essential to write out the range of values for each function, in order not to be led astray.

Now suppose instead that X and Y have a uniform distribution on the square $-1 < x < 1$ and $-1 < y < 1$, so that

$$f_{X,Y}(x,y) = \begin{cases} \ell & -1 < x < 1,\ -1 < y < 1 \\ 0 & \text{otherwise} \end{cases}.$$

Since the space $(-1,1) \times (-1,1)$ is a square box with lengths of side 2, it is easy to see that its area is 4, so $\ell = 1/4$.

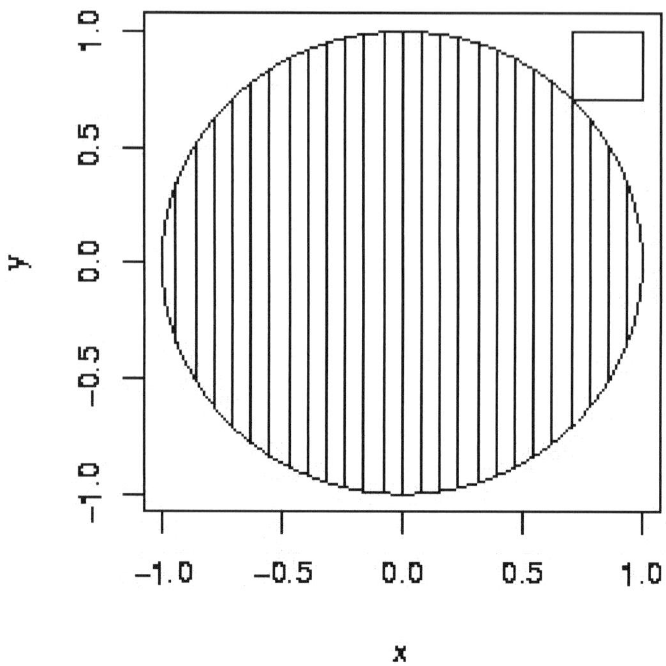

Figure 4.1: Area of positive density in example is shaded. The box in the upper right corner is a region of zero probability.

```
Commands:
s=((-50:50)/50)*pi  # gives 101 points between -pi and pi
x=cos(s)
y=sin(s) # x and y define the circle
plot(x,y,type="n") # draws the coordinates of the plot
polygon(x,y,density =10,angle=90) # shades the circle
w=1/sqrt(2)
lines(c(w,w),c(w,1),lty=1) # these draw the four lines
# of the box in the upper right corner
lines(c(w,1),c(w,w),lty=1)
lines(c(w,1),c(1,1),lty=1)
lines(c(1,1),c(w,1),lty=1)
```

Also

$$\begin{aligned} f_X(x) &= \int_{-\infty}^{\infty} f_{X,Y}(x,y)dy \\ &= \int_{-1}^{1} \frac{1}{4}dy = \frac{y}{4} \Big|_{-1}^{1} = \frac{1}{2} \quad \text{for} \quad -1 < x < 1. \end{aligned}$$

Hence

$$f_X(x) = \begin{cases} \frac{1}{2} & -1 < x < 1 \\ 0 & \text{otherwise} \end{cases}.$$

By symmetry,

$$f_Y(y) = \begin{cases} \frac{1}{2} & -1 < y < 1 \\ 0 & \text{otherwise} \end{cases}.$$

Therefore

$$f_X(x)f_Y(y) = \begin{cases} \frac{1}{4} & -1 < x < 1, -1 < y < 1 \\ 0 & \text{otherwise} \end{cases}.$$

Hence $f_{X,Y}(x,y) = f_X(x)f_Y(y)$ so X and Y are independent.

Thus, in the circle, uniform distributions are not independent, but in the square, they are independent.

Now reconsider the problem 3 of section 4.2.2. Here X and Y have the probability density function

$$f_{X,Y}(x,y) = \begin{cases} k \mid x+y \mid & -1 < x < 1, -2 < y < 1 \\ 0 & \text{otherwise} \end{cases}.$$

While this density is positive over the rectangle $-1 < x < 1, -2 < y < 1$, the function $\mid x + y \mid$ does not factor into a function of x times a function of y. Hence X and Y are not independent in this case.

4.3.1 Summary

The conditional density of Y given X (where both X and Y are continuous) is given by

$$f_{Y|X}(y \mid x) = \frac{f_{X,Y}(x,y)}{f_X(x)}.$$

X and Y are independent if

$$f_{Y|X}(y \mid x) = f_Y(y).$$

4.3.2 Exercises

1. Vocabulary: State in your own words, the meaning of:
   (a) the conditional density of Y given X.
   (b) independence of continuous random variables.
2. Reconsider problem 2 of section 4.2.
   (a) Find the conditional probability density of Y given X: $f_{Y|X}(y \mid x)$.
   (b) Find the cumulative conditional probability density of Y given X: $F_{Y|X}(y \mid x)$.
   (c) use your answer to (a) and (b) to address the question of whether X and Y are independent.
3. Reconsider problem 3 of section 4.2.
   (a) Find the conditional probability density of X given Y, $f_{X|Y}(x \mid y)$.
   (b) Find the cumulative probability density of X given Y, $F_{X|Y}(x \mid y)$.
   (c) Use your answer to (a) and (b) to address the question of whether X and Y are independent.

4.4 Existence and properties of expectations

The expectation of a random variable X with probability density function (pdf) $f_X(x)$ is defined as

$$E(X) = \int_{-\infty}^{\infty} x f_X(x) dx. \tag{4.27}$$

It should come as no surprise that this expectation is said to exist only when

$$E(\mid X \mid) = \int_{-\infty}^{\infty} \mid x \mid f_X(x) dx < \infty. \tag{4.28}$$

The reason for this is the same as that explored in Chapter 3, namely that where (4.28) is violated, the value of (4.27) would depend on the order in which segments of the real line are added together. This is an unacceptable property for an expectation to have.

If (4.28) is violated, then

$$\begin{aligned} \infty = \int_{-\infty}^{\infty} \mid x \mid f_X(x) dx = \int_{-\infty}^{0} \mid x \mid f_X(x) dx + \int_{0}^{\infty} \mid x \mid f_X(x) dx \\ = \int_{-\infty}^{0} (-x) f_X(x) dx + \int_{0}^{\infty} x f_X(x) dx. \end{aligned} \tag{4.29}$$

Hence, at least one of the integrals in (4.29) must be infinity. Suppose first that

$$\int_{0}^{\infty} x f_X(x) dx = \infty.$$

Then if $g(x) \geq x f_X(x)$, for all $x \epsilon (0, \infty)$, then $\int_0^\infty g(x) dx = \infty$. Thus no function greater than or equal to $x f_X(x)$ can have a finite integral on $(0, \infty)$.

Therefore the Riemann strategy, approximating the integrand above and below by piecewise constant functions, and showing that the difference between the approximations goes to zero as the grid gets finer, fails when (4.28) does not hold. A similar statement applies to approximating $-xf(x)$ from above, and hence approximating $xf(x)$ from below. Consequently we accept (4.28) as necessary for the existence of the expectation (4.27).

I now show that each of the properties given in section 3.4 (except the fourth, whose proof is postponed to section 4.7) for expectations of discrete random variables holds for continuous ones as well. The proofs are remarkably similar in many cases.

1. Suppose X is a random variable having an expectation, and let k be any constant. Then kX is a random variable that has an expectation, and $E(kX) = kE(X)$.

   Proof. We divide this according to whether k is zero, positive or negative.
   Case 1: If $k = 0$, then kX is a trivial random variable, take the value 0 with probability one. Its expectation exists, and is zero. Therefore

   $$E(kX) = 0 = kE(X).$$

   Case 2: $k > 0$.
   Then $Y = kX$ has cdf

   $$\begin{aligned} F_Y(y) = P\{Y \leq y\} = P\{kX \leq y\} = P\{X \leq y/k\} \\ = F_X(y/k). \end{aligned}$$

Differentiating both sides with respect to y,

$$f_Y(y) = \frac{f_X(y/k)}{k},$$

so Y has pdf $\frac{1}{k} f_X(y/k)$.
Therefore

$$E(\mid Y \mid) = \int_{-\infty}^{\infty} \frac{\mid y \mid}{k} f_X(y/k) dy.$$

Let $x = y/k$. Then

$$E(\mid Y \mid) = \int_{-\infty}^{\infty} k \mid x \mid f_X(x) dx = kE(\mid X \mid) < \infty.$$

Therefore the expectation of Y exists. Also, using the same substitution,

$$\begin{aligned} E(Y) &= \int_{-\infty}^{\infty} \frac{y}{k} f_X(y/k) dy = \int_{-\infty}^{\infty} kx f_X(x) dx \\ &= kE(X). \end{aligned}$$

Case 3: $k < 0$.
Now $Y = kX$ has cdf

$$\begin{aligned} F_Y(y) &= P\{Y \leq y\} = P\{kX \leq y\} = P\{X > y/k\} \\ &= 1 - F_X(y/k). \end{aligned}$$

Again differentiating, $f_Y(y) = -f_X(y/k)/k$, so Y has pdf $-\frac{1}{k} f_X(y/k)$.
Then the expectation of $\mid Y \mid$ is

$$E(\mid Y \mid) = \int_{-\infty}^{\infty} \mid y \mid f_Y(y) dy = -\frac{1}{k} \int_{-\infty}^{\infty} \mid y \mid f_X(y/k) dy.$$

Again, let $x = y/k$, but because $k < 0$ this reverses the sense of the integral. Hence

$$\begin{aligned} E(\mid Y \mid) &= -\frac{1}{k} \int_{\infty}^{-\infty} \mid kx \mid f_X(x) k dx \\ &= \int_{-\infty}^{\infty} \mid kx \mid f_X(x) dx = \mid k \mid \int_{-\infty}^{\infty} \mid X \mid f_X(x) dx \\ &= \mid k \mid E \mid X \mid < \infty. \end{aligned}$$

Therefore Y has an expectation, and it is

$$\begin{aligned} E(Y) &= \int_{-\infty}^{\infty} -\frac{y}{k} f_X(y/k) dy = \int_{\infty}^{-\infty} -\frac{k}{k} kx f_X(x) dx \\ &= k \int_{-\infty}^{\infty} x f_X(x) dx = kE(X). \end{aligned}$$

□

2. If $E(\mid X \mid) < \infty$ and $E(\mid Y \mid) < \infty$, then $X + Y$ has an expectation and $E(X + Y) = E(X) + E(Y)$.

Proof.

$$\begin{aligned}
E \mid X+Y \mid &= \int_{-\infty}^{\infty}\int_{-\infty}^{\infty} \mid x+y \mid f_{X,Y}(x,y)dxdy \\
&\leq \int_{-\infty}^{\infty}\int_{-\infty}^{\infty} (\mid x \mid + \mid y \mid) f_{X,Y}(x,y)dxdy \\
&= \int_{-\infty}^{\infty} \mid x \mid \int_{-\infty}^{\infty} f_{X,Y}(x,y)dydx \\
&+ \int_{-\infty}^{\infty} \mid y \mid \int_{-\infty}^{\infty} f_{X,Y}(x,y)dxdy \\
&= \int_{-\infty}^{\infty} \mid x \mid f_X(x)dx + \int_{-\infty}^{\infty} \mid y \mid f_Y(y)dy \\
&= E(\mid X \mid) + E(\mid Y \mid) < \infty \\
E(X+Y) &= \int_{-\infty}^{\infty}\int_{-\infty}^{\infty} (x+y) f_{X,Y}(x,y)dxdy \\
&= \int_{-\infty}^{\infty} x f_{X,Y}(x,y)dydx + \int_{-\infty}^{\infty} y f_{X,Y}(x,y)dxdy \\
&= \int_{-\infty}^{\infty} x f_X(x)dx + \int_{-\infty}^{\infty} y f_Y(y)dy = E(X) + E(Y)
\end{aligned}$$

□

Of course, again by induction, if $X_1, \ldots, X_k$ are random variables having expectations, then $X_1 + \ldots + X_k$ has an expectation whose value is

$$E(X_1 + \ldots + X_k) = \sum_{i=1}^{k} E(X_i).$$

3. Let $\min X = \max\{x|F(x) = 0\}$ and $\max X = \min\{x|F(x) = 1\}$, which may, respectively, be $-\infty$ and ∞. Also suppose X is non-trivial. Then

$$\min X < E(X) < \max X.$$

Proof.

$$\begin{aligned}
-\infty \leq \min X = \int_{-\infty}^{\infty} (\min X) f(x)dx &< \int_{-\infty}^{\infty} x f(x)dx = E(X) \\
&< \int_{-\infty}^{\infty} (\max X) f(x)dx = \max X \leq \infty.
\end{aligned}$$

□

4. Let X be non-trivial and have expectation c. Then there is some positive probability $\epsilon > 0$ that X exceeds c by a fixed amount $\eta > 0$, and positive probability $\epsilon > 0$ that c exceeds X by a fixed amount $\eta > 0$.
   The proof of this property is postponed to section 4.7.
5. Let X and Y be continuous random variables. Suppose that $E[X]$ and $E[X \mid Y]$ exist. Then

$$E[X] = EE[X \mid Y].$$

Proof.

$$
\begin{aligned}
E[X \mid Y] &= \int_{-\infty}^{\infty} x f_{X|Y}(x \mid y) dx \\
&= \int_{-\infty}^{\infty} x \frac{f_{X,Y}(x,y)}{f_Y(y)} dx \\
EE[X \mid Y] &= \int_{-\infty}^{\infty} \int_{-\infty}^{\infty} x \frac{f_{X,Y}(x,y)}{f_Y(y)} dx f_Y(y) dy \\
&= \int_{-\infty}^{\infty} \int_{-\infty}^{\infty} x f_{X,Y}(x,y) dy dx \\
&= \int_{-\infty}^{\infty} x f_X(x) dx \\
&= E[X].
\end{aligned}
$$

□

6. If g is a real valued function, $Y = g(x)$ and Y has an expectation, then

$$E(Y) = \int_{-\infty}^{\infty} g(x) f_X(x) dx.$$

Proof. We apply 5, reversing the roles of X and Y, so we write 5 as

$$E(Y) = E_X E[Y \mid X].$$

Now $Y \mid X = g(X)$. So $E[Y \mid X] = g(X)$.
Hence $E_X E[Y \mid X] = E_X g(X) = \int_{-\infty}^{\infty} g(x) f_X(x) dx$.
But $E_X E[Y \mid X] = E(Y)$. □

7. If X and Y are independent random variables, then $E[g(X)h(Y)] = E[g(X)]E[h(Y)]$, provided these expectations exist.

Proof.

$$
\begin{aligned}
E[g(X)h(Y)] &= \int_{-\infty}^{\infty} \int_{-\infty}^{\infty} g(x)h(y) f_{X,Y}(x,y) dx dy \\
&= \int_{-\infty}^{\infty} \int_{-\infty}^{\infty} g(x)h(y) f_X(x) f_Y(y) dx dy \\
&= \int_{-\infty}^{\infty} g(x) f_X(x) dx \int_{-\infty}^{\infty} h(y) f_X(y) dy \\
&= E[g(X)]E[h(Y)].
\end{aligned}
$$

□

8. Suppose $E \mid X \mid^k < \infty$ from some k. Let $j < k$. Then $E \mid X \mid^j < \infty$.

Proof.

$$
\begin{aligned}
E \mid X \mid^j &= \int \mid x \mid^j f_X(x) dx \\
&= \int \mid x \mid^j f_X(x) I(\mid x \mid \leq 1) dx + \int \mid x \mid^j f_X(x) I(\mid x \mid > 1) dx \\
&\leq 1 + \int \mid x \mid^k f_X(x) I(\mid x \mid > 1) dx \\
&\leq 1 + E(\mid X \mid^k) < \infty.
\end{aligned}
$$

□

9. All the properties of covariances and correlations given in section 2.11 hold for all continuous random variables as well, provided the relevant expectations exist.

4.4.1 *Summary*

The expectation of a continuous random variable X is defined to be

$$E(X) = \int_{-\infty}^{\infty} x f_X(x) dx$$

and is said to exist provided $E \mid X \mid < \infty$. It has many of the properties found in Chapter 3 of expectations of discrete random variables.

4.4.2 *Exercises*

1. Reconsider problem 2 of section 4.2, continued in problem 2 of section 4.3.
   (a) Find the conditional expectation and the conditional variance of Y given X.
   (b) Find the covariance of X and Y.
   (c) Find the correlation of X and Y.
2. Reconsider problem 3 of section 4.2, continued in problem 3 of section 4.3.
   (a) Find the conditional expectation and the conditional variance of Y given X.
   (b) Find the covariance of X and Y.
   (c) Find the correlation of X and Y.

4.5 Extensions

It should be obvious that there are very strong parallels between the discrete and continuous cases, between sums and integrals. Indeed the integral sign, "$\int$" was originally an elongated "S," for sum. There are senses of integral, particularly the Riemann-Stieltjes integral introduced in section 4.8, that unite these two into a single theory.

Many applications rely on the extension of the ideas of this chapter to vectors of random variables. Thus, for example, we can have $\boldsymbol{X} = (X_1, \ldots, X_k)$, which is just the random variable $(X_1, \ldots, X_k)$ considered together. If $\boldsymbol{x} = (x_1, \ldots, x_k)$ is a point in k-dimensional real space, we can write

$$F_{\boldsymbol{X}}(\boldsymbol{x}) = P\{\boldsymbol{X} \leq \boldsymbol{x}\} = P\{X_1 \leq x_1, X_2 \leq x_2, \ldots, X_k \leq x_k\}.$$

Similarly there can be a multivariate density function $f_{\boldsymbol{X}}(\boldsymbol{x})$, with marginal and conditional densities defined just as before.

This generalization is crucial to the rest of this book. Open your mind to it.

4.5.1 *An interesting relationship between cdf's and expectations of continuous random variables*

Suppose X is a continuous variable on $[0, \infty)$.

Then

$$E(X) = \int_0^{\infty} (1 - F_X(x)) dx$$

provided the expectation of X exists.

Proof.

$$\begin{aligned} E(X) &= \int_0^\infty x f_X(x)dx = \int_0^\infty \int_0^x dy f_X(x)dx \\ &= \int_{0<y<x<\infty} f_X(x)dxdy = \int_0^\infty \int_y^\infty f_X(x)dxdy \\ &= \int_0^\infty (1 - F_X(y))dy = \int_0^\infty (1 - F_X(x))dx. \end{aligned}$$

□

Not only is the result similar to the discrete case set forth in section 3.10.2, but also the steps in the proof are the same, with sums replaced by integrals.

4.6 Chapter retrospective so far

Many of the difficult issues involved in moving beyond random variables taking only finitely many values occur in Chapter 3, which concentrates on random variables taking at most countably many values. The further extension, in this chapter, to continuous random variables, mainly just recapitulates Chapter 3, substituting integrals for infinite sums, for each of the properties we have taken up so far. However, the story is more complex for the dominated and bounded convergence theorems, which are studied next.

4.7 Bounded and dominated convergence for Riemann integrals

The purpose of this section is to explore how close one can come to the results of section 3.11 on dominated and bounded convergence using Riemann integration. The answer is that one can get nearly, but not quite, all the way. To make precise exactly how close one can come requires a series of lemmas and theorems of increasing strength. But first it is necessary to introduce further material on limits, leading to a useful tool in studying convergence, namely Cauchy's criterion, and to be precise about what is meant by a Riemann integral. These are the subjects of the next two supplements.

4.7.1 A supplement about limits of sequences and Cauchy's criterion

Up to this point it has been possible to discuss limits of sequences directly from the definition. For the purpose of the remainder of this chapter, however, it is necessary to go more deeply into this concept.

Before doing so, it is useful to give some guidance about quantified expressions. Consider for example, the definition of continuity of a function f at a point x_0: for all $\epsilon > 0$, there exists a $\delta > 0$, such that for all x, if $| x - x_0 | < \delta$, then $| f(x) - f(x_0) | < \epsilon$. How should such an expression be handled?

If a quantified expression is given as an assumption, then you get to choose each of the "for all" variables, but your opponent gets to choose each of the "there exists" variables. On the other hand, to prove a quantified expression, your opponent chooses each of the "for all" variables, while you choose each of the "there exists" variables. This principle is used repeatedly in the material to come.

The order of quantifiers in a quantified expression matters. For example, when I am trying to prove that a function f is continuous at a point x_0, my choice of $\delta > 0$ can depend on f, x_0 and ϵ. And whatever choice of δ I make, my opponent's choice of x can depend on my choice of δ.

The first new idea to introduce is that of a point of accumulation: An infinite set of numbers $a_1, a_2, \ldots$ has a point of accumulation ξ if, for every $\epsilon > 0$ no matter how small, the interval $(\xi - \epsilon, \xi + \epsilon)$ contains infinitely many a_i's.

Theorem 4.7.1. *Let $a_1, a_2, \ldots$ be a bounded set of numbers. Then it has a point of accumulation.*

Proof. Suppose first that the numbers $a_1, a_2, \ldots$ are in the interval [0,1]. Consider all numbers whose decimal expansion is of the form $0.0, 0.1, \ldots, 0.9$. There are ten sets of numbers, at least one of which has infinitely many a_i's. Suppose that each member of that set has the decimal expansion $0.b_1$. Now consider the ten sets of numbers whose decimal expansions are $0.b_10, 0.b_11, 0.b_12, \ldots, 0.b_19$. Again at least one of these ten sets has infinitely many a's, say those with decimal expansion $0.b_1b_2$. This process leads to a number $\xi = 0.b_1b_2, \ldots$ that is a point of accumulation, because, no matter what $\epsilon > 0$ is taken, there are infinitely many a's within ϵ of ξ.

If the interval is not $[0,1]$, but instead $[c, c+d]$, then the point $\xi = c + d(0.b_1b_2, \ldots)$ suffices, whose points x in $[c, c+d]$ have been transformed into points on $[0,1]$ with the transformation $(x-c)/d$. □

Applied to sequences of points, a_n, we say that it has a point of accumulation ξ if for every $\epsilon > 0$, infinitely many values of n satisfy $\mid \xi - a_n \mid < \epsilon$. This, then, includes the possibility that infinitely many a_n's equal ξ.

With that definition, we have the following:

Theorem 4.7.2. *A bounded sequence a_n has a limit if and only if it has exactly one point of accumulation.*

Proof. We know from Theorem 4.7.1 that a bounded sequence has at least one accumulation point ξ. Suppose first that ξ is the only accumulation point. We will show that it is the limit of the a_n's. Let $\epsilon > 0$ be given, and consider the points a_n outside the set $(\xi - \epsilon, \xi + \epsilon)$. If there are infinitely many of them, then the subsequence of a_n's outside $(\xi - \epsilon, \xi + \epsilon)$ has an accumulation point, which is an accumulation point of the a_n's. This contradicts the hypothesis that the a_n's have only one accumulation point. Therefore there are only finitely many values of n such that a_n is outside the interval $(\xi - \epsilon, \xi + \epsilon)$. But this is the same as the existence of an N such that, for all n greater than or equal to N, $\mid \xi - a_n \mid < \epsilon$. Thus ξ is the limit of the a_n's.

Now suppose that the sequence a_n has at least two points of accumulation, ξ and η. Then let $\mid \xi - \eta \mid = a$. By choosing $\epsilon < a/3$, no point will have all but a finite number of the a_n's within ϵ of it, so there is no limit. This completes the proof. □

Perhaps it is useful to give some examples at this point. The sequence $a_n = 1/n$ has the limit 0, which is, of course, its only accumulation point. Similarly the sequence $b_n = 1 - 1/n$ has limit 1. Now consider the sequence c_n that, for even n, that is, n's of the form $n = 2m$ (where m is an integer) takes the value $1/m$, and for odd n, that is, n's of the form $n = 2m + 1$, takes the value $1 - 1/m$. This sequence has two accumulation points, 0 and 1, and hence no limit. In all three cases, the accumulation point is not an element of the sequence.

Up to this point, checking whether a sequence of real numbers converges to a limit has required knowing what the limit is. The Cauchy criterion for convergence of a sequence allows discussion of whether a sequence has a limit (*i.e.*, convergence) without specification of what that limit is. The Cauchy criterion can be stated as follows:

A sequence $a_1, a_2, \ldots,$ satisfies the Cauchy criterion for convergence if, for every $\epsilon > 0$, there is an N such that

$$\mid a_n - a_m \mid < \epsilon$$

if n and m are both greater than or equal to N.

The importance of the Cauchy criterion lies in the following theorem:

Theorem 4.7.3. *A sequence satisfies the Cauchy criterion if and only if it has a limit.*

Proof. Suppose $a_1, a_2, \ldots,$ is a sequence that has a limit ℓ. Let $\epsilon > 0$ be given. Then there is an N such that for all n greater than or equal to N, $| a_n - \ell | < \epsilon/2$. Then for all n and m greater than or equal to N,

$$| a_n - a_m | \leq | a_n - \ell | + | \ell - a_m | < \epsilon/2 + \epsilon/2 = \epsilon.$$

Therefore the sequence $a_1, a_2, \ldots$ satisfies the Cauchy criterion.

Now suppose $a_1, a_2, \ldots$ satisfies the Cauchy criterion. Then, choose some $\epsilon > 0$. There exists an N such that $| a_n - a_m | < \epsilon$ if n and m are greater than or equal to N. Hold a_n fixed. Then except possibly for $a_1, \ldots, a_{N-1}$, all the a_m's are within ϵ of a_n. Therefore the a's are bounded. Hence Theorem 4.7.1 applies, and says that the sequence a_n has a limit point ξ. Suppose it has a second limit point η. Let $a = | \xi - \eta |$ and choose $\epsilon < a/3$. Then there are infinitely many n's such that $| \xi - a_n | < \epsilon$ and infinitely many m's such that $| \eta - a_m | < \epsilon$. For those choices of n and m, we have $| a_n - a_m | > a/3$, which contradicts the assumption that the a's satisfy the Cauchy criterion. Therefore there is only one limit point ξ, and $\lim_{n\to\infty} a_n = \xi$. □

If, in the proof of Theorem 4.7.1, had the largest b been chosen when several b's corresponded to the decimal expansion of an infinite number of a's, the resulting ξ would be the largest point of accumulation of the bounded sequence a_n. This largest accumulation point is called the limit superior, and is written $\overline{\lim}_{n\to\infty} a_n$. Similarly always choosing the smallest leads to the smallest accumulation point, called the limit inferior, and written $\underline{\lim}_{n\to\infty} a_n$. A bounded sequence a_n has a limit if and only if $\overline{\lim}_{n\to\infty} a_n = \underline{\lim}_{n\to\infty} a_n$.

An interval of the form $a \leq x \leq b$ is a closed interval; an interval of the form $a < x < b$ is an open interval. Intervals of the form $a < x \leq b$ or $a \leq x < b$ are called half-open.

Lemma 4.7.4. *A closed interval I has the property that it contains every accumulation point of every sequence $\{a_n\}$ whose elements satisfy $a_n \in I$ for all n.*

Proof. Suppose that $I = \{x \mid a \leq x \leq b\}$, and let a_n be a sequence of elements in I. Let $b^* = \overline{lim}_{n\to\infty} a_n$.

If $b^* \leq b$ we are done. Therefore suppose that $b^* > b$. Let $\epsilon = (b^* - b)/2$. Then because $a_n \leq b$ for all n, $| b^* - a_n | > \epsilon$ for all n, so b^* is not the $\overline{lim}_{n\to\infty} a_n$, a contradiction. Hence $b^* \leq b$.

A similar argument applies to $a^* = \underline{\lim}_{n\to\infty} a_n$, and shows $a \leq a^*$. Consequently $a \leq a^* \leq b^* \leq b$, so if c is an arbitrary accumulation point of a_n, we have $a \leq a^* \leq c \leq b^* \leq b$, so $c \in I$, as claimed. □

Open and half-open intervals do not have this property. For example, if $I = \{x \mid a < x < a + 2\}$, the sequence $a_n = a + 1/n$ satisfies $a_n \in I$ for all n, but $\lim_{n\to\infty} a_n = a \notin I$.

A second lemma shows that bounded non-decreasing sequences have a limit:

Lemma 4.7.5. *Suppose a_n is a non-decreasing bounded sequence. Then a_n has a limit.*

Proof. We have that there is a b such than $a_n \leq b$ for all n. Also we have $a_{n+1} \geq a_n$ for all n. Let $x \leq b$ be chosen to be $\overline{\lim}\ a_n$ and suppose, contrary to the hypothesis, that $y = \underline{\lim}\ a_n$ satisfies $y < x$. Let $\epsilon = (x - y)/2 > 0$. Then by definition of the $\overline{\lim}$, there are an infinite number of n's such that $x - a_n < \epsilon$. Take any such n. Because the a_n's are non-decreasing, $x - a_{n+1} < \epsilon$, $x - a_{n+2} < \epsilon$, etc. Thus for all $m \geq n$, $x - a_m < \epsilon$. But then there cannot be infinitely many n's such that $| y - a_n | < \epsilon$. A contradiction to the definition of $\overline{\lim}$. Hence $x = y$, and $\{a_n\}$ has a limit. □

Lemma 4.7.6. *Suppose G_n is a non-increasing sequence of non-empty closed subsets of $[a, b]$, so $G_n \supseteq G_{n+1}$ for all n. Then $G = \cap_{n=1}^{\infty} G_n$ is non-empty.*

Proof. Let $x_n = \inf G_n$. The point x_n exists because G_n is non-empty and bounded. Furthermore, $x_n \epsilon G_n$, because G_n is closed. The sequence $\{x_n\}$ is non-decreasing, because $G_n \supseteq G_{n+1}$. It is also bounded above by b. Therefore by Lemma 4.7.5, x_n converges to a limit x. Choose an n and $k > n$. Then $x_k \in G_k \subseteq G_n$. Then $x \epsilon G_n$ because G_n is closed. Since $x \epsilon G_n$ for all n, $x \epsilon G$. □

4.7.2 *Exercises*

1. Vocabulary. Explain in your own words:
   (a) accumulation point of a set
   (b) accumulation point of a sequence
   (c) Cauchy criterion
   (d) limit superior
   (e) limit inferior
2. Consider the three examples given just after the proof of Theorem 4.7.2. For each of them, identify the limit superior and the limit inferior.
3. Prove the following: Suppose b_n is a non-increasing bounded sequence. Then b_n has a limit.
4. Let $U \geq L$. Let $x_1, x_2, \ldots$ be a sequence that is convergent but not absolutely convergent. Show that there is a reordering of the x's such that U is the limit superior of the partial sums of the x's, and so that L is the limit inferior. Hint: Study the proof of Riemann's Rainbow Theorem 3.3.5.
5. Consider the following two statements about a space $\mathcal{X}$:
   (a) For every $x \epsilon \mathcal{X}$, there exists a $y \epsilon \mathcal{X}$ such that $y = x$.
   (b) There exists a $y \epsilon \mathcal{X}$ such that for every $x \epsilon \mathcal{X}, y = x$.
      i. For each statement, find a necessary and sufficient condition on $\mathcal{X}$ such that the statement is true.
      ii. If one statement is harder to satisfy than the other (*i.e.*, the $\mathcal{X}$'s satisfying it are a narrower class), explain why.

4.7.3 *References*

The approach used in this section is from Courant (1937, pp. 58-61).

4.7.4 *A supplement on Riemann integrals*

To understand the material to come, it is useful to be more precise about a concept considered only informally up to this point, Riemann integration, the ordinary kind of integral we have been using.

A cell is a closed interval $[a, b]$ such that $a < b$, so the interior (a, b) is not empty. A collection of cells is non-overlapping if their interiors are disjoint. A partition of a closed interval $[a, b]$ is a finite set of couples (ξ_k, I_k) such that the I_k's are non-overlapping cells, such that $\cup_{k=1}^{n} I_k = [a, b]$, and ξ_k is a point such that $\xi_k \epsilon I_k$.

If $\delta > 0$, then a partition $\pi = (\xi_i, [u_i, v_i]; i = 1, \ldots, n)$ for which, for all $i = 1, \ldots, n$

$$\xi_i - \delta < u_i \leq \xi_i \leq v_i < \xi_i + \delta$$

is a δ-fine partition of $[a, b]$.

If f is a real-valued function on $[a, b]$, then a partition π has a Riemann sum

$$\sum_{\pi} f = \sum_{i=1}^{n} f(\xi_i)(v_i - u_i). \tag{4.30}$$

Definition: A number A is the Riemann integral of f on $[a, b]$ if for every $\epsilon > 0$ there is a $\delta > 0$ such that, for every δ-fine partition π,

$$|\sum_{\pi} f - A| < \epsilon.$$

Many of the treatments of the Riemann integral use an equivalent formulation that looks at the $\underline{\lim}$ of the Riemann sums of functions at least as large as f and the $\overline{\lim}$ of the Riemann sums of functions no larger than f. If these two numbers are equal, then the Riemann integral of f is defined and equal to both of them.

A function such as (4.30), which is constant on a finite number of intervals, is called a step function. Riemann integrals are limits of areas under step functions as the partition that defines them gets finer.

As practice using the formal definition of Riemann integration, suppose $g(x)$ and $h(x)$ are Riemann integrable functions. Then we'll show that $g(x) + h(x)$ is integrable, and that

$$\int \Big(g(x) + h(x) \Big) dx = \int g(x)dx + \int h(x)dx.$$

Proof. Let $a = \int g(x)dx$ and $b = \int h(x)dx$. Let $\epsilon > 0$ be given. Then there is a $\delta_g > 0$ such that, for every δ_g-fine partition π_g,

$$|\sum_{\pi_g} g - a| < \epsilon/2.$$

Similarly there is a $\delta_h > 0$ such that, for every δ_h-fine partition π_h,

$$|\sum_{\pi_h} h - b| < \epsilon/2.$$

Let $\delta = \min(\delta_g, \delta_h) > 0$, and let π be an arbitrary δ-fine partition. Then π is both a δ_g-fine and δ_h-fine partition. Then

$$|\sum_{\pi} \Big(g(x) + h(x) \Big) - (a + b)| \leq |\sum_{\pi} g(x) - a| + |\sum_{\pi} h(x) - b| < \epsilon/2 + \epsilon/2 = \epsilon.$$

Since this is true for all δ-fine partitions π, and for all $\epsilon > 0$, $g(x) + h(x)$ has a Riemann integral, and it equals $a + b$.

4.7.5 Summary

This supplement makes more precise exactly what is meant by the Riemann integral of a function.

4.7.6 Exercises

1. Vocabulary: state in your own words what is meant by:
   (a) Riemann sum
   (b) Riemann integral
   (c) Step function
2. Use the definition of Riemann integral to find $\int_0^1 x dx$. Hint: You may find it helpful to review section 1.2.2.

4.7.7 Bounded and dominated convergence for Riemann integrals

Having introduced the Cauchy criterion and given a rigorous definition of the Riemann integral, along with some of its properties, I am now ready to proceed to the goal of this section, bounded and dominated convergence for Riemann integration. I do so in a series of steps, proving a rather restricted result, and then gradually relaxing the conditions. We start with some facts about some special sets called elementary sets.

A subset of $\mathbb{R}$ is said to be elementary if it is the finite union of bounded intervals (open, half-open or closed). Two important properties of elementary subsets are:

(i) if F is an elementary set and if $\mid g(x) \mid \leq K$ for all $x \epsilon F$, then $\mid \int_F g(x) \mid \leq K \mid F \mid$, where $\mid F \mid$ is the sum of the lengths of the bounded intervals comprising F, and is called the measure of F.

(ii) if F is an elementary set and $\epsilon > 0$, there is a closed elementary subset H of F such that $\mid H \mid > \mid F \mid - \epsilon$.

The first is obvious. To show the second, if F is elementary, it is the finite union of intervals, say $I_1, \ldots, I_N$. Choose $\epsilon > 0$. Suppose the endpoints of I_i are a_i and b_i, where $a_i \leq b_i$, and I_i is open or closed at each end. If $a_i = b_i$, I_i must be $\{a_i\}$ and is closed. If $a_i < b_i$, then, choose ϵ_i' so that $0 < \epsilon_i' < \min\{\epsilon/2N, (b_i - a_i)/2\}$.

Consider $I_i' = [a_i + \epsilon_i', b_i - \epsilon_i'] \subset I_i$. Let $H = \cup_{i=1}^n I_i', a_i' = a_i + \epsilon_i'$ and $b_i' = b_i - \epsilon_i'$.

H is closed because it is a finite union of closed intervals, and

$$\mid H \mid = \sum_{i=1}^N (b_i' - a_i') = \sum_{i=1}^N [(b_i - a_i) - 2\epsilon_i'] = \mid F \mid - 2\sum_{i=1}^N \epsilon_i' > \mid F \mid - \epsilon.$$

□

Definition: A sequence A_n is **contracting** if and only if $A_1 \supseteq A_2 \supseteq \ldots$.

Lemma 4.7.7. *Suppose A_n is a contracting sequence of bounded subsets of $\mathbb{R}$, with an empty intersection. For each n, define*

$$\alpha_n = \sup\{\mid E \mid \mid E \text{ is an elementary subset of } A_n\}.$$

Then $\alpha_n \to 0$ as $n \to \infty$.

Proof. The sequence α_n is non-increasing. Suppose the lemma is false. Then there is some $\delta > 0$ such that $\alpha_n > \delta$ for all n. For each n, let F_n be a closed elementary subset of A_n such that

$$\mid F_n \mid > \alpha_n - \delta/2^n,$$

and let

$$H_n = \cap_{i=1}^n F_n.$$

Now $H_n \subseteq A_n$ and H_n's are a decreasing sequence of closed intervals. To show that each H_n is not empty, consider

(a) for every n, if F is an elementary subset of $A_n \backslash F_n$, then $\mid F \mid + \mid F_n \mid = \mid F \cup F_n \mid \leq \alpha_n$ and $\mid F_n \mid > \alpha_n - \delta/2^n$. Consequently $\mid F \mid < \delta/2^n$.

(b) For every n, if G is an elementary subset of $A_n \backslash H_n$, then since

$$G = (G \backslash F_1) \cup (G \backslash F_2) \cup \ldots \cup (G \backslash F_n),$$

it follows that $\mid G \mid \leq \sum_{i=1}^n \mid G \backslash F_i \mid \leq \sum_{i=1}^n \delta/2^i < \delta$.

For every n, because $\alpha_n > \delta$, the set A_n must have an elementary subset G_n such that $\mid G_n \mid > \delta$ so it follows that each H_n is non-empty. Then H_n is a decreasing sequence of non-empty closed intervals, and $H_n \subseteq A_n$. It follows from Lemma 4.7.6 that $\cap_{n=1}^{\infty} H_n$ is non-empty. Therefore $\cap_{n=1}^{\infty} A_n$ is non-empty, a contradiction. □

Theorem 4.7.8. *Suppose f_n is a sequence of Riemann integrable functions, suppose $f_n \to f$ point-wise, that f is Riemann integrable, and that for some constant $K > 0$ we have $\mid f_n \mid \leq K$ for every n. Then*

$$\int_a^b f_n \to \int_a^b f.$$

Proof. Let $g_n = \mid f_n - f \mid$. Then $g_n \geq 0$ for all n and $g_n \to 0$ point-wise. Therefore there is no loss of generality in supposing $f_n \geq 0$ and $f = 0$. Let $\epsilon > 0$ and for each n, define

$$A_n = \{x\epsilon[a,b] \mid f_i(x) \geq \frac{\epsilon}{2(b-a)} \text{ for at least one } i \geq n\}.$$

Now Lemma 4.7.7 applied to A_n says that there is an N such that for all n greater than or equal to N, and F being an elementary subset of A_n, we have $\mid F \mid < \epsilon/2K$. Now we must show that for all n greater than or equal to N, we have $\int_a^b f_n \leq \epsilon$. Fix $n \geq N$. It suffices to show that when s is a step function and $0 \leq s \leq f_n$ we have $\int_a^b s \leq \epsilon$. Let s be such a step function, let

$$F = \{x\epsilon[a,b] \mid s(x) \geq \frac{\epsilon}{2(b-a)}\}, \text{ and } G = [a,b]\backslash F.$$

Then F and G are elementary sets, and since $F \subseteq A_n$ we have $\mid F \mid < \epsilon/2K$.

Then

$$\begin{aligned} \int_a^b s &= \int_F s + \int_G s \leq \int_F K + \int_G \tfrac{\epsilon}{2(b-a)} \leq \int_F K + \int_a^b \tfrac{\epsilon}{2(b-a)} \\ &= K \mid F \mid + \tfrac{\epsilon}{2(b-a)}(b-a) < \epsilon. \end{aligned}$$

□

Now this bounded convergence theorem does not quite generalize Theorem 3.11.1, since it assumes that the limit function f is integrable. What happens if this assumption is not made?

Corollary 4.7.9. *Suppose f_n is a sequence of Riemann integrable functions, suppose $f_n \to f$ point-wise, and, for some constant $K > 0$, we have $\mid f_n \mid \leq K$ for every n. Then*

(a) $\int_a^b f_n$ is a sequence that satisfies the Cauchy criterion.

(b) If f is Riemann integrable, then $\int_a^b f_n \to \int_a^b f$.

Proof. In light of the theorem, only (a) requires proof.

Let $h_{n,m} = \mid f_n - f_m \mid$. Then $h_{n,m} \geq 0$ for all n and m. We may suppose without loss of generality that $m \geq n$. Then $\lim_{n\to\infty} h_{n,m} = 0$.

Now the proof of the theorem applies to $h_{n,m}$, showing that $\lim_{n\to\infty, m\geq n} \int h_{n,m}(x)dx = \lim_{n\to\infty, m\geq n} \int \mid f_n - f_m \mid = 0$. Thus $\int f_n$ satisfies the Cauchy criterion. □

To show what the issue is about whether f is integrable, consider the following example.

Example 1. (Dirichlet): In this example we consider rational numbers p/q, where p and q are natural numbers having no common multiple except one. Thus $2/4$ is to be reduced to $1/2$.

Let

$$f_n(x) = \begin{cases} 1 & \text{if } x = p/q \text{ and } n \leq q, 0 < x < 1 \\ 0 & \text{otherwise, } 0 < x < 1. \end{cases}$$

So then $f_2(x) = 1$ at $x = 1/2$ and is zero elsewhere on the unit interval. Similarly $f_3(x) = 1$ at $x = 1/3, 1/2, 2/3$ and is zero elsewhere, etc. Each such $f_n(x)$ is continuous except at a finite number of points, and hence is Riemann integrable. Indeed the integral of each f_n is zero.

Now let's look at $f(x) = \lim_{n\to\infty} f_n(x)$. This function is 1 at each rational number, and zero otherwise. The $\overline{\lim}$ of each Riemann sum of f is 1, and the $\underline{\lim}$ of each Riemann sum is zero. Hence f is not Riemann integrable. □

Finally, I wish to extend the result from bounded convergence to dominated convergence. To this end, I wish to substitute for the assumption $\mid f_n \mid\leq K$ for all n, the weaker assumption that $\mid f_n(x) \mid\leq k(x)$ where $k(x)$ is integrable. To do this, I find, for every $\epsilon > 0$, a constant K big enough that $\int g \leq \int \min(g, K) + \epsilon$. In particular,

Lemma 4.7.10. *Let k be a non-negative function with $\int k < \infty$ and let $\epsilon > 0$ be given. Then there exists a constant K so large that $\int g \leq \int \min(g, K) + \epsilon$ for all non-negative integrable functions g satisfying $g(x) \leq k(x)$.*

Proof. Define a lower sum for g any number of the form $\sum_{i=1}^{r} y_i \mid I_i \mid$, where $I_i(i = 1, \ldots, r)$ are a partition of $[a, b]$ and $g(x) \geq y_i$ for all $x\epsilon I_i$. $\int g$ is the least upper bound of all lower sums of g.

Let $\epsilon > 0$ be given, and let $\pi = (y_i, I_i, i = 1, \ldots, r)$ be a lower sum for k such that $\sum_{i=1}^{r} y_i \mid I_i \mid> \int k - \epsilon$. Let $K = \max\{y_1, \ldots, y_r\}$. Let g satisfy the assumptions of the lemma. Additionally, let $\eta = (x_j, J_j, j = 1, \ldots, s)$ be a lower sum for $g - \min(g, K)$. Let $H_{ij} = I_iJ_j$. I claim that $\sum_{i,j}(x_j + y_i) \mid H_{ij} \mid$ is a lower sum for k. Since the H_{ij}'s are a partition of $[a, b]$, what has to be shown is $k(x) \geq x_j + y_i$ for all $x\epsilon H_{ij}$.

(a) If $g(x) \leq K$, then $\min(g(x), K) = g(x)$. Hence $g(x) - \min(g(x), K) = g(x) - g(x) = 0$. Then $x_j \leq 0$, and $y_i + x_j \leq y_i \leq g(x) \leq k(x)$.

(b) If $g(x) > K$ then $\min(g(x), K) = K$. Therefore $g(x) - \min(g(x), K) = g(x) - K$. Then $y_i + x_j \leq K + g(x) - K = g(x) \leq k(x)$.

Therefore $\int k - \sum_{i=1}^{r} y_i \mid I_i \mid$ is an upper bound for $\sum x_jJ_j$, which is a lower sum of $\int g - \min(g, K)$. Since $(x_j, J_j, j = 1, \ldots, s)$ is an arbitrary such lower sum, $\int k - \sum_{i=1}^{r} y_i \mid I_i \mid$ is an upper bound for all such lower sums of $\int g - \min(g, K)$, so it is an upper bound for $\int g - \min(g, K)$. Since $\int k - \sum_{i=1}^{r} y_i \mid I_i \mid< \epsilon$, this proves the lemma. □

Now $\min\{f_n(x), K\} \leq K$ so the theorem applies to $\min\{f_n(x), K\}$, and the result is a contribution of less than ϵ, for any $\epsilon > 0$, to the resulting integrals. Hence we have

$$\Big| \int f_n - \int \min\{f_n, K\}\Big| < \epsilon, \text{ so } \int f_n \leq 2\epsilon$$

as a consequence of the proof of Theorem 4.7.8. Since this is true for all $\epsilon > 0$, we have $\int f_n \to 0$. This derives the final form of the result:

Theorem 4.7.11. *Suppose $f_n(x)$ is a sequence of Riemann-integrable functions satisfying*

(i) $f_n(x) \to f(x)$

(ii) $\mid f_n(x) \mid\leq k(x)$ where k is Riemann integrable.

Then

(a) $\int f_n(x)dx$ is a sequence satisfying the Cauchy criterion.

(b) If $f(x)$ is Riemann integrable, then $\int f_n(x)dx \to \int f(x)dx$.

Theorem 4.7.12. *Suppose $f_n(x)$ and $g_n(x)$ are two sequences of Riemann-integrable functions satisfying conditions (i) and (ii) of Theorem 4.7.11 with respect to the same limiting function $f(x)$. Then*

$$\lim_{n\to\infty} \int f_n(x)dx = \lim_{n\to\infty} \int g_n(x)dx.$$

Proof. Consider the sequence of functions $f_1, g_1, f_2, g_2, \ldots$. Let the nth member of the sequence be denoted h_n. I claim that the sequence of functions h_n satisfies conditions (i) and (ii) of Theorem 4.7.11, with respect to f.

(i) Let $\epsilon > 0$ be given. Since $f_n(x) \rightarrow f(x)$, there is some number N_f such that, for all $n \geq N_f$,

$$| f_n(x) - f(x) |< \epsilon.$$

Similarly there is an N_g such that for all $n \geq N_g$,

$$| g_n(x) - f(x) |< \epsilon.$$

Let $N = 2\text{Max}\{\text{N}_\text{f}, \text{N}_\text{g}\} + 1$. Then, for all $n \geq N$,

$$| h_n(x) - f(x) |< \epsilon.$$

(ii) Let $k_f(x)$ be Riemann integrable and such that $| f_n(x) |\leq k_f(x)$ for all x. Similarly, let $k_g(x)$ be Riemann integrable and such that $| g_f(x) |\leq k_g(x)$. Then $| h_f(x) |\leq k_f(x) + k_g(x)$, and $k_f(x) + k_g(x)$ is Riemann integrable.

Therefore, Theorem 4.7.11 applies to h_n, so $\int h_n(x)dx$ is a Cauchy sequence, and therefore has a limit. Since $\int f_n(x)dx$ and $\int g_n(x)dx$ are also Cauchy sequences, they have limits, which we'll call a and b, respectively. Then a and b are accumulation points of the set $\{\int h_n(x)dx\}$, so by Theorem 4.7.2, we must have $a = b$. □

Theorem 4.7.12 suggests that when conclusion (a) of Theorem 4.7.11 applies, we know what the value of $\int f(x)dx$ "ought" to be, namely $\lim_{n\rightarrow\infty} \int f_n(x)dx$, (which limit exists because it satisfies the Cauchy criterion). Theorem 4.7.12 shows that this extension of Riemann integration is well-defined, by showing that if, instead of choosing the sequence $f_n(x)$ of Riemann-integrable functions one chose any other sequence $g_n(x)$ also converging to f, the limit of the sequence of integrals would be the same. Nonetheless, this would be a messy theory, because each use would require distinguishing the two cases of Theorem 4.7.11. Instead, I will soon introduce a generalized integral, the McShane integral, that satisfies a strong dominated convergence theorem and does so in a unified way.

4.7.8 Summary

This section gives a sequence of increasingly more general results on bounded and dominated convergence, culminating in Theorem 4.7.11.

4.7.9 Exercises

1. Vocabulary. Explain in your own words:
   (a) Riemann integrability
   (b) bounded convergence
   (c) dominated convergence
2. In Example 1, what is $\int f_n(x)dx$? Show that it is a Cauchy sequence. What is its limiting value?

4.7.10 References

The first bounded convergence theorem (without uniform convergence) for Riemann integration is due to Arzela. A useful history is given by Luxemburg (1971). Lemma 4.7.7, Theorem 4.7.8 and Corollary 4.7.9 are from Lewin (1986). Lemma 4.7.10 and Theorem 4.7.11 follow Cunningham (1967). Other useful material includes Kestelman (1970) and Bullen and Vyborny (1996).

4.7.11 A supplement on uniform convergence

The disappointment that Theorem 4.7.11 does not permit the conclusion that $f(x)$ is Riemann integrable leads to the thought that either the assumptions should be made stronger or that the notion of integral should be strengthened. While most of the rest of this chapter is devoted to the second possibility, this supplement explores a strengthening of the notion of convergence.

The kind of convergence in assumption (i) of Theorem 4.7.11 is pointwise in $x\epsilon[a, b]$. It says that for each x, $f_n(x)$ converges to $f(x)$. Formally this is translated as follows: for every $x\epsilon[a, b]$ and for every $\epsilon > 0$, there exists an $N(x, \epsilon)$ such that, for all $n \geq N(x, \epsilon)$, $\mid f_n(x) - f(x) \mid < \epsilon$. In this supplement, we consider a stronger sense of convergence, called uniform convergence: for every $\epsilon > 0$ there exists an $N(\epsilon)$ such that for all $x\epsilon[a, b]$, $\mid f_n(x) - f(x) \mid < \epsilon$. Thus every sequence of functions that converges uniformly also converges pointwise, by taking $N(x, \epsilon) = N(\epsilon)$. However, the converse is not the case, as the following example shows: Consider the sequence of functions $f_n(x) = x^n$ in the interval $x\epsilon[0, 1]$. This sequence converges pointwise to the function $f(x) = 0$ for $0 \leq x < 1$, and $f(1) = 1$. Choose, however, an $\epsilon > 0$, and an N. Then for all $n \geq N$, we have $x^n - 0 > \epsilon$ if $1 > x > \epsilon^{1/n}$. Hence the convergence is not uniform.

The distinction between pointwise and uniform convergence is an example in which it matters in what order the quantifiers come, as explained in section 4.7.1. (See also problem 3 in section 4.7.2.)

Now we explore what happens to Theorem 4.7.11 if instead of assuming (i) we assume instead that $f_n(x) \to f(x)$ uniformly for $x\epsilon[a, b]$. First we have the following lemma:

Lemma 4.7.13. *Let $f_n \to f(x)$ uniformly for $x\epsilon[a, b]$, and suppose $f_n(x)$ is Riemann integrable. Then f is Riemann integrable.*

Proof. Let j and J be respectively the supremum of the lower sums and the infimum of the upper sums of the Riemann approximations to f. Let $\epsilon_n = \sup_{x\epsilon[a,b]} \mid f_n(x) - f(x) \mid$. The definition of uniform convergence is equivalent to the assertion that $\epsilon_n \to 0$. Then for all $x\epsilon[a, b]$ and $n \geq N$, $f_n(x) - \epsilon_n \leq f(x) \leq f_n(x) + \epsilon_n$.

Integrating, this implies, for all $n \geq N$

$$\int f_n(x)dx - \epsilon_n(b - a) \leq j \leq J \leq \int f_n(x)dx + \epsilon_n(b - a).$$

Then $0 \leq J - j \leq 2\epsilon_n(b - a)$.

As $n \to \infty$, the right-hand side goes to 0, so $j = J$ and f is Riemann integrable. □

Next, we see what happens to assumption (ii) of Theorem 4.7.11 when $f_n(x) \to f(x)$ uniformly:

Lemma 4.7.14. *If $f_n(x)$ is a sequence of Riemann-integrable functions satisfying $f_n(x) \to f(x)$ uniformly in $x\epsilon[a, b]$, then $\mid f_n(x) \mid \leq k(x)$ where k is Riemann integrable on $[a, b]$.*

Proof. Choose an $\epsilon > 0$. Then by uniform convergence there exists an $N(\epsilon)$ such that, for all $n \geq N \mid f_n(x) - f(x) \mid < \epsilon$.

Let $k(x) = \sum_{i=1}^{N} \mid f_i(x) \mid + \mid f(x) \mid + \epsilon$. Then

$$\mid f_i(x) \mid \leq k(x) \text{ for } i = 1, \ldots, N.$$

For $n \geq N, \mid f_n(x) \mid \leq \mid f(x) \mid + \epsilon \leq k(x)$. Therefore,

$$\mid f_n(x) \mid \leq k(x) \text{ for all } n.$$

To show that k is integrable,

$$\int_a^b \left[\sum_{i=1}^{N} | f_1(x) | + | f(x) | + \epsilon \right] dx$$

$$= \sum_{i=1}^{N} \int_a^b | f_i(x) | dx + \int | f(x) | dx + \epsilon(b-a),$$

which displays the integral of k as the sum of $N+2$ terms, each of which is finite (using Lemma 4.7.13). Hence k is Riemann integrable. □

Thus if $f_n(x)$ converges to $f_n(x)$ uniformly for $x \epsilon [a,b]$, Theorem 4.7.11 part (b) applies, and we can conclude that

$$\int f_n(x)dx \to \int f(x)dx.$$

Hence, we have the following corollary to Theorem 4.7.11.

Corollary 4.7.15. *Suppose $f_n(x)$ is a sequence of Riemann-integrable functions converging uniformly to a function f. Then*

(a) f is Riemann integrable

(b) $\int f_n(x)dx \to \int f(x)dx$.

It turns out that the assumption of uniform convergence is a serious restriction, which is why the modern emphasis is on generalizing the idea of the integral. The development of such an integral begins in section 4.8.

4.7.12 Bounded and dominated convergence for Riemann expectations

We now specialize our considerations to expectations of random variables, where the expectation is understood to be a Riemann integral. There are two ways in which these expectations are special cases of the integrals considered in section 4.7.7:

(A) There is an underlying probability density $h(x)$ satisfying

  (i) $h(x) \geq 0$ for all x

  (ii) $\int h(x)dx = 1$

(B) A random variable $y(X)$ is considered to have an expectation only when $E \mid y(X) \mid = \int \mid y(x) \mid h(x)dx < \infty$ for reasons discussed in Chapter 3.

Additionally, there is one respect in which these expectations are more general than the integrals of section 4.7.7: we want the domain of integration to be the whole real line, and not just a closed interval $[a,b]$. As it will turn out, the restrictions (A) and (B) permit this extension without further assumptions. To be clear, we mean by an integral over the whole real line, that

$$\int_{-\infty}^{\infty} f(x)dx = \lim_{a \to -\infty} \lim_{b \to \infty} \int_a^b f(x)dx.$$

That all our integrands are absolutely integrable assures us that the order in which limits are taken is irrelevant.

Theorem 4.7.16. *Let T be the set of $x \epsilon \mathbb{R}$ such that $h(x) > 0$. Let $Y_n(X)$ be a sequence of random variables converging to $Y(x)$ in the sense that*

$$Y_n(x) \to Y(x) \text{ for all } x \epsilon T.$$

Additionally, suppose there is a random variable $g(x)$ such that

$$| Y_n(x) | \leq g(x)$$

and

$$\int_{\mathbb{R}} g(x)h(x)dx < \infty.$$

Then

(a) the sequence $E(Y_n)$ satisfies the Cauchy criterion and

(b) if $E(Y)$ exists, then $E(Y) = \lim_{n\to\infty} E(Y_n)$.

Proof. The only aspect of this result not included in Theorem 4.7.11 is the extension of the integrals to an infinite range. We address that issue as follows:

Let $\epsilon > 0$ be given. Necessarily, $g(x) \geq 0$ and $h(x) \geq 0$. By assumption $\int_{-\infty}^{\infty} g(x)h(x) < \infty$. Then there is an a such that

$$\int_{-\infty}^{a} g(x)h(x) < \epsilon/6. \tag{4.31}$$

Also there is a b such that

$$\int_{b}^{\infty} g(x)h(x) < \epsilon/6. \tag{4.32}$$

On the interval $[a,b], g(x)h(x)$ satisfies the conditions of Theorem 4.7.11, so there is an N such that

$$|\int_a^b y_n(x)h(x)dx - \int_a^b y_m(x)h(x)dx| < \epsilon/3 \tag{4.33}$$

for all n and m satisfying $n, m \geq N$. Then

$$\begin{aligned}
&|\int_{-\infty}^{\infty} y_n(x)h(x)dx - \int_{-\infty}^{\infty} y_m(x)h(x)dx| \\
&\leq |\int_{-\infty}^{a} (y_n(x) - y_m(x))h(x)dx| + |\int_a^b y_n(x)h(x)dx - \int_a^b y_m(x)h(x)dx| \\
&\quad + |\int_b^{\infty} (y_n(x) - y_m(x)h(x)dx| \\
&\leq \int_{-\infty}^{a} 2g(x)h(x)dx + \epsilon/3 + \int_b^{\infty} 2g(x)h(x)dx \\
&\leq 2(\epsilon/6) + \epsilon/3 + 2(\epsilon/6) = \epsilon.
\end{aligned}$$

This proves part (a).

The proof of part (b) is the same, substituting $y(x)$ for $y_m(x)$ throughout. □

Example 1 of section 4.7 applies to expectations, where $[a,b] = [0,1]$ and $h(x) = I_{[0,1]}(x)$.

The result of this analysis is that, under the assumptions made, part (a) implies that the sequence $E[Y_n]$ has a limit. However, Example 1 shows that the limiting random variable Y is not necessarily integrable in the Riemann sense. However, when it is Riemann integrable, then part (b) shows that

$$\lim_{n\to\infty} E(Y_n) = E(\lim_{n\to\infty} Y_n),$$

which is the goal. Thus it may fairly be concluded that the barrier to achieving the goal lies in a weakness in the Riemann sense of integration. Hence in section 4.9 I seek a more general integral, one that coincides with Riemann integration when it is defined, but that allows other functions to be integrated.

I am now in a position to address the sense in which Riemann probabilities are countably additive. I distinguish between two senses of countable additivity, as follows:

Weak Countable Additivity:

If $A_1, \ldots$ are disjoint events such that $P\{A_i\}$ is defined, and if $P\{\cup_{i=1}^{\infty} A_i\}$ is defined, then

$$\sum_{i=1}^{\infty} P\{A_i\} = P\{\cup_{i=1}^{\infty} A_i\}.$$

Strong Countable Additivity:

If $A_1, \ldots$ are disjoint events such that $P\{A_i\}$ is defined, then $P\{\cup_{i=1}^{\infty} A_i\}$ is defined and

$$\sum_{i=1}^{\infty} P\{A_i\} = P\{\cup_{i=1}^{\infty} A_i\}.$$

The distinction between weak and strong countable additivity lies in whether $\cup_{i=1}^{\infty} A_i$ has a defined probability. Riemann probabilities are not strongly countably additive, as the following example shows:

Example 2, a continuation of Example 1: We start with a special case, and then show that the construction is general. Consider the uniform density on $(0, 1)$, so $f(x) = 1$ if $0 < x < 1$ and $f(x) = 0$ otherwise. Consider the (countable) set Q of rational numbers. Let A_i be the set consisting of the ith rational number (in any order you like). Then $\int I_{A_i} f(x)dx$ exists and equals 0. Now $Q = \cup_{i=1}^{\infty} A_i$, but $I_Q(x)f(x)$ is a function that is 1 on each rational number $x, 0 < x < 1$, and zero otherwise. It is not Riemann integrable. Hence strong countable additivity fails.

Now suppose $f(x)$ is an arbitrary density satisfying $f(x) \geq 0$ and $\int_{-\infty}^{\infty} f(x) = 1$. Let $F(x) = \int_{-\infty}^{x} f(x)dx$ be the cumulative distribution function. Then F is differentiable with derivative $f(x)$, non-decreasing, and satisfies $F(-\infty) = 0$ and $F(\infty) = 1$.

Let $A_i = \{x \mid F(x) = q_i\}$. Then $P\{A_i\} = 0$ (and exists). However, consider the set $A = \cup_{i=1}^{\infty} A_i = \{x \mid F(x) \epsilon Q\}$, so $F^{-1}(A) = Q$. Suppose, contrary to the hypothesis, that A is integrable, so that $\int_{-\infty}^{\infty} I_A(x)f(x)dx$ exists. Consider the transformation $y = F(x)$, whose differential is $dy = f(x)dx$. Then $\int_{-\infty}^{\infty} I_A(x)f(x)dx = \int_0^1 I_{F^{-1}(y)}(y)dy = \int_0^1 I_Q(y)dy$. Since the latter integral does not exist in the Riemann sense, A is not integrable with respect to the density $f(x)$. Hence the Riemann probabilities defined by the density $f(x)$ are not strongly countably additive. □

Thus the most that we can hope for Riemann probabilities is weak countable additivity.

Theorem 4.7.17. *Let $f(x)$ be a density function, and let $A_1, \ldots,$ be a countable sequence of disjoint sets whose Riemann probability is defined. If $\cup_{i=1}^{\infty} A_i$ has a Riemann probability, then*

$$P\{\cup_{i=1}^{\infty} A_i\} = \sum_{i=1}^{\infty} P\{A_i\}.$$

Proof. Consider the random variables

$$Y_n(x) = \sum_{i=1}^{n} I_{A_i}(x).$$

We know that $Y_n(x)$ converges point-wise to the random variable

$$Y(x) = \sum_{i=1}^{\infty} I_{A_i}(x) = I_{\cup_{i=1}^{\infty} A_i}(x).$$

Also $| Y(x) | \leq 1$, which satisfies

$$\int_{\mathbb{R}} 1 f(x)dx = 1 < \infty.$$

Therefore Theorem 4.7.16 applies. Since we have assumed that $\cup_{i=1}^{\infty} A_i$ has a Riemann probability, it satisfies

$$\begin{aligned} P\{\cup_{i=1}^{\infty} A_i\} = EY = \lim_{n\to\infty} E(Y_n) &= \lim_{n\to\infty} E \sum_{i=1}^{n} I_{A_i}(x) = \\ \lim_{n\to\infty} (\sum_{i=1}^{n} P\{A_i\}) &= \sum_{i=1}^{\infty} P\{A_i\}. \end{aligned}$$

□

Theorem 4.7.17 shows that Riemann probabilities are weakly countably additive.

Finally, we postponed the proof of the following result; which is property 4 from section 4.4.

Theorem 4.7.18. *Let X be non-trivial and have expectation c. Then there is some positive probability $\epsilon > 0$ that X exceeds c by a fixed amount $\eta > 0$, and positive probability $\epsilon > 0$ that c exceeds X by a fixed amount $\eta > 0$.*

Proof. Let $A_i = \{x \mid \frac{1}{i} > x - c \geq \frac{1}{i+1}\}, i = 0, 1, \ldots, \infty$ where $\frac{1}{0}$ is taken to be infinity. The A_i's are disjoint and $\cup_{i=1}^{\infty} A_i = \{x - c > 0\}$. Similarly let $B_j = \{\frac{1}{j} > c - x \geq \frac{1}{j+1}\}, j = 0, \ldots, \infty$, so the B_j's are disjoint and

$$\cup_{i=1}^{\infty} B_j = \{c - x > 0\}.$$

Since X is non-trivial, $P\{X \neq c\} > 0$. All three sets, $\{x \mid x > c\}$, $\{x \mid x < c\}$ and $\{x \mid x \neq c\}$ have Riemann probabilities. Hence by weak countable additivity their probabilities are respectively the sum of the probabilities of countable disjoint sets $\{A_1, \ldots\}$, $\{B_1, \ldots\}$ and $\{A_1, B_1, \ldots\}$. But

$$\begin{aligned} 0 < P\{X \neq c\} &= P\{X > c\} + P\{X < c\} \\ &= \sum_{i=0}^{\infty} P\{A_i\} + \sum_{j=0}^{\infty} P\{B_j\}. \end{aligned}$$

By exactly the same argument as in section 3.4, there is both an i and a j such that $P\{A_i\} > 0$ and $P\{B_j\} > 0$. Then taking

$$\epsilon = \min(P\{A_i\}, P\{B_j\}) > 0 \text{ and } \eta = \min\{\frac{1}{i+1}, \frac{1}{j+1}\} \text{ suffices.}$$

□

4.7.13 Summary

Theorem 4.7.16 gives a dominated convergence theorem for Riemann probabilities. Theorem 4.7.17 uses this result to show that Riemann probabilities are weakly countably additive, while Example 2 shows that they are not strongly countably additive.

4.7.14 Exercises

1. Vocabulary. Explain in your own words:
   (a) Riemann probability
   (b) Riemann expectation
   (c) Weak countable additivity
   (d) Strong countable additivity
2. In section 3.9 the following example is given:
   Let X_n take the value n with probability $1/n$, and otherwise take the value 0. Then $E(X_n) = 1$ for all n. However $\lim_{n\to\infty} P\{X_n = 0\} = 1$, so the limiting distribution puts all its mass at 0, and has mean 0.
   (a) Does this example contradict the dominated convergence theorem? Explain your reasoning.
   (b) Let Y_n take the value $\sqrt{n}$ with probability $1/n$, and otherwise take the value 0. Answer the same question.
3. Example 1 after Corollary 4.7.9 displays a sequence of functions $f_n(x)$ that converge to a limiting function $f(x)$.
   (a) Use the definition of uniform convergence to examine whether this convergence is uniform.
   (b) If this convergence were uniform, what consequence would it have for the integration of the limiting function f? Why?

4.7.15 Discussion

Riemann probabilities are a convenient way to specify an uncountable number of probabilities simultaneously, by specifying a density. The results of this chapter so far show that the probabilities thus specified are coherent, weakly but not strongly countably additive, and satisfy a dominated convergence theorem, but not the strongest version of a dominated convergence theorem. There is nothing wrong with such a specification, because it is coherent and therefore avoids sure loss. However, it suggests that you could say just a bit more by accepting the same density with respect to a stronger sense of integral than Riemann's. This would mean that you are declaring bets on more sets, which you may or may not be comfortable doing. But the reward for doing so is that stronger mathematical results become available. Section 4.8 introduces the Riemann-Stieltjes integral, which unifies the material on expectations found in Chapters 1, 3 and earlier in Chapter 4. In turn, the Riemann-Stieltjes integral forms a basis for understanding the McShane-Stieltjes integral, the subject of section 4.9.

4.8 A first generalization of the Riemann integral: The Riemann-Stieltjes integral

When two mathematical systems have similar or identical properties, there is usually a reason for it. Indeed, much of modern mathematics can be understood as finding generalizations that explain such apparent coincidences. In our case, we have expectations in Chapter 1 defined on finite discrete probabilities, extended in Chapter 3 to discrete probabilities on countable sets and separately in this chapter to continuous probabilities. The properties of these expectations found in sections 1.6, 3.4 and 4.4 are virtually identical. Indeed the only notable distinction comes in the countable case discussed in Chapter 3, where we find that we must have the condition that the sum of the absolute values must be finite in order to avoid having the sum depend on the order of addition. There should be a reason, a generalization, that explains why the discrete and continuous cases are so similar. Explaining that generalization is the purpose of this section.

4.8.1 Definition of the Riemann-Stieltjes integral

Recall from 4.7.1 that the Riemann integral is defined as follows: a number A is the Riemann integral of g on $[a, b]$ if for every $\epsilon > 0$ there is a $\delta > 0$ such that, for every δ-fine partition π,

$$\Big|\sum_{\pi} g - A\Big| < \epsilon \tag{4.34}$$

where

$$\sum_{\pi} g = \sum_{i=1}^{n} g(\xi_i)(\nu_i - u_i), \tag{4.35}$$

and where the partition $\pi = (\xi_i, [u_i, \nu_i], i = 1, \ldots, n)$ satisfies

$$\xi_i - \delta < u_i \leq \xi_i \leq \nu_i < \xi_i + \delta, \tag{4.36}$$

the condition for π to be δ-fine.

Suppose $\alpha(x)$ is a non-decreasing function on $[a, b]$. Then the *Riemann-Stieltjes integral of g with respect to α* satisfies (4.34), where (4.35) is modified to read

$$\sum_{\pi,\alpha} g = \sum_{i=1}^{n} g(\xi_i)(\alpha(\nu_i) - \alpha(u_i)). \tag{4.37}$$

Thus the Riemann integral is the special case of the Riemann-Stieltjes integral, where $\alpha(x) = x$. Intuitively, the function α allows the integral to put extra emphasis on some parts of the interval $[a, b]$, and less on others.

The definition of the Riemann-Stieltjes integral can also apply to functions α that are non-increasing, and to functions that are the difference of two functions, one non-increasing and the other non-decreasing. Such functions are called functions of bounded variation (see Jeffreys and Jeffreys (1950), pp. 24-25). This book will use Riemann-Stieltjes integration with respect to cumulative distribution functions, which are non-decreasing.

The Riemann-Stieltjes integral of g with respect to α is written

$$\int_a^b g(x)d\alpha(x). \tag{4.38}$$

Conditions for the existence of the Riemann-Stieltjes integral are given by Dresher (1981) and Jeffreys and Jeffreys (1950). The leading case when it does not exist is when $g(x)$ and $\alpha(x)$ have a common point of discontinuity. For example, let $a = 0, b = 1$ and suppose

$$\begin{aligned} g(x) &= \alpha(x) = 0 \text{ for } 0 \leq x < 1/2 \\ g(x) &= \alpha(x) = 1 \text{ for } 1/2 \leq x \leq 1. \end{aligned} \tag{4.39}$$

In every partition π there will be one index i for which $\alpha(x_i) - \alpha(x_{i-1}) = 1$, while the rest are zero. Then $g(\xi_i) = 0$ or 1 depending on whether $\xi_i < 1/2$ or $\xi_i \geq 1/2$. Thus the value of (4.35) depends on π, so the integral does not exist.

4.8.2 The Riemann-Stieltjes integral in the finite discrete case

We start with the integral with respect to an indicator function. Thus suppose

$$\alpha(x) = \begin{cases} 1 & x \geq c \\ 0 & x < c \end{cases} = I_{\{x \geq c\}}(x) \tag{4.40}$$

and that $g(x)$ is continuous at c. I now show that

$$\int_a^b g(x)d\alpha(x) = g(c), \tag{4.41}$$

where $a \leq c \leq b$.

Proof. Suppose that $\pi = (\xi_i, [u_i, \nu_i], i = 1, \ldots, n)$. There is one value of the index, say $i = j$, where

$$\alpha(\nu_j) - \alpha(u_j) = 1,$$

while $\alpha(\nu_i) - \alpha(u_i) = 0$ for $i \neq j$.

Put another way, $\alpha(v_i) - \alpha(u_j) = \begin{cases} 1 & \text{if } i = j \\ 0 & \text{if } i \neq j \end{cases}.$

Hence

$$\sum_\pi g = \sum_{i=1}^n g(\xi_i)[\alpha(\nu_i) - \alpha(u_i)] = g(\xi_j). \tag{4.42}$$

Because of the continuity of g at c, it follows that

$$\lim_{\substack{n\to\infty \\ \delta\to 0}} g(\xi_j) = g(c).$$

Hence

$$\lim_{n\to\infty} \sum_{i=1}^n g(\xi_i)[\alpha(\nu_i) - \alpha(u_i)] = g(c) \tag{4.43}$$

for all δ-fine partitions π, so

$$\int_a^b g d\alpha = g(c).$$

□

The expression in (4.40) is the cumulative distribution function of a random variable that puts probability 1 at $x = c$.

Now suppose that $\alpha_1(\cdot)$ and $\alpha_2(\cdot)$ are two non-decreasing functions. Then if $g(\cdot)$ has a Riemann-Stieltjes integral with respect to each, with respective values A_1 and A_2, then it has a Riemann-Stieltjes integral A with respect to $\alpha_1(\cdot) + \alpha_2(\cdot)$, and $A = A_1 + A_2$.

The proof of this follows essentially from the fact that (4.37) can be written in this case as

$$\begin{aligned}\sum_\pi g &= \sum_{i=1}^n g(\xi_i)[\alpha_1(\nu_i) + \alpha_2(\nu_i) - \alpha_1(u_i) - \alpha_2(u_i)] \\ &= \sum_{i=1}^n g(\xi_i)[\alpha_1(\nu_i) - \alpha_1(u_i)] + \sum_{i=1}^n g(\xi_i)[\alpha_2(\nu_i) - \alpha_2(u_i)]. \end{aligned} \tag{4.44}$$

By induction, if $\alpha_1(\cdot), \ldots, \alpha_n(\cdot)$ are non-decreasing functions, and g has a Riemann-Stieltjes integral with respect to each, with respective values $A_1, \ldots, A_n$, then g has a Riemann-Stieltjes integral A with respect to $\sum_{i=1}^n \alpha_i$, and its value is $A = \sum_{i=1}^n A_i$.

Similarly, if k is a constant, and if g has Riemann-Stieltjes integral A with respect to

α, then it has Riemann-Stieltjes integral kA with respect to $k\alpha$. This follows again from (4.37), because in this case

$$\sum_{\pi} g = \sum_{i=1}^{n} g(\xi_i)[k\alpha(\nu_i) - k\alpha(u_i)]$$
$$= k \sum_{i=1}^{n} g(\xi_i)[\alpha(v_i) - \alpha(u_i)]. \tag{4.45}$$

Now consider a random variable X that takes a finite number of values $x_1, \ldots, x_n$, where

$$P\{X = x_i\} = p_i$$

and $\sum_{i=1}^{n} p_i = 1$. Let $F_X(x)$ be the cdf of X.

Then I claim

$$P\{X \leq x\} = F_X(x) = \sum_{i=1}^{n} p_i I_{\{x \geq x_i\}}(x), \tag{4.46}$$

since the summation is over all p_i's for which $x \geq x_i$.

Now using (4.44) (4.45) and (4.46), we have

$$\int x dF_X(x) = \int x d\Big(\sum_{i=1}^{n} p_i I_{\{x \geq x_i\}}(x)\Big)$$
$$= \sum_{i=1}^{n} p_i \int x dI_{\{x \geq x_i\}}(x)$$
$$= \sum_{i=1}^{n} p_i x_i = E(X). \tag{4.47}$$

Hence the Riemann-Stieltjes integral with respect to the cdf is the expectation for discrete random variables, such as those of Chapter 1.

4.8.3 The Riemann-Stieltjes integral in the countable discrete case

This subsection considers the case in which X is a random variable with values $x_1, x_2, \ldots$ such that

$$P\{X = x_i\} = p_i \quad i = 1, \ldots \tag{4.48}$$

and

$$\sum_{i=1}^{\infty} p_i = 1. \tag{4.49}$$

Again, the first goal is to show

$$\int_a^b x dF(x) = \sum_{i=1}^{\infty} x_i p_i \tag{4.50}$$

when

$$a \leq x_i \leq b \text{ for all } i. \tag{4.51}$$

Toward this end, the following simple fact is useful: if $h(x) \leq g(x)$ for all x, and both h and g have Riemann-Stieltjes integrals with respect to α, then

$$\int h(x) d\alpha(x) \leq \int g(x) d\alpha(x). \tag{4.52}$$

The demonstration of this fact again relies on the same fact for the sums: for all partitions π,

$$\sum_{\pi} h = \sum_{i=1}^{n} h(\xi_i)(\alpha(\nu_i) - \alpha(u_i)) \leq \sum_{i=1}^{n} g(\xi_i)(\alpha(\nu_i) - \alpha(u_i)) = \sum_{\pi} g. \qquad (4.53)$$

Now we have the result.

Theorem 4.8.1. *Assume* (4.48) *and* (4.49), *and* (4.52). *Then* (4.50) *holds.*

Proof. Let $\epsilon > 0$ be given. Then there exists an n such that $\sum_{i=n+1}^{\infty} p_i < \epsilon/2K$ where $K = \max\{|a|, |b|\}$. Then, letting $F_n(x) = \sum_{i=1}^{n} p_i I_{\{x \geq x_i\}}(x)$, we have

$$\begin{aligned} \Big| \int_a^b x dF(x) - \sum_{i=1}^{\infty} p_i x_i \Big| \leq & \\ & \Big| \int_a^b x dF(x) - \int_a^b x dF_n(x) \Big| + \Big| \int_a^b x dF_n(x) - \sum_{i=1}^{n} p_i x_i \Big| \\ & + \Big| \sum_{i=1}^{n} p_i x_i - \sum_{i=1}^{\infty} p_i x_i \Big| . \end{aligned} \qquad (4.54)$$

I now address each of these terms in turn. The first term admits the following approximation:

$$\begin{aligned} \Big| \int_a^b x dF(x) - \int_a^b x dF_n(x) \Big| &= \Big| \int_a^b x d(F(x) - F_n(x)) \Big| \\ &\leq \int_a^b |x| d(F(x) - F_n(x)) \leq K\epsilon/2K = \epsilon/2 \end{aligned} \qquad (4.55)$$

since $x \leq |x| \leq K$ and $F(x) - F_n(x)$ has rise $\sum_{i=n+1}^{\infty} p_i < \epsilon/2K$.

The second term requires division because $\sum_{i=1}^{n} p_i < 1$:

$$\Big| \int_a^b x dF_n(x) - \sum_{i=1}^{n} p_i x_i \Big| \leq \Big| \int_a^b \frac{x dF_n(x)}{\sum_{i=1}^{n} p_i} - \frac{\sum_{i=1}^{n} p_i x_i}{\sum_{i=1}^{n} p_i} \Big| = 0 \qquad (4.56)$$

by (4.47) and the fact that $\frac{F_n(x)}{\sum_{i=1}^{n} p_i}$ is a cumulative density function. Finally the third term,

$$\Big| \sum p_i x_i - \sum_{i=1}^{\infty} p_i x_i \Big| = \Big| \sum_{i=n+1}^{\infty} p_i x_i \Big| < K\epsilon/2K = \epsilon/2. \qquad (4.57)$$

Therefore, putting together (4.54), (4.55), (4.56) and (4.57),

$$\Big| \int_a^b x dF(x) - \sum_{i=1}^{\infty} p_i x_i \Big| < \epsilon/2 + 0 + \epsilon/2 = \epsilon. \qquad (4.58)$$

Since $\epsilon > 0$ is arbitrary, we have

$$\int_a^b x dF(x) = \sum_{i=1}^{\infty} p_i x_i \qquad (4.59)$$

□

It is noteworthy that in the above discussion, the condition $\sum |x_i|p_i < \infty$ did not occur. Because of the condition $a \leq x_i \leq b$, we have $|x_i| \leq K$, where $K = \max\{|a|, |b|\} < \infty$ and therefore $\sum |x_i|p_i \leq \sum Kp_i = K \sum p_i = K < \infty$. Thus we automatically have the condition in question. That we cannot casually let $K \to \infty$ is hinted at by the observation that in the proof of Theorem 4.8.1, we choose n so that $\sum_{i=n+1}^{\infty} p_i < \epsilon/2K$. This division by K is only a hint, however, as there is no reason to deny that some other proof of Theorem 4.8.1 might be found that does not require division by K.

So now we wish to explore what happens if $a \to -\infty$ and $b \to \infty$, to see under what circumstances we can write

$$\int_{-\infty}^{\infty} xdF(x) = \sum_{i=1}^{\infty} p_i x_i. \tag{4.60}$$

Since (4.60) does not involve a and b, it makes sense to write (4.60) only when the order in which $a \to -\infty$ and $b \to \infty$ doesn't matter. To examine this, let

$$x_i^*(a, b) = \text{ median } \{a, x_i, b\}. \tag{4.61}$$

The median of three numbers is the middle number. Since $b > a, x_i^*(a, b) = x_i$ if $a \leq x_i \leq b, x_i^*(a, b) = a$ if $x_i < a$, and $x_i^*(a, b) = b$ if $x_i > b$. Thus $x_i^*(a, b)$ truncates x_i to live in the interval $[a, b]$. Also let $F^*(a, b)$ be the cdf of the numbers $x_i^*(a, b)$. Then we may use Theorem 4.8.1 to write, for each finite a and b such that $b > a$.

$$\int_a^b xdF^*_{(a,b)}(x) = \sum_{i=1}^{\infty} p_i x_i^*(a, b). \tag{4.62}$$

Now consider the consequence if we hold a fixed, say $a = 0$, and allow b to get arbitrarily large. Then the right-hand side of (4.62) approaches s^+, the sum of the positive terms in the right-hand side of (4.62). Similarly if $b = 0$ and $a \to -\infty$, the right-hand side of (4.62) approaches s^-, the sum of the negative terms in the right-hand side of (4.62). The limiting value is finite and independent of the order of these two operations if and only if both s^+ and s^- are finite. But this is exactly the condition that

$$\sum_{i=1}^{\infty} p_i |x_i| < \infty. \tag{4.63}$$

Thus we write (4.60) only where (4.63) holds.

Consequently the Riemann-Stieltjes integral has as a special case, the material of Chapter 3 concerning expectations of discrete random variables that take a countable number of possible values.

4.8.4 The Riemann-Stieltjes integral when F has a derivative

This subsection considers the case introduced in section 4.1 in which the cdf $F(x)$ has a derivative $f(x)$ (called the density function) so that

$$F(x) = \int_{-\infty}^{x} f(y)dy \tag{4.64}$$

and

$$F'(x) = f(x), \tag{4.65}$$

where the integral in (4.64) is understood in the Riemann sense.

We wish to show first that in this case,

$$\int_a^b xdF(x) = \int_a^b xf(x)dx \tag{4.66}$$

providing both integrals exist.

Let $[u_i, \nu_i], i = 1, \ldots, n$ be a set of closed intervals, not overlapping except at the end-points, whose union is $[a, b]$. For each i, by the mean-value theorem there is a point $\xi_i \epsilon [u_i, \nu_i]$ such that

$$F(\nu_i) - F(u_i) = F'(\xi_i)(\nu_i - u_i) = f(\xi_i)(\nu_i - u_i). \tag{4.67}$$

We now consider the partition $\pi = (\xi_i, [u_i, \nu_i])$. Now

$$\sum_{\pi,F} x = \sum_{i=1}^{n} \xi_i(F(\nu_i) - F(u_i)) = \sum_{i=1}^{n} \xi_i f(\xi_i)(\nu_i - u_i) = \sum_{\pi} xf. \tag{4.68}$$

Thus (4.66) holds in the Riemann sense for all δ-fine partitions π if and only if it holds in the Riemann-Stieltjes sense on xf for all δ-fine partitions π.

We now consider the extension to the whole real line, letting $a \to -\infty$ and $b \to \infty$. Once again we seek a condition so that the result does not depend on the order in which these limits are approached. Again, we consider the uncertain quantity (also known as a random variable)

$$X^*(a, b) = \text{ median } \{a, X, b\} \tag{4.69}$$

and let $F^*_{a,b}$ be the cdf of X^*. Then for each value of a and b, we have, applying (4.66),

$$\int_a^b xdF^*_{a,b}(x) = \int_a^b xf(x)dx + aP\{x < a\} + bP\{x > b\}. \tag{4.70}$$

Again holding $a = 0$ and letting $b \to \infty$, the limit is

$$I^+ = \int_0^\infty xdF^*_{0,\infty}(x), \tag{4.71}$$

while holding $b = 0$ and letting $a \to -\infty$, the limit is

$$I^- = \int_{-\infty}^0 xdF^*_{-\infty,0}(x). \tag{4.72}$$

Then $\int_{-\infty}^\infty xdF(x)$ exists independent of the order in which $a \to -\infty$ and $b \to \infty$ when and only when both I^+ and I^- are finite, so when

$$\int |x|f(x)dx < \infty.$$

Hence the Riemann-Stieltjes theory finds the same condition for the existence of an expectation as was found in section 4.4.

4.8.5 Other cases of the Riemann-Stieltjes integral

The Riemann-Stieltjes integral is not limited to the discrete and absolutely continuous cases. To give one example, consider a person's probability p of the outcome of the flip of a coin. This person puts probability 1/2 on the coin being fair (*i.e.*, $p = 1/2$) and probability 1/2 on a uniform distribution on $[0, 1]$ for p. Thus this distribution is a $1/2 - 1/2$ mixture of a discrete distribution and a continuous one. The cdf of these two parts are respectively an indicator function for $p = 1/2$, and the function $F(p) = p$ $(0 \leq p \leq 1)$. The cdf for the mixture is the convex combination of these with weights $1/2$ each, and therefore equals

$$\frac{1}{2}I_{\{p=1/2\}}(p) + \frac{1}{2}p. \tag{4.73}$$

The Riemann-Stieltjes integral gracefully handles expectations, with respect to this cdf, of functions not having a discontinuity at $p = 1/2$.

A second kind of example of Riemann-Stieltjes integrals that are neither discrete nor continuous are expectations with respect to cdf's that are continuous but not differentiable. The most famous of these is an example due to Cantor. While it is good mathematical fun, it is not essential to the story of this book, and therefore will not be further discussed here.

The next section introduces a generalization of the Riemann-Stieltjes integral and establishes the (now usual) properties of expectation for the generalization. Since each Riemann-Stieltjes uncertain quantity (random variable) has a McShane-Stieltjes expectation, it is not necessary to establish them for Riemann-Stieltjes expectations.

4.8.6 Summary

The Riemann-Stieltjes integral unites the discrete uncertain quantities (random variables) of Chapters 1 and 3 with the Riemann continuous case discussed in the first part of this chapter.

4.8.7 Exercises

1. (a) Vocabulary. Explain in your own words what the Riemann-Stieltjes integral is.
   (b) Why is it useful to think about?
2. Consider the following distribution for the uncertain quantity P, that indicates my probability that a flipped coin will come up heads. With probability $2/3$, I believe that the coin is fair, $(P = 1/2)$. With probability $1/3$, I believe that P is drawn from the density $3p^2, 0 < p < 1$.
   (a) Find the cdf F of P. Is F non-decreasing?
   (b) Use the Riemann-Stieltjes integral to find

   $$\int_0^1 pdF(p) \text{ and } \int_0^1 p^2 dF(p).$$

   (c) Use the results of (b) to find Var (P).

4.9 A second generalization: The McShane-Stieltjes integral

The material presented in section 4.7 makes it clear that to have a strong dominated convergence theorem and probabilities that are strongly countably additive a stronger integral than Riemann's might be convenient. This section introduces such an integral, the McShane-Stieltjes Integral. It is a mild generalization, having the following properties:

(i) A Riemann-Stieltjes integrable function is McShane-Stieltjes integrable, and the integrals are equal.

(ii) McShane-Stieltjes probabilities are strongly countably additive.

(iii) McShane-Stieltjes expectations satisfy a strong dominated (and bounded) convergence theorem: the limiting function is always McShane-Stieltjes-integrable. (For those readers familiar with abstract integration theory, it turns out that the McShane-Stieltjes integral is the Lebesgue integral on the real line. For those readers to whom the last sentence is meaningless or frightening, don't let it bother you).

For short, we'll call the McShane-Stieltjes integral the McShane integral, as does most of the literature. The basic idea of the McShane integral is surprisingly similar to that of the Riemann integral. The only change is to replace the positive number δ with a positive function $\delta(x)$, or, to put it a different way, to replace Riemann's uniformly-fine δ with McShane's locally-fine $\delta(x)$. To see why this might be a good idea, consider the following integral:

$$\int_{0.002}^{0.2} \left(\frac{1}{x}\right) sin \left(\frac{1}{x}\right) dx. \tag{4.74}$$

As illustrated in Figure 4.2, the integrand swings more and more widely as $x \to 0.002$. Indeed Figure 4.2 is a ragged mess close to the origin. This happens because the 100 equally spaced points used to make Figure 4.2 are sparse (relative to the amount of fluctuation in $(1/x)sin(1/x)$) for small x, and thick (relative to the amount of fluctuation) for large x.

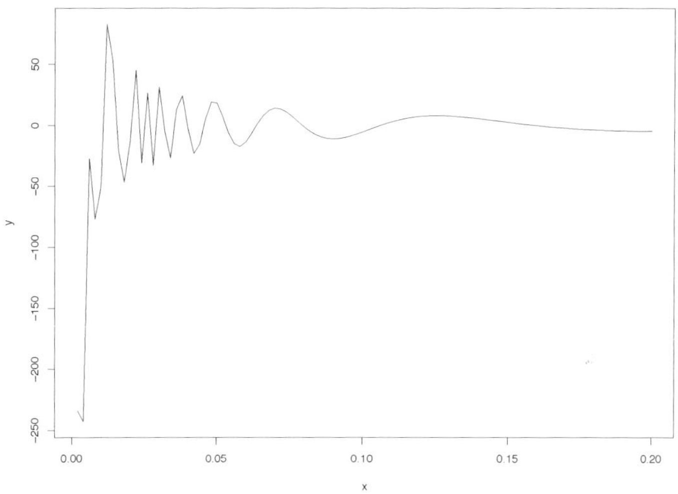

Figure 4.2: Plot of $y = (1/x)sin(1/x)$ with uniform spacing.

```
Commands:
x=(1:100)/500
y=(1/x) * sin (1/x)
plot(x,y,type="l")
```

To remedy this, it makes sense to evaluate the function at points that are bunched closer to the origin, which is to the left in Figure 4.2. For comparison, suppose I replot the function with points proportional to $1/x$, in Figure 4.3. This figure is plotted with the same number of points over the same domain, ($[0.002, 0.2]$), as Figure 4.2, but reveals much more of the structure of the function. To appreciate how different Figures 4.2 and 4.3 are, compare their vertical axes.

Finding an integral of a function is much like plotting the function. In both cases, the function is evaluated at a set of points. When the function is plotted, those points are connected (by straight lines). When the integral is evaluated, a point in the interval between points is taken as representative, and the integral is approximated by the area (in the one-dimensional case) found by multiplying the value of the function at the point by the length of the interval. Both methods rely for accuracy on the relative constancy of the function over the interval.

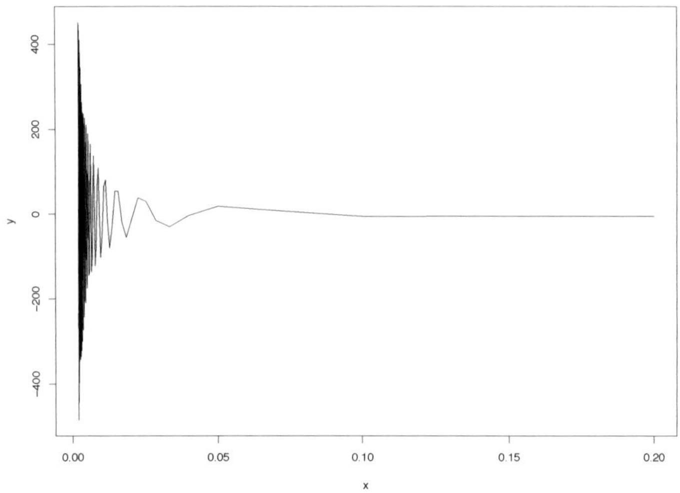

Figure 4.3: Plot of $y = (1/x)sin(1/x)$ with non-uniform spacing.

```
Commands:
x=(0.2)/(1:100)
y=(1/x) *  sin (1/x)
plot(x,y,type="l")
```

This is a heuristic argument intended to suggest that allowing locally-fine $\delta(x)$ may be a good idea. Because the function $y = (1/x)sin(1/x)$ is continuous on the bounded interval $[0.002, 0.2]$ it is Riemann integrable, and therefore this example does not settle the question of whether using the McShane locally-fine $\delta(x)$ allows one to integrate functions that are not Riemann integrable. Such an example is coming, just after the formal introduction of the McShane integral. Since the approach here is rigorous, I will define several terms before defining the McShane integral itself.

Recall from section 4.7.1 that a **cell** is a closed interval $[a, b]$ such that $a < b$, so the interior (a, b) is not empty. A collection of cells is **non-overlapping** if their interiors are disjoint. If $[a, b]$ is a cell, $\lambda([a, b]) = b - a > 0$ is the **length** of the cell $[a, b]$. More generally, if α is a non-decreasing function on the cell $A = [a, b]$, then $\alpha(A) = \alpha(b) - \alpha(a) \geq 0$.

A **partition** of a cell A is a collection $\pi = \{(A_1, x_1), \ldots, (A_p, x_p)\}$ where $A_1, \ldots, A_p$ are non-overlapping cells whose union is A, and $x_1, \ldots, x_p$ are points in $\mathbb{R}$ (the real numbers). The point x_i is called the **evaluation point** of cell A_i. Let δ be a positive function defined on a set $E \subset \mathbb{R}$. A partition $\{(A_1, x_1), \ldots, (A_p, x_p)\}$, with $x_i \epsilon E$ for all $i = 1, \ldots, p$, is called **δ-fine** if

$$A_i \subset (x_i - \delta(x_i), x_i + \delta(x_i)) \text{ for all } i = 1, \ldots, p. \tag{4.75}$$

When (4.75) holds for some i, A_i is said to be within a $\delta(x_i)$-neighborhood of x_i.

This is where the distinction between a Riemann and a McShane integral comes in.

In the Riemann case, a δ-fine partition is defined for a real number $\delta > 0$, while in the McShane case, a δ-partition is defined for a positive function $\delta(x) > 0$. While seemingly a trivial distinction, this difference has important implications, as will now be explained.

First, the following lemma will be useful later:

Lemma 4.9.1. *Suppose $\delta(x)$ and $\delta'(x)$ are positive functions on $\mathbb{R}$ satisfying $\delta(x) \le \delta'(x)$. Then every δ-fine partition is δ'-fine.*

Proof. Suppose a partition $\{(A_1, x_1), \ldots, (A_p, x_p)\}$ is a δ-fine partition of A. Then, for all $i = 1, \ldots, p$

$$A_i \subset (x_i - \delta(x_i), x_i + \delta(x_i)) \subseteq (x_i - \delta'(x_i), x_i + \delta'(x_i)),$$

so $\{(A_1, x_1), \ldots, (A_p, x_p)\}$ is δ'-fine. □

Let $\pi = \{(A_1, x_1), \ldots, (A_p, x_p)\}$ be a partition and let A be a cell. If $\{x_1, \ldots, x_p\}$ and $\cup_{i=1}^{p} A_i$ are subsets of A, then π is a **partition in** $\boldsymbol{A}$. If in addition, $\cup_{i=1}^{p} A_i = A$, then π is a **partition of A**.

It is not obvious whether there always is a δ-fine partition of a cell. That there is, constitutes the following lemma:

Lemma 4.9.2. *(Cousin) For each positive function δ on a cell A, there is a δ-fine partition π of A.*

Proof. Let $A = [a, b]$ with $a < b$, and let $c \,\epsilon\, (a, b)$. If π_a and π_b are δ-fine partitions of the cells $[a, c]$ and $[c, b]$, respectively, then $\pi = \pi_a \cup \pi_b$ is a δ-fine partition of A.

Now assume the lemma is false. Then we can construct cells $A = A_0 \supset A_1 \supset \ldots$ such that for $n = 0, 1, \ldots,$ no δ-fine partition of A_n exists and $\lambda(A_n) = (b - a)/2^n$. Since the sequence $A_0, A_1, A_2 \ldots$ is a non-increasing sequence of non-empty closed intervals, the intersection of them is non-empty, using Lemma 4.7.6. Thus there is some number z such that

$$z \in \cap_{n=0}^{\infty} A_n, \text{ where } z \epsilon A.$$

Since $\delta(z) > 0$, there is an integer $k \ge 0$ such that $\lambda(A_k) < \delta(z)$. Then $\{(A_k, z)\}$ is a δ-fine partition of A_k, which is a contradiction. □

A partition $\{(A_1, x_1), \ldots, (A_p, x_p)\}$ is said to be *anchored* in a set $B \subset A$ if $x_i \epsilon B$, $i = 1, \ldots, p$.

Corollary 4.9.3. *For each positive function δ on a cell A, there is a δ-fine partition of π of A anchored in A.*

Proof. The proof is the same as that of Cousin's Lemma, with the additional observation that $\{(A_k, z)\}$ is anchored in A, because $z \epsilon A$. □

Corollary 4.9.4. *Let δ be a positive function on a cell A. Each δ-fine partition π in A is a subset of a δ-fine partition η of A.*

Proof. Let $\pi = \{(A_1, x_1), \ldots, (A_p, x_p)\}$ and let $B_1, \ldots, B_k$ be cells such that $\{A_1, \ldots, A_p, B_1, \ldots, B_k\}$ is a non-overlapping family whose union is A. By Cousin's Lemma, there are δ-fine partitions π_j of B_j, for $j = 1, \ldots, k$. Then $\eta = \pi_{\cup}(\cup_{j=1}^{k} \pi_j)$ is the desired δ-fine partition of A. □

I now define a Stieltjes sum, which is the fundamental quantity in the definition of the McShane integral. Let α be a non-decreasing function on a cell A, and let $\pi =$

$\{(A_1, x_1), \ldots, (A_p, x_p)\}$ be a partition in A. For any function f on $\{x_1, \ldots, x_p\}$, the **α-Stieltjes sum of A associated with f** is

$$\sigma(f, \pi; \alpha) = \sum_{i=1}^{p} f(x_i)\alpha(A_i). \tag{4.76}$$

Finally, I am in a position to give the definition of the McShane integral.

Let α be a non-decreasing function on a cell A. A function f on A is said to be **McShane integrable over A with respect to α** if there is a real number I such that: given $\epsilon > 0$, there is a positive function δ on A such that

$$\left|\sigma(f, \pi; \alpha) - I\right| < \epsilon \tag{4.77}$$

for each δ-fine partition π of A.

Before discussing the properties of this integral, we must assure ourselves that it is well defined, which means that the number I is uniquely defined in this way. Suppose that the number $J \neq I$ also satisfies the definition. Let $\epsilon = \left|I - J\right| /2 > 0$. From the definition of the McShane integral, there are positive functions δ_I and δ_J on A so that $\left|\sigma(f, \pi; \alpha) - I\right| < \epsilon$ for each δ_I-fine partition of A, and $\left|\sigma(f, \pi; \alpha) - J\right| < \epsilon$ for each δ_J-fine partition P of A. Let $\delta = \min\{\delta_I, \delta_J\}$, and apply Cousin's Lemma to find a δ-fine partition π of A. Then this partition π is both δ_I-fine and δ_J-fine using Lemma 4.9.1. Thus I may write

$$\left|I - J\right| \leq \left|I - \sigma(f, \pi; \alpha)\right| + \left|\sigma(f, \pi; \alpha) - J\right| < 2\epsilon = \left|I - J\right|,$$

a contradiction. □

Having assured ourselves that the McShane integral is well defined, we may now observe that it is a generalization of the Riemann integral because of the simple fact that a special case of a positive function $\delta(x)$ is the constant function δ. Therefore when a Riemann integral exists, the McShane integral exists and gives the same value, which is the first property of McShane integrals stated in the introduction to this section.

A bit of extra notation will be useful in what follows. Let $M(A, \alpha)$ be the family of all McShane-integrable functions over A with respect to α.

We now need to reassure ourselves that the McShane integral is in fact more general than the Riemann integral, as otherwise this whole development would lose its point. We already know about a function that is not Riemann integrable (and that kept coming up as the canonical counterexample in section 4.7), namely the Dirichlet function

$$f(x) = \begin{cases} 1 \text{ if } x \text{ is rational} \\ 0 \text{ if } x \text{ is irrational.} \end{cases} \tag{4.78}$$

I will now show that $f \epsilon M([0, 1], \lambda)$ and $\int f d\lambda = 0$.

Choose $\epsilon > 0$ and let $\{r_1, r_2, \ldots\}$ be an enumeration of the rational numbers in $[0, 1]$. Define the positive function δ on $[0, 1]$ as follows:

$$\delta(x) = \begin{cases} \epsilon 2^{-n-1} & \text{if } x = r_n \text{ and } n = 1, 2, \ldots \\ 1 & \text{if } x \text{ is irrational.} \end{cases} \tag{4.79}$$

Let $\pi = \{(A_1, x_1), \ldots, (A_p, x_p)\}$ be a δ-fine partition of $[0, 1]$, which we know exists by Cousin's Lemma. Suppose the points $x_{i_1}, x_{i_2}, \ldots, x_{i_k}$ are equal to r_n. Then

$$\cup_{j=1}^{k} A_{i_j} \subset (r_n - \delta(r_n), r_n + \delta(r_n)), \text{ so} \tag{4.80}$$

$$\sum_{j=1}^{k} f(x_{i_j})\lambda(A_{i_j}) \leq \sum_{j=1}^{k} \lambda(A_{i_j}) < \epsilon 2^{-n}. \tag{4.81}$$

Since $f(x) = 0$ when x is irrational, irrational evaluation points do not contribute to the Stieltjes sum. Therefore we have

$$0 \leq \sigma(f, \pi; \lambda) < \sum_{n=1}^{\infty} \epsilon 2^{-n} = \epsilon. \tag{4.82}$$

Therefore $\int_0^1 f d\lambda$ exists and equals 0. □

This example has two important implications. The first, already mentioned, is that it shows that the McShane integral is strictly more powerful than the Riemann integral. The second implication is that it opens the possibility that the McShane integral supports a strong dominated convergence theorem and strong countable additivity. It does, as is shown below, but it requires some effort to prove.

4.9.1 Extension of the McShane integral to unbounded sets

So far, the theory of the McShane integral as presented has been limited to cells $[a, b]$, where $a < b$ and both a and b are real numbers. However, for our purposes we need to define integrals over $(-\infty, \infty)$. One way to do this is to mimic what is done for Riemann integrals, namely to let

$$\int_{-\infty}^{\infty} f(x)dx = \lim_{a \to -\infty} \lim_{b \to \infty} \int_a^b f(x)dx,$$

provided that the limiting value does not depend on the order in which the integrals are taken. In principle, however, this extended Riemann integral is a new object, for which the properties of the Riemann integral on a bounded set would have to be reexamined. Perhaps some of its properties would hold and others not. In the case of the McShane integral, however, a second more elegant strategy is available. By extending the definitions to include $-\infty$ and ∞, the McShane integral can be defined so that it applies directly to unbounded sets such as $(-\infty, \infty), (-\infty, b], (-\infty, b), (a, \infty)$ and $[a, \infty)$. The purpose of this subsection is to show the steps in this extension.

To do this, we need to establish notation and conventions for handling ∞ and $-\infty$. First, let $\overline{\mathbb{R}} = \mathbb{R} \cup \{\infty\} \cup \{-\infty\}$. We have the ordering $-\infty < x < \infty$ for all $x \epsilon \mathbb{R}$. We also have some rules for extending arithmetic to $\mathbb{R}$:

$$\begin{aligned}
&\infty + x = x + \infty = \infty \text{ unless } x = -\infty \\
&-\infty + x = x + -\infty = -\infty \text{ unless } x = \infty \\
&\text{If } c > 0, \text{ then } c\infty = \infty c = \infty \text{ and } c(-\infty) = (-\infty)c = -\infty \\
&\text{If } c < 0, \text{ then } c\infty = \infty c = -\infty \text{ and } c(-\infty) = (-\infty)c = \infty \\
&0 \cdot \infty = \infty \cdot 0 = 0
\end{aligned}$$

It is also useful to write $[(a, b)]$ to indicate the four sets $(a, b), [a, b], (a, b]$ and $[a, b)$.

We also need to establish the topology on $\overline{R}$, which means a specification of which sets are open. All sets of the form $(a, b) = \{x \mid a < x < b\}$ are open, where $a, b \epsilon \overline{\mathbb{R}}$. Additionally, sets of the form $[-\infty, a), (a, \infty]$ and $[-\infty, \infty]$ are open, as is $\varnothing$. A closed set is the complement of an open set.

If A is a non-empty set in $\overline{\mathbb{R}}$, the *interior* of A, denoted A^o, is the largest open interval of $\overline{\mathbb{R}}$ that is contained in A. The *closure* of A, denoted A^c, is the smallest closed interval that contains A. Thus if $-\infty < a < b < \infty$, the closure of the sets $[(a, b)]$ is $[a, b]$, and the interior of these sets is (a, b). The sets $[a, \infty], [-\infty, b], \varnothing$ and $[-\infty, \infty]$ are their own interiors and closures.

Finally, we clarify distances from ∞ and $-\infty$ as follows: for x positive, the $x-$neighborhood of $-\infty$ is $(-\infty, -1/x)$ and that of ∞ is $(1/x, \infty)$.

With these definitions and conventions, we now review the results leading to the definition of the McShane integral. The purpose is to show which definitions results require change and which do not in the shift from an integral defined on a bounded cell $[a, b]$, $-\infty < a < b < \infty$ to one defined on a possibly unbounded cell $-\infty \leq a < b \leq \infty$.

Redefine a *partition* of $A = [a, b]$ to be a collection $\pi = \{(A_1, x_1), \ldots, (A_p, x_p)\}$ where $A_1, \ldots, A_p$ are non-overlapping cells whose union is A, and $x_1, \ldots, x_p$ are points in $\overline{\mathbb{R}}$. Let δ be a positive function defined on a set $E \subset \overline{\mathbb{R}}$. A partition $\{(A_1, x_1), \ldots, (A_p, x_p)\}$ with $x_i \epsilon E$ for all $i = 1, \ldots, p$ is called *δ-fine* if A_i is contained in a $\delta(x_i)$ neighborhood of x_i.

The evaluation point for a cell $[(-\infty, a)]$ must be $-\infty$, since if x is any other possible evaluation point, $-\infty < x$, the neighborhood $(x - \delta(x), x + \delta(x))$ is bounded, and hence cannot contain the cell. Similarly the evaluation point for the cells $[(b, \infty)]$ must be ∞.

Next, I must show that Lemmas 4.9.1 and 4.9.2 and Corollary 4.9.4 extend to cells in $\overline{\mathbb{R}}$.

Lemma 4.9.1*. *Suppose $\delta(x)$ and $\delta'(x)$ are positive functions on $\overline{\mathbb{R}}$ satisfying $\delta(x) \leq \delta'(x)$. Then every δ-fine partition is δ'-fine.*

Proof. Suppose a partition $\{(A_1, x_1), \ldots, (A_p, x_p)\}$ is δ-fine. There can be at most one set A_i of the form $[(-\infty, x)]$ because the A's have disjoint interiors. For that set,

$$A_i = [(-\infty, x)] \subset [(-\infty, -1/\delta(\infty))] \subseteq [(-\infty, -1/\delta'(\infty))]$$

because $-1/\delta(\infty) \leq -1/\delta'(\infty)$.

Similarly there can be at most one set A_j of the form $[(x, \infty)]$. For that set

$$A_j = [(x, \infty)] \subset [(1/\delta(x), \infty)] \subseteq [(1/\delta'(x), \infty)]$$

because $1/\delta(x) \geq 1/\delta'(x)$.

The space $[-1/\delta(-\infty), 1/\delta(\infty)]$ is bounded, and hence Lemma 4.9.1 applies to it. □

Lemma 4.9.2*. *(Cousin) For each positive function δ on a cell A, there is a δ-fine partition π of A.*

Proof. In addition to the δ-fine partition π of $[-1/\delta(-\infty), 1/\delta(\infty)] \cap A$ assured by Lemma 4.9.2, the partition

$$\pi^* = \pi \cup \{[-\infty, -1/\delta(-\infty)] \cap A\} \cup \{[1/\delta(\infty), \infty)] \cap A\}$$

suffices. □

Corollaries 4.9.3* and 4.9.4* have the same statement and proof as Corollaries 4.9.3 and 4.9.4, so need not be repeated.

The functions f to be integrated have to be defined on all of $\overline{\mathbb{R}}$, and in particular for $-\infty$ and ∞. It is important to choose $f(-\infty) = f(\infty) = 0$ for this purpose. Having done so, we now consider that the contribution of the cells $A_i = [(-\infty, x_i)$ and $A_j = [(x_j, \infty)]$ to the Stieltjes sum (4.76) is $f(-\infty)\alpha(A_i) + f(\infty)\alpha(A_j)$. Because $f(-\infty) = f(\infty) = 0$, for every value of $\alpha(A_i)$ and $\alpha(A_j)$ (including ∞), we have

$$f(-\infty)\alpha(A_i) + f(\infty)\alpha(A_j) = 0 + 0 = 0.$$

(This is the reason for the otherwise possibly mysterious convention that $\infty \cdot 0 = 0$.)

Hence the Stieltjes sum (4.76) is unchanged by consideration of cells in $\overline{\mathbb{R}}$.

With these conventions, then, the definition of the McShane integral, and the proof that it is well defined, extend word-for-word.

4.9.2 Properties of the McShane integral

Our first task is to show some simple properties of M, namely the sense in which it is additive with respect to each of its inputs.

Lemma 4.9.5. *Let A be a cell, let f and g be elements of $M(A, \alpha)$ and let c be a real number. Then $f+g$ and cf belong to $M(A, \alpha)$ and*

$$\int_A (f+g)d\alpha = \int_A f d\alpha + \int_A g d\alpha; \tag{4.83}$$

and

$$\int_A cf\ d\alpha = c\int_A f d\alpha. \tag{4.84}$$

If, in addition, $f \leq g$, the

$$\int_A f d\alpha \leq \int_A g d\alpha. \tag{4.85}$$

Proof. For each partition π of A, we have

$$\sigma(f+g, \pi, \alpha) = \sigma(f, \pi, \alpha) + \sigma(g, \pi, \alpha). \tag{4.86}$$

Let $\epsilon > 0$ be given. Since f is McShane integrable over A with respect to α, there is a positive function δ_f and a number I_f such that

$$\mid \sigma(f, \pi, \alpha) - I_f \mid < \epsilon/2 \tag{4.87}$$

for all δ_f-fine partitions π of A. Similarly there is a positive function δ_g and a number I_g such that

$$\mid \sigma(g, \pi, \alpha) - I_g \mid < \epsilon/2 \tag{4.88}$$

for all δ_g-fine partitions π of A. Let $\delta = \min(\delta_f, \delta_g)$, a positive function on A. Using Lemma 4.9.1*, a partition π that is δ-fine is both δ_f-fine and δ_g-fine. Let π be a δ-fine partition. Then

$$\begin{aligned} &\mid \sigma(f+g, \pi, \alpha) - (I_f + I_g) \mid \\ &= \mid \sigma(f, \pi, \alpha) - I_f + \sigma(g, \pi, \alpha) - I_g \mid && \text{(using (4.86))} \\ &\leq \mid \sigma(f, \pi, \alpha) - I_f \mid + \mid \sigma(g, \pi, \alpha) - I_g \mid \\ &< \epsilon/2 + \epsilon/2 = \epsilon. && \text{(uses (4.87) and (4.88))} \end{aligned}$$

Therefore $f+g$ is McShane integrable over A with respect to α, and its integral is $I_f + I_g$. This proves (4.83).

The proofs for cf, and for $f \leq g$ are similar, using

$$\sigma(cf, \pi, \alpha) = c\sigma(f, \pi, \alpha) \tag{4.89}$$

and, if $f \leq g$,

$$\sigma(f, \pi, \alpha) \leq \sigma(g, \pi, \alpha), \tag{4.90}$$

respectively. □

Lemma 4.9.6. *The following both hold:*

a) Let α and β be non-decreasing functions on a cell A, and suppose f is McShane integrable with respect to both α and β on A. Then f is McShane integrable with respect to $\alpha + \beta$ and

$$\int_A f d(\alpha+\beta) = \int_A f d\alpha + \int_A f d\beta. \tag{4.91}$$

b) Let $c \geq 0$ be a non-negative constant. If f is McShane integrable with respect to α, a non-decreasing function on a cell A, it is also McShane integrable with respect to $c\alpha$ on A, and

$$\int_A fd(c\alpha) = c\int_A fd\alpha. \tag{4.92}$$

Proof. a) For each partition π of A, we have

$$\sigma(f,\pi,\alpha+\beta) = \sigma(f,\pi,\alpha) + \sigma(f,\pi,\beta). \tag{4.93}$$

Let $\epsilon > 0$ be given. Since f is McShane integrable with respect to α on A, there is a positive function δ_α and a number I_α such that

$$\mid \sigma(f,\pi,\alpha) - I_\alpha \mid < \epsilon/2, \tag{4.94}$$

for all δ_α-fine partitions π of A. Similarly, there is a positive function δ_β and a number I_β such that

$$\mid \sigma(f,\pi,\beta) - I_\beta \mid < \epsilon/2, \tag{4.95}$$

for all δ_β-fine partitions π of A. Let $\delta = \min(\delta_\alpha, \delta_\beta)$, a positive function on A. Let π be a δ-fine partition of A. Again using Lemma 4.9.1*, a partition that is δ-fine is both δ_α-fine and δ_β-fine. Hence in particular, π is both δ_α-fine and δ_β-fine. Then

$$\begin{aligned}
&\mid \sigma(f,\pi,\alpha+\beta) - (I_\alpha + I_\beta) \mid = \\
&\mid \sigma(f,\pi,\alpha) - I_\alpha + \sigma(f,\pi,\beta) - I_\beta \mid && \text{(using (4.93))} \\
&\leq \mid \sigma(f,\pi,\sigma - I_\alpha) \mid + \mid \sigma(f,\pi,\beta - I_\beta) \mid && \text{(using (4.94) and (4.95))} \\
&< \epsilon/2 + \epsilon/2 = \epsilon.
\end{aligned}$$

Therefore f is McShane integrable over A with respect to $\alpha+\beta$, and its integral is $I_\alpha + I_\beta$. This proves a).

The proof for b) similarly relies on the equality

$$\sigma(f,\pi,c\alpha) = c\sigma(f,\pi,\alpha) \tag{4.96}$$

for all partitions π of A. □

The proofs of Lemma 4.9.5 and 4.9.6 are similar. Both rely fundamentally on Lemma 4.9.1*, a principle used repeatedly in the proofs to follow.

The Cauchy criterion for sequences, introduced in section 4.7.1, has a useful analog for McShane integrals. Like the result for sequences, it can be applied without knowing the value of the limit.

Theorem 4.9.7. *(Cauchy's Test) A function f on a cell A is McShane integrable with respect to α on A if and only if for each $\epsilon > 0$, there is a positive function δ on A such that*

$$\mid \sigma(f,\pi,\alpha) - \sigma(f,\xi,\alpha) \mid < \epsilon \tag{4.97}$$

for all δ-fine partitions π and ξ of A.

Proof. Suppose first that for each $\epsilon > 0$, there is such a positive function δ on A. For $n = 1, 2, \ldots$ choose $\epsilon_n = 1/n$. Then by assumption there is a positive function δ_n satisfying (4.97). Let $\delta_n^* = \min\{\delta_1, \delta_2, \ldots, \delta_n\}$. Then every δ_n^*-fine partition is δ_i-fine, for $i = 1, \ldots, n$, (using Lemma 4.9.1*) and $\delta_1^* \geq \delta_2^* \ldots$. Let π_n be a δ_n^*-fine partition for each n. I claim that $\sigma(f,\pi_n;\alpha)$ is a sequence satisfying the Cauchy criterion. To see this, choose $\epsilon > 0$, and let $N > 1/\epsilon$. Let n and m be chosen so that $n \geq m \geq N$. Then π_n and π_m are δ_N^*-fine. By (4.97),

$$\mid \sigma(f,\pi_n;\alpha) - \sigma(f,\pi_m;\alpha) \mid < 1/N < \epsilon. \tag{4.98}$$

Hence $\sigma(f, \pi_n; \alpha)$ satisfies the Cauchy criterion as a sequence of real numbers. Using Theorem 4.7.3, it then follows that this sequence has a limit I.

Now choose a (possibly different) number $\epsilon > 0$. There is an integer $k > 2/\epsilon$ such that $\mid \sigma(f, \pi_k; \alpha) - I \mid < \epsilon/2$. Let $\delta = \delta_k^*$. If π is a δ-fine partition of A, then

$$\mid \sigma(f, \pi; \alpha) - I \mid \leq \mid \sigma(f, \pi; \alpha) - \sigma(f, \pi_k; \alpha) \mid + \mid \sigma(f, \pi_k; \alpha) - I \mid < \frac{1}{k} + \frac{\epsilon}{2} < \epsilon. \tag{4.99}$$

This proves that f is McShane integrable on A with respect to α.

In the second part of the proof, I suppose that f is McShane integrable on A with respect to α, and prove that it satisfies (4.97). To show this, choose $\epsilon > 0$. By definition of the McShane integral, there is a positive function δ and a number I such that

$$\mid \sigma(f, \pi; \alpha) - I \mid < \epsilon/2 \tag{4.100}$$

for all δ-fine partitions π.

Let π and ξ be δ-fine partitions. Then

$$\begin{aligned} \mid \sigma(f, \pi; \alpha) - \sigma(f, \xi; \alpha) \mid &= \mid \sigma(f, \pi; \alpha) - I - (\sigma(f, \xi; \alpha) - I) \mid \\ &\leq \mid \sigma(f, \pi; \alpha) - I \mid + \mid \sigma(f, \xi; \alpha) - I \mid \\ &< \epsilon/2 + \epsilon/2 = \epsilon. \end{aligned} \tag{4.101}$$

This proves (4.97) and hence the theorem. □

The proof of the next lemma uses Cauchy's test twice.

Lemma 4.9.8. *If A is a cell, and f is McShane integrable on A with respect to α then f is McShane integrable on B with respect to α for every cell $B \subseteq A$.*

Proof. Let $\epsilon > 0$ be given. Because f is McShane integrable on A with respect to α, there is a positive function δ on A and a number I such that

$$\mid \sigma(f, \pi; \alpha) - I \mid < \epsilon \tag{4.102}$$

for every δ-fine partition π on A. By Cauchy's test, we have

$$\mid \sigma(f, \pi; \alpha) - \sigma(f, \xi; \alpha) \mid < \epsilon \tag{4.103}$$

for every δ-fine partitions π and ξ on A. If $B = A$, there is nothing to prove. If $B \subset A$, then A can be represented as

$$A = B \cup C \cup D$$

where C is a cell, and D is either a cell or is the null set. By Cousin's Lemma 4.9.2* there is a δ-fine partition π_c of C, and, if D is a cell, a δ-fine partition π_D of D as well. Let π_B and ξ_B be δ-fine partitions of B. Then $\pi = \pi_B \cup \pi_C \cup \pi_D$ and $\xi = \xi_B \cup \pi_c \cup \pi_D$ are δ-fine partitions of A. (Of course, take $\pi_D = \varnothing$ if $D = \varnothing$.) Now

$$\sigma(f, \pi; \alpha) = \sigma(f, \pi_B; \alpha) + \sigma(f, \pi_C; \alpha) + \sigma(f, \pi_D; \alpha) \tag{4.104}$$

and

$$\sigma(f, \xi; \alpha) = \sigma(f, \xi_B; \alpha) + \sigma(f, \pi_C; \alpha) + \sigma(f, \pi_D; \alpha) \tag{4.105}$$

where again $\sigma(f, \pi_D; \alpha) = 0$ if $D = \varnothing$.

Therefore

$$\epsilon > \mid \sigma(f, \pi; \alpha) - \sigma(f, \xi; \alpha) \mid = \mid \sigma(f, \pi_B; \alpha) - \sigma(f, \xi_B; \alpha) \mid . \tag{4.106}$$

Applying Cauchy's test, we conclude that f is McShane integrable on B with respect to α, which completes the proof. □

Lemma 4.9.8 shows that if f is McShane integrable on a cell $[a,b]$, then it is integrable on a smaller cell contained in $[a,b]$. The next lemma shows the reverse, that if f is McShane integrable on $[a,c]$ and on $[c,b]$, then it is McShane integrable on $[a,b]$ and the integrals add. More formally,

Lemma 4.9.9. *Let f be a function on a cell $[a,b]$ and let $c\epsilon(a,b)$. If f is McShane integrable with respect to α on both $[a,c]$ and $[c,b]$, then it is McShane integrable with respect to α on $[a,b]$ and*

$$\int_a^b f d\alpha = \int_a^c f d\alpha + \int_c^b f d\alpha. \tag{4.107}$$

Proof. Let $I = \int_a^c f d\alpha + \int_c^b f d\alpha$, and let $\epsilon > 0$ be given.

Then by definition of the McShane integral, there are positive functions δ_a and δ_b on the cells $[a,c]$ and $[c,b]$, respectively, such that

$$\mid \sigma(f, \pi_a; \alpha) - \int_a^c f d\alpha \mid < \epsilon/2 \tag{4.108}$$

and

$$\mid \sigma(f, \pi_b; \alpha) - \int_c^b f d\alpha \mid < \epsilon/2 \tag{4.109}$$

for every δ_a-fine partition π_a of $[a,c]$ and for every δ_b-fine partition π_b of $[c,b]$. The key to the proof is the following definition of the positive function δ. Let $\delta(x)$ be defined as follows:

$$\delta(x) = \begin{cases} \min\{\delta_a(x), c-x\} \text{ if } x < c \\ \min\{\delta_b(x), x-c\} \text{ if } x > c \\ \min\{\delta_a(x), \delta_b(x)\} \text{ if } x = c \end{cases} . \tag{4.110}$$

Crucially, $\delta(x) > 0$ for all $x\epsilon[a,b]$.

Now choose a δ-fine partition $\pi = \{(A_1, x_1), \ldots, (A_p, x_p)\}$ of $[a,b]$. Because of the choice of the function δ, we have:

$$\begin{array}{ll} (i) & \text{if } A_i \subset [a,c], \text{ then } x_i\epsilon[a,c] \\ (ii) & \text{if } A_i \subset [c,b], \text{ then } x_i\epsilon[c,b] \\ (iii) & \text{if } c\epsilon A_i, \text{ then } x_i = c. \end{array} \tag{4.111}$$

There are now two cases to consider:

(a) Each A_i is contained in either $[a,c]$ or $[c,b]$. In this case $\pi = \pi_a \cup \pi_b$, where π_a is a δ_a-fine partition of $[a,c]$ and π_b is a δ_b-fine partition of $[c,b]$. Since

$$\sigma(f, \pi; \alpha) = \sigma(f, \pi_a; \alpha) + \sigma(f, \pi_b; \alpha), \tag{4.112}$$

we can conclude that

$$\begin{aligned} &|\sigma(f, \pi; \alpha) - I| \\ &= \mid \sigma(f, \pi_a; \alpha) - \int_a^c f d\alpha + \sigma(f, \pi_b; \alpha) - \int_c^b f d\alpha \mid \\ &\leq \mid \sigma(f, \pi_a; \alpha) - \int_a^c f d\alpha| + |\sigma(f, \pi_b; \alpha) - \int_c^b f d\alpha| \\ &< \epsilon/2 + \epsilon/2 = \epsilon. \end{aligned} \tag{4.113}$$

(b) There is an A_i contained in neither $[a, c]$ nor $[c, b]$. In this case $c \epsilon A_i$. Then the partition

$$\xi = \{(A_1, x_1), \ldots, (A_i \cap [a, c], x_i), (A_i \cap [c, b], x_i), \ldots, (A_p, x_p)\} \tag{4.114}$$

satisfies the condition of case (a), so, using (4.113),

$$|\sigma(f, \xi; \alpha) - I| < \epsilon. \tag{4.115}$$

But

$$\sigma(f, \xi; \alpha) = \sigma(f, \pi; \alpha), \text{ so} \tag{4.116}$$

$$|\sigma(f, \pi; \alpha) - I| < \epsilon. \tag{4.117}$$

This establishes the lemma.

□

The next series of results are aimed at showing that the McShane integral is "absolute," which means that if $f \epsilon M(A, \alpha)$, then $|f| \epsilon M(A, \alpha)$. A few lemmas are necessary to get there.

The first lemma looks a lot like Cauchy's test, but shows that the partitions involved can be limited to those that have common cells:

Lemma 4.9.10. *A function f on a cell A belongs to $M(A, \alpha)$ if and only if for each $\epsilon > 0$, there is a positive function δ such that*

$$|\sigma(f, \pi; \alpha) - \sigma(f, \xi; \alpha)| < \epsilon \tag{4.118}$$

for all partitions $\pi = \{(A_1, x_1), (A_2, x_2), \ldots, (A_p, x_p)\}$ and $\xi = \{(A_1, y_1), (A_2, y_2), \ldots, (A_p, y_p)\}$ of A that are δ-fine.

Proof. If $f \epsilon M(A, \alpha)$, then Cauchy's test applies to π and ξ to yield the result. The work in the proof then, is proving the converse, namely that restricting π and ξ to have the same cells still allows one to prove that f is McShane integrable.

Choose an $\epsilon > 0$, and let δ be a positive function such that (4.118) holds for all partitions π and ξ as stated in the Lemma. Let $\gamma = \{(B_1, u_1), \ldots, (B_p, u_p)\}$ and $\eta = \{(C_1, v_1), \ldots, (C_q, v_q)\}$ be δ-fine partitions of A. (We know that Cauchy's criterion applies to γ and η.)

For $i = 1, \ldots, p$ and $j = 1, \ldots, q$, let

$$A_{i,j} = B_i \cap C_j, x_{i,j} = u_i \text{ and } y_{i,j} = v_j,$$

and let $N = \{(i, j)$ such that $A_{i,j}$ is a cell$\}$.

Now let

$$\pi = \{(A_{i,j}, x_{i,j}) : (i, j) \epsilon N\} \text{ and}$$
$$\xi = \{(A_{i,j}, y_{i,j}) : (i, j) \epsilon N\}.$$

Both π and ξ are δ-fine partitions of A, because γ and η, respectively, are. Now we have

$$\begin{aligned}
\sigma(f,\pi;\alpha) &= \sum_{(i,j)\epsilon N} f(x_{i,j})\alpha(A_{i,j}) \\
&= \sum_{i=1}^{p}\sum_{j=1}^{q} f(x_{i,j})\alpha(A_{i,j}) \\
&\text{(uses the convention that } \alpha(D)=0 \text{ if D is not a cell)} \\
&= \sum_{i=1}^{p} f(u_i)\sum_{j=1}^{q}\alpha(B_i\cap C_j) \\
&= \sum_{i=1}^{p} f(u_i)\alpha(B_i) \\
&= \sigma(f,\gamma,\alpha).
\end{aligned} \tag{4.119}$$

In the same way,

$$\sigma(f,\xi;\alpha) = \sigma(f,\eta;\alpha). \tag{4.120}$$

Therefore

$$|\sigma(f,\pi;\alpha) - \sigma(f,\xi;\alpha)| = |\sigma(f,\gamma;\alpha) - \sigma(f,\eta;\alpha)| < \epsilon,$$

so $f\epsilon M(A,\alpha)$ by Cauchy's test. □

The next lemma allows even greater control over the partitions and over the sums:

Lemma 4.9.11. *A function f on a cell A belongs to $M(A,\alpha)$ if and only if for each $\epsilon > 0$ there is a positive function δ in A such that*

$$\sum_{i=1}^{n} |f(x_i) - f(y_i)|\alpha(A_i) < \epsilon \tag{4.121}$$

for all partitions $\pi = \{(A_1,x_1),\ldots,(A_n,x_n)\}$ and

$$\xi = \{(A_1,y_1),\ldots,A_n,y_n)\}$$

in A that are δ-fine.

Remark: Lemma 4.9.11 differs from Lemma 4.9.10 in two ways. Obviously (4.121) is not the same as (4.118), but in addition the partitions in 4.9.11 are *in* A, where those in 4.9.10 are *of* A.

Proof. First suppose that for each $\epsilon > 0$ there is a positive function δ in A such that (4.121) holds. Because each partition in A is a subset of a partition of A, the condition of Lemma 4.9.10 holds. Then

$$\begin{aligned}
|\sigma(f,\pi;A) - \sigma(f,\xi;A)| &= |\sum_{i=1}^{n} f(x_i)\alpha(A_i) - \sum_{i=1}^{n} f(y_i)\alpha(A_i)| \\
= |\sum_{i=1}^{n}[f(x_i) - f(y_i)]\alpha(A_i)| &\le \sum_{i=1}^{n} |f(x_i) - f(y_i)|\alpha(A_i) < \epsilon
\end{aligned} \tag{4.122}$$

so Lemma 4.9.10 applies and shows that $f\epsilon M(A,\alpha)$.

So now suppose that $f\epsilon M(A,\alpha)$, and we seek to prove (4.121). Using the construction of Lemma 4.9.10, we may consider δ-fine partitions π and ξ of A, having the same sets $A_1, \ldots, A_n$. Reordering the index as needed, there is an integer $k, 0 \leq k \leq n$ such that $f(x_i) \geq f(y_i)$ for $i = 1, 2, \ldots, k$ and $f(x_i) < f(y_i)$ for $i = k+1, \ldots, n$. Then the partitions

$$\gamma = \{(A_1, x_1), \ldots, (A_k, x_k), (A_{k+1}, y_{k+1}), \ldots, (A_n, y_n)\}$$

and

$$\eta = \{(A_1, y_1), \ldots, (A_k, y_k), (A_{k+1}, x_{k+1}), \ldots, (A_n, x_n)\}$$

are δ-fine partitions. Hence, by Lemma 4.9.10,

$$\begin{aligned} \epsilon > & |\sigma(f, \gamma; \alpha) - \sigma(f, \eta; \alpha)| \\ & = |\sum_{i=1}^{k} f(x_i)\alpha(A_i) + \sum_{i=k+1}^{n} f(y_i)\alpha(A_i) \\ & - \sum_{i=1}^{k} f(y_i)\alpha(A_i) - \sum_{i=k+1}^{n} f(x_i)\alpha(A_i)| \\ & = |\sum_{i=1}^{k}(f(x_i) - f(y_i))\alpha(A_i) + \sum_{i=k+1}^{n}(f(y_i) - f(x_i))\alpha(A_i)|. \end{aligned} \tag{4.123}$$

Now each of these terms is non-negative, so the absolute value of the sum is the sum of the absolute values.

Hence

$$\begin{aligned} \epsilon > & \sum_{i=1}^{k} |f(x_i) - f(y_i)|\alpha(A_i) + \sum_{i=k+1}^{n} |f(y_i) - f(x_i)|\alpha(A_i) \\ = & \sum_{i=1}^{n} |f(x_i) - f(y_i)|\alpha(A_i), \end{aligned} \tag{4.124}$$

which is (4.121). □

Corollary 4.9.12. *Let A be a cell. If $f \epsilon M(A, \alpha)$ then $|f| \epsilon M(A, \alpha)$ and*

$$|\int_A f d\alpha| \leq \int_A |f| d\alpha. \tag{4.125}$$

Proof. Using Lemma 4.9.11, let $\epsilon > 0$ be given. Then there is a positive function δ on A such that (4.121) holds. Then

$$\begin{aligned} \epsilon > & \sum_{i=1}^{n} |f(x_i) - f(y_i)|\alpha(A_i) \\ \geq & \sum_{i=1}^{n} ||f(x_i)| - |f(y_i)||\alpha(A_i). \end{aligned} \tag{4.126}$$

Applying Lemma 4.9.11, this implies that $|f| \epsilon M(A, \alpha)$. (4.125) then follows from (4.85). □

Corollary 4.9.12 establishes that the McShane integral is absolute, which, as we saw in Chapter 3, is vital for our purposes.

Corollary 4.9.13. *Let A be a cell. If f and g are in $M(A,\alpha)$, then so are* $\max\{f,g\}$ *and* $\min\{f,g\}$.

Proof.

$$\max\{f,g\} = \frac{1}{2}(f+g+|f-g|)$$

$$\min\{f,g\} = \frac{1}{2}(f+g-|f-g|)$$

hold pointwise. Then the result follows from Corollary 4.9.12 and Lemma 4.9.5. □

Now we are ready to consider a sequence of results culminating in a dominated convergence theorem.

Lemma 4.9.14. *(Henstock) Let A be a cell and let $f\epsilon M(A,\alpha)$. For every $\epsilon > 0$, there is a positive function δ on A such that*

$$\sum_{i=1}^{p} |f(x_i)\alpha(A_i) - \int_{A_i} f d\alpha| < \epsilon \tag{4.127}$$

for every δ-fine partition $\{(A_1,x_1),\ldots,(A_p,x_p)\}$ in A.

Proof. Let $\epsilon > 0$ be given. Since $f\epsilon M(A,\alpha)$, there is a positive function δ on A such that $|\sigma(f,\pi;\alpha) - \int_A f d\alpha| < \epsilon/3$ for all δ-fine partitions π of A. Because of Corollary 4.9.4, we may consider a δ-fine partition $\{(A_1,x_1),\ldots,(A_p,x_p)\}$ of A. After reordering if necessary, there is an integer k, $0 \leq k \leq p$ such that $f(x_i)\alpha(A_i) - \int_{A_i} f d\alpha$ is non-negative for $i = 1,\ldots,k$ and negative for $i = k+1,\ldots,p$. Using Cousin's Lemma 4.9.2* and Lemma 4.9.8, there is a δ-fine partition π_i of A_i such that

$$|\sigma(f,\pi_i;\alpha) - \int_{A_i} f d\alpha| < \epsilon/3p \text{ for } i = 1,\ldots,p.$$

Define two new partitions as follows:

$$\xi = \{(A_1,x_1),\ldots,(A_k,x_k)\} \cup_{i=k+1}^{p} \pi_i \tag{4.128}$$

$$\eta = \{(A_{k+1},x_{k+1}),\ldots,(A_p,x_p)\} \cup_{i=1}^{k} \pi_i. \tag{4.129}$$

Both of these are δ-fine partitions of A. Then

$$\begin{aligned}
\epsilon/3 &> |\sigma(f,\xi;\alpha) - \int_A f d\alpha| \\
&\geq \sum_{i=1}^{k} [f(x_i)\alpha(A_i) - \int_{A_i} f d\alpha] - |\sum_{i=k+1}^{p} [\sigma(f,\pi_i;\alpha) - \int_{A_i} f d\alpha]| \\
&\geq \sum_{i=1}^{k} |f(x_i)\alpha(A_i) - \int_{A_i} f d\alpha| - (p-k)\epsilon/3p.
\end{aligned} \tag{4.130}$$

Also

$$\begin{aligned}
\epsilon/3 &> |\sigma(f,\eta;\alpha) - \int_A f d\alpha| \\
&\geq \sum_{i=k+1}^{p} [\int_{A_i} f d\alpha - f(x_i)\alpha(A_i)] - |\sum_{i=1}^{k} \sigma(f,\pi_i;\alpha) - \int_{A_i} f d\alpha| \\
&\geq \sum_{i=k+1}^{p} |f(x_i)\alpha(A_i) - \int_{A_i} f d\alpha| - k\epsilon/3p.
\end{aligned} \tag{4.131}$$

Adding (4.130) and (4.131) yields

$$2\epsilon/3 \geq \sum_{i=1}^{p} |f(x_i)\alpha(A_i) - \int_{A_i} f d\alpha| - p(\epsilon/3p), \tag{4.132}$$

so $\epsilon > \sum_{i=1}^{p} |f(x_i)\alpha(A_i) - \int_{A_i} f d\alpha|$. □

The heart of the issue of dominated convergence is found in monotone convergence. A sequence of functions f_n is non-decreasing (or non-increasing) if $f_n \leq f_{n+1}$(or $f_n \geq f_{n+1}$) for $n = 1, 2, \ldots$. If a non-decreasing (non-increasing) sequence converges to a function f, we write $f_n \nearrow f (f_n \searrow f)$.

Theorem 4.9.15. *(Monotone Convergence) Let f be a function on a cell A, and let f_n be a sequence of functions in $M(A, \alpha)$ such that $f_n \nearrow f$. If $\lim_{n\to\infty} \int_A f_n d\alpha$ is finite, then $f \epsilon M(A, \alpha)$ and*

$$\int_A f d\alpha = \lim \int_A f_n d\alpha. \tag{4.133}$$

Proof. Let $\epsilon > 0$ be given. For each $n, n = 1, 2, \ldots$, by Henstock's Lemma (4.9.14), there is a positive function δ_n on A such that

$$\sum_{i=1}^{q} |f_n(y_i)\alpha(B_i) - \int_{B_i} f_n d\alpha| < \epsilon 2^{-n} \tag{4.134}$$

for each δ_n-fine partition $\{(B_1, y_1), \ldots, (B_q, y_q)\}$ in A.

Let $I = \lim \int_A f_n d\alpha$. By assumption $I < \infty$. Therefore there is a positive integer r with

$$\int_A f_r d\alpha > I - \epsilon. \tag{4.135}$$

Because $f_n(x) \to f(x)$ for each $x \epsilon A$, there is an integer $n(x) \geq r$ such that

$$|f_{n(x)}(x) - f(x)| < \epsilon. \tag{4.136}$$

Now the function δ on A is defined as follows:

$$\delta(x) = \delta_{n(x)}(x) \tag{4.137}$$

for each x. That $\delta(x) > 0$ for all x follows from the fact that $\delta_n(x) > 0$ for all n and x.

The theorem is now proved by showing that

$$|\sigma(f, \pi; \alpha) \quad I| < \epsilon[2 + \alpha(A)] \tag{4.138}$$

for any δ-fine partition $\pi = \{(A_1, x_1), \ldots, (A_p, x_p)\}$ of A.

We do this in three steps. To begin, we have

$$\begin{aligned} |\sigma(f, \pi; \alpha) - \sum_{i=1}^{p} f_{n(x_i)}(x_i)\alpha(A_i)| &= |\sum_{i=1}^{p} f(x_i)\alpha(A_i) - \sum_{i=1}^{p} f_{n(x_i)}(x_i)\alpha(A_i)| \\ &\leq \sum_{i=1}^{p} |f(x_i) - f_{n(x_i)}(x_i)|\alpha(A_i) \\ &\leq \epsilon \sum_{i=1}^{p} \alpha(A_i) \quad \text{(uses (4.136))} \\ &= \epsilon\alpha(A) \end{aligned} \tag{4.139}$$

which is the first step.

To establish the second step we may eliminate all A_i that are of the form $[(-\infty, a)]$ or $[(b, \infty)]$, as they do not contribute to the Stieltjes sum. The integers $n(x_1), \ldots, n(x_p)$ need not be distinct. However, there is a (possibly less numerous) set that includes each of them. Let $k_1 < \ldots < k_s$ be k distinct integers such that

$$\{n(x_1), \ldots, n(x_p)\} = \{k_1, \ldots, k_s\}, \tag{4.140}$$

where $s \leq p$. Then $\{1, \ldots, p\}$ is the disjoint union of the sets $T_j = \{i | n(x_i) = k_j\}$ for $j = 1, \ldots, s$. For each $i \epsilon T_j$,

$$\begin{aligned} A_i \subset & \{x | x_i - \delta(x_i) < x < x_i + \delta(x_i)\} \\ = & \{x | x_i - \delta_{n(x_i)}(x_i) < x < x_i + \delta_{n(x_i)}(x_i)\} \\ = & \{x | x_i - \delta_{k_j}(x_i) < x < x_i + \delta_{k_j}(x_i)\}. \end{aligned} \tag{4.141}$$

It follows that $\{(A_i, x_i) : i \epsilon T_j\}$ is a δ_{k_j}-fine partition in A. Hence

$$\begin{aligned} & |\sum_{i=1}^{p} f_{n(x_i)}(x_i)\alpha(A_i) - \sum_{i=1}^{p} \int_{A_i} f_{n(x_i)} d\alpha| \\ = & |\sum_{j=1}^{s} \sum_{i \epsilon T_j} (f_{n(x_i)}(x_i)\alpha(A_i) - \int_{A_i} f_{n(x_i)} d\alpha)| \\ \leq & \sum_{j=1}^{s} \sum_{i \epsilon T_j} |f_{n(x_i)}(x_i)\alpha(A_i) - \int_{A_i} f_{n(x_i)} d\alpha| \\ \leq & \sum_{j=1}^{s} \epsilon 2^{-k_j} < \epsilon \sum_{k=1}^{\infty} 2^{-k} = \epsilon, \end{aligned} \tag{4.142}$$

using (4.134). This completes the second step.

To establish the third step, we show that I is within ϵ of $\sum_{i=1}^{p} \int_{A_i} f_n(x) d\alpha$ as follows:

$$\begin{aligned} I - \epsilon < & \int_A f_r d\alpha && \text{(uses (4.135))} \\ = & \sum_{i=1}^{p} \int_{A_i} f_r d\alpha && \text{(uses 4.9.9)} \\ \leq & \sum_{i=1}^{p} \int_{A_i} f_{n(x_i)} d\alpha && \text{(since } r \leq n(x_i), f_r \leq f_{n(x_i)} \text{ and (4.84) applies)} \\ \leq & \sum_{i=1}^{p} \int_{A_i} f_{k_s} d\alpha && \text{(since } n(x_i) \leq k_s \text{, the same reasoning applies)} \\ = & \int_A f_{k_s} d\alpha && \text{(from (4.107))} \\ \leq & I && \text{(because } f_{k_s} \leq f \text{, and apply (4.85))} \\ < & I + \epsilon. \end{aligned}$$

Then

$$|I - \sum_{i=1}^{p} \int_{A_i} f_{n(x_i)} d\alpha| < \epsilon, \tag{4.143}$$

completing the third step.

Summarizing, we have

$$
\begin{aligned}
|\sigma(f,\pi;\alpha) - I)| &\leq |\sigma(f,\pi;\alpha) - \sum_{i=1}^{p} f_{n(x_i)}(x_i)\alpha(A_i)| \\
&+ |\sum_{i=1}^{p} f_{n_i}(x_i)\alpha(A_i) - \sum_{i=1}^{p}\int_{A_i} f_{n_i(x_i)} d\alpha| \\
&+ |\sum_{i=1}^{p}\int_{A_i} f_{n_i(x_i)} d\alpha - I| \\
&< \epsilon\alpha(A) + \epsilon + \epsilon = \epsilon(\alpha(A) + 2)
\end{aligned}
$$

using (4.139), (4.142) and (4.143). This establishes (4.138), and hence the theorem. □

Next, I give two lemmas that extend the result from monotone functions.

Lemma 4.9.16. *Let A be a cell, and let f_n and g be McShane integrable on A with respect to α, and satisfy $f_n \geq g$ for $n = 1, \ldots,$. Then inf f_n is McShane integrable on A with respect to α.*

Proof. Let $g_n = \min\{f_1, \ldots, f_n\}$ for $n = 1, 2, \ldots$. Then g_n is McShane integrable by 4.9.13. Also g_n is monotone decreasing, and approaches $\inf f_n$. Also $g \leq g_n$ for all n. Then

$$\int_A g d\alpha \leq \lim \int_A g_n d\alpha \leq \int_A g_1 d\alpha, \tag{4.144}$$

using (4.85) once again.

Therefore the functions $-g_n$ are McShane integrable on A with respect to α. The sequence $\{-g_n\}$ is monotone increasing, and approaches $\sup -f_n$. By (4.144), $\lim \int_A(-g_n)d\alpha$ is finite. Therefore $\{-g_n\}$ satisfies the conditions of the Monotone Convergence Theorem 4.9.15, so $\sup\{-f_n\}$ is McShane integrable. But $\sup\{-f_n\} = -\inf\{f_n\}$, so $\inf f_n$ is McShane integrable. □

Lemma 4.9.17. *(Fatou) Suppose f, g, and $f_n(n = 1, 2, \ldots)$ are functions on a cell A such that $f_n \geq g$ for $n = 1, 2, \ldots$ and $f = \liminf f_n$. Also suppose that f_n and g are McShane integrable on A with respect to α. If $\liminf \int_A f_n d\alpha$ is finite, then f is McShane integrable on A with respect to α and*

$$\int_A f d\alpha \leq \liminf \int_A f_n d\alpha. \tag{4.145}$$

Proof. Let $g_n = \inf_{k \geq n} f_k$ for $n = 1, 2, \ldots$. Then by Lemma 4.9.16, g_n is McShane integrable, and $g_n \nearrow f$. Since $g_n \leq f_n$ for all n,

$$\int_A g_1 d\alpha \leq \lim \int_A g_n d\alpha \leq \lim inf \int_A f_n d\alpha. \tag{4.146}$$

Now apply the monotone convergence theorem, and conclude that f is McShane integrable and

$$\int_A f d\alpha = \lim \int_A g_n d\alpha. \tag{4.147}$$

But (4.147) and (4.146) imply (4.145). □

Corollary 4.9.18. *(Lebesgue Dominated Convergence Theorem) Let A be a cell and suppose that f_n and g are McShane integrable on A with respect to α. If $|f_n| \leq g$ for $n = 1, 2, \ldots$ and if $f = \lim f_n$, then (i) f is McShane integrable on A with respect to α and (ii)*

$$\int_A f d\alpha = \lim \int_A f_n d\alpha. \tag{4.148}$$

Proof. Fatou's Lemma implies (i). To obtain (ii), we have

$$\begin{aligned}
&\int_A f d\alpha \\
&= \int_A \liminf f_n d\alpha && \text{(because } f = \lim f_n\text{)} \\
&\leq \liminf \int_A f_n d\alpha && \text{(Fatou's Lemma applied to } \{f_n\}\text{).} \\
&\leq \limsup \int_A f_n d\alpha && \text{(property of lim sup and lim inf)} \\
&\leq \int_A \limsup f_n && \text{(Fatou's Lemma applied to } \{-f_n\}\text{).} \\
&= \int_A f d\alpha && \text{(because } f = \lim f_n\text{).}
\end{aligned}$$

Now (4.148) follows immediately. □

Example 2: Let

$$f(x) = (-1)^{i+1}/(i+1) \quad i - 1 < x < i$$

defined for $x \epsilon (0, \infty)$. Thus $f(x)$ is a step function, constant on intervals of unit length. It is an open question whether to consider that this function has a Riemann integral. Courant (1937, p. 249) would say that it does, because

$$\lim_{A \to \infty} \int_0^A f(x) dx$$

exists (and was shown in equation (3.10) to have the value $\log 2$). However, Taylor (1955, p. 652) would insist that $f(x)$ be absolutely integrable, which is not the case for this example, since $\sum_{i=1}^{\infty} 1/(i+1) = \infty$. From the McShane viewpoint, according to Corollary 4.9.12, if $f \epsilon M(A, \alpha)$ then $|f| \epsilon M(A, \alpha)$. Hence, f is not McShane integrable. Thus the statement that all functions that are Riemann integrable are McShane integrable holds only if one takes the Taylor, and not the Courant view. The extension of Riemann integrals to the whole real line introduced just before section 4.7 is restricted to the expectations of functions such that $E|X| < \infty$, thus excluding functions like f above.

4.9.3 McShane probabilities

Suppose X is an uncertain quantity with F as cdf. Then F is non-decreasing, and satisfies

$$P\{A\} = F(x) = \int I_A(x) dF(x) \tag{4.149}$$

where $A = (-\infty, x]$. In greater generality, let A be any set for which the McShane integral

$$\int I_A(x) dF(x) \tag{4.150}$$

exists, and define $P\{A\}$ to be equal to that integral. Then $P\{A\}$ are McShane probabilities. If $P\{A\}$ is what you would pay for a ticket that pays \$1 if A occurs and nothing otherwise, then these McShane probabilities are your probabilities.

Theorem 4.9.19. *(Strong Countable Additivity) Let $A_1, \ldots$ be a countable sequence of disjoint events having McShane probabilities with respect to a cdf F. Then $A = \cup_{i=1}^{\infty} A_i$ has a McShane probability with respect to F and*

$$P\{A\} = \sum_{i=1}^{\infty} P\{A_i\}. \tag{4.151}$$

Proof. Let $f_n(x) = \sum_{i=1}^{n} I_{A_i}(x)$. Then $f_n(x) \nearrow f = \sum_{i=1}^{\infty} I_{A_i}(x) = I_A(x)$ and $|f| \leq 1$. Now the constant function 1 has McShane integral 1. Then the dominated convergence theorem applies, so A has a McShane probability with respect to F, and

$$\begin{aligned} P\{A\} &= \int I_A(x)dF(x) = \lim_{n\to\infty} \int f_n(x)d\alpha \\ &= \lim_{n\to\infty} \int \sum_{i=1}^{n} I_{A_i}(x)dF = \sum_{i=1}^{\infty} \int I_{A_i}(x)dF \\ &= \sum_{i=1}^{\infty} P\{A_i\}. \end{aligned} \tag{4.152}$$

□

4.9.4 Comments and relationship to other literature

The material in this section on McShane integrals relies heavily on Pfeffer (1993, Chapters 1 and 2). Indeed my 4.9.2 to 4.9.18 are respectively his 1.2.4, 1.2.5, 2.1.3, 2.1.5, 2.1.8-2.1.10, 2.2.1-2.2.4, 2.3.1 and 2.3.4-2.3.7.

There is an elegant abstract theory of integration, using measure theory and the Lebesgue integral, that applies to integration on general spaces (see Billingsley (1995), for example). It turns out that the McShane integral is the Lebesgue integral (Pfeffer (1993, Chapter 4) and McShane (1983)). Because the McShane integral is only slightly more complicated than the Riemann integral, a number of senior mathematicians have suggested that it be used instead of the Riemann integral in elementary courses (see Bartle et al. (1997)).

A further generalization of the Riemann integral is found by restricting partitions to those satisfying $x_i \epsilon A_i$ for $i = 1, \ldots, p$. This leads to the Henstock-Kurzweil approach to the Denjoy-Perron integral. Because this integral is not absolute, it is not suitable for our purposes. For more about this integral, see Henstock (1963), Pfeffer (1993) and Yee and Vyborny (2000).

From the perspective of this book, it is coherent for a person to specify a density and only Riemann probabilities. Indeed a person could specify any number in the interval $[0, 1]$ for the Dirichlet example. Advanced methods (fundamentally the Hahn-Banach theorem) show that each such choice is coherent, see Bhaskara Rao and Bhaskara Rao (1983). Only the choice of 0 is countably additive. Thus whether to specify a density and Riemann probabilities, or to specify a density and McShane probabilities, or to make some other choice, is a personal matter that is not to be coerced. Each choice has certain mathematical consequences, but other than lack of coherence, none of them is "wrong".

4.9.5 Summary

This section introduces the McShane integral. Three promises were made at the beginning of this section, namely

(i) The McShane integral is a generalization of the Riemann integral (see the discussion after Corollary 4.9.4).

(ii) The McShane integral has a strong dominated convergence theorem (see Corollary 4.9.18).

(iii) McShane probabilities are strongly countably additive (see Theorem 4.9.19).

Thus all three promises have been fulfilled.

4.9.6 Exercises

1. Vocabulary. Explain in your own words:
   (a) partition
   (b) δ-fine partition
   (c) cell
   (d) McShane-Stieltjes (or McShane) integral
   (e) Cousin's Lemma
2. Why is Cousin's Lemma important? If it were not true, what consequences would that have?
3. (a) Prove (4.89).
   (b) Use (4.89) to show that $cf\epsilon M(A, \alpha)$ and that (4.84) holds.
4. (a) Prove (4.90).
   (b) Use (4.90) to show that, if $f \leq g$, (4.85) holds.
5. (a) Prove (4.96).
   (b) Use (4.96) to prove part b) of Lemma 4.9.6.
6. Prove (4.111).

4.10 The road from here

The McShane integral (equivalently, the Lebesgue integral) can be extended to vectors of length k, and indeed to infinite dimensional spaces. There's lots of excellent probability that lies this way. To explore it further, however, would take this book too far from its main goal, which is to understand uncertainty. Hence I leave advanced probability to other books.

There is one matter, however, that does come up later, namely the strong law of large numbers. Consequently the next section is devoted to that subject.

4.11 The strong law of large numbers

Where there is a weak law of large numbers (see section 2.13), there must be a strong law. This section proves the strong law and also shows the sense in which the strong law is stronger than the weak law. To do so requires some more precision in notation, to which I now turn.

4.11.1 Random variables (otherwise known as uncertain quantities) more precisely

Up to now, it has not been necessary to have notation for the sample space, the space of uncertain outcomes. For example, in a single flip of a coin, this space can be thought of as $S = \{H, T\}$, because the coin will show either a head or a tail.

For a countably additive probability, the set of subsets of S over which the probability is defined is a σ-field, $\mathcal{F}$, satisfying the following conditions:

1. $\emptyset \epsilon \mathcal{F}$
2. if $A_1, A_2, \ldots, \epsilon \mathcal{F}$ then $\cup_{i=1}^{\infty} A_i \epsilon \mathcal{F}$
3. if $A \epsilon \mathcal{F}$ then $\overline{A} \epsilon \mathcal{F}$.

The countably additive probability P is then defined as a function from $\mathcal{F}$ to $\mathcal{R}$ satisfying assumptions (1.1) (1.2) and (3.2). A probability space is then defined as the triple $(S, \mathcal{F}, P)$.

Let $A_1, A_2 \ldots,$ be a sequence of events, so $A_i \epsilon \mathcal{F}$ for all i. Define

$$B_n = \cup_{m=n}^{\infty} A_m \text{ and } C_n = \cap_{m=n}^{\infty} A_m. \tag{4.153}$$

Obviously

$$C_n \leq A_n \leq B_n,$$

and the sequence C_n increases in n, while the sequence B_n decreases in n.

Let

$$B = \lim_{n \to \infty} B_n = \cap_n B_n = \cap_n \cup_{m \geq n} A_m. \tag{4.154}$$

Similarly, let

$$C = \lim_{n \to \infty} C_n = \cup_n C_n = \cup_n \cap_{m \geq n} A_m. \tag{4.155}$$

Lemma 4.11.1.

(a) $B = \{w \epsilon S : w \epsilon A_n$ *for infinitely many values of* $n\}$.

(b) $C = \{w \epsilon S : w \epsilon A_n$ *for all but a finite number of* n's$\}$.

Proof. (a) $w \epsilon B \iff w \epsilon \cap_n \cup_{m \geq n} A_m \iff$ for all $n, w \epsilon \cup_{m \geq n} A_m$. Hence no matter how large n is, there is an $m \geq n$ such that $w \epsilon A_m$. Hence $w \epsilon A_n$ for infinitely many values of n.

Conversely, if $w \epsilon A_n$ for infinitely many values of n, then for all $n, w \epsilon \cup_{m \geq n} A_m$, so $w \epsilon B$.

(b) $w \epsilon C \iff w \epsilon \cup_n \cap_{m \geq n} A_m$. Then there is some n such that $w \epsilon \cap_{m \geq n} A_m$. Therefore $w \epsilon A_m$ for all $m \geq n$, so $w \epsilon A_m$ for all but a finite number of values of n.

Conversely, if $w \epsilon A_m$ for all but a finite number of values of n, then there is some n such that $w \epsilon A_m$ for all $m \geq n$, so $w \epsilon \cup_{m \geq n} A_m$, so $w \epsilon \cup_n \cap_{m \geq n} A_m = C$.

□

The sets B and C are respectively called the limit superior and limit inferior for the sequence of sets $A_1, A_2, \ldots$

Lemma 4.11.2. *(Borel-Cantelli)*

$$P\{B\} = 0 \quad \textit{if} \quad \sum_{i=1}^{\infty} P\{A_n\} < \infty$$

Proof.

$$B = \cap_n \cup_{m=n}^{\infty} A_n \subseteq \cup_{m=n}^{\infty} A_m \quad \text{for all } n.$$

Therefore

$$P\{B\} \leq \sum_{m=n}^{\infty} P\{A_m\} \to 0 \quad \text{as } n \to \infty.$$

□

Lemma 4.11.3.

(a) Let $A_1, A_2, \ldots$ be a non-decreasing sequence of events, so $A_1 \subseteq A_2 \subseteq \ldots$ and let

$$A = \cup_{i=1}^{\infty} A_i = \lim_{i \to \infty} A_i.$$

Then $P\{A\} = \lim_{i\to\infty} P\{A_i\}$.

(b) Let $B_1, B_2, \ldots$ be a non-increasing sequence of events, so $B_1 \supseteq B_2 \supseteq \ldots$ and let

$$B = \cap_{i=1}^{\infty} B_i = \lim_{i \to \infty} B_i.$$

Then

$$P\{B\} = \lim_{i \to \infty} P\{B_i\}.$$

Proof. (a) $A = A_1 \cup A_2\overline{A}_1 \cup A_3\overline{A}_2 \cup \ldots$ is the union of a disjoint family of events. Then

$$\begin{aligned} P\{A\} &= P\{A_1\} + \sum_{i=1}^{\infty} P\{A_{i+1}\overline{A}_i\} \\ &= P\{A_1\} + \lim_{n\to\infty} \sum_{i=1}^{n-1} [P\{A_{i+1}\} - P\{A_i\}] \\ &= \lim_{n\to\infty} P\{A_n\}. \end{aligned} \tag{4.156}$$

(b) Let $A_i = \overline{B}_i$. Then the A_i's are non-decreasing, so *(a)* applies.

$$A = \cup_{i=1}^{\infty} A_i = \cup_{i=1}^{\infty} \overline{B}_i = \left(\overline{\cap_{i=1}^{\infty} B_i}\right) = \overline{B}. \tag{4.157}$$

$$\begin{aligned} 1 - P\{B\} = P\{\overline{B}\} = P\{A\} &= \lim_{i\to\infty} P\{A_i\} = \lim_{i\to\infty} [1 - P\{B_i\}] \\ = 1 - \lim_{i\to\infty} P\{B_i\}. \end{aligned} \tag{4.158}$$

Hence

$$P\{B\} = \lim_{i\to\infty} P\{B_i\}.$$

□

4.11.2 Modes of convergence of random variables

There are several different senses in which a sequence of random variables might be said to approach a limiting random variable. This section deals with only two:

(a) Convergence in probability:
X_n converges in probability to X $\Longleftrightarrow$

$$P\{\mid X_n - X \mid > \epsilon\} \to 0 \quad \text{for all } \epsilon > 0.$$

This case is denoted $X_n \overset{p}{\to} X$.

(b) Convergence almost surely:
X_n converges to X almost surely (written $X_n \overset{a.s.}{\to} X$) $\Longleftrightarrow$

$$P\{w\epsilon S : X_n(w) \to X(w)\} = 1.$$

The weak law of large numbers (section 2.13) can be rephrased to say that if $X_1, \ldots$ are independent and identically distributed with mean μ, then

$$\overline{X}_n = \sum_{i=1}^{n} X_i/n \xrightarrow{p} \mu,$$

or, more properly $\overline{X}_n$ converges in probability to the random variable that takes the value μ with probability 1.

Let $A_n(\epsilon) = \{w :\mid X_n(w) - X(w) \mid > \epsilon\}$, and let

$$B_m(\epsilon) = \cup_{n \geq m} A_n(\epsilon). \tag{4.159}$$

Lemma 4.11.4.

$$P\{B_m(\epsilon)\} \to 0 \text{ as } m \to \infty \text{ if and only if } X_n \xrightarrow{a.s.} X.$$

Proof. Fix $\epsilon > 0$.

$(B_m(\epsilon), m \geq 1)$ is a non-increasing sequence of sets whose limit is

$$A(\epsilon) = \cap_m B_m(\epsilon) = \{w \epsilon S : w \epsilon A_n(\epsilon) \text{ for infinitely many values of } n\}. \tag{4.160}$$

Therefore $P\{B_m(\epsilon)\} \to 0$ as $m \to \infty$ if and only if $P\{A(\epsilon)\} = 0$.

Let

$$\begin{aligned} C =& \{w \epsilon S : X_n(w) \to X(w) \text{ as } n \to \infty\} \\ P\{\overline{C}\} =& P\{\cup_{\epsilon > 0} A(\epsilon)\} = P\{\cup_{m=1}^{\infty} A(m^{-1})\} \\ \leq& \sum_{m=1}^{\infty} P\{A(m^{-1})\} = 0 \text{ if } P\{A(\epsilon)\} = 0 \text{ for all } \epsilon > 0. \end{aligned} \tag{4.161}$$

So $P\{C\} = 1$ in this case, and hence $X_n(w) \xrightarrow{a.s.} X(w)$.

Now suppose $P\{A(\epsilon)\} \neq 0$ for some $\epsilon > 0$. Then $P\{\overline{C}\} > 0$, so X_n does not almost surely approach X, and $P\{B_m(\epsilon)\}$ does not approach 0 as $m \to \infty$. □

Lemma 4.11.5. *If $\sum_n P\{A_n(\epsilon)\} < \infty$ for all $\epsilon > 0$, then $X_n \xrightarrow{a.s.} X$.*

Proof. Fix $\epsilon > 0$.

$$P\{B_m(\epsilon)\} = P\{\cup_{n \geq m} A_n(\epsilon)\} \leq \sum_{n=m}^{\infty} P\{A_n(\epsilon)\} \to 0. \tag{4.162}$$

Application of Lemma 4.11.4 now completes the result. □

Lemma 4.11.6. *If $X_n \xrightarrow{a.s.} X$ then $X_n \xrightarrow{p} X$.*

Proof. If $X_n \xrightarrow{a.s.} X$ then by Lemma 4.11.4, $B_m(\epsilon) \to 0$. But $A_n(\epsilon) \leq B_n(\epsilon)$, so $A_n(\epsilon) \to 0$. Hence $P\{\mid X_n - X \mid > \epsilon\} \to 0$, so $X_n \xrightarrow{p} X$. □

The following example shows that almost sure convergence is stronger than convergence in probability, by displaying a sequence of random variables that converge in probability, but not almost surely.

Example:

Let X_n be a sequence of independent random variables such that

$$X_n = \begin{cases} 1 & \text{with probability } 1/n \\ 0 & \text{otherwise.} \end{cases}$$

Obviously $X_n \xrightarrow{p} 0$, the random variable taking the value of 0 with probability 1.

Let $0 < \epsilon < 1$. Then

$$A_n(\epsilon) = \{w \mid X_n(w) - 0 \mid > \epsilon\} = \{w \mid X_n(w) = 1\}.$$

Hence $B_m(\epsilon) = \cup_{n \geq m} A_n(\epsilon)$ is the event that at least one

$$X_n(w) = 1, \text{where } n \geq m.$$

Hence

$$\begin{aligned}
P\{B_m(\epsilon)\} &= 1 - \lim_{r \to \infty} P\{X_n = 0 \text{ for all } m \leq n \leq r\} \\
&= 1 - \lim_{M \to \infty} \left(1 - \frac{1}{m}\right)\left(1 - \frac{1}{m+1}\right) \cdots \left(\frac{M}{M+1}\right) \qquad \text{(uses independence)} \\
&= 1 - \lim_{M \to \infty} \left\{\frac{m-1}{m} \cdot \frac{m}{m+1} \cdots \frac{M}{M+1}\right\} \\
&= 1 - \lim_{M \to \infty} \frac{m-1}{M+1} = 1. \qquad (4.163)
\end{aligned}$$

Therefore X_n does not converge almost surely to 0. □

Having shown that almost sure convergence is stronger than convergence in probability, and having been reminded that the weak law of large numbers shows that $\overline{X}_n$ converges in probability to μ provided the X_i's are independent, identically distributed and have mean μ, the reader may not be astonished to learn that the strong law of large numbers is the same result, under the same conditions, with respect to almost sure convergence.

4.11.3 Four algebraic lemmas

It will not be obvious why the four lemmas in this subsection are interesting or important. However, they are each used in the proof of the strong law in the next section.

For the purposes of this section and much of the rest, $\alpha > 1$ is a constant.

Lemma 4.11.7. *Let $\alpha > 1$. There exists a $K > 0$ such that, for all $k \geq K, \alpha^{k-1} \leq \alpha^k - 1$.*

Proof. The inequality is equivalent to

$$1 \leq \alpha^k - \alpha^{k-1} = \alpha^{k-1}(\alpha - 1).$$

Since $\alpha > 1, (\alpha - 1) > 0$, and $\alpha^{k-1} \to \infty$. Hence there is some K such that for all $k \geq K, \alpha^{k-1}(\alpha - 1) > 1$. □

It is now necessary to introduce the floor function, $\lfloor x \rfloor$, which is the largest integer no larger than x.

Lemma 4.11.8. *Let $\beta_k = \lfloor \alpha^k \rfloor$. Then there is a finite constant A such that*

$$\sum_{k=m}^{\infty} \frac{1}{\beta_k^2} \leq \frac{A}{\beta_m^2} \textit{ for all } m \geq 1.$$

Remark: What makes the lemma a bit tricky to prove is the operation of the floor function. So for practice and to make this lemma plausible, I prove it first without the floor function. Thus (within this remark only) I redefine $\beta_k = \alpha^k$. Then

$$\begin{aligned}
\sum_{k=m}^{\infty} \frac{1}{\beta_k^2} &= \sum_{k=m}^{\infty} \frac{1}{\alpha^{2k}} = \frac{1}{\alpha^{2m}} \sum_{k=0}^{\infty} \frac{1}{\alpha^{2k}} \\
&= \frac{1}{\beta_m^2} \frac{1}{1 - 1/\alpha^2} = \frac{1}{\beta_m^2} \left(\frac{\alpha^2}{\alpha^2 - 1}\right) \qquad (4.164)
\end{aligned}$$

so $A = \alpha^2/(\alpha^2-1)$ suffices. The intuition of the lemma is that $\lfloor \alpha^k \rfloor$ is "almost" like α^k, so something like this proof should work, at least for large m.

Proof. I first prove the result for all large m, specifically for all $m \geq K$, where K is the number found in Lemma 4.11.7, as follows:

Let $m \geq K$. Then

$$\begin{aligned}
\beta_m^2 \sum_{k=m}^{\infty} \frac{1}{\beta_k^2} &\leq \alpha^{2m} \sum_{k=m}^{\infty} \frac{1}{\alpha^{2(k-1)}} && \text{(uses Lemma 4.11.7)} \\
&= \alpha^{2m+2} \sum_{k=m}^{\infty} \left(\frac{1}{\alpha^2}\right)^k \\
&= \frac{\alpha^{2m+2}}{\alpha^{2m}} \sum_{k=0}^{\infty} \left(\frac{1}{\alpha^2}\right)^k \\
&= \alpha^2 \frac{1}{1-1/\alpha^2} \\
&= \alpha^4/(\alpha^2-1). && (4.165)
\end{aligned}$$

Hence $A_1 = \alpha^4/(\alpha^2-1)$ is sufficient for all $m \geq K$.

Now let $\beta_m^* = \begin{cases} \beta_m & m \geq K \\ \beta_K & m \leq K \end{cases}$.

Then $\beta_m^* \geq \beta_m$ for all m.

Using this, for $m \leq K$

$$\begin{aligned}
\beta_m^2 \sum_{k=m}^{\infty} \frac{1}{\beta_k^2} &\leq \beta_m^{*2} \sum_{k=m}^{\infty} \frac{1}{\beta_k^2} \\
&= \beta_K^{*2} \sum_{k=m}^{K} \frac{1}{\beta_k^2} + \sum_{k=K+1}^{\infty} \frac{\beta_m^2}{\beta_k^2} \\
&\leq A_1 + \beta_K^2 \sum_{k=m}^{K} \frac{1}{\beta_k^2} \\
&\leq A_1 + \beta_K^2 \sum_{k=1}^{K} \frac{1}{\beta_k^2}. && (4.166)
\end{aligned}$$

Hence $A = A_1 + \beta_K^2 \sum_{k=1}^{K} \frac{1}{\beta_K^2}$ suffices for all m. □

(The key point in the above proof is that once it is proved for all large $m \geq K$, the finite initial part is easily bounded.)

Lemma 4.11.9. $\lim_{k\to\infty} \left(\frac{\beta_{K+1}}{\beta_K}\right) = \alpha$.

$$\frac{\beta_{k+1}}{\beta_k} = \frac{\lfloor \alpha^{k+1} \rfloor}{\lfloor \alpha^k \rfloor} \leq \frac{\alpha^{k+1}}{\alpha^{k-1}} = \frac{\alpha}{1-1/\alpha^k}. \tag{4.167}$$

Hence

$$\limsup_{k\to\infty} \left(\frac{\beta_{k+1}}{\beta_k}\right) = \alpha. \tag{4.168}$$

Similarly

$$\frac{\beta_{k+1}}{\beta_K} \geq \frac{\alpha^{k+1}-1}{\alpha_K} = \alpha - 1/\alpha^k, \tag{4.169}$$

so

$$\lim_{k\to\infty} \inf \left(\frac{\beta_{k+1}}{\beta_k}\right) = \alpha. \tag{4.170}$$

Hence $\lim_{k\to\infty} \frac{\beta_{K+1}}{\beta_K} = \alpha$. □

There's one additional lemma that comes up in the proof of the strong law.

Lemma 4.11.10. *If* $\lim_{n\to\infty} x_n = c$ *then* $\lim \frac{\sum_{i=1}^n x_i}{n} = c$.

Proof. Choose $\epsilon > 0$.

There is an N_1 such that for all $n \geq N_1$, $| x_n - c | < \epsilon/2$. Now $\sum_{i=1}^{N_1} | x_i - c |$ is a fixed number, so there is some N_2 such that, for all $n \geq N_2$, $\frac{\sum_{i=1}^{N_1} |x_i - c|}{n} < \epsilon/2$.

Let $N = \max\{N_1, N_2, 2\}$. Then for all $n \geq N$,

$$\begin{aligned}\left|\frac{\sum_{i=1}^n x_i}{n} - c\right| &= \left|\frac{\sum_{i=1}^n (x_i - c)}{n}\right| \leq \frac{\sum_{i=1}^n |x_i - c|}{n} \\ &\leq \sum_{i=1}^{N_1} \frac{|x_i - c|}{n} + \sum_{i=N_1+1}^{n} \frac{|x_i - c|}{n} < \epsilon/2 + \epsilon/2n < \epsilon. \end{aligned} \tag{4.171}$$

Hence $\lim \sum_{i=1}^n x_i/n = c$. □

4.11.4 The strong law of large numbers

Finally, the stage is now set for a proof of the strong law:

Theorem 4.11.11. *Let* $X_1, X_2, \ldots,$ *be a sequence of independent and identically distributed random variables such that* $E \mid X_1 \mid < \infty$, *and let* $E(X_1) = \mu$. *Then*

$$\overline{X}_n = \sum_{i=1}^n X_i/n \overset{a.s.}{\to} \mu.$$

Proof. First suppose that X_1 (and hence all the other X's) are non-negative. (This restriction is removed at the end of the proof).

Let

$$Y_n = \begin{cases} X_n & \text{if } X_n < n \\ 0 & \text{otherwise} \end{cases}.$$

(The Y_n's are still independent, but no longer identically distributed.)

Now

$$\begin{aligned}
&\sum_{n=1}^{\infty} P\{X_n \neq Y_n\} \\
=&\sum_{n=1}^{\infty} P\{X_n \geq n\} && \text{(definition of } Y_n) \\
=&\sum_{n=1}^{\infty} P\{X_1 \geq n\} && (X\text{'s identically distributed}) \\
\leq&\sum_{n=1}^{\infty} P\{\lfloor X_1 \rfloor \geq n\} && \lfloor X_1 \rfloor \leq X_1 \\
\leq& E(\lfloor X_1 \rfloor) && \text{by 3.10.2} \\
\leq& E(X_1) && \lfloor X_1 \rfloor \leq X_1 \\
<&\infty. && \text{by assumption} \qquad (4.172)
\end{aligned}$$

Applying the Borel-Cantelli Lemma 4.11.2

$$P\{X_n \neq Y_n \text{ for infinitely many values of } n\} = 0. \tag{4.173}$$

Therefore

$$\frac{1}{n}\sum_{i=1}^{n}(X_i - Y_i) \overset{a.s.}{\rightarrow} 0 \text{ as } n \rightarrow \infty. \tag{4.174}$$

Hence it suffices to show $\sum_{i=1}^{n} Y_i/n \overset{a.s.}{\rightarrow} \mu$ as $n \rightarrow \infty$.

The substitution of the Y's for the X's is called truncation, and is widely used in probability theory.

Let $S'_n = \sum_{i=1}^{n} Y_i$, let $\alpha > 1$ and $\epsilon > 0$ be given.

Then

$$P\left\{\frac{1}{\beta_n}\left|S'_{\beta_n} - E(S'_{\beta_n})\right| > \epsilon\right\} \leq \frac{1}{\epsilon^2}\cdot\frac{1}{\beta_n^2}Var(S'_{\beta_n}) \tag{4.175}$$

by Tchebychev's Inequality (see section 2.13).

Consequently

$$\begin{aligned}
&\sum_{n=1}^{\infty} P\{\frac{1}{\beta_n} \mid S'_{\beta_n} - E(S'_{\beta_n}) \mid > \epsilon\} \\
\leq&\frac{1}{\epsilon^2}\sum_{n=1}^{\infty}\frac{1}{\beta_n^2}Var(S'_{\beta_n}) \\
=&\frac{1}{\epsilon^2}\sum_{n=1}^{\infty}\frac{1}{\beta_n^2}\sum_{i=1}^{\beta_n}Var(Y_i) \\
=&\frac{1}{\epsilon^2}\sum_{n=1}^{\infty}\sum_{i=1}^{\infty}\frac{1}{\beta_n^2}Var(Y_i)I_{\{i \leq \beta_n\}} \\
=&\frac{1}{\epsilon^2}\sum_{i=1}^{\infty}Var(Y_i)\sum_{n=1}^{\infty}\frac{1}{\beta_n^2}I\{i \leq \beta_n\} \\
\leq&\frac{1}{\epsilon^2}\sum_{i=1}^{\infty}E(Y_i^2)\sum_{n:\beta_n \geq i}\frac{1}{\beta_n^2}.
\end{aligned} \tag{4.176}$$

Let $m = \min\{n \mid \beta_n \geq i\}$.

Then

$$\begin{aligned}
&\sum_{n=1}^{\infty} P\left\{\frac{1}{\beta_n} \mid S'_{\beta_n} - E(S'_{\beta_n}) \mid > \epsilon\right\} \\
&\leq \frac{1}{\epsilon^2}\sum_{i=1}^{\infty} E(Y_i^2) \sum_{n=m}^{\infty} \frac{1}{\beta_n^2} \\
&\leq \frac{1}{\epsilon^2}\sum_{i=1}^{\infty} E(Y_i^2) \cdot \frac{A}{\beta_m^2} && \text{(uses Lemma 4.11.8)} \\
&\leq \frac{A}{\epsilon^2}\sum_{i=1}^{\infty} \frac{1}{i^2} E(Y_i^2). && \text{(definition of } m) \qquad (4.177)
\end{aligned}$$

Next, we bound $\sum_{i=1}^{\infty} \frac{1}{i^2} E(Y_i^2)$ as follows:

Let $B_{i_j} = \{j - 1 \leq X_i < j\}$ and note $P\{B_{ij}\} = P\{B_{1j}\}$ for all i and j.

Then

$$\begin{aligned}
\sum_{i=1}^{\infty} \frac{1}{i^2} E(Y_i^2) &= \sum_{i=1}^{\infty} \frac{1}{i^2} \sum_{j=1}^{i} E(Y_i^2 I_{B_{ij}}) \\
&\leq \sum_{i=1}^{\infty} \frac{1}{i^2} \sum_{j=1}^{i} j^2 P\{B_{ij}\} && \text{(on } B_{ij}, X_i \text{ is no larger than } j) \\
&= \sum_{i=1}^{\infty}\sum_{j=1}^{\infty} \frac{j^2}{i^2} P\{B_{ij}\} I_{j \leq i} \\
&= \sum_{j=1}^{\infty} j^2 \sum_{i=j}^{\infty} \frac{1}{i^2} P\{B_{ij}\} \\
&= \sum_{j=1}^{\infty} j^2 P\{B_{1j}\} \sum_{i=j}^{\infty} \frac{1}{i^2}. && (4.178)
\end{aligned}$$

Now to bound $\sum_{i=j}^{\infty} \frac{1}{i^2}$, think of this as a step function, less than $1/x^2$ if $x < i$. It is necessary to separate out the case of $j = 1$, as follows:

$$\sum_{i=1}^{\infty} \frac{1}{i^2} = 1 + \sum_{i=2}^{\infty} \frac{1}{i^2} \leq 1 + \int_1^{\infty} \frac{1}{x^2} dx = 1 - 1/x \mid_1^{\infty} = 2 = 2/j.$$

If $j \geq 2$,

$$\sum_{i=j}^{\infty} \frac{1}{i^2} \leq \int_{j-1}^{\infty} \frac{1}{x^2} dx = 1/(j-1) \leq 2/j.$$

Hence for all $j \geq 1, \sum_{i=j}^{\infty} \frac{1}{i^2} \leq 2/j$.

Therefore

$$\begin{aligned}
&\sum_{j=1}^{\infty} j^2 P\{B_{1j}\} \sum_{i=j}^{\infty} \frac{1}{i^2} \\
\leq& \sum_{j=1}^{\infty} j^2 P\{B_{1j}\} \cdot 2/j \\
=& 2 \sum_{j=1}^{\infty} j P\{B_{1j}\} \\
=& 2 \sum_{j=1}^{\infty} [(j-1)+1] P\{B_{1j}\} \\
\leq& 2(E(X)+1) < \infty.
\end{aligned} \tag{4.179}$$

Hence

$$\sum_{n=1}^{\infty} P\left\{\frac{1}{\beta_n} \mid S'_{\beta_n} - E(S'_{\beta_n}) \mid > \epsilon\right\} < \infty, \tag{4.180}$$

using (4.177), (4.178) and (4.179).

Therefore, by Lemma 4.11.5,

$$\frac{1}{\beta_n}[S'_{\beta_n} - E(S'_{\beta_n})] \overset{a.s.}{\rightarrow} 0 \text{ as } n \to \infty. \tag{4.181}$$

We now turn to evaluating the expectation:

$$E(Y_n) = E(X_n I_{\{X_n < n\}}) = E(X_1 I_{\{X_i < n\}}) \to E(X_1) = \mu \tag{4.182}$$

as $n \to \infty$ by monotone convergence. Hence, applying Lemma 4.11.10,

$$\frac{1}{\beta_n} E(S'_{B_n} = \frac{1}{\beta_n} \sum_{i=1}^{\beta_n} E(Y_i) \to \mu \text{ as } n \to \infty. \tag{4.183}$$

Therefore we may conclude

$$\frac{1}{\beta_n} S'_n \overset{a.s.}{\rightarrow} \mu \text{ as } n \to \infty. \tag{4.184}$$

This proves the result, but only for particular β_n's, not for all n. Now, because the Y_i's are non-negative, the sequence S'_n is non-decreasing. Therefore, if $\beta_n \leq m \leq \beta_{n+1}$,

$$\frac{1}{\beta_{n+1}} S'_{\beta_n} \leq \frac{S'_m}{m} \leq \frac{1}{\beta_n} S'_{\beta_{n+1}}. \tag{4.185}$$

Now

$$\frac{\beta_n}{\beta_{n+1}} \frac{S'_{\beta_n}}{\beta_n} \leq \frac{S'_m}{m} \leq \frac{\beta_{n+1}}{\beta_n} \frac{S'_{\beta_{n+1}}}{\beta_{n+1}}. \tag{4.186}$$

Let $m \to \infty$ and apply Lemma 4.11.9 to obtain

$$\alpha^{-1}\mu \leq \liminf \frac{S'_m}{m} \leq \limsup \frac{S'_m}{m} \leq \alpha\mu \text{ almost surely.} \tag{4.187}$$

Since this holds for all $\alpha > 1$, we may now let $\alpha \to 1$, and find

$$\lim \frac{S'_m}{m} = \mu \text{ almost surely as } m \to \infty \tag{4.188}$$

when the X_is are non-negative. The last step is to remove this constraint.

For general X_i's, define

$$X_n^+(w) = \max\{X_n(w), 0\} \text{ and } X_n^-(w) = -\min\{X_n(w), 0\}.$$

Then X_n^+ and X_n^- are non-negative, and $X_n = X_n^+ - X_n^-$.

Since $X_n^+ \leq\mid X_n \mid$ and $X_n^- \leq\mid X_n \mid$, both $E(X_n^+)$ and $E(X_n^-)$ exist, and $E(X_n) = E(X_n^+) - E(X_n^-)$.

Therefore

$$\frac{1}{n}S_n = \frac{1}{n}\left(\sum_{i=1}^{n} X_i^+ - \sum_{i=1}^{n} X_i^-\right) \xrightarrow{a.s.} E(X_1^+) - E(X_1^-) = E(X_1) = \mu \tag{4.189}$$

as $n \to \infty$.

This completes the proof of the theorem. □

4.11.5 *Summary*

This section states and proves the strong law of large numbers, and contrasts it with the weak law of large numbers.

4.11.6 *Exercises*

1. Vocabulary. Explain in your own words:
   (a) σ-field
   (b) convergence in probability
   (c) almost sure convergence
   (d) weak law of large numbers
   (e) strong law of large numbers
2. Consider the sequence of independent random variables defined by

$$X_n = \begin{cases} n & \text{with probability } 1/n \\ 0 & \text{with probability } 1 - 1/n \end{cases}.$$

   (a) Does X_n converge almost surely? If so, to what random variable does it converge? Explain your answer.
   (b) Does X_n converge in probability? If so, to what random variable does it converge? Explain your answer.
3. Consider the sequence of independent random variables defined by

$$X_n = \begin{cases} n & \text{with probability } 1/2n\log n \\ 0 & \text{with probability } 1 - 1/n\log n \\ -n & \text{with probability } 1/2n\log n \end{cases}.$$

   Answer the same questions as in problem 2.

4.11.7 *Reference*

Many probability books have proofs of the strong law of large numbers. This one is due to Grimmett and Stirzaker (2001); in general I can recommend this book as being both clear and concise.

Chapter 5

Transformations

5.1 Introduction

Transformations of random variables are essential tools. If X is a random variable, and g is a function, then $Y = g(X)$ is a new random variable. If I know the distribution of X and I know the function g, how do I find the distribution of Y?

Section 5.2 addresses this question when X is discrete. The continuous univariate case, both linear and non-linear, is the subject of 5.3. To deal with the continuous multivariate case requires the development of some matrix algebra, which begins in section 5.4, and culminates in 5.8. For a one-to-one transformation, one substitutes the function value into the density of X, and rescales locally so that the density of Y integrates to one. Then 5.9 shows the derivation of the absolute value of the determinant of the Jacobian matrix as the necessary scaling factor in the multivariate case, linear or non-linear.

The chapter concludes with a discussion of the Borel-Kolmogorov paradox in section 5.10.

5.2 Transformations of discrete random variables

Suppose X is a discrete random variable, such that

$$P\{X = x_i\} = p_i > 0, \quad i = 1, 2, \ldots \tag{5.1}$$

where $\sum_{i=1}^{\infty} p_i = 1$.

Let g be a function such that $g(x_i) \neq g(x_j)$ if $x_i \neq x_j$. Such a function is called one-to-one. Each one-to-one function g has a one-to-one inverse g^{-1} such that $g^{-1}g(x_i) = x_i$ for all i. We seek the distribution of $Y = g(X)$.

$$P\{Y = y_j\} = P\{g^{-1}(Y) = g^{-1}(y_j)\} = P\{X = g^{-1}(y_j)\} = p_j \tag{5.2}$$

if $g^{-1}(y_j) = x_j$.

It is easy to tell whether a function g is one-to-one. The way to tell is to find the inverse function g^{-1}. If you can solve for g^{-1} uniquely, then the function is one-to-one. For example, suppose $g(x) = x^2$. Then we might have $g^{-1}(x) = \pm\sqrt{x}$, so if the random variable X can take both positive and negative values, g would not be one-to-one in general. However if X is restricted to be positive, then g is one-to-one, and $g^{-1}(x) = \sqrt{x}$.

To make this concrete, let's look at an example. Suppose X has a Poisson distribution with parameter λ, *i.e.*,

$$P\{X = k\} = \begin{cases} \frac{e^{-\lambda}\lambda^k}{k!} & k = 0, 1, 2, \ldots \\ 0 & \text{otherwise} \end{cases}.$$

Let $g(x) = 2x$. Then we seek the distribution of $Y = 2X$. Clearly Y has positive values on only the even integers. Also clearly $g^{-1}(y) = y/2$ so g is one-to-one. Then

$$P\{Y = j\} = P\{X = j/2\} = \begin{cases} \frac{e^{-\lambda}\lambda^{(j/2)}}{(j/2)!} & j = 0, 2, 4, \ldots \\ 0 & \text{otherwise} \end{cases}. \tag{5.3}$$

Suppose now that $\mathbf{X} = (X_1, X_2, \ldots, X_k)$ is a vector of discrete random variables, satisfying $P\{\mathbf{X} = \mathbf{x}_j\} = p_j, j = 1, 2, \ldots,$ and $\sum_{j=1}^{\infty} p_j = 1$ and that $\mathrm{g}(\mathbf{x}) = (\mathrm{y}_1, \mathrm{y}_2, \ldots, \mathrm{y}_\mathrm{k})$ is a one-to-one function. We seek the distribution of $\boldsymbol{Y} = g(\boldsymbol{X})$, where now $\boldsymbol{Y}$ is a k-dimensional vector. Again, to check whether the function g is one-to-one, we compute the inverse function g^{-1}. If g^{-1} can be solved for uniquely, the function g is one-to-one. In this case

$$P\{\mathbf{Y} = \mathbf{y}_j\} = P\{g^{-1}(\mathbf{Y}) = g^{-1}(\mathbf{y}_j)\} = P\{X = g^{-1}(\mathbf{y}_j)\} = p_j \quad \text{if } g^{-1}(\mathbf{y}_j) = \mathbf{x}_j. \tag{5.4}$$

Thus the multivariate case works exactly the way the univariate case does.

Of course, marginal distributions are found from joint distributions by summing, and conditional distributions are found by application of Bayes Theorem.

As an example, let X_1 and X_2 have the joint distribution

$$P\{X_1 = x_1, X_2 = x_2\} = x_1x_2/60 \text{ for } x_1 = 1, 2, 3 \text{ and } x_2 = 1, 2, 3, 4.$$

(a) Find the joint distribution of $Y_1 = X_1X_2$ and $Y_2 = X_2$.

(b) Find the marginal distribution of Y_2.

(c) Find the conditional distribution of Y_1 given Y_2.

Solution:

(a) Let $g(x_1, x_2) = (x_1x_2, x_2)$. Let $y_1 = x_1x_2$ and $y_2 = x_2$. Then $x_1 = y_1/y_2$ and $x_2 = y_2$. Since this inverse function exists, the function g is one-to-one.
Hence, applying (5.4),

$$P\{Y_1 = y_1, Y_2 = y_2\} = y_1/60 \text{ for } (y_1, y_2)\epsilon\{(1,1), (2,1), (3,1), (2,2), (4,2), (6,2), (3,3), (6,3), (9,3), (4,4)(8,4), (12,4)\} = D$$

and

$$P\{Y_1 = y_1, Y_2 = y_2\} = 0 \quad \text{otherwise.}$$

(b)

$$\begin{aligned} P\{Y_2 = y_2\} &= \sum_{(y_1,y_2)\epsilon D} P\{Y_1 = y_1, Y_2 = y_2\} \\ &= \sum_{(y_1,y_2)\epsilon D} y_1/60 = y_2 \cdot 6/60 = y_2/10, \quad y_2 = 1, 2, 3, 4 \end{aligned}$$

and $P\{Y_2 = y_2\} = 0$ otherwise.

(c)

$$P\{Y_1 = y_1 \mid Y_2 = y_2\} = \frac{P\{Y_1 = y_1, Y_2 = y_2\}}{P\{Y = y_2\}} = \frac{y_1/60}{y_2/10} = (y_1/y_2) \cdot \frac{1}{6}$$

$$\text{for } y_1\epsilon\{y_2, 2y_2, 3y_2\}$$

$$\text{and } P\{Y_1 = y_1 \mid Y_2 = y_2\} = 0 \quad \text{otherwise.}$$

As can be seen from this example, keeping the domain straight is an important part of the calculation.

Now suppose that g is not necessarily one-to-one. Then fix a value for $Y = g(X)$, say y_j, and let S_j be the set of values x_i of X such that $g(x_i) = y_j$, *i.e.*, $S_j = \{X_i \mid p_i > 0 \text{ and } g(x_i) = y_j\}$. Also let Z_j be an indicator function for y_j. Then, applying property 6 of section 3.5,

$$P\{Y = y_j\} = E[Z] = EE[Z \mid X].$$

Now

$$E[Z \mid X = x_i] = \begin{cases} 1 & Y = g(x_i) \\ 0 & \text{otherwise} \end{cases}.$$

Hence

$$P\{Y = y_j\} = \sum_{x_i \epsilon S_j} p_i. \tag{5.5}$$

This demonstration applies equally well to univariate and multivariate random variables and transformations. Also note that in the special case that S_j consists of only a single element, (5.5) coincides with (5.2) in the univariate case and (5.4) in the multivariate case.

5.2.1 *Summary*

To transform a discrete random variable with a function g, one must check to see if the function is one-to-one. This may be done by calculating the inverse of the function, g^{-1}. If there is an inverse, the function is one-to-one. In this case, probabilities that the transformed random variable take particular values can be computed using (5.2) in the univariate case, or (5.4) in the multivariate case.

When g is not one-to-one, (5.5) applies.

5.2.2 *Exercises*

1. Let X have a Poisson distribution with parameter λ. Suppose $Y = X^2$. Find the distribution of Y. Is g one-to-one?
2. Let X_1 and X_2 be independent random variables each having the distribution

$$P\{X_i = i\} = \begin{cases} 1/6 & i = 1, 2, 3, 4, 5, 6 \\ 0 & \text{otherwise} \end{cases}.$$

   (a) Find the joint distribution of $Y_1 = X_1 + X_2$ and $Y_2 = X_1$.
   (b) Find the marginal distribution of Y_1.
   (c) Find the conditional distribution of Y_1 given Y_2.
   [Y_1 is the distribution of the sum of two fair dice X_1 and X_2 on a single throw.]

5.3 Transformation of univariate continuous distributions

Suppose X is a random variable with cdf $F_X(x)$ and density $f_X(x)$, so that

$$F_X(x) = \int_{-\infty}^{x} f_X(y)dy.$$

Suppose also that g is a real valued function of real numbers. Then $Y = g(X)$ is a new random variable. The purpose of this section is to discuss the distribution of Y, which depends on g and the distribution of X.

Suppose X is a continuous variable on $[-1, 1]$, and let $Y = X^2$, so $g(x) = x^2$, as illustrated in Figure 5.1. Consider the set $S = [0.25, 0.81]$. Then the event $Y \in S$ corresponds to $X \in [-0.9, -0.5] \cup [0.5, 0.9]$, as illustrated in Figure 5.2.

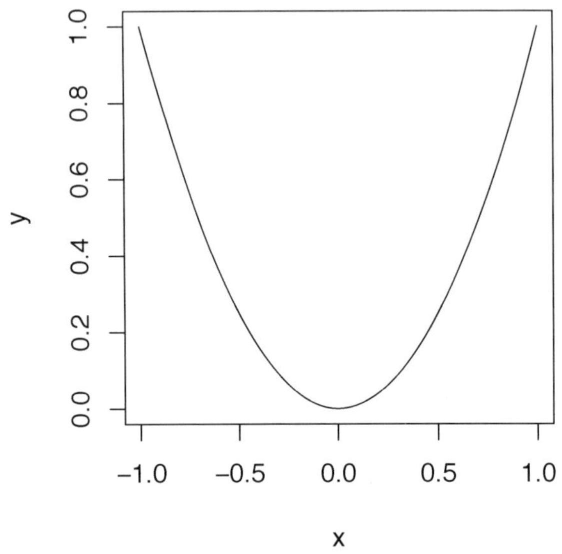

Figure 5.1: Quadratic relation between X and Y.

```
Commands:
x=((-100:100)/100)
y=(x**2)
plot(x,y,type="l")          # type="l" draws a line
```

Then we are asking about the probability that X falls in the two intervals marked in Figure 5.2.

Of course, the probability that X falls in the union of these two intervals is the sum of the probability that X falls in each. So if we can understand how to analyze each piece separately, they can be put together to find probabilities in the more general case. What distinguishes each piece is that within the relevant range of values for y, g is one-to-one.

It is geometrically obvious that a continuous one-to-one function on the real line can't double back on itself, *i.e.*, if it is increasing it has to go on increasing, and if it is decreasing it has to go on decreasing. (Such functions are called monotone increasing and monotone decreasing, respectively.) So we'll consider those two cases, at first separately, and then together.

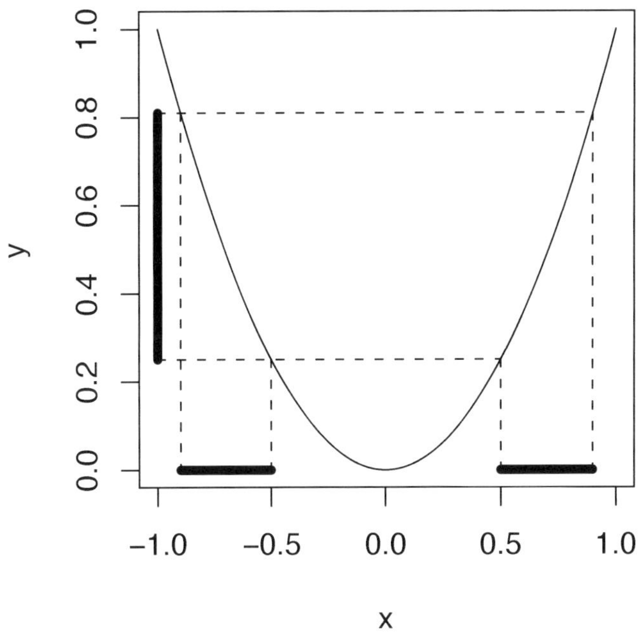

Figure 5.2: The set $[0.25, 0.81]$ for Y is the transform of two intervals for X.

```
Commands:
x=((-100:100)/100)
y=(x**2)
plot(x,y,type="l")
                                  #segments draws a line
                                  #from the (x,y) coordinates
                                  #listed first to the (x,y)
                                  #coordinates listed second
segments (-1,0.25,0.5,0.25,lty=2) #lty=2 gives a dotted line
segments (-0.9,0.81,-0.9,0,lty=2)
segments (-0.5,0.25,-0.5,0,lty=2)
segments (0.5,0.25,0.5,0,lty=2)
segments (0.9,0.81,0.9,0,lty=2)

segments(-0.9,0,-0.5,0,lwd=5)     #lwd=5 gives a line width
                                  #5 times the usual line

segments(0.5,0,0.9,0,lwd=5)
segments(-1,0.25,-1,0.81,lwd=5)
```

Suppose, then, that g is a monotone increasing function on an interval in the real line. We'll also suppose that it is not only continuous, but has a derivative. Then we can compute the c.d.f. of $Y = g(X)$ as follows:

$$\begin{aligned} F_Y(y) &= P\{Y \leq y\} = P\{g(X) \leq y\} \\ &= P\{X \leq g^{-1}(y)\} = F_X(g^{-1}(y)). \end{aligned} \tag{5.6}$$

Differentiating with respect to y, the density of Y is

$$f_Y(y) = \frac{dF_Y(y)}{dy} = f_X(g^{-1}(y))\frac{dg^{-1}(y)}{dy} \tag{5.7}$$

using the chain rule. Since g is monotone increasing, so is g^{-1}, so $\frac{dg^{-1}(y)}{dy}$ is positive.

Now suppose that g is a monotone decreasing differentiable function on an interval in the real line. Then the c.d.f. of $Y = g(X)$ is

$$\begin{aligned} F_Y(y) &= P\{Y \leq y\} = P\{g(X) \leq y\} \\ &= 1 - P\{X < g^{-1}(y)\} = 1 - F_X(g^{-1}(y)). \end{aligned} \tag{5.8}$$

Again (5.8) can be differentiated to give

$$f_Y(y) = \frac{dF_Y(y)}{dy} = -f_X(g^{-1}(y))\frac{dg^{-1}(y)}{dy}. \tag{5.9}$$

Because g is monotone decreasing, so is g^{-1}. Therefore in this case $\frac{dg^{-1}(y)}{dy}$ is negative, but the result for $f_Y(y)$ is positive, as it must be.

Formulae (5.7) and (5.8) can be summarized as follows: If g is one-to-one, then $Y = g(X)$ has density

$$f_Y(y) = f_X(g^{-1}(y)) \mid \frac{dg^{-1}(y)}{dy} \mid . \tag{5.10}$$

Let's see how this works in the case of a linear transformation, *i.e.*, a function $g(x)$ of the form $g(x) = ax + b$ for some a and b. The first step is to compute g^{-1}. If $y = ax + b$, then

$$g^{-1}(y) = x = (y - b)/a. \tag{5.11}$$

From (5.11) we learn some important things. The most important is that in order for g to be one-to-one, we must have $a \neq 0$. Indeed, if $a > 0$, then g is monotone increasing. If $a < 0$, then g is monotone decreasing. The derivative of g^{-1} is now easy to compute:

$$\frac{dg^{-1}(y)}{dy} = 1/a \tag{5.12}$$

so the absolute value is available:

$$\mid \frac{dg^{-1}(y)}{dy} \mid = \frac{1}{\mid a \mid}. \tag{5.13}$$

Thus for a linear $g(x) = ax + b$, $Y = g(X)$ has density

$$f_Y(y) = f_X(\frac{y-b}{a}) \cdot \frac{1}{\mid a \mid}. \tag{5.14}$$

Suppose, for example, that X has a uniform density on $[0, 1]$, which is to say

$$f_X(x) = \begin{cases} 1 & 0 < x < 1 \\ 0 & \text{otherwise} \end{cases}. \tag{5.15}$$

The corresponding c.d.f. is $F_X(x) = \begin{cases} 0 & x \leq 0 \\ x & 0 < x < 1 \\ 1 & x \geq 1 \end{cases}.$

Then, with $g(x) = ax + b$ and $a > 0$, Y will have values in the interval $(b, a + b)$, and its density is

$$f_Y(y) = \begin{cases} f_X(\frac{y-b}{a}) \cdot \frac{1}{|a|} = \frac{1}{|a|} & b < y < a + b \\ 0 & \text{otherwise} \end{cases}.$$

The corresponding c.d.f. is $F_Y(y) = \begin{cases} 0 & y \leq b \\ \frac{y-b}{a} & b < y < a + b \\ 1 & y \geq a + b \end{cases}.$

If $a < 0$, then Y has values in the interval $(a + b, b)$, and its density is

$$f_Y(y) = \begin{cases} f_X(\frac{y-b}{a}) \cdot \frac{1}{|a|} = \frac{1}{|a|} & a + b < y < b \\ 0 & \text{otherwise} \end{cases}.$$

The corresponding c.d.f. here is $F_Y(y) = \begin{cases} 0 & y \leq a + b \\ \frac{y-b}{|a|} & a + b < y < b \\ 1 & y \geq b \end{cases}.$

Therefore, in both cases y has a uniform distribution on an interval of length $|a|$. Thus the role of the factor $\frac{dg^{-1}(y)}{dy}$ is to compensate for the fact that the length of the interval has been changed by the transformation from 1 (because X is uniform on $(0, 1)$), to $|a|$. And this, in turn, is because the derivative of a function (here a c.d.f.) depends on the scale of the variable the derivative is being taken with respect to, which is what the chain rule is all about.

Now consider an example of a linear transformation of a non-uniform random variable. Suppose X has the density

$$f_X(x) = \begin{cases} |\, x \,| & -1 < x < 1 \\ 0 & \text{otherwise} \end{cases}. \tag{5.16}$$

First, we'll check that this is a legitimate density. It is certainly non-negative. Its integral is

$$\begin{aligned} \int_{-\infty}^{\infty} f_X(x)dx &= \int_{-1}^{1} |\, x \,|\, dx = \int_{-1}^{0} (-x)dx + \int_{0}^{1} xdx \\ &= \frac{-x^2}{2}\Big|_{-1}^{0} + \frac{x^2}{2}\Big|_{0}^{1} = -(-\frac{1}{2}) + \frac{1}{2} = 1, \end{aligned}$$

so (5.16) is a legitimate density. Its cumulative distribution function is

$$F_X(x) = \begin{cases} 0 & x \leq -1 \\ \frac{1}{2} - \frac{x^2}{2} & -1 < x < 0 \\ \frac{1}{2} + \frac{x^2}{2} & 0 \leq x < 1 \\ 1 & x > 1 \end{cases}.$$

Then if $g(x) = ax + b$ with $a > 0$, the random variable $Y = g(X)$ has positive density in the range $b - a$ to $b + a$, and has density

$$f_Y(y) = \begin{cases} |\, \frac{y-b}{a^2} \,| & b - a < y < b + a \\ 0 & \text{otherwise} \end{cases}$$

and has c.d.f.

$$F_Y(y) = \begin{cases} 0 & y \leq b-a \\ \frac{1}{2} - \frac{1}{2}(\frac{y-b}{a})^2 & b-a < y \leq b \\ \frac{1}{2} + \frac{1}{2}(\frac{y-b}{a})^2 & b < y < b+a \\ 1 & y \geq b+a \end{cases}.$$

A similar derivation can be found for $a < 0$, and is offered below as an exercise.

Finally, let's look at some examples of non-linear functions. Suppose X again has a uniform density on $[0, 1]$, *i.e.*, its density satisfies (5.15), and now let $g(x) = x^2$. Thus $Y = g(X)$ has positive density also only on the space $(0, 1)$. Computing $g^{-1}(x)$, we find $g^{-1}(y) = \pm\sqrt{y}$. Hence it appears that g is not 1 to 1, because both $\sqrt{x}$ and $-\sqrt{x}$ are possible values of the inverse. However the inverse $-\sqrt{x}$ is irrelevant here, because X takes values only on $(0, 1)$. Hence g is one-to-one as a function from $(0, 1)$ to $(0, 1)$, with inverse $g^{-1}(y) = \sqrt{y}$. Then the derivative is

$$\frac{dg^{-1}(y)}{dy} = \frac{d}{dy}y^{1/2} = \frac{1}{2}y^{-1/2} = \frac{1}{2\sqrt{y}}.$$

Hence Y has density

$$f_Y(y) = f_X(g^{-1}(y)) \mid \frac{dg^{-1}(y)}{dy} \mid = \begin{cases} \frac{1}{2\sqrt{y}} & 0 < y < 1 \\ 0 & \text{otherwise} \end{cases}$$

and c.d.f.

$$F_Y(y) = \begin{cases} 0 & y \leq 0 \\ \sqrt{y} & 0 < y < 1 \\ 1 & y \geq 1 \end{cases}.$$

This example illustrates an important point. Given an arbitrary function (especially a non-linear one), it may not be obvious whether it is one-to-one. Computing the inverse is an excellent way to check. If the inverse is found and is unique, then the function is indeed one-to-one. In the example, the function $g(x) = x^2$ is one-to-one as a function from $(0, 1)$ to $(0, 1)$, but not as a function, say, from $(-1, 1)$ to $(0, 1)$. Thus, in that example, had X had a uniform distribution on $(-1, 1)$, it would have been necessary to do separate analyses for X in $(-1, 0)$ and $(0, 1)$ and then to put them together, because $g(x) = x^2$ is one-to-one as a function from $(-1, 0)$ to $(0, 1)$ (with inverse $-\sqrt{x}$) and from $(0, 1)$ to $(0, 1)$ (with inverse $\sqrt{x}$).

5.3.1 *Summary*

If X has a continuous distribution with pdf $f_X(x)$, and g is a differentiable one-to-one function on the set where $f_X(x) > 0$, then the density of $Y = g(x)$ is given by (5.8). Whether g is one-to-one can be checked by computing its inverse.

5.3.2 *Exercises*

1. Vocabulary. State in your own words the meaning of:
   (a) one-to-one function
   (b) inverse function
   (c) monotone increasing (decreasing) function
2. Let X have the density specified by (5.16) and let $g(x) = ax + b$ with $a < 0$. Find the p.d.f. and the c.d.f. of $Y = g(X)$.

5.3.3 A note to the reader

The purpose of the remainder of this chapter is to develop the multivariate analog of the results of section 5.3. This requires what may seem like a long digression into linear algebra, but it also provides tools we'll need for the rest of the book. An alternative would be to retreat to "it can be shown that...," but that makes the book less self-contained, and by not showing the proofs, obscures the force of the assumptions made.

A reader whose grasp of matrix and vector notation is not solid might benefit from rereading section 2.12.1 at this point.

5.4 Linear spaces, inner products and orthogonality

This section introduces some of the tools needed to understand linear transformations, and then non-linear transformations, in many dimensions.

Definition: A linear space (also called a vector space) is a set of elements $\mathcal{M}$ closed in the following sense: if $\mathbf{x} \epsilon \mathcal{M}$, $\mathbf{y} \epsilon \mathcal{M}$ and α and β are real numbers, then

$$\alpha \mathbf{x} + \beta \mathbf{y} \ \epsilon \ \mathcal{M}.$$

If $\alpha = \beta = 0$, then $\mathbf{0} \ \epsilon \mathcal{M}$. Also, of course, by induction, if $\mathbf{x}_1, \ldots, \mathbf{x}_n \epsilon \mathcal{M}$ and $\alpha_1, \alpha_2, \ldots, \alpha_n$ are real numbers, then

$$\sum_{i=1}^{n} \alpha_i \mathbf{x}_i \ \epsilon \ \mathcal{M}.$$

Consider the following examples:

(i) $S_1 = \{(a, a, 0)' \ , -\infty < a < \infty\}$.

(ii) $S_2 = \{(a, a, 0)' \ , a \leq 0\}$.

(iii) $S_3 = \{(a, b, 0)' \ , -\infty < a < \infty, \ -\infty < b < \infty\}$.

(iv) $S_4 = \{(a, b, c)', -\infty < a < \infty, -\infty < b < \infty, -\infty < c < \infty, c \neq 0\}$.

S_1 and S_3 are linear spaces, but S_2 is not, since if $\alpha = -1$, $(-a, -a, 0)' \notin S_2$ if $a < 0$. Also S_4 is not a linear space because $(0, 0, 0)$ is excluded. However, each of these examples has more than finitely many elements.

Definition: A set of vectors $\mathbf{x}_1, \mathbf{x}_2, \ldots, \mathbf{x}_n$ is said to *span* a linear space $\mathcal{M}$ if every element $\mathbf{x} \epsilon \mathcal{M}$ can be expressed as a linear combination of $\{\mathbf{x}_1, \mathbf{x}_2, \ldots, \mathbf{x}_n\}$, *i.e.*, if there exist numbers $\alpha_1, \alpha_2, \ldots, \alpha_n$ such that

$$\mathbf{x} = \sum_{i=1}^{n} \alpha_i \mathbf{x}_i.$$

Definition: A set of vectors $\{\mathbf{x}_1, \ldots, \mathbf{x}_n\}$ is said to be *linearly independent* if the only numbers $\alpha_1, \ldots, \alpha_n$ satisfying

$$\sum_{i=1}^{n} \alpha_i \mathbf{x}_i = \mathbf{0}$$

are $\alpha_1 = \alpha_2 = \ldots = \alpha_n = 0$. Otherwise they are said to be linearly dependent.

If the vectors $\{\mathbf{x}_1, \mathbf{x}_2, \ldots, \mathbf{x}_n\}$ are linearly independent and $\sum_{i=1}^{n} \alpha_i \mathbf{x}_i = \sum_{i=1}^{n} \alpha_i' \mathbf{x}_i$ for some numbers α_i and α_i', then $\alpha_i = \alpha_i'$. The reason for this is $\mathbf{0} = \sum_{i=1}^{n} \alpha_i \mathbf{x}_i - \sum_{i=1}^{n} \alpha_i' \mathbf{x}_i = \sum_{i=1}^{n} (\alpha_i - \alpha_i') \mathbf{x}_i$ By definition of linear independence, we must have $\alpha_i - \alpha_i' = 0$ for all i, so $\alpha_i = \alpha_i'$.

Lemma 5.4.1. *A set of non-zero vectors* $\{\mathbf{x}_1, \mathbf{x}_2, \ldots, \mathbf{x}_n\}$ *is linearly dependent if and only if some* $\mathbf{x}_k, 2 \leq k \leq n$ *is a linear combination of the preceding ones.*

Proof. A single vector $\mathbf{x}_1$ is automatically linearly independent. Let k be the first integer for which $\mathbf{x}_1, \ldots, \mathbf{x}_k$ are linearly dependent, $2 \leq k \leq n$. Then there are numbers $\alpha_1, \alpha_2, \ldots, \alpha_k$, not all zero, such that $\alpha_1\mathbf{x}_1 + \alpha_2\mathbf{x}_2 + \ldots + \alpha_k\mathbf{x}_k = \mathbf{0}$.

Also $\alpha_k \neq 0$ by definition of k. Then

$$\mathbf{x}_k = -(\alpha_1/\alpha_k)\mathbf{x}_1 + \ldots + (-\alpha_{k-1}/\alpha_k)\mathbf{x}_{k-1}.$$

Conversely, if $\mathbf{x}_k$ is a linear combination of $\mathbf{x}_1, \ldots, \mathbf{x}_{k-1}$, then obviously the set $\{\mathbf{x}_1, \ldots, \mathbf{x}_n\}$ is linearly dependent.

We need two more definitions before we can begin to prove something:

Definition: A set of vectors $\{\mathbf{x}_1, \ldots, \mathbf{x}_n\}$ is said to be a *basis* for a linear space $\mathcal{M}$ if they are linearly independent and span $\mathcal{M}$.

Definition: A linear space $\mathcal{M}$ is finite dimensional if it has a finite basis.

Theorem 5.4.2. *If* $\mathcal{M}$ *is finite dimensional every linearly independent set can be extended to be a basis.*

Proof. Let $\{\mathbf{y}, \ldots \mathbf{y}_m\}$ be a linearly independent set. Since $\mathcal{M}$ is finite dimensional, it has a finite basis $\{\mathbf{x}, \ldots, \mathbf{x}_n\}$. Now consider the vectors

$$\mathbf{y}_1, \ldots, \mathbf{y}_m, \mathbf{x}_1, \ldots, \mathbf{x}_n$$

in that order. This set is linearly dependent, since the $\mathbf{x}$'s form a basis, so each y_i may be expressed as a linear combination of the $\mathbf{x}$'s. Applying the lemma, there is a first element that is a linear combination of the others. Furthermore, since the y's are linearly independent, this first element must be an $\mathbf{x}$, say $\mathbf{x}_i$. Now consider

$$\mathbf{y}_1, \ldots, \mathbf{y}_m, \mathbf{x}_1, \ldots, \mathbf{x}_{i-1}, \mathbf{x}_{i+1}, \ldots, \mathbf{x}_n.$$

Every vector in $\mathcal{M}$ is a linear combination of vectors in this set, since $\mathbf{x}_i$ is a linear combination of them, and $\mathbf{x}_1, \ldots, \mathbf{x}_n$ are a basis for $\mathcal{M}$. If this set is linearly independent, the theorem is proved. If not, the lemma is applied recursively until it is. Thus we obtain a linearly independent set that includes $\mathbf{y}_1, \ldots, \mathbf{y}_m$ and that spans $\mathcal{M}$, and is therefore a basis for $\mathcal{M}$. □

Theorem 5.4.3. *If* $\{\mathbf{x}_1, \ldots, \mathbf{x}_n\}$ *and* $\{\mathbf{y}_1, \ldots, \mathbf{y}_m\}$ *are both bases of a linear space* $\mathcal{M}$, *then* $n = m$.

Proof. First, we use the properties of a basis that $\mathbf{y}_1, \ldots, \mathbf{y}_m$ are independent and $\mathbf{x}_1, \ldots, \mathbf{x}_n$ span $\mathcal{M}$. Apply the lemma to

$$\mathbf{y}_m, \mathbf{x}_1, \ldots, \mathbf{x}_n.$$

As before, one of the $\mathbf{x}$'s, say $\mathbf{x}_i$, is the first linearly dependent element, so

$$\mathbf{y}_m, \mathbf{x}_1, \ldots, \mathbf{x}_{i-1}, \mathbf{x}_{i+1}, \ldots, \mathbf{x}_n$$

are linearly independent, and span $\mathcal{M}$. Now apply the same argument to

$$\mathbf{y}_{m-1}, \mathbf{y}_m, \mathbf{x}_1, \ldots, \mathbf{x}_{i-1}, \mathbf{x}_{i+1}, \ldots, \mathbf{x}_n.$$

After using this argument m times, we obtain a set $\mathbf{y}_1, \ldots, \mathbf{y}_m$ followed by $n - m$ x's. Hence $n \geq m$. Reversing the roles of the x's and y's, we also have $m \geq n$. Hence $n = m$. □

Definition: The number of elements in the basis of a finite dimensional linear space is called the dimension of the space.

5.4.1 A mathematical note

There's a very elegant theory of linear spaces, also called vector spaces. The theorems above are from Halmos (1958), one of the most elegant of the expositions. However, we don't need the generality of the abstract theory, so we won't explore it further here.

5.4.2 Inner products

Definition: The inner product of two vectors of the same dimension, $\mathbf{x} = (x_1, \ldots, x_n)$ and $\mathbf{y} = (y_1, \ldots, y_n)$ is denoted $< \mathbf{x}, \mathbf{y} >$, and equals

$$< \mathbf{x}, \mathbf{y} >= \sum_{i=1}^{n} x_i y_i.$$

Some simple properties of the inner product are:

(a) $< \mathbf{x}, \mathbf{y} >=< \mathbf{y}, \mathbf{x} >$

(b) $< a\mathbf{x}, \mathbf{y} >= a < \mathbf{x}, \mathbf{y} >$ for any number a

(c) $< \mathbf{x}, \mathbf{y} + \boldsymbol{z} >=< \mathbf{x}, \mathbf{y} > + < \mathbf{x}, \boldsymbol{z} >$

Additionally, the length of a vector $\mathbf{x}$ is defined to be

$$| \mathbf{x} |=< x, x >^{1/2} .$$

The notation for length looks like the notation for absolute value. There is no harm in this double use of parallel vertical lines, since if $\mathbf{x} = (x)$, a vector of length one,

$$| \mathbf{x} |= + < x, x >^{1/2}= \left(\sum_{i=1}^{1} x_i^2 \right)^{1/2} = (x^2)^{1/2} =| x | .$$

The distance between two vectors $\mathbf{x}$ and $\mathbf{y}$, is $< x-y, x-y >^{1/2}$. This leads to an important geometrical interpretation. Think about a triangle (in n-dimensions) with vertices $\mathbf{x}, \mathbf{y}$ and $\mathbf{0}$. Recall the Pythagorean Theorem, which says that for a right triangle, the square of the length of the hypotenuse equals the sum of squares of the lengths of the other two sides. Then, for a right triangle,

$$\begin{aligned} 0 &= < \mathbf{x}, \mathbf{x} > + < \mathbf{y}, \mathbf{y} > - < \mathbf{x} - \mathbf{y}, \mathbf{x} - \mathbf{y} > \\ &= < \mathbf{x}, \mathbf{x} > + < \mathbf{y}, \mathbf{y} > -\{< \mathbf{x}, \mathbf{x} > + < \mathbf{y}, \mathbf{y} > - < \mathbf{x}, \mathbf{y} > - < \mathbf{y}, \mathbf{x} >\} \\ &= < \mathbf{x}, \mathbf{y} > + < \mathbf{y}, \mathbf{x} >= 2 < \mathbf{x}, \mathbf{y} > . \end{aligned}$$

Therefore $\mathbf{x}$ and $\mathbf{y}$ form a right triangle if and only if $< \mathbf{x}, \mathbf{y} >= 0$. In this case $\mathbf{x}$ and $\mathbf{y}$ are said to be orthogonal. Similarly a set of vectors $\{\mathbf{x}_1, \mathbf{x}_2, \ldots, \mathbf{x}_n\}$ is said to be an orthogonal set if each pair of them is orthogonal, and to be orthonormal if in addition each $\mathbf{x}_i$ satisfies $< \mathbf{x}_i, \mathbf{x}_i >= 1$.

Theorem 5.4.4. *If $\mathbf{x}_1, \ldots, \mathbf{x}_n$ are linearly independent vectors, there are numbers c_{ij}, $1 \leq j < i \leq n$ such that the vectors $\mathbf{y}_1, \ldots, \mathbf{y}_n$ given by*

$$\begin{aligned} \mathbf{y}_1 &= \mathbf{x}_1 \\ \mathbf{y}_2 &= c_{21}\mathbf{x}_1 + \mathbf{x}_2 \\ &\vdots \\ \mathbf{y}_n &= c_{n1}\mathbf{x}_1 + c_{n2}\mathbf{x}_2 + \ldots + c_{n,n-1}\mathbf{x}_{n-1} + \mathbf{x}_n \end{aligned}$$

form an orthogonal set of non-zero vectors.

Proof. Consider $\mathbf{y}_1, \ldots, \mathbf{y}_n$ defined by

$$\begin{aligned}
\mathbf{y}_1 &= \mathbf{x}_1 \\
\mathbf{y}_2 &= \mathbf{x}_2 - \frac{< \mathbf{y}_1, \mathbf{x}_2 >}{< \mathbf{y}_1, \mathbf{y}_1 >}\mathbf{y}_1 \\
&\vdots \\
\mathbf{y}_n &= \mathbf{x}_n - \frac{< \mathbf{y}_1, \mathbf{x}_n >}{< \mathbf{y}_1, \mathbf{y}_1 >}\mathbf{y}_1 - \ldots - \frac{< \mathbf{y}_{n-1}, \mathbf{x}_n >}{< \mathbf{y}_{n-1}, \mathbf{y}_{n-1} >}\mathbf{y}_{n-1}.
\end{aligned}$$

We claim first that $\mathbf{y}_k \neq 0$ for all k, by induction. When $k = 1$, we have $\mathbf{y}_1 \neq 0$. Suppose that $\mathbf{y}_1, \ldots, \mathbf{y}_{k-1}$ are all non-zero. Then $\mathbf{y}_k$ is well defined (*i.e.*, no zero division), and $\mathbf{y}_k$ is a linear combination of $\mathbf{x}_1, \ldots, \mathbf{x}_k$, where $\mathbf{x}_k$ has the coefficient 1. Since the $\mathbf{x}$'s are linearly independent, we have $\mathbf{y}_k \neq 0$. Hence $\mathbf{y}_1, \ldots, \mathbf{y}_n$ are all non-zero.

Next we claim that the $\mathbf{y}$'s are orthogonal, and again proceed by induction. A single vector is trivially orthogonal. Assume that, for $k \geq 2$, $\mathbf{y}_1, \ldots, \mathbf{y}_{k-1}$ are an orthogonal set. Then

$$\mathbf{y}_k = \mathbf{x}_k - \sum_{i=1}^{k-1} \frac{< \mathbf{y}_i, \mathbf{x}_k >}{< \mathbf{y}_i, \mathbf{y}_i >}\mathbf{y}_i.$$

Choose some $j < k$, and form the inner product $< \mathbf{y}_j, \mathbf{y}_k >$.

Then $< \mathbf{y}_j, \mathbf{y}_k > = < \mathbf{y}_j, \mathbf{x}_k > - \sum_{i=1}^{k-1} \frac{<\mathbf{y}_i, \mathbf{x}_k>}{<\mathbf{y}_i, \mathbf{y}_i>} < \mathbf{y}_j, \mathbf{y}_i >$.

Since $\mathbf{y}_1, \ldots, \mathbf{y}_{k-1}$ are an orthogonal set by the inductive hypothesis, $< \mathbf{y}_j, \mathbf{y}_i > = 0$ if $i \neq j$. Therefore

$$\begin{aligned}
< \mathbf{y}_j, \mathbf{y}_k > &= < \mathbf{y}_j, \mathbf{x}_k > - \frac{< \mathbf{y}_j, \mathbf{x}_k >}{< \mathbf{y}_j, \mathbf{y}_j >} < \mathbf{y}_j, \mathbf{y}_j > \\
&= \mathbf{0}
\end{aligned}$$

Now the c's can be deduced from the definition of the $\mathbf{y}$'s. □

This process is known as Gram-Schmidt orthogonalization.

Theorem 5.4.5. *The set of vectors spanned by the* $\mathbf{x}$*'s in Theorem 5.4.4 is the same as the set of vectors spanned by the* $\mathbf{y}$*'s.*

Proof. Any vector that is a linear combination of the $\mathbf{y}$'s is a linear combination of the $\mathbf{x}$'s by substitution. Hence the set spanned by the $\mathbf{y}$'s is contained in or equal to the set spanned by the $\mathbf{x}$'s.

To prove the opposite inclusion, we proceed by induction on n. If $n = 1$ the statement is trivial. Suppose it is true for $n - 1$. Let $\mathbf{z} = \sum_{i=1}^{n} d_i \mathbf{x}_i$ for some set of coefficients $d_1, \ldots, d_n$.

Then

$$\begin{aligned}
\mathbf{z} &= d_n \mathbf{x}_n + \sum_{i=1}^{n-1} d_i x_i \\
&= d_n(\mathbf{y}_n - c_{n1}\mathbf{x}_1 - \ldots - c_{n,n-1}\mathbf{x}_{n-1}) + \sum_{i=1}^{n-1} d_i \mathbf{x}_i \\
&= d_n \mathbf{y}_n + \sum_{i=1}^{n-1} (d_i - d_n c_{ni})\mathbf{x}_i.
\end{aligned}$$

By the inductive hypothesis, there are coefficients $e_1, \ldots, e_{n-1}$ such that

$$\sum_{i=1}^{n-1}(d_i - d_n c_{ni})\mathbf{x}_i = \sum_{i=1}^{n-1} e_i \mathbf{y}_i.$$

Hence

$$\mathbf{z} = d_n \mathbf{y}_n + \sum_{i=1}^{n-1} e_i \mathbf{y}_i,$$

so $\mathbf{z}$ is in the space spanned by $\mathbf{y}_1, \ldots, \mathbf{y}_n$. This completes the proof. □

A set of orthogonal non-zero vectors $\mathbf{x}_1, \ldots, \mathbf{x}_n$ can be turned into a set of orthonormal non-zero vectors as follows: let

$$\mathbf{z}_i = \frac{\mathbf{x}_i}{|\ \mathbf{x}_i\ |}, \text{ for all } i = 1, \ldots, n. \tag{5.17}$$

Theorem 5.4.6. *Let $\mathbf{x}_1, \ldots, \mathbf{x}_p$ be an orthonormal set in a linear space $\mathcal{M}$ of dimension n. There are additional vectors $\mathbf{x}_{p+1}, \ldots, \mathbf{x}_n$ such that $\mathbf{x}_1, \ldots, \mathbf{x}_n$ are an orthonormal basis for $\mathcal{M}$.*

Proof. An orthonormal set of vectors is linearly independent, since if not, there is a non-trivial linear combination for them that is zero, i.e., there are constants $c_1, \ldots, c_p$, not all zero, such that

$$\sum_{i=1}^{p} c_i \mathbf{x}_i = \mathbf{0}.$$

But then

$$0 =< \sum_{i=1}^{p} c_i \mathbf{x}_i, \mathbf{x}_j >= \sum_{i=1}^{p} c_i < \mathbf{x}_i, \mathbf{x}_j >= c_j$$

for $j = 1, \ldots, p$, which is a contradiction.

By Theorem 5.4.2 such a linearly independent set can be extended to be a basis. By Theorem 5.4.4 such a basis can be orthogonalized. By Theorem 5.4.5 it is a basis. And it can be made into an orthonormal basis using (5.17), without changing its functioning as a basis. □

Theorem 5.4.7. *Suppose $\mathbf{u}_1, \ldots, \mathbf{u}_n$ are an orthonormal basis. Then any vector $\mathbf{v}$ can be expressed as*

$$\mathbf{v} = \sum_{i=1}^{n} < \mathbf{u}_i, \mathbf{v} > \mathbf{u}_i.$$

Proof. Because $\mathbf{u}_1, \ldots, \mathbf{u}_n$ span the space, there are numbers $\alpha_1, \ldots, \alpha_n$ such that $\mathbf{v} = \sum_{j=1}^{n} \alpha_j \mathbf{u}_j$. If I show $\alpha_i =< \mathbf{u}_i, \mathbf{v} >$, I will be done.

Now

$$\begin{aligned}\sum_{i=1}^{n} < \mathbf{u}_i, \mathbf{v} > \mathbf{u}_i &= \sum_{i=1}^{n} < \mathbf{u}_i, \sum_{j=1}^{n} \alpha_j \mathbf{u}_j > \mathbf{u}_i \\ &= \sum_{i=1}^{n}\sum_{j=1}^{n} \alpha_j < \mathbf{u}_i, \mathbf{u}_j > \mathbf{u}_i.\end{aligned}$$

We use the notation δ_{ij} (Kronecker's delta) which is 1 if $i = j$ and 0 otherwise and note that

$$< \mathbf{u}_i, \mathbf{u}_j >= \delta_{ij}.$$

Then we have

$$\sum_{i=1}^{n} < \mathbf{u}_i, \mathbf{v} > \mathbf{u}_i = \sum_{i=1}^{n} \sum_{j=1}^{n} \alpha_j \delta_{ij} \mathbf{u}_i = \sum_{i=1}^{n} \alpha_i \mathbf{u}_i.$$

Therefore $\sum_{i=1}^{n} (< \mathbf{u}_i, \mathbf{v} > -\alpha_i)\mathbf{u}_i = \mathbf{0}$.

Since the u's are independent, $\alpha_i =< \mathbf{u}_i, \mathbf{v} >$, which concludes the proof. □

The linear space of vectors of the form $\mathbf{x} = (x_1, x_2, \ldots, x_n)$ where x_i are unrestricted real numbers has dimension n. To see this, consider the basis consisting of the unit vectors $\mathbf{e}_i$, with a 1 in the ith position and zero otherwise. The vectors $\mathbf{e}_i$ are linearly independent (indeed they are orthonormal), and span the space, since every vector $\mathbf{x} = (x_1, \ldots, x_n)$ satisfies

$$\mathbf{x} = \sum_{i=1}^{n} x_i \mathbf{e}_i.$$

Since there are n vectors $\mathbf{e}_i$, the dimension of the space is n.

There are many orthonormal sets of n vectors in this space. Indeed Theorem 5.4.6 applies to say that one can start with an arbitrary vector of length 1, and find $n - 1$ additional vectors such that together they form an orthonormal set of n vectors. These observations show that there are many examples of the following definition:

A real $n \times n$ matrix is called orthogonal if and only if its columns (and therefore rows) form an orthonormal set of vectors. It might seem reasonable to call such a matrix "orthonormal" instead of "orthogonal," but such is not the traditional usage.

Suppose A is an orthogonal matrix. The $(i, j)^{th}$ element of AA' is

$$\sum_{k=1}^{n} a_{ik} a_{jk} =< \mathbf{a}_i, \mathbf{a}_j >= \delta_{ij}, \text{ where } \mathbf{a}_i = (a_{i1}, \ldots, a_{in}).$$

Therefore we have

$$AA' = I.$$

Additionally $A'A = I$, shown by taking the transpose of both sides. Therefore an orthogonal matrix always has an inverse, and orthogonality can also be characterized by the relation

$$A^{-1} = A'.$$

Having defined an orthogonal matrix, we can now state a simple Corollary to Theorem 5.4.6: A unit vector $\mathbf{x}$ is a vector such that $< \mathbf{x}, \mathbf{x} >= 1$.

Corollary 5.4.8. *Let $\mathbf{x}_1$ be a unit vector. Then there exists an orthogonal matrix A with $\mathbf{x}_1$ as first column (row).*

Also it is obvious that if A is orthogonal, so is A^{-1}, because $AA' = I$ implies $(A')'A' = I$. Similarly if A and B are orthogonal, so is AB, because

$$(AB)'AB = B'A'AB = B'IB = B'B = I.$$

Our next target is to characterize orthogonal matrices among all square matrices. To do so, we need a simple lemma first:

Lemma 5.4.9. *Suppose B is a symmetric matrix. Then $\mathbf{y}'B\mathbf{y} = 0$ for all $\mathbf{y}$ if and only if $B = 0$.*

Proof. First let $\mathbf{y} = \mathbf{e}_i$. Then

$$0 = \mathbf{e}_i' B \mathbf{e}_i = b_{ii} \text{ for all } i. \tag{5.18}$$

Now let $\mathbf{y} = \mathbf{e}_i + \mathbf{e}_j$. Then

$$\begin{aligned} 0 &= (\mathbf{e}_i + \mathbf{e}_j)' B(\mathbf{e}_i + \mathbf{e}_j) = b_{ii} + b_{jj} + b_{ij} + b_{ji} \\ &= b_{ij} + b_{ji} = 2b_{ij} \text{ by symmetry for all } i \text{ and } j \neq i. \end{aligned}$$

Then $b_{ij} = 0$ for $i \neq j$. Putting this together with (5.18), $b_{ij} = 0$ for all i and j, *i.e.*, $B = 0$.
However, if $B = 0$, obviously $\mathbf{y}' B y = 0$ for all $\mathbf{y}$. □

Theorem 5.4.10. *The following are equivalent:*
(i) A is orthogonal.
(ii) A preserves length, i.e., $\mid A\mathbf{x} \mid = \mid \mathbf{x} \mid$ for all $\mathbf{x}$.
(iii) A preserves distance, i.e., $\mid A\mathbf{x} - A\mathbf{y} \mid = \mid \mathbf{x} - \mathbf{y} \mid$ for all $\mathbf{x}$ and $\mathbf{y}$.
(iv) A preserves inner products, i.e., $< A\mathbf{x}, A\mathbf{y} > = < \mathbf{x}, \mathbf{y} >$ for all $\mathbf{x}$ and $\mathbf{y}$.

Proof. *(i)* $\leftrightarrow$ *(ii)*
For all $\mathbf{x}$,

$$\begin{aligned} \mid A\mathbf{x} \mid = \mid \mathbf{x} \mid &\text{ if and only if} \\ \mid A\mathbf{x} \mid^2 = \mid \mathbf{x} \mid^2 &\text{ if and only if} \\ \mathbf{x}' A' A \mathbf{x} = \mathbf{x}'\mathbf{x} &\text{ if and only if} \\ \mathbf{x}'(A'A - I)\mathbf{x} = 0. & \end{aligned}$$

Using the lemma and the symmetry of $A'A$, this is equivalent to

$$A'A = I,$$

i.e., A is orthogonal.

$$\begin{aligned} (ii) \to (iii): &\quad \mid A\mathbf{x} - A\mathbf{y} \mid = \mid A(\mathbf{x} - \mathbf{y}) \mid = \mid \mathbf{x} - \mathbf{y} \mid \text{ for all } \mathbf{x} \text{ and } \mathbf{y}. \\ (iii) \to (ii): &\quad \text{Take } \mathbf{y} = \mathbf{0}. \\ (i) \to (iv): &\quad < A\mathbf{x}, A\mathbf{y} > = (A\mathbf{y})' A\mathbf{x} = \mathbf{y}' A' A \mathbf{x} = \mathbf{y}'\mathbf{x} = < \mathbf{x}, \mathbf{y} > \\ &\quad \text{for all } \mathbf{x} \text{ and } \mathbf{y}. \\ (iv) \to (ii): &\quad \text{Take } \mathbf{y} = \mathbf{x}. \text{ Then } < A\mathbf{x}, A\mathbf{x} > = < \mathbf{x}, \mathbf{x} >, \\ &\quad i.e., \mid A\mathbf{x} \mid = \mid \mathbf{x} \mid \text{ for all } \mathbf{x}. \end{aligned}$$

□

We now do something more ambitious, and characterize orthogonal matrices among all transformations: Mirsky (1990), Theorem 8.1.11, p. 228.

Theorem 5.4.11. *Let f be a transformation of the space of n-dimensional vectors to the same space.*
If $f(\mathbf{0}) = \mathbf{0}$ and for all $\mathbf{x}$ and $\mathbf{y}$,

$$\mid f(\mathbf{x}) - f(\mathbf{y}) \mid = \mid \mathbf{x} - \mathbf{y} \mid$$

then $f(\mathbf{x}) = A\mathbf{x}$ where A is an orthogonal matrix.

Remark: Such a function f preserves origin and distance.

Proof.

$$| f(\mathbf{x}) |=| f(\mathbf{x}) - f(\mathbf{0}) |=| \mathbf{x} - \mathbf{0} |=| \mathbf{x} | \quad \text{for all } \mathbf{x}.$$

Thus $< f(\mathbf{x}), f(\mathbf{x}) >=< \mathbf{x}, \mathbf{x} >$.

Also for all $\mathbf{x}$ and $\mathbf{y}$ by hypothesis,

$$< f(\mathbf{x}) - f(\mathbf{y}), f(\mathbf{x}) - f(\mathbf{y}) >=< \mathbf{x} - \mathbf{y}, \mathbf{x} - \mathbf{y} > .$$

Therefore

$$< f(\mathbf{x}), f(\mathbf{y}) >=< \mathbf{x}, \mathbf{y} >, \quad \text{for all } \mathbf{x} \text{ and } \mathbf{y}.$$

This is the fundamental relationship to be exploited. Now let $\mathbf{x} = \mathbf{e}_i$ and $\mathbf{y} = \mathbf{e}_j$. Then

$$< f(\mathbf{e}_i), f(\mathbf{e}_j) >=< \mathbf{e}_i, \mathbf{e}_j >= \delta_{ij},$$

which shows that the vectors $f(\mathbf{e}_i)$ form an orthonormal set. Since there are n of them, they form a basis. Let A be the orthogonal matrix with $f(\mathbf{e}_i)$ in the ith row, so that

$$f(\mathbf{e}_i) = A\mathbf{e}_i \quad i = 1, \ldots, n.$$

Using Theorem 5.4.7 with $\mathbf{v} = f(\mathbf{x})$. we have

$$\begin{aligned} f(\mathbf{x}) &= \sum_{i=1}^{n} < f(\mathbf{e}_i), f(\mathbf{x}) > f(\mathbf{e}_i) \\ &= \sum_{i=1}^{n} < \mathbf{e}_i, \mathbf{x} > A\mathbf{e}_i \\ &= A \sum_{i=1}^{n} < \mathbf{e}_i, \mathbf{x} > \mathbf{e}_i = A\mathbf{x}. \end{aligned}$$

□

Corollary 5.4.12. *Let f be a transformation of the space of n-dimensional vectors to itself. If*

$$| f(\mathbf{x}) - f(\mathbf{y}) |=| \mathbf{x} - \mathbf{y} |,$$

then

$$f(\mathbf{x}) = A\mathbf{x} + \mathbf{c}$$

where A is orthogonal and $\mathbf{c}$ is a fixed vector.

Proof. Let $g(\mathbf{x}) = f(\mathbf{x}) - f(\mathbf{0})$. Then $g(\mathbf{0}) = 0$ and $| g(\mathbf{x}) - g(\mathbf{y}) |=| \mathbf{x} - \mathbf{y} |$, so Theorem 5.4.11 applies to g. Then $g(\mathbf{x}) = A\mathbf{x}$ where A is orthogonal. Hence

$$f(\mathbf{x}) = A\mathbf{x} + f(\mathbf{0}).$$

□

This result allows us to understand distance-preserving transformations in n-dimensional space. The simplest such transformation adds a constant to each vector. Geometrically this is called a translation. It simply moves the origin, shifting each vector by the same amount. The orthogonal transformations are more interesting. They amount to a rotation of the axes, changing the co-ordinate system but preserving distances (and hence volumes). They include transformations like

$$\begin{pmatrix} 1 & 0 \\ 0 & -1 \end{pmatrix}$$

which leaves the first co-ordinate unchanged, but reverses the sense of the second (this is sometimes called a reflection). Thus a distance (and volume) preserving transformation consists only of a translation, a reflection and a rotation.

5.4.3 Summary

Orthogonal matrices satisfy $A' = A^{-1}$. Transformations preserve distances if and only if they are of the form $f(x) = Ax + b$, where A is orthogonal.

5.4.4 Exercises

1. Vocabulary. Explain in your own words:
   (a) linear space
   (b) span
   (c) linear independence
   (d) basis
   (e) finite dimensional linear space
   (f) inner product, length, distance
   (g) orthogonal vectors
   (h) orthonormal vectors
   (i) orthogonal matrix
   (j) Graham-Schmidt orthogonalization
   (k) Kronecker's delta
   (l) A preserves length
   (m) A preserves separation
   (n) A preserves inner products
2. Prove the following about inner products:
   (a) $< \mathbf{x}, \mathbf{y} >=< \mathbf{y}, \mathbf{x} >$
   (b) $< a\mathbf{x}, \mathbf{y} >= a < \mathbf{x}, \mathbf{y} >$ for any number a
   (c) $< \mathbf{x}, \mathbf{y} + \mathbf{z} >=< \mathbf{x}, \mathbf{y} > + < \mathbf{x}, \mathbf{z} >$

5.5 Permutations

An assignment of n letters to n envelopes can be thought of as assigning to each envelope i a letter numbered $\beta(i)$, such that $\beta(i) \neq \beta(j)$ if $i \neq j$ (i.e., different envelopes $(i \neq j)$ get different letters $(\beta(i) \neq \beta(j))$). Such a $\boldsymbol{\beta}$ is called a permutation of $\{1, 2, \ldots, n\}$, and is written $\boldsymbol{\beta} \epsilon \mathcal{A}\{1, 2, \ldots, n\}$. Two (and hence more) permutations β_1, β_2 can be performed in succession. The permutation $\beta_2\beta_1$ of β_1 followed by β_2 takes the value

$$\beta_2\beta_1(i) = \beta_2(\beta_1(i)).$$

Permutations have the following properties:

(i) if $\boldsymbol{\beta}_1 \epsilon \mathcal{A}$ and $\boldsymbol{\beta}_2 \epsilon \mathcal{A}$, then $\boldsymbol{\beta}_2\boldsymbol{\beta}_1 \epsilon \mathcal{A}$

(ii) there is an identity permutation, $\mathbf{1}$, satisfying

$$\boldsymbol{\beta}\mathbf{1} = \mathbf{1}\boldsymbol{\beta} = \boldsymbol{\beta}$$

(iii) if $\boldsymbol{\beta} \epsilon \mathcal{A}$, there is a $\beta^{-1} \epsilon \mathcal{A}$ such that $\boldsymbol{\beta}\boldsymbol{\beta}^{-1} = \boldsymbol{\beta}^{-1}\boldsymbol{\beta} = \mathbf{1}$.

Any set $\mathcal{A}$ together with an operation (here the composition of permutations) satisfying these properties is called a group.

We now use the group structure on permutations to prove a result that is useful in the development to follow:

Result 1: Let $\boldsymbol{\beta}_1$ be fixed, and $\boldsymbol{\beta}_2$ vary over all permutations of $\{1, 2, \ldots, n\}$. Then $\boldsymbol{\beta}_2\boldsymbol{\beta}_1$ and $\boldsymbol{\beta}_1\boldsymbol{\beta}_2$ vary over all permutations of $\{1, 2, \ldots, n\}$.

Proof. Let $\boldsymbol{\gamma}$ be an arbitrary permutation. Then $\boldsymbol{\beta}_2 = \boldsymbol{\gamma}\boldsymbol{\beta}_1^{-1}$ has the property that $\boldsymbol{\beta}_2\boldsymbol{\beta}_1 = \boldsymbol{\gamma}\boldsymbol{\beta}_1^{-1}\boldsymbol{\beta}_1 = \boldsymbol{\gamma}$. Also $\boldsymbol{\beta}_2 = \boldsymbol{\beta}_1^{-1}\boldsymbol{\gamma}$ has the property that $\boldsymbol{\beta}_1\boldsymbol{\beta}_2 = \boldsymbol{\beta}_1\boldsymbol{\beta}_1^{-1}\boldsymbol{\gamma} = \boldsymbol{\gamma}$. □

Result 2: Each permutation can be obtained from any other permutation by a series of exchanges of adjacent elements.

The proof of this is obvious by induction on n. Find n among the $\beta(i)$'s. Move it to last place by a sequence of adjacent exchanges. Now the induction hypothesis applies to the $n-1$ remaining elements. □

For any real number x, let sgn (x) (pronounced "signature") be defined as

$$\text{sgn } (x) = \begin{cases} 1 & \text{if } x > 0 \\ 0 & \text{if } x = 0 \\ -1 & \text{if } x < 0 \end{cases}. \tag{5.19}$$

It follows that sgn $(xy) =$ sgn (x) sgn (y). This function definition is now extended to permutations as follows:

$$\text{sgn } (\boldsymbol{\beta}) = \text{ sgn } \left(\prod_{1 \le i < j \le n} (\beta(j) - \beta(i)) \right). \tag{5.20}$$

For example, if $n = 2$, there are two possible permutations: $\boldsymbol{\beta}_1 = (1,2)$, which leaves both elements in place, and $\boldsymbol{\beta}_2 = (2,1)$, which switches them. Thus $\beta_1(1) = 1$, $\beta_2(1) = 2$, $\beta_1(2) = 2$, and $\beta_2(2) = 1$. Applying (5.20),

$$\text{sgn}(\boldsymbol{\beta}_1) = \text{sgn}(\beta_1(2) - \beta_1(1)) = \text{sgn}(2-1) = \text{sgn}(1) = 1$$

and

$$\text{sgn}(\boldsymbol{\beta}_2) = \text{sgn}(\beta_2(2) - \beta_2(1)) = \text{sgn}(1-2) = \text{sgn}(-1) = -1.$$

Because we are discussing the permutation of distinct integers, $\beta(j) \neq \beta(i)$, so sgn$(\boldsymbol{\beta}) \neq 0$, for all $\boldsymbol{\beta}$. This extension has the following properties:

(i) Let $1 \le r < s \le n$. Then

$$\text{sgn } (1, \dots, r-1, s, r+1, \dots, s-1, r, s+1, \dots, n) = -1.$$

Proof. Let $\boldsymbol{\alpha}$ be the permutation resulting from switching elements r and s, leaving all the others alone. Among the $n(n-1)/2$ factors in the definition of sgn $(\boldsymbol{\alpha})$, the only ones that are negative are those involving r or s, and numbers between r and s, specifically

$$\begin{array}{ll} (r+1)-s, (r+2)-s, \dots, & (s-1)-s \\ r-(r+1), r-(r+2), \dots, & r-(s-1), \\ \text{and } r-s. & \end{array}$$

There are exactly $2(s-r+1)+1$ of these. Therefore sgn $(\boldsymbol{\alpha}) = (-1)^{2(s-r+1)+1} = -1$. □

The same argument shows that if $\boldsymbol{\beta}$ is an arbitrary permutation, and $\boldsymbol{\alpha}$ is related to $\boldsymbol{\beta}$ by switching the rth and s^{th} elements of $\boldsymbol{\beta}$, then

$$\text{sgn } (\boldsymbol{\beta}) = - \text{ sgn } (\boldsymbol{\alpha}).$$

(ii) Therefore, of the many ways of moving from one permutation to another by a sequence of transpositions, all of these sequences have either an even number or an odd number of transpositions.

(iii) Let $\boldsymbol{\alpha}$ be a permutation of the integers $(1, 2, \ldots, n-1)$. Let $\boldsymbol{\beta} = (\boldsymbol{\alpha}, n)$ be the permutation defined by

$$\beta(i) = \alpha(i) \qquad i = 1, \ldots, n-1$$
$$\beta(n) = n.$$

Then $\text{sgn}\ (\boldsymbol{\beta}) = \text{sgn}\ (\boldsymbol{\alpha})$.

Proof. Consider a sequence of exchanges of pairs of elements that changes $\boldsymbol{\alpha}$ to the identity permutation on $\{1, 2, \ldots, n-1\}$. Each such sequence has either an odd number of exchanges (if $\text{sgn}\ (\boldsymbol{\alpha}) = -1$) or an even number (if $\text{sgn}\ (\boldsymbol{\alpha}) = 1$). Each such sequence also changes $\boldsymbol{\beta}$ to the identity permutation on $\{1, 2, \ldots, n\}$. Therefore $\text{sgn}\ (\boldsymbol{\beta}) = \text{sgn}\ (\boldsymbol{\alpha})$. □

(iv) $\text{sgn}\ (\beta_2\beta_1) = \text{sgn}\ (\beta_2)\text{sgn}\ (\beta_1)$.

Now consider $\text{sgn}\ (\boldsymbol{\beta}_2\boldsymbol{\beta}_1)$. Consider the number of exchanges of adjacent elements required to change $(1, 2, \ldots, n)$ to $\beta_1(1), \beta_1(2), \ldots, \beta_1(n)$. That number is odd if and only if $\text{sgn}\ (\boldsymbol{\beta}_1) = -1$ and even if and only if $\text{sgn}\ (\boldsymbol{\beta}_1) = 1$. Next consider the number of exchanges needed to transform $(\beta_1(1), \ldots, \beta_1(n))$ to $(\beta_2\beta_1(1), \beta_2\beta_1(2), \ldots, \beta_2\beta_1(n))$. Again, that number is either odd or even according to the $\text{sgn}\ (\boldsymbol{\beta}_2)$. But the composition of these two sequences changed $(1, 2, \ldots, n)$ to $(\beta_2\beta_1(1), \ldots, \beta_2\beta_1(n))$.

Hence we have the result

$$\text{sgn}\ (\boldsymbol{\beta}_2\boldsymbol{\beta}_1) = \text{sgn}\ (\boldsymbol{\beta}_2)\ \text{sgn}\ (\boldsymbol{\beta}_1),$$

which is property (iv). □

Every permutation β has an inverse permutation β^{-1} such that

$$\boldsymbol{\beta}^{-1}\boldsymbol{\beta} = \mathbf{1},$$

where $\mathbf{1}$ is the identity permutation.

Since $\text{sgn}\ (\mathbf{1}) = 1$, we must have

$$1 = \text{sgn}\ (\boldsymbol{\beta}^{-1}\boldsymbol{\beta}) = (\text{sgn}\ \boldsymbol{\beta}^{-1})(\text{sgn}\ \boldsymbol{\beta}).$$

Hence

$$\text{sgn}\ (\boldsymbol{\beta}^{-1}) = \text{sgn}\ (\boldsymbol{\beta}). \tag{5.21}$$

This shows that the subset of permutations $\boldsymbol{\beta}$ with $\text{sgn}(\boldsymbol{\beta}) = 1$ also satisfies the conditions for a group, and is called a subgroup. There is a large (and beautiful) literature on group theory.

5.5.1 *Summary*

A permutation of the first n integers is a rearrangement of them. The function sgn is defined on numbers, and then on permutations. It satisfies

$$\text{sgn}\ (\boldsymbol{\beta}_2\boldsymbol{\beta}_1) = \text{sgn}\ (\boldsymbol{\beta}_2)\ \text{sgn}\ (\boldsymbol{\beta}_1)$$

for all n and all permutations $\boldsymbol{\beta}_1$ and $\boldsymbol{\beta}_2$.

5.5.2 Exercises

1. Vocabulary. State in your own words the meaning of:
   (a) permutation
   (b) signature of a number
   (c) signature of a permutation
2. For $n = 3$,
   (a) What are the six possible permutations?
   (b) For each of them, apply (5.20) to find its signature.
3. Let $n = 3$, and let $\beta_1(1,2,3) = (1,3,2)$ and $\beta_2(1,2,3) = (2,1,3)$.
   (a) Compute $\boldsymbol{\beta}_1\boldsymbol{\beta}_2$.
   (b) Compute $\boldsymbol{\beta}_2\boldsymbol{\beta}_1$.
   (c) Show that $\boldsymbol{\beta}_1\boldsymbol{\beta}_2 \neq \boldsymbol{\beta}_2\boldsymbol{\beta}_1$.
4. Using the same setup as problem 3
   (a) compute sgn $(\boldsymbol{\beta}_1\boldsymbol{\beta}_2)$ directly
   (b) compute sgn $(\boldsymbol{\beta}_2\boldsymbol{\beta}_1)$ directly
   (c) show sgn $(\boldsymbol{\beta}_1\boldsymbol{\beta}_2) =$ sgn $(\boldsymbol{\beta}_2\boldsymbol{\beta}_1)$
5. Determine whether the following sets and operations form a group:
   (a) the positive integers under addition
   (b) all integers (positive, negative and zero) under addition
   (c) all integers under multiplication
   (d) all rational numbers (positive, negative and zero) under addition
   (e) the same under multiplication
   (f) all real numbers under multiplication
6. Prove or disprove: The set of permutations $\boldsymbol{\beta}$ of $\{1, 2, \ldots, n\}$ such that $\text{sgn}(\boldsymbol{\beta}) = -1$ form a subgroup.

5.6 Number systems; DeMoivre's Formula

"A rose, by any other name, would smell as sweet."

W. Shakespeare, Romeo and Juliet

This is a good point at which to explore systems of numbers. First, I review different kinds of numbers and their traditional names. Then I discuss those names from the viewpoint of modern mathematics, and then move on to the specific theorem we need.

The *natural* numbers are the numbers $1, 2, 3, \ldots$. The *integers* include the natural numbers, and also $0, -1, -2, \ldots$. *Rational* numbers are zero and ratios of non-zero integers. The *real* numbers include the rational numbers and limits of them. The real numbers that are not rational numbers are called *irrational* numbers. The imaginary numbers are $i = \sqrt{-1}$ (meaning $i^2 = -1$) and real multiples of i. Finally, the *complex* numbers are of the form $x + yi$ where x and y are real numbers.

These are scary names. They reflect, historically, the reluctance of mathematicians to expand their horizons to admit the possibility of more general views of what numbers are legitimate. Each of these sets of numbers has its own set of rules, but there's nothing "irrational" about irrational numbers. Complex numbers are no less "real" than real numbers, nor are they more complex. Every number is "imaginary" in a certain sense. Each set of numbers has its uses. So far in this book we have used only real numbers. But now we'll need to use a result from complex numbers.

The reason for studying complex numbers is to find solutions to polynomial equations. For example, the equation

$$x^2 + 1 = 0$$

cannot be solved if x is restricted to the real numbers. However, it can be solved using complex numbers, and indeed $x = \pm\, i$, where $i = \sqrt{-1}$, are those solutions. We begin with some results on Taylor series for real numbers.

5.6.1 *A supplement with more facts about Taylor series*

Recall the general form of Taylor series around $x_0 = 0$:

$$f(x) = \sum_{k=0}^{\infty} \frac{f^{(k)}(0) \cdot x^k}{k!}.$$

The particular case of this we have used most heavily is the series for the exponential function $f(x) = e^x$. Since $f^{(1)} = e^x$, it follows (use induction for a formal proof), that $f^{(k)}(x) = e^x$ for all k. Combined with the observation that $e^0 = 1$, we have $f^{(k)}(0) = 1$ for all k. Then the Taylor series for e^x is $\sum_{k=0}^{\infty} x^k/k!$. Since this series converges absolutely for all x, we may write

$$e^x = \sum_{k=0}^{\infty} x^k/k!.$$

The goal here is to apply the same kind of reasoning to $\sin x$ and $\cos x$. First, we explore the derivatives of $\sin x$ at 0. We have

$$\begin{aligned} f(x) &= \sin x & f(0) &= 0 \\ f^{(1)}(x) &= \cos x & f^{(1)}(0) &= 1 \\ f^{(2)}(x) &= -\sin x & f^{(2)}(0) &= 0 \\ f^{(3)}(x) &= -\cos x & f^{(3)}(0) &= -1 \\ f^{(4)}(x) &= \sin x & f^{(4)}(0) &= 0 \end{aligned}$$

After the 4th derivative, we are back where we started, so it is clear that the sequence $(0, 1, 0, -1)$ will be repeated indefinitely. Also all even powers of x will have coefficient 0 in the Taylor series for $\sin x$. Thus the series for $\sin x$ is

$$x - \frac{x^3}{3!} + \frac{x^5}{5!} - \frac{x^7}{7!} + \ldots = \sum_{k=0}^{\infty} \frac{(-1)^k x^{2k+1}}{(2k+1)!}.$$

This series also converges absolutely, so we may write

$$\sin x = \sum_{k=0}^{\infty} \frac{(-1)^k x^{2k+1}}{(2k+1)!}.$$

Now let's examine $\cos x$. Again, we explore its derivatives at $x = 0$:

$$\begin{aligned} f(x) &= \cos x & f(0) &= 1 \\ f^{(1)}(x) &= -\sin x & f^{(1)}(0) &= 0 \\ f^{(2)}(x) &= -\cos x & f^{(2)}(0) &= -1 \\ f^{(3)}(x) &= \sin x & f^{(3)}(0) &= 0 \\ f^{(4)}(x) &= \cos x & f^{(4)}(0) &= 1 \end{aligned}$$

Again after four derivatives we're back where we started, but now the pattern is $(1, 0, -1, 0)$ repeated indefinitely. Also all the odd powers of x will have coefficient 0 in the Taylor series for $\cos x$. Thus the series for $\cos x$ is

$$1 - \frac{x^2}{2!} + \frac{x^4}{4!} - \frac{x^6}{6!} + \ldots = \sum_{k=0}^{\infty} \frac{(-1)^k x^{2k}}{(2k)!}.$$

Again the series converges absolutely, so we may write

$$\cos x = \sum_{k=0}^{\infty} \frac{(-1)^k x^{2k}}{(2k)!}.$$

5.6.2 DeMoivre's Formula

In order to progress, we must define what is meant by the exponential function e^z where z is a complex number. The standard way to do this is by use of the Taylor series, thus:

$$e^z = \sum_{k=0}^{\infty} \frac{z^k}{k!} \tag{5.22}$$

where z is now in general a complex number. Of course in the special case that z is a real number, the definition coincides with the usual exponential function of a real variable. This series converges absolutely for all complex numbers z for the same reason that it does for all real z: it is dominated by the geometric series (see Courant (1937) Vol. 1, p. 413, and Vol. 2, p. 529).

It is now important to show that the definition given for complex exponentials works the same way as it does for real numbers, namely that

$$e^{z_1+z_2} = e^{z_1} e^{z_2} \tag{5.23}$$

where z_1 and z_2 are arbitrary complex numbers.

Proof. The proof is nothing more than the Binomial Theorem and a change of variable:

$$\begin{aligned} e^{z_1+z_2} &= \sum_{j=0}^{\infty} \frac{(z_1+z_2)^j}{j!} = \sum_{j=0}^{\infty}\sum_{k=0}^{j} \binom{j}{k, j-k} \cdot \frac{1}{j!} z_1^k z_2^{j-k} \\ &= \sum_{0 \le k \le j \le \infty} \frac{z_1^k}{k!} \frac{z_2^{j-k}}{(j-k)!}. \end{aligned}$$

Now let $\ell = j - k$. Then the range of summation is $0 \le j \le \infty$, $0 \le \ell \le \infty$, and

$$e^{z_1+z_2} = \sum_{j=0}^{\infty} \frac{z_1^j}{j!} \sum_{\ell=0}^{\infty} \frac{z_2^\ell}{\ell!} = e^{z_1} e^{z_2}.$$

□

Now consider z of the form $z = it$, where t is a real number, and, of course, $i = \sqrt{-1}$. Substituting $z = it$ into (5.22) yields

$$e^{it} = \sum_{k=0}^{\infty} \frac{(it)^k}{k!} = \sum_{k=0}^{\infty} \frac{i^k t^k}{k!}. \tag{5.24}$$

Everything here is familiar, except for powers of i. So let's examine those.

We have $i^0 = 1, i^1 = i, i^2 = -1, i^3 = -i, i^4 = 1$ and then it starts over again. So once more the powers of i have a repeating pattern of length four, with the pattern $(1, i, -1, -i)$. Comparing the pattern to those of sin and cos found in section 5.6.2, we see that

$$(1, i, -1, -i) = (1, 0, -1, 0) + i(0, 1, 0, -1),$$

so the pattern for powers of i equals the pattern for $\cos x$ plus i times the pattern for $\sin x$. Writing out the full expression for e^{it} yields

$$\begin{aligned} e^{it} &= \sum_{k=0}^{\infty} \frac{i^k t^k}{k!} = \sum_{j=0}^{\infty} \frac{(-1)^j t^{2k}}{(2j)!} + i \sum_{j=0}^{\infty} \frac{(-1)^j t^{2j+1}}{(2j+1)!} \\ &= \cos t + i \sin t. \end{aligned} \tag{5.25}$$

This standard formula in complex variables is known as Euler's Formula.

Formulas (5.23) and (5.25) can be combined to prove some important trigonometric identities as follows:

First suppose $z_1 = it_1$ and $z_2 = it_2$. Then

$$e^{z_1+z_2} = \cos(t_1 + t_2) + i \sin(t_1 + t_2), \tag{5.26}$$

using Euler's Formula (5.25).

Using (5.23), we have

$$\begin{aligned} e^{z_1+z_2} = e^{z_1} e^{z_2} =& (\cos t_1 + i \sin t_1)(\cos t_2 + i \sin t_2) \\ =& (\cos t_1 \cos t_2 - \sin t_1 \sin t_2) \\ &+ i(\cos t_1 \sin t_2 + \sin t_1 \cos t_2). \end{aligned} \tag{5.27}$$

Equating (5.26) and (5.27), and separately displaying the real and purely imaginary parts of the result yields

$$\cos(t_1 + t_2) = \cos t_1 \cos t_2 - \sin t_1 \sin t_2 \tag{5.28}$$

and

$$\sin(t_1 + t_2) = \cos t_1 \sin t_2 + \sin t_1 \cos t_2. \tag{5.29}$$

Formulae (5.28) and (5.29) are standard trigonometric identities for the sine and cosine of the sums of angles.

The formula

$$(\cos t_1 + i \sin t_1)(\cos t_2 + i \sin t_2) = \cos(t_1 + t_2) + i \sin(t_1 + t_2) \tag{5.30}$$

is known as DeMoivre's Formula. Taking $t - t_1 = t_2$ and multiplying n times yields

$$(\cos t + i \sin t)^n = \cos nt + i \sin nt. \tag{5.31}$$

5.6.3 *Complex numbers in polar co-ordinates*

All complex numbers can be written in the form $c = x + yi$ where $i = \sqrt{-1}$. Transforming to polar co-ordinates, suppose $x = r\cos\theta$ and $y = r\sin\theta$. Then $r = \sqrt{x^2 + y^2}$ is the distance of the point (x, y) from the origin. The complex number c can now be written as

$$c = r(\cos\theta + i \sin\theta). \tag{5.32}$$

The angle θ is called the amplitude of c, and r is called the absolute value or modulus of c. When c is a real number, so when $y = 0$, $|\, c \,|$ is the absolute value of the real number c. Also if c and c' are two complex numbers, $|\, c - c' \,|$ is the distance from c to c' in the plane.

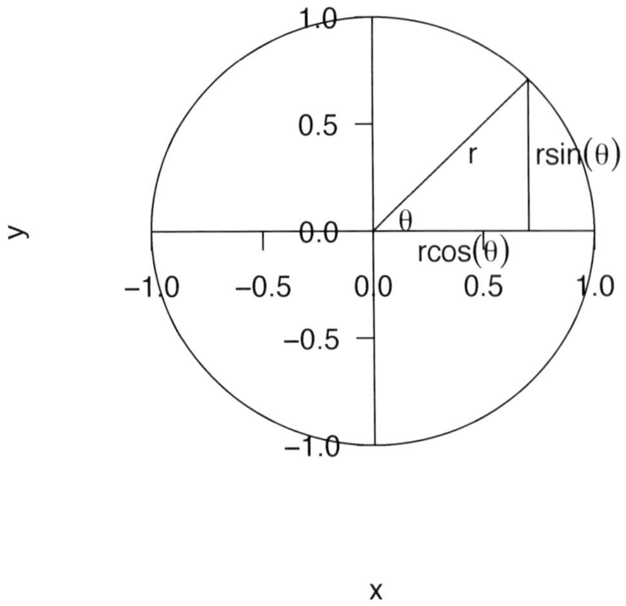

Figure 5.3: The geometry of polar co-ordinates for complex numbers.

```
Commands:
s=(-100:100)/100 * pi
x=cos(s)
y=sin(s)
w=1/sqrt(2)
plot(x,y,axes=F,type="l",xlab=" ",ylab=" ")
segments(0,0,w,w)
segments(0,0,w,0)
segments(w,w,w,0)
text(w/2+0.1,w/2,"r",adj=0.5)
text(0.2,-0.1,expression(rcos(theta)),adj=0)
text(w+0.03,w/2,expression(rsin(theta)),adj=0,xpd=T)
text(.15,.05,expression(theta))
```

Figure 5.3 illustrates the geometry of the transformation of a complex number to polar co-ordinates. The point $x + iy$ is represented as $r\cos\theta + ir\sin\theta$, where the real axis is horizontal and the imaginary axis is vertical.

Now suppose that a second complex number is written

$$c' = r'(\cos\theta' + i\sin\theta'). \tag{5.33}$$

Multiplying c and c' together yields

$$\begin{aligned} cc' &= r(\cos\theta + i\sin\theta)r'(\cos\theta' + i\sin\theta') \\ &= rr'(\cos(\theta+\theta') + i\sin)(\theta+\theta') \end{aligned} \tag{5.34}$$

using DeMoivre's Formula (5.30). Thus the result of multiplying two complex numbers is that their absolute values multiply and their amplitudes add.

5.6.4 *The fundamental theorem of algebra*

Now we are in a position to tackle this famous result. The issue is solutions to polynomial equations. As shown in section 5.6 the equation

$$x^2 + 1 = 0$$

has roots $x = \pm i$, so the equation can be factored as $0 = x^2 + 1 = (x - i)(x + i)$. The result that we are after, called the Fundamental Theorem of Algebra, is that every polynomial of the type

$$f(x) = x^m + \alpha_{m-1}x^{m-1} + \ldots + \alpha_0 = 0 \tag{5.35}$$

where the α's are real or complex, can be factored as

$$x^m + \alpha_{m-1}x^{m-1} + \ldots + \alpha_0 = \prod_{i=1}^{m}(x - \lambda_i) \tag{5.36}$$

where the λ_i's are, in general, complex.

The key step in proving the Fundamental Theorem of Algebra is a lemma known as Gauss's Theorem, because he proved it first in his doctoral thesis in 1799. It says:

Gauss's Theorem: Consider a polynomial of the form (5.35), where m is a positive integer and the α's are real or complex numbers. Then there is a complex number β such that $f(\beta) = 0$.

Proof. Suppose to the contrary that the polynomial $f(x)$ in (5.35) has no complex root, so that $f(x) \neq 0$ for all complex numbers x. Then in particular $f(0) = \alpha_0 \neq 0$. We now study the number of times $f(x)$ makes circuits around 0 for various values of r as θ goes from 0 to 2π, which we'll call $g(r)$. When $r = 0$, $g(r) = 0$ because $f(x) = f(0) = \alpha_0$ for all θ. We next show that for large $r, g(r) = m$. But g is constant in r because otherwise $f(x)$ would have a complex root, so this is a contradiction.

I interrupt the proof to give a visual image of what's going on: It should come as no surprise that for large $r, f(x)$ behaves very similarly to x^m, and hence that $f(x)$ winds around the origin m times. I imagine the path taken by $f(x)$ as if it were a string in the complex plane, that loops back on itself. Think of a spike at the origin, preventing $f(x)$ from passing through the origin. Also I think of diminishing r as pulling the string tighter and tighter. Prevented from passing through the origin by the spike, as $r \to 0$ the string would be wound m times around the spike, so $f(0)$ would be 0, a contradiction.

Figure 5.4 illustrates this mental picture. The spike at zero is the large dot. The curve represents some function that winds around zero twice. As r shrinks, but the curve is not allowed to pass through the origin, the string is wound more and more tightly around zero.

I now resume the formal proof.

The first part of this demonstration is to show that for large $r, f(x)$ behaves like x^m. (This should come as no surprise, since the lower order terms in the polynomial matter less and less for large r.) In particular, let r be larger than $r_0 = \max\{\sum_{i=0}^{m-1} \mid \alpha_i \mid, 1\}$. Then

$$\begin{aligned}
\mid f(x) - x^m \mid &= \mid \alpha_{m-1}x^{m-1} + \ldots + \alpha_0 \mid \\
&\leq \mid \alpha_{m-1} \mid\mid x \mid^{m-1} + \ldots + \mid \alpha_0 \mid \\
&= r^{m-1}[\mid \alpha_{m-1} \mid + \frac{\mid \alpha_{m-2} \mid}{r} + \ldots + \frac{\mid \alpha_0 \mid}{r^{m-1}}] \\
&\leq r^{m-1}[\mid \alpha_{m-1} \mid + \mid \alpha_{m-2} \mid + \ldots + \mid \alpha_0 \mid] \\
&\leq r^m = \mid x \mid^m = \mid x - 0 \mid^m .
\end{aligned}$$

Hence for all complex numbers x whose absolute value is larger than r_0, $f(x)$ is closer to

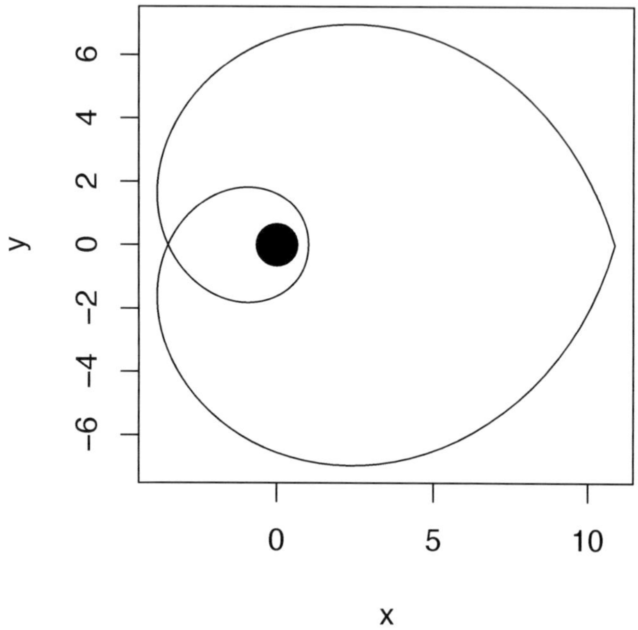

Figure 5.4: Illustration of a curve $f(x)$ winding twice around the origin.

```
Commands:
s=(-100:100)/100 * pi
x=(1+(s**2))*cos(2*s)
 y=(1+(s**2))*sin(2*s)
  plot(x,y,type="l")
 points(0,0,cex=3,pch=16)
```

x^m than is the origin. Hence $f(x)$ can be continuously stretched or shrunk to x^m without passing through the origin. Hence for $| x |> r_0$, $g(r)$ is the same as the number of times x^m makes circuits around the origin. But DeMoivre's Formula shows that x^m makes m circuits around the origin.

Hence $g(r) = m$ if $r > r_0$, and $g(0) = 0$, but $g(r)$ is constant. This contradiction completes the proof of Gauss's Theorem. □

We proceed to prove the Fundamental Theorem, that is, equation (5.36), by induction on m. When $m = 1$ the result is obvious. Suppose it is true for $m - 1$.

We use the following identity:

$$x^k - \beta^k = (x - \beta)(x^{k-1} + \beta x^{k-2} + \ldots + \beta^{k-2}x + \beta^{k-1}). \quad (5.37)$$

Using Gauss's Theorem, we know there is some number β such that $f(\beta) = 0$. Then

$$f(x) = f(x) - f(\beta) = (x^n - \beta^n) + a_{n-1}(x^{n-1} - \beta^{n-1}) + \ldots + a_1(x - \beta).$$

Each of these summands has a factor $(x - \beta)$, using (5.37). Hence

$$f(x) = (x - \beta)g(x)$$

where $g(x)$ is a polynomial of degree $m-1$, with leading coefficient 1, so g can be written

$$g(x) = x^{m-1} + \gamma_{m-2}x^{m-2} + \ldots + \gamma_0$$

for some numbers γ. Now the inductive hypothesis applies to g, so there are complex numbers $\lambda_1, \ldots, \lambda_{m-1}$ such that

$$g(x) = \prod_{i=1}^{m-1} (x - \lambda_i).$$

Therefore

$$f(x) = \prod_{i=1}^{m} (x - \lambda_i)$$

where $\lambda_m = \beta$. □

Complex numbers and real numbers operate the same way with respect to addition, subtraction, multiplication and division. (Technically both the complex and real numbers form what is called a field.) The differences between real and complex numbers occur mainly when it comes to continuity and other limiting procedures.

The next section, on determinants, uses only addition, subtraction, multiplication and division. As a result, the Theorems derived apply to both the real and complex fields. The neutral word "number" in the work to come, means simultaneously a real and a complex number, as we're proving theorems for both simultaneously.

5.6.5 *Summary*

Complex numbers work just like real numbers with respect to addition, division, multiplication and subtraction, remembering that $i^2 = -1$.

The Fundamental Theorem of Algebra says that every polynomial of degree m can be factored into m linear factors with m roots, possibly complex and not necessarily distinct.

5.6.6 *Exercises*

1. Let $x = a + bi$ and $y = c + di$, where a, b, c, and d are real numbers. Prove that $xy = 0$ if and only if at least one of x and y is zero.
2. Again suppose x and y are complex numbers. Show that $x + y = y + x$ and $xy = yx$.

5.6.7 *Notes*

This proof is based on that in Courant and Robbins (1958, pp. 269-271 and p. 102). Other proofs can be found in Hardy (1955, pp. 492-497).

For more on the names and history of number systems, see Asimov (1977, pp. 97-108).

5.7 Determinants

The determinant of a square $n \times n$ matrix A may be defined as follows:

$$\det(A) = \mid A \mid = \sum_{\boldsymbol{\beta} \epsilon \mathcal{A}} (\text{sgn } \boldsymbol{\beta}) a_{1,\beta(1)} a_{2,\beta(2)} \cdots a_{n,\beta(n)} \tag{5.38}$$

where the sum extends over all $n!$ permutations $\boldsymbol{\beta} \epsilon \mathcal{A}$ of the integers $\{1, 2, \ldots, n\}$.

Some special cases will help to explain the notation. When $n = 1$, the matrix A consists of a single number, *i.e.*,

$$A = [a],$$

and there is only the identity permutation to consider. Hence

$$| A |= a.$$

Now suppose $n = 2$. Then

$$A = \begin{bmatrix} a_{11} & a_{12} \\ a_{21} & a_{22} \end{bmatrix}, \text{ and}$$

$$\begin{aligned} | A | &= \text{ sgn } (1,2)a_{11}a_{22} + \text{ sgn } (2,1)a_{12}a_{21} \\ &= a_{11}a_{22} - a_{12}a_{21}. \end{aligned}$$

Finally, if $n = 3$, then

$$A = \begin{bmatrix} a_{11} & a_{12} & a_{13} \\ a_{21} & a_{22} & a_{23} \\ a_{31} & a_{32} & a_{33} \end{bmatrix}, \text{ and}$$

$$\begin{aligned} | A | &= \text{ sgn } (1,2,3)a_{11}a_{22}a_{33} + \text{ sgn } (1,3,2)a_{11}a_{23}a_{32} \\ &+ \text{ sgn } (2,1,3)a_{12}a_{21}a_{33} + \text{ sgn } (2,3,1)a_{12}a_{23}a_{31} \\ &+ \text{ sgn } (3,1,2)a_{13}a_{21}a_{32} + \text{ sgn } (3,2,1)a_{13}a_{22}a_{31} \\ &= a_{11}a_{22}a_{33} - a_{11}a_{23}a_{32} - a_{12}a_{21}a_{33} \\ &+ a_{12}a_{23}a_{31} + a_{13}a_{21}a_{32} - a_{13}a_{22}a_{31}. \end{aligned}$$

While the definition of the determinant may seem grossly complicated the first time a person sees it, determinants turn out to have many useful properties.

It simplifies the notation in the work to follow to write $\boldsymbol{\beta} = (\beta_1, \beta_2, \ldots, \beta_n)$ where before we were writing $\boldsymbol{\beta} = (\beta(1), \beta(2), \ldots, \beta(n))$.

As defined, the determinant appears to treat the rows and columns of a matrix differently. The next result shows that this is not the case.

Theorem 5.7.1. *The following both hold:*

(i) If $\boldsymbol{\beta} = (\beta_1 \ldots, \beta_n)$ is a fixed permutation of $(1, 2, \ldots, n)$ then

$$| A |= sgn\ (\boldsymbol{\beta}) \sum_{\boldsymbol{\mu}} sgn\ (\boldsymbol{\mu}) a_{\beta_1,\mu_1} a_{\beta_2,\mu_2} \cdots a_{\beta_n,\mu_n},$$

where the sum is over all permutations $\boldsymbol{\mu}$ of $\{1, 2, \ldots, n\}$.

(ii) If $\boldsymbol{\mu} = (\mu_1, \ldots, \mu_n)$ is a fixed permutation of $(1, 2, \ldots, n)$, then

$$| A |= sgn\ (\boldsymbol{\mu}) \sum_{\boldsymbol{\beta}} sgn\ (\boldsymbol{\beta}) a_{\beta_1,\mu_1} a_{\beta_2,\mu_2}, \cdots a_{\beta_n,\mu_n},$$

where the sum is over all permutations $\boldsymbol{\beta}$ of $\{1, 2, \ldots, n\}$.

Proof. (i) $| A |= \sum_{\boldsymbol{\nu} \epsilon \mathcal{A}} (\text{ sgn } \boldsymbol{\nu}) a_{1,\nu_1}, \ldots, a_{n,\nu_n}$.

Let $\boldsymbol{\mu} = \boldsymbol{\nu\beta}$. Then

$$a_{\beta_1,\mu_1} a_{\beta_2,\mu_2}, \ldots, a_{\beta_n,\mu_n} = a_{\beta_1,\nu\beta_1} a_{\beta_2,\nu\beta_2}, \ldots, a_{\beta_n,\nu\beta_n}.$$

For each i, $i = 1, \ldots, n$, there is an integer j, $1 \leq j \leq n$ such that $\beta(i) = j$. Then $a_{\beta_i,\nu(\beta_i)} = a_{j,\nu_j}$, so

$$a_{\beta_i,v_{\beta_i}} \epsilon \{a_{1,\nu_1}, a_{2,\nu_2}, \ldots a_{n,\nu_n}\}.$$

Also for each j, $j = 1, \ldots, n$, there is an integer i, $1 \leq i \leq n$ such that $\beta(i) = j$. Then $a_{j,\nu_j} = a_{\beta_i,\nu_{\beta_i}}$, so the sets

$$\{a_{1,\nu_1}, a_{2,\nu_2}, \ldots, a_{n,\nu_n}\} \text{ and } \{a_{\beta_1,\nu\beta_1}, \ldots, a_{\beta_n,\nu\beta_n}\}$$

comprise the same n numbers, rearranged. And so

$$a_{1,\nu_1}a_{2,\nu_2},\ldots,a_{n,\nu_n} = a_{\beta_1,\nu\beta_1},\ldots,a_{\beta_n,\nu\beta_n}. \tag{5.39}$$

Also sgn $(\boldsymbol{\beta})$ sgn $(\boldsymbol{\mu})$ = sgn $(\boldsymbol{\beta})$ sgn $(\boldsymbol{\nu})$ sgn $(\boldsymbol{\beta})$ = sgn $(\boldsymbol{\nu})$. Finally, using result 1 of section 5.5,

$$| A |= (\text{sgn } \boldsymbol{\beta}) \sum_{\boldsymbol{\mu}} (\text{sgn } \boldsymbol{\mu}) a_{\beta_1 \mu_1},\ldots,a_{\beta_n,\mu_n},$$

proving (i).

The proof of (ii) is similar. Let $\boldsymbol{\nu} = \boldsymbol{\mu}\boldsymbol{\beta}^{-1}$, so $\boldsymbol{\mu} = \boldsymbol{\nu}\boldsymbol{\beta}$. The above argument applies, again proving (5.36). In addition, sgn $(\boldsymbol{\beta})$ sgn $(\boldsymbol{\mu})$ = sgn $(\boldsymbol{\nu})$, so

$$| A |= (\text{sgn } \boldsymbol{\mu}) \sum_{\boldsymbol{\beta}} (\text{sgn } \boldsymbol{\beta}) a_{\beta_1,\nu_1},\ldots,a_{\beta_n,\nu_n},$$

using result 1 of section 5.5 again. This proves (ii). □

Theorem 5.7.1 shows that $| A |$ can be written in a fully symmetric form as follows:

$$| A |= \frac{1}{n!} \sum_{\boldsymbol{\alpha}} \sum_{\boldsymbol{\beta}} (\text{sgn } \alpha)(\text{sgn } \beta) a_{\alpha_1\beta_1} a_{\alpha_2\beta_2},\ldots,a_{\alpha_n\beta_n}. \tag{5.40}$$

This is the sum of $(n!)^2$ terms, $n!$ groups of $n!$ identical terms. While not very useful for computation, this expression has one obvious and convenient consequence:

$$| A |=| A' | \tag{5.41}$$

where A' is the transpose of A.

Theorem 5.7.2. *If two rows (or columns) of a matrix A are interchanged, the determinant of the resulting matrix, A^*, is given by*

$$| A^* |= (-1) \, | A | \, .$$

Proof. Let $1 \leq r < s \leq n$, and suppose the rth and s^{th} rows of A are interchanged. Then $A^* = [a^*_{ij}]$, where

$$a^*_{ij} = \begin{cases} a_{ij} & \text{if } i \neq r, s \\ a_{sj} & \text{if } i = r \\ a_{rj} & \text{if } i = s \end{cases} .$$

Then

$$\begin{aligned} | A^* | &= \sum_{\boldsymbol{\beta}} \text{sgn } (\boldsymbol{\beta}) a^*_{1\beta_1} \ldots a^*_{n\beta_n} \\ &= \sum_{\boldsymbol{\beta}} \text{sgn}(\boldsymbol{\beta}) a_{1\beta_1} \ldots a_{s\beta_r} \ldots a_{r\beta_s} \ldots a_{n\beta_n}. \end{aligned}$$

Let $\boldsymbol{\phi} = \boldsymbol{\gamma\beta}$, where γ is the permutation that switches r and s, and leaves all other elements unchanged. By property (i) of section 5.5, sgn $(\gamma) = -1$.

$$\begin{aligned} | A^* | &= \sum_{\boldsymbol{\phi}} \frac{(\text{sgn } \boldsymbol{\phi})}{(\text{sgn } (\boldsymbol{\gamma}))} a_{1\beta_1} \ldots a_{r\beta_r} \ldots a_{s\beta_s} \ldots a_{n\beta_n} \\ &= (-1) \sum_{\boldsymbol{\phi}} (\text{sgn } \boldsymbol{\phi}) a_{1\beta_1} \ldots a_{n\beta_n} = (-1) \, | A | \, . \end{aligned}$$

□

Corollary 5.7.3. *If a matrix A has two identical rows (columns), its determinant is zero.*

Proof. Switching the identical rows does not change the matrix.

Hence

$$\mid A \mid = - \mid A \mid,$$

whence

$$\mid A \mid = 0.$$

□

Theorem 5.7.4. *If each element of a row (or column) of a matrix is multiplied by a constant k, the determinant of the matrix is also multiplied by that constant.*

Proof. Let $[a_{ij}]$ be the starting matrix, and suppose the rth row is multiplied by k. Then

$$\begin{vmatrix} a_{11} & \dots & a_{1n} \\ ka_{r1} & \dots & ka_{rn} \\ \vdots & & \\ a_{n1} & & a_{nn} \end{vmatrix} = \sum_{\beta} \text{sgn } (\boldsymbol{\beta}) a_{1\beta_1} \dots (ka_{r\beta_r}) \dots a_{n\beta_n}$$

$$= k \sum_{\beta} \text{sgn } (\boldsymbol{\beta}) a_{1\beta_1} \dots a_{r\beta_r} \dots a_{n\beta_n} = k \mid A \mid .$$

□

Corollary 5.7.5. *If a row (or column) of a matrix is the zero vector, the determinant of the matrix is zero. (Take $k = 0$ above.)*

Theorem 5.7.6. *Suppose A and B are two square $n \times n$ matrices that are identical except for the rth row (column). Let C be a matrix that is the same as A and B on all rows (columns) except the rth and whose r^{th} row (column) is the sum of the rth row (column) of A and the rth row (column) of B. Then*

$$\mid C \mid = \mid A \mid + \mid B \mid .$$

Proof. Suppose $A = [a_{ij}]$ and $B = [b_{ij}]$. Then $C = [c_{ij}]$, where $c_{ij} = a_{ij}$ for $i \neq r, j = 1, \dots, n$ and $c_{rj} = a_{rj} + b_{rj}, j = 1, \dots, n$.

Then

$$\begin{aligned} \mid C \mid &= \sum_{\beta} (\text{sgn } \boldsymbol{\beta}) c_{1\beta_1} c_{2\beta_2} \cdots c_{n\beta_n} \\ &= \sum_{\beta} (\text{sgn } \boldsymbol{\beta}) c_{1\beta_1} c_{2\beta_2} \cdots c_{r-1,\beta_{r-1}} (a_{r\beta_r} + b_{r\beta_r}) \dots c_{n\beta_n} \\ &= \sum_{\beta} (\text{sgn } \boldsymbol{\beta}) c_{1\beta_1} c_2 \beta_2 \dots c_{r-1,\beta_{r-1}} a_{r\beta_n} \cdots c_{n\beta_n} \\ &+ \sum_{\beta} (\text{sgn } \boldsymbol{\beta}) c_{1\beta_1} c_{2\beta_2} \cdots c_{r-1\beta_{r-1}} \cdots c_{r-1,\beta_{r-1}} b_{r\beta_n} \cdots c_{n\beta_n} \\ &= \sum_{\beta} (\text{sgn } \boldsymbol{\beta}) a_{1\beta_1} a_{2\beta_2} \cdots a_{r\beta_r} \cdots a_{n\beta_n} \\ &+ \sum_{\beta} (\text{sgn } \boldsymbol{\beta}) b_{1\beta_1} b_{2\beta_2} \dots b_{r\beta_n} b_{n\beta_n} \\ &= \mid A \mid + \mid B \mid . \end{aligned}$$

□

Theorem 5.7.7. *Let $A = [a_{ij}]$ and $B = [b_{jk}]$ be two $n \times n$ matrices. Also let $C = [c_{ik}]$ be the matrix product of A and B, i.e., $C = AB$, where*

$$c_{ik} = \sum_{j=1}^{n} a_{ij} b_{jk}.$$

Then $|C| = |A||B|$.

Proof.

$$\begin{aligned}
|C| &= \sum_{\boldsymbol{\lambda}} \text{sgn } (\boldsymbol{\lambda}) c_{1\lambda_1} \cdots c_{n\lambda_n} \\
&= \sum_{\boldsymbol{\lambda}} (\text{sgn } \boldsymbol{\lambda}) (\sum_{\mu_1=1}^{n} a_{1\mu_1} b_{\mu_1 \lambda_1}) (\sum_{\mu_2=1}^{n} a_{2\mu_2} b_{\mu_2,\lambda_2}) \cdot (\sum_{\mu_n=1}^{n} a_{n\mu_n} b_{\mu_n,\lambda_n}) \\
&= \sum_{\mu_1=1}^{n} \cdots \sum_{\mu_n=1}^{n} a_{1\mu_1} \cdots a_{n\mu_n} \sum_{\boldsymbol{\lambda}} \text{sgn } (\boldsymbol{\lambda}) b_{\mu_1\lambda_1} \cdots b_{\mu_n\lambda_n}
\end{aligned}$$

The inner sum is determinant, *i.e.*,

$$\begin{vmatrix} b_{\mu_1,1} & \cdots & b_{\mu_1 n} \\ \vdots & & \vdots \\ b_{\mu_n,1} & \cdots & b_{\mu_n,n} \end{vmatrix}$$

and is zero if any two μ's are equal, by Corollary 5.7.1. Therefore, out of the n^n terms in the summation over the μ's, only $n!$ remain, namely those in which the μ's are all different, *i.e.*, those that comprise a permutation. Hence

$$\begin{aligned}
|C| &= \sum_{\boldsymbol{\mu}} a_{1\mu_1} \cdots a_{n\mu_n} \sum_{\boldsymbol{\lambda}} \text{sgn } (\boldsymbol{\lambda}) b_{\mu_1\lambda_1} \cdots b_{\mu_n\lambda_n} \\
&= \sum_{\boldsymbol{\mu}} (\text{sgn } \boldsymbol{\mu}) a_{1\mu_1} \cdots a_{n\mu_n} \sum_{\boldsymbol{\lambda}} (\text{sgn } \boldsymbol{\mu})(\text{ sgn } \boldsymbol{\lambda}) b_{\mu_1\lambda_1} \cdots b_{\mu_n\lambda_n} \\
&= |A||B|.
\end{aligned}$$

□

Theorem 5.7.8. *Let A be an $n \times n$ matrix, and let A^* be a matrix which has each row (column) the same as A except that a constant multiple of one row (column) is added to another. Then*

$$|A^*| = |A|.$$

Proof. Suppose k times the sth row is added to the rth row. Then

$$| A^* | = \begin{vmatrix} a_{11} & \dots & a_{1n} \\ \vdots & & \vdots \\ a_{r1} + ka_{s1} & \dots & a_{rn} + ka_{sn} \\ \vdots & & \vdots \\ a_{s1} & \dots & a_{sn} \\ \vdots & & \vdots \\ a_{n1} & \dots & a_{rn} \end{vmatrix}$$

$$= \begin{vmatrix} a_{11} & \dots & a_{1n} \\ \vdots & & \vdots \\ a_{r1} & \dots & a_{rn} \\ \vdots & & \vdots \\ a_{s1} & \dots & a_{sn} \\ \vdots & & \vdots \\ a_{n1} & \dots & a_{nr} \end{vmatrix} + \begin{vmatrix} a_{11} & \dots & a_{1n} \\ \vdots & & \vdots \\ ka_{s1} & \dots & ka_{sn} \\ \vdots & & \vdots \\ a_{s1} & \dots & a_{sn} \\ \vdots & & \vdots \\ a_{n1} & \dots & a_{nn} \end{vmatrix}$$

$$= | A | + k \begin{vmatrix} a_{11} & \dots & a_{1n} \\ \vdots & & \vdots \\ a_{s1} & \dots & a_{sn} \\ \vdots & & \vdots \\ a_{s1} & \dots & a_{sn} \\ \vdots & & \vdots \\ a_{n1} & \dots & a_{nn} \end{vmatrix} = | A | + k \ \ 0 = | A |$$

□

Lemma 5.7.9. *Suppose A is an $n \times n$ matrix having the structure*

$$A = \begin{bmatrix} B & \mathbf{0} \\ \mathbf{b}' & a \end{bmatrix}$$

where B is $(n-1) \times (n-1), \mathbf{0}$ and $\mathbf{b}$ are $1 \times (n-1)$ column vectors, and a is a number. $| A | = a | B |$.

Proof. In the expression for $| A |$ given in (5.35), of the $n!$ summands, each has exactly one element from the last column. Each of them, excepting those containing a, have a factor of zero, and hence are zero. Each of those containing a is a product of a permutation in B, multiplied by sgn $(\boldsymbol{\beta})$, where $\boldsymbol{\beta}$ has the form $\boldsymbol{\beta} = (\boldsymbol{\alpha}, n)$. Using result (iii) of section 5.5, sgn $(\boldsymbol{\beta}) =$ sgn $(\boldsymbol{\alpha})$. Therefore

$$| A | = a | B | .$$

□

We now study vectors $\mathbf{x}$ satisfying $A\mathbf{x} = \mathbf{0}$. One such $\mathbf{x}$ is always $\mathbf{x} = \mathbf{0}$, called the trivial solution. The question is whether there are non-trivial solutions $\mathbf{x} \neq 0$.

Theorem 5.7.10. *There exists a non-trivial $\mathbf{x}$ such that $A\mathbf{x} = 0$ if and only if $| A | = 0$.*

Proof. Suppose first that there is such a non-trivial $\mathbf{x}$. I will show that $| A |= 0$.

If A has a zero row, then $| A |= 0$ by Corollary 2. Since $\mathbf{x}$ is non-trivial, there is some i, $1 \leq i \leq n$ such that $x_i \neq 0$. Let $\mathbf{y} = \mathbf{x}/x_i$. Then $y_i = 1$ and $A\mathbf{y} = 0$.

Now let the non-zero elements of y be indexed by a set I, where $\phi \subset I \subseteq \{1, 2, \ldots, n\}$. By Theorem 5.4.7, the rows of A may be multiplied by y_j, for $j \epsilon I$, and added to row i, without changing $| A |$. This results in a matrix whose i^{th} row is zero, and has the same determinant as A. Hence by Corollary 5.7 2, $| A |= 0$.

To complete the proof of the theorem, I now assume that $| A |= 0$ and prove the existence of a non-trivial vector $\mathbf{x}$ such that $A\mathbf{x} = 0$. The proof proceeds by induction on n. For $n = 1$, the statement is obvious. Suppose then that it is true for $n - 1$.

If $a_{ni} = 0$ for all i, $1 \leq i \leq n$, then the vector $\mathbf{x} = (0, \ldots, 0, 1)$ suffices. Suppose then, that there is a non-zero element in the nth row of A. Without changing the determinant of A, the columns can be rearranged so that $a_{nn} \neq 0$ (see Theorem 5.7.2). Now subtract a_{ni}/a_{nn} from the ith row of A, to obtain the matrix

$$\begin{bmatrix} B & \mathbf{0} \\ b' & a_{nn} \end{bmatrix}$$

where B is $(n-1) \times (n-1)$, and b and $\mathbf{0}$ are column vectors of length $n-1$. By Theorem 5.7.8, this matrix has the same determinant as A. Using the lemma, we then have

$$0 =| A |= a_{nn} \mid B \mid .$$

Since $a_{nn} \neq 0$, we have $0 =| B |$, where B is an $(n - 1) \times (n - 1)$ matrix. Consequently the inductive hypothesis applies to B, where

$$b_{ij} = a_{ij} - \frac{a_{in}a_{nj}}{a_{nn}} \quad i, j = 1, 2, \ldots, n - 1.$$

Therefore there are numbers $x_1, \ldots, x_{n-1}$, not all zero, such that

$$0 = \sum_{j=1}^{n-1} b_{ij}x_j = \sum_{j=1}^{n-1} \left(a_{ij} - \frac{a_{in}a_{nj}}{a_{nn}} \right) x_j \quad i = 1, \ldots, n - 1. \tag{5.42}$$

Let $x_n = -1/a_{nn} \left(\sum_{j=1}^{n-1} a_{nj}x_j \right)$, so that

$$\sum_{j=1}^{n} a_{nj}x_j = 0. \tag{5.43}$$

Substituting (5.43) into (5.42),

$$\begin{aligned} 0 &= \sum_{j=1}^{n-1} (a_{ij} - \frac{a_{in}a_{nj}}{a_{nn}})x_j = \sum_{j=1}^{n-1} a_{ij}x_j - \frac{a_{in}}{a_{nn}} \sum_{j=1}^{n-1} a_{nj}x_j \\ &= \sum_{j=1}^{n-1} a_{ij}x_j + a_{in}x_n = \sum_{j=1}^{n} a_{ij}x_j \ , \end{aligned} \tag{5.44}$$

for $i = 1, \ldots, n - 1$.

Now (5.44) and (5.43) together yield

$$\sum_{j=1}^{n} a_{ij}x_j = 0 \qquad n_n i = 1, \ldots, n, \text{ and } \mathbf{x} \neq \mathbf{0}.$$

□

By the same proof, using the symmetry between rows and columns, we have $| A |= 0$ if and only if there is a non-trivial $\mathbf{x}$ such that $\mathbf{x}'A = 0$.

There is a nice geometric interpretation of the determinant. However, that discussion must be postponed until further linear algebra has been developed later in this chapter.

5.7.1 *Summary*

The determinant is defined in (5.35) as a function from square matrices to numbers, either real or complex. Among its important properties are:

$$| AB |=| A | \, | B |$$

and $| A |= 0$ if and only if there exists a non-trivial $\mathbf{x}$ such that $A\mathbf{x} = \mathbf{0}$.

5.7.2 *Exercises*

1. We know from Theorem 5.4.10 that if an $n \times n$ matrix A satisfies $| A |= 0$, then there is some vector $\mathbf{x}, \mathbf{x} \neq 0$ such that $A\mathbf{x} = \mathbf{0}$. We also know from Corollary 5.7 2 that if matrix A has a row of zeros, say the ith row, then $| A |= 0$. Find $\mathbf{x} \neq \mathbf{0}$ such that $A\mathbf{x} = \mathbf{0}$.
2. From Corollary 5.7 1 we know that if a matrix A has two identical rows, say rows i and j, then $| A |= 0$. As in exercise 1, find $\mathbf{x} \neq \mathbf{0}$ such that $A\mathbf{x} = \mathbf{0}$.

5.7.3 *Real matrices*

We return for a moment to real matrices, to notice that there are two kinds of real matrices for which it is easy to calculate a determinant:

(a) Suppose D is a diagonal matrix, $D_{\boldsymbol{\lambda}}$. Then $| D |= \prod_{i=1}^{n} \lambda_i$.

(b) Suppose P is an orthogonal matrix. Then $1 =| I |=| P' | \, | P |=| P |^2$. Therefore $| P |= \pm 1$.

5.7.4 *References*

There are many fine books on aspects of linear algebra. Two that I have found especially helpful are Mirsky (1990) and Schott (2005).

5.8 Eigenvalues, eigenvectors and decompositions

We now study numbers λ (just what sort of numbers is part of the story), that satisfy the following determinental equation:

$$| \lambda I - A |= 0$$

and we restrict ourselves to symmetric matrices A.

A *polynomial* is a function that can be written as

$$f(x) = a_m x^m + a_{m-1} x^{m-1} + \ldots + a_1 x + a_0.$$

If $a_m \neq 0$, f is said to have *degree m*.

Lemma 5.8.1. *If A is $n \times n$ real and symmetric, there are n real numbers λ_j (not necessarily distinct) such that*

$$| \lambda I - A |= \prod_{j=1}^{n} (\lambda - \lambda_j).$$

Proof. Consider $| \lambda I - A |$ as a function of λ. It is a polynomial of degree n, and the coefficient of λ^n is 1, since the highest power of λ comes from the diagonal of $\lambda I - A$, and is $\prod_{i=1}^{n}(\lambda - a_{ii})$. Hence $| \lambda I - A |$ may be written as

$$| \lambda I - A |= \lambda^n + \alpha_{n-1}\lambda^{n-1} + \ldots + \alpha_0.$$

Therefore by the Fundamental Theorem of Algebra, this polynomial has n roots, which may be complex numbers. It remains to show that, in this case, the roots are real. Let β be one of them. Then we know that

$$| \beta I - A |= 0.$$

Now applying Theorem 5.7.10 of section 5.7, there is a complex vector $\mathbf{x} \neq 0$ such that

$$(\beta I - A)\mathbf{x} = 0,$$

so $\beta\mathbf{x} = A\mathbf{x}$. Let $\beta = r + is$, where r and s are real numbers, and let $\mathbf{x} = \mathbf{w} + i\mathbf{z}$ where $\mathbf{w}$ and $\mathbf{z}$ are real vectors.

Then we have

$$A(\mathbf{w} + i\mathbf{z}) = (r + is)(\mathbf{w} + i\mathbf{z}).$$

Now multiply this equation on the left by the complex vector $(\mathbf{w} - i\mathbf{z})'$, to get

$$(\mathbf{w} - i\mathbf{z})'A(\mathbf{w} + i\mathbf{z}) = (r + is)(\mathbf{w} - i\mathbf{z})'(\mathbf{w} + i\mathbf{z}).$$

Because A is symmetric, $\mathbf{w}'A\mathbf{z} = \mathbf{z}'A\mathbf{w}$. Then

$$\mathbf{w}'A\mathbf{w} + \mathbf{z}'A\mathbf{z} = (r + is)(\mathbf{w}'\mathbf{w} + \mathbf{z}'\mathbf{z}).$$

Now since $\mathbf{x} \neq 0$, $\mathbf{w}'\mathbf{w} + \mathbf{z}'\mathbf{z} > 0$. Therefore we must have $s = 0$, so β is real. □

The numbers λ_j are called the eigenvalues of A (also called characteristic values). When A is symmetric, we showed above that the λ_j's are real numbers. Hence as real numbers,

$$| \lambda_j I - A |- 0,$$

so Theorem 5.7.10 of section 5.7 applies, and assures us that there is a real vector $\mathbf{x}_j \neq \mathbf{0}$ such that

$$\lambda_j \mathbf{x}_j = A\mathbf{x}_j.$$

Without loss of generality, we may take $| \mathbf{x}_j |= 1$. Such a vector $\mathbf{x}_j$ is called the eigenvector associated with λ_j (also called a characteristic vector associated with λ_j). When the λ_j's are not necessarily distinct, all that Theorem 5.7.10 gives us is a single vector $\mathbf{x}_j$ associated with possibly many equal λ_j's.

Theorem 5.8.2. *(Spectral Decomposition of a Symmetric Matrix) Let A be an $n \times n$ symmetric matrix. Then there exists an orthogonal matrix P and a diagonal matrix D such that*

$$A = PDP'.$$

Proof. By induction on n. The theorem is obvious when $n = 1$. Suppose then, that it is true for $n - 1$, where $n \geq 2$. We will then show that it is true for n.

Let λ_1 be an eigenvalue of A. From Lemma 5.8.1, we know that λ_1 is real, because A is symmetric. We also know that there is a real eigenvector associated with λ_1 such that

$$A\mathbf{x}_1 - \lambda_1\mathbf{x}_1.$$

Let S be an orthogonal matrix with $\mathbf{x}_1$ as first column. Such an S is shown to exist by

Theorem 5.4.6 of section 5.4. In the calculation that follows, the ith row of a matrix B is denoted B_{i*}; similarly the jth column of B is denoted B_{*j}.

Now for $r = 1, \ldots, n$,

$$\begin{aligned}
(S^{-1}AS)_{r1} &= (S^{-1})_{r*}AS_{*1} = S^{-1}_{r*}A\mathbf{x}_1 && (S_{*1} = \mathbf{x}_1 \text{ by construction})\\
&= \lambda_1(S^{-1})_{r*}\mathbf{x}_1 && \text{(eigenvector)}\\
&= \lambda_1(S^{-1})_{r*}S_{*1} && \text{(by construction)}\\
&= \lambda_1(S^{-1}S)_{r1} = \lambda_1 I_{r1} = \lambda_1\delta_{r1}.
\end{aligned}$$

Since A is symmetric, so is $S^{-1}AS = S'AS$. Therefore $(S^{-1}AS)_{1r} = \lambda_1\delta_{r1} \quad r = 1, \ldots, n$. Then the matrix $B = S^{-1}AS$ has the form

$$B = \begin{pmatrix} \lambda_1 & 0_1^{n-1} \\ 0_{n-1}^1 & B_1 \end{pmatrix}$$

where B_1 is a symmetric $(n-1) \times (n-1)$ matrix. The inductive hypothesis applies to B_1. Therefore there is an orthogonal matrix C_1 and a diagonal matrix D_1, both of order $n-1$, such that $B_1C_1 = C_1D_1$. Therefore

$$\begin{pmatrix} \lambda_1 & 0 \\ 0 & B_1 \end{pmatrix}\begin{pmatrix} 1 & 0 \\ 0 & C_1 \end{pmatrix} = \begin{pmatrix} 1 & 0 \\ 0 & C_1 \end{pmatrix}\begin{pmatrix} \lambda_1 & 0 \\ 0 & D_1 \end{pmatrix}.$$

Let $C = \begin{pmatrix} 1 & 0 \\ 0 & C_1 \end{pmatrix}$ and $D = \begin{pmatrix} \lambda_1 & 0 \\ 0 & D_1 \end{pmatrix}$. Then D is diagonal. Also

$$\begin{aligned}
C'C &= \begin{pmatrix} 1 & 0 \\ 0 & C_1 \end{pmatrix}'\begin{pmatrix} 1 & 0 \\ 0 & C_1 \end{pmatrix} = \begin{pmatrix} 1 & 0 \\ 0 & C_1' \end{pmatrix}\begin{pmatrix} 1 & 0 \\ 0 & C_1 \end{pmatrix}\\
&= \begin{pmatrix} 1 & 0 \\ 0 & C_1'C_1 \end{pmatrix} = \begin{pmatrix} 1 & 0 \\ 0 & I_{n-1} \end{pmatrix} = I.
\end{aligned}$$

Therefore C is orthogonal. Let $P = SC$. P is orthogonal, as it is the product of two orthogonal matrices.

Also $S^{-1}ASC = CD$, or $A = SCD(SC)^{-1} = PDP^{-1} = PDP'$. □

Before we proceed to the next decomposition theorem, we need one more lemma:

Lemma 5.8.3. *Let T be an $n \times n$ real matrix such that $\mid T \mid \neq 0$. Then $T'T$ has n positive eigenvalues.*

Proof. Since $T'T$ is symmetric, we know from Lemma 5.8.1 that it has n real eigenvalues. It remains to show that they are positive.

Let $y = Tx$. Then

$$x'T'Tx = y'y = \sum_{i=1}^{n} y_i^2 \geq 0.$$

Because $\mid T \mid \neq 0$, Theorem 5.4.10 of section 5.4 applies, and says that if $\mathbf{x} \neq \mathbf{0}$ then $\mathbf{y} \neq \mathbf{0}$. Therefore, for $\mathbf{x} \neq \mathbf{0}$, $\mathbf{x}'T'T\mathbf{x} > 0$.

Now let λ_j be an eigenvalue of $T'T$, and $\mathbf{x}_j \neq 0$ an associated eigenvector. Then

$$0 < \mathbf{x}_j'T'T\mathbf{x}_j = \lambda_j\mathbf{x}_j'\mathbf{x}_j = \lambda_j.$$

□

Theorem 5.8.4. *(Singular Value Decomposition of a Matrix) Let A be an $n \times n$ matrix such that $| A |\neq 0$. There exist orthogonal matrices P and Q and a diagonal matrix D with positive diagonal elements such that $A = PDQ$.*

Proof. From Lemma 5.8.3, we know that $A'A$ has positive eigenvalues. Let D^2 be an $n \times n$ diagonal matrix whose diagonal elements are those n positive eigenvalues, and let D be the diagonal matrix whose diagonal elements are the positive square roots of the diagonal elements of D^2.

Since $A'A$ is symmetric, by Theorem 1, there is an orthogonal matrix Q such that

$$QA'AQ' = D^2.$$

Let $P = AQ'D^{-1}$. Then P is orthogonal, because

$$P'P = D^{-1}QA'AQ'D^{-1} = D^{-1}D^2D^{-1} = I.$$

Also

$$\begin{aligned} P'AQ' &= D^{-1}QA'AQ' = D^{-1}D^2 = D, \text{ or} \\ A &= PDQ. \end{aligned}$$

□

Corollary 5.8.5. *A has an inverse matrix if and only if $| A |\neq 0$.*

Proof. If $| A |\neq 0$, then Theorem 5.8.4 shows that, defining $A^{-1} = Q'D^{-1}P'$, we have

$$\begin{aligned} AA^{-1} &= PDQQ'D^{-1}P' = PDD^{-1}P' = PP' = I \\ A^{-1}A &= Q'D^{-1}P'PDQ = Q'D^{-1}DQ = Q'Q = I. \end{aligned}$$

Suppose $| A |= 0$. Then Theorem 5.7.10 applies, and says that there is a vector $\mathbf{x} \neq 0$ such that $A\mathbf{x} = \mathbf{0}$. Suppose A^{-1} existed, contrary to hypothesis. Then $\mathbf{0} = A^{-1}A\mathbf{x} = \mathbf{x}$, a contradiction. Therefore A has no inverse if $| A |= 0$. □

When A has an inverse, $| A |= 1/ | A^{-1} |$, because $1 =| I |= | AA^{-1} |=| A | | A^{-1} |$.

Theorem 5.8.4 offers a geometric interpretation of the absolute value of the determinant of a non-singular matrix A. We know that such an A can be written as $A = PDQ$, where P and Q are orthogonal. We also know $| A |=| P | | D | | Q |$, and that $|| P ||= 1$ (meaning the absolute value of the determinant of P), and $|| Q ||= 1$, while $|| D ||$ is the product of the numbers down the diagonal of D.

Consider a unit cube. What happens to its volume when operated on by A? First, we have the orthogonal matrix Q. From Theorem 5.4.10, we know that an orthogonal matrix rotates the cube, but it is still a unit cube after operation by Q. Now what does D do to it? D stretches or shrinks each dimension by a factor d_i, so the volume of the cube (in n-space) is now $\prod_{i=1}^n d_i$. The resulting figure is no longer a cube, but rather a rectangular solid. Finally P again rotates the rectangular solid, but does not change its volume. Hence the volume of the cube is multiplied by $\prod_{i=1}^n d_i$, which is $|| A ||$.

You may recall the following result from section 5.3: Suppose X has a continuous distribution with pdf $f_X(x)$. Let $g(x) = ax + b$ with $a \neq 0$. Then $Y = g(X)$ has the density

$$f_Y(y) = f_X\left(\frac{y-b}{a}\right) \cdot \frac{1}{| a |}. \tag{5.45}$$

The time has come to state the multivariate generalization of this result. Suppose $\mathbf{X}$ has a continuous multivariate distribution with pdf $f_{\mathbf{X}}(\mathbf{x})$. Let $g(\mathbf{x}) = A\mathbf{x} + \mathbf{b}$, with $\mid A \mid \neq 0$. Then $\mathbf{Y} = g(\mathbf{X})$ has the density

$$f_{\mathbf{Y}}(y) = f_{\mathbf{x}}(A^{-1}(\mathbf{y} - b)) \cdot \frac{1}{\| A \|} = f_{\mathbf{X}}(A^{-1}(\mathbf{y} - b)) \| A^{-1} \| .$$

Thus $\| A \|$ is the appropriate multivariate generalization of $\mid a \mid$ in the univariate case.

The next decomposition theorem useful as an alternative way of decomposing a positive-definite matrix. (Recall the definition of positive-definite in section 2.12.4.) A few preliminary facts are useful to establish:

Lemma 5.8.6. *If A is symmetric and positive definite, every submatrix whose diagonal is a subset of the diagonal of A is also positive definite.*

Proof. Let A_1 be such a submatrix. Without loss of generality, we may reorder the rows and columns of A so that A_1 is the upper left-hand corner of A, and then write

$$A = \begin{pmatrix} A_1 & A_2 \\ A_2' & A_3 \end{pmatrix}.$$

Let A_1 be $m \times m$, and $\mathbf{x}$ a vector of length m, $\mathbf{x} \notin 0$. If A is $n \times n$, append a vector of 0's of length $n - m$ to x, and let $y = (\mathbf{x}, \mathbf{0})'$. Then

$$0 < y'Ay = x'A_1x.$$

So A_1 is positive definite. □

A lower triangular matrix T has zeros above the main diagonal. Its determinant is the product of its diagonal elements. If those diagonal elements are not zero, T is non-singular, and therefore has an inverse.

Theorem 5.8.7. *(Schott) Let A be an $n \times n$ positive definite matrix. Then there exists a unique lower triangular matrix T with positive diagonal elements such that*

$$A = TT'.$$

Proof. To shorten what is written, let "ltmwpde" stand for "lower triangular matrix with positive diagonal elements." The proof proceeds by induction on n. When $n = 1$, A consists of a single positive number a. Then the 1×1 matrix T consisting of $\sqrt{a}$ is *ltmwpde*. Now assume the theorem is true for all $(n-1) \times (n-1)$ positive definite matrices. Let A be an $n \times n$ positive definite matrix. Then A can be partitioned as

$$A = \begin{pmatrix} A_{11} & a_{12} \\ a_{12}' & a_{22} \end{pmatrix}$$

where A_{11} is $(m-1) \times (m-1)$ and positive definite. So the induction hypothesis applies to A_{11}, yielding the existence of T_{11}, a *ltmwpde*, which is $(n-1) \times (n-1)$. Now the relation $A = TT'$ where T is *ltmwpde*, holds if and only if

$$\begin{pmatrix} A_{11} & a_{12} \\ a_{12}' & a_{22} \end{pmatrix} = \begin{pmatrix} T_{11}^* & 0 \\ t_{12}' & t_{22} \end{pmatrix} \begin{pmatrix} T_{11}^{*\prime} & t_{12} \\ 0' & t_{22} \end{pmatrix}$$
$$= \begin{pmatrix} T_{11}^* T_{11}^{*\prime} & T_{11}^* t_{12} \\ t_{12}' T_{11}^{*\prime} & t_{12}' t_{12} + t_{22}^2 \end{pmatrix}.$$

Which yields three necessary and sufficient equations:

1. $A_{11} = T_{11}^* T_{11}^{*\prime}$

2. $a_{12} = T^*_{11} t_{12}$
3. $a_{22} = t'_{12} t_{12} + t^2_{22}$

Now because the inductive hypothesis, T_{11} is unique, so $T^*_{11} = T_{11}$ from (1). Because T_{11} is *ltmwpde*, it is non-singular and has an inverse. Then the only solution to (2) is $\mathbf{t}_{12} = T_{11}^{-1}\mathbf{a}_{12}$. Using (3),

$$\begin{aligned} t^2_{22} = a_{22} - t'_{12}t_{12} &= a_{22} - a'_{12}T_{11}^{-1\prime}T_{11}^{-1}a_{12} \\ &= a_{22} - a'_{12}(T_{11}T'_{11})^{-1}a_{12} \\ &= a_{22} - a'_{12}A_{11}^{-1}a_{12}. \end{aligned}$$

Now we check that the last will be positive: Because A is positive definite, $x'Ax > 0$ for all $x \neq 0$. Consider x of the form $x = (a'_{12}A_{11}^{-1}, -1)'$. Because of its last element, $x \neq 0$. Then

$$\begin{aligned} 0 < x'Ax &= a'_{12}A_{11}^{-1}A_{11}A_{11}^{-1}a_{12} - 2a'_{12}A_{11}^{-1}a'_{12} + a_{22} \\ &= a_{22} - a'_{12}A_{11}^{-1}a_{12}. \end{aligned}$$

Thus the only solution is

$$t_{22} = (a_{22} - a'_{12}A_{11}^{-1}a_{12})^{1/2}.$$

Thus these solutions are unique. This completes the inductive step, and the proof. □

The uniqueness part of Theorem 5.4.4 proves important in its application in Chapter 8.

5.8.1 *Projection matrices*

In Chapter 2, the projection matrix $\bar{P}_X = I - X(X'X)^{-1}X'$ was found to be useful in proving that $\hat{\beta} = (X'X)^{-1}X'y$ minimizes the sum of squares. Now that the material about eigenvectors and eigenvalues is available, it is possible to return to the subject of projection.

There is one additional property of $\bar{P}_X$ that is necessary. The matrix $\bar{P}_X$ is idempotent, which means that $\bar{P}_X\bar{P}_X = \bar{P}_X$. To show this, consider

$$\begin{aligned} \bar{P}_X \cdot \bar{P}_X &= [I - X(X'X)^{-1}X'][I - X(X'X)^{-1}X'] \\ &= I - X(X'X)^{-1}X' - X(X'X)^{-1}X' + X(X'X)^{-1}X'X(X'X)^{-1}X' \\ &= I - 2(X(X'X)^{-1}X') + X(X'X)^{-1}X' \\ &= I - X(X'X)^{-1}X' \\ &= \bar{P}_X. \end{aligned}$$

Because $\bar{P}_X$ is symmetric (see proof in Chapter 2), we know from Theorem 5.8.2 that it can be written in the form

$$\bar{P}_X = PDP'$$

where P is orthogonal and D is diagonal. Now consider the consequence of the idempotent property:

$$PDP' = PDP'PDP' = PD^2P'.$$

Therefore, $D = D^2$. Since D is diagonal, this means $d_i^2 = d_i$ for each $i = 1, \ldots, n$, i.e., d_i is either 0 or 1. So, rearranging P as necessary, one can write

$$\bar{P}_X = P\begin{pmatrix} I & 0 \\ 0 & 0 \end{pmatrix}P'.$$

Furthermore, it is then clear that

$$\begin{aligned} P_X \equiv I - \bar{P}_X &= I - (I - X(X'X)^{-1}X') \\ &= X(X'X)^{-1}X' \end{aligned}$$

can be written

$$P_X = P\begin{pmatrix} 0 & 0 \\ 0 & I \end{pmatrix}P'.$$

So what is happening here is that the sum of squares $y'y$ is being decomposed as

$$y'y = y'(P_X + \bar{P}_X)y = y'P_Xy + y'\bar{P}_Xy.$$

The second summand, $y'\bar{P}_Xy$ is exactly the value of the minimized sum of squares found in Chapter 2. Thus, a sum of squares, $y'y$ has been shown to be the sum of two sums of squares, $y'P_Xy$ and $y'\bar{P}_Xy$. Furthermore, $\bar{P}_XP_X = 0$, which is the condition for orthogonality. So, in fact, this is just a multivariate version of Pythagoras's Theorem, with $\bar{P}P_X = 0$ showing the (multivariate) right angle needed.

5.8.2 *Generalizations*

An infinite dimensional linear space with an inner product and a completeness assumption is called a Hilbert Space. The equivalent of a symmetric matrix in infinite dimensions is called a self-adjoint operator. There is a spectral theorem for such operators in Hilbert Space (see Dunford and Schwartz, 1988).

There is also a singular value decomposition theorem for non-square matrices of not-necessarily full rank (see Schott, 2005, p. 140).

5.8.3 *Summary*

This section gives three decompositions that are fundamental to multivariate analysis: the spectral decomposition of a symmetric matrix, the singular value decomposition of a non-singular matrix, and the triangular decomposition of a positive definite matrix.

5.8.4 *Exercises*

1. Let A be a symmetric 2×2 matrix, so A can be written

$$A = \begin{pmatrix} a_{11} & a_{12} \\ a_{12} & a_{22} \end{pmatrix}.$$

   Find the spectral decomposition of A.

2. Let B be a non-singular 2×2 matrix, so B can be written

$$B = \begin{pmatrix} b_{11} & b_{12} \\ b_{21} & b_{22} \end{pmatrix},$$

   where $b_{11}b_{22} \neq b_{12}b_{21}$.
   Find the singular value decomposition of B.

3. Let C be a positive definite 2×2 matrix, so C can be written

$$C = \begin{pmatrix} c_{11} & c_{12} \\ c_{21} & c_{22} \end{pmatrix},$$

   where $c_{11} > 0, c_{22} > 0$ and $c_{11}c_{22} - c_{21}c_{12} > 0$.
   Find the triangular decomposition of C.

5.9 Non-linear transformations

It may seem that the jump from linear to non-linear transformations is a huge one, because of the variety of non-linear transformations that might be considered. Such is not the case, however, because locally every non-linear transformation is linear, with the matrix governing the linear transformations being the matrix of first partial derivatives of the function. Thus we have done the hard work already in section 5.8 (and the sections that led to it).

Theorem 5.9.1. *Suppose* $\mathbf{X}$ *has a continuous multivariate distribution with pdf* $f_{\mathbf{X}}(\mathbf{x})$ *in* n*-dimensions. Suppose there is some subset* S *of* $\mathbb{R}^n$ *such that* $P\{\mathbf{X}\epsilon S\} = 1$. *Consider new random variables* $\mathbf{Y} = (Y_1, \ldots, Y_n)$ *related to* $\mathbf{X}$ *by the function* $g(\mathbf{X}) = \mathbf{Y}$, *so there are* n *functions*

$$\begin{aligned} y_1 &= g_1(\mathbf{x}) = g_1(x_1, x_2, \ldots, x_n) \\ y_2 &= g_2(\mathbf{x}) = g_2(x_1, x_2, \ldots, x_n) \\ &\vdots \\ y_n &= g_n(\mathbf{x}) = g_n(x_1, x_2, \ldots, x_n). \end{aligned}$$

Let T be the image of S under g, that is, T is the set (in $\mathbb{R}^n$) such that there is an $\mathbf{x}\epsilon S$ such that $g(\mathbf{x})\epsilon T$. (This is sometimes written $g(S) = T$.)

We also assume that g is one-to-one as a function from S to T, that is, if $g(\mathbf{x}_1) = g(\mathbf{x}_2)$ then $\mathbf{x}_1 = \mathbf{x}_2$. If this is the case, then there is an inverse function u mapping points in T to points in S such that

$$x_i = u_i(\mathbf{y}) \text{ for } i = 1, \ldots, n.$$

Now suppose that the functions $\mathbf{g}$ and $\mathbf{u}$ have continuous first partial derivatives, that is, the derivatives $\partial u_i/\partial y_j$ and $\partial g_i/\partial x_j$ exist and are continuous for all $i = 1, \ldots, n$ and $j = 1, \ldots, n$. Then the following matrices can be defined:

$$J = \begin{bmatrix} \frac{\partial u_1}{\partial y_1} & \cdots & \frac{\partial u_1}{\partial y_n} \\ \vdots & & \vdots \\ \frac{\partial u_n}{\partial y_1} & \cdots & \frac{\partial u_n}{\partial y_n} \end{bmatrix} \text{ and } J^* = \begin{bmatrix} \frac{\partial g_1}{\partial x_1} & \cdots & \frac{\partial g_1}{\partial x_n} \\ \vdots & & \vdots \\ \frac{\partial g_n}{\partial x_1} & \cdots & \frac{\partial g_n}{\partial x_n} \end{bmatrix}.$$

The matrices J and J^* are called Jacobian matrices.

Then

$$\begin{aligned} f_{\mathbf{Y}}(\mathbf{y}) &= \begin{cases} f_{\mathbf{X}}(u(\mathbf{y})) \mid\mid J \mid\mid & \text{if } \mathbf{y}\epsilon T \\ 0 & \text{otherwise} \end{cases} \\ &= \begin{cases} f_{\mathbf{X}}(u(\mathbf{y}))\,(1/ \mid\mid J^* \mid\mid) & \text{if } \mathbf{y}\epsilon T \\ 0 & \text{otherwise.} \end{cases} \end{aligned}$$

Proof. Let $\epsilon > 0$ be a given number. (Of course, toward the end of this proof, we'll be taking a limit as $\epsilon \to 0$.) There are bounded subsets $S_\epsilon \subset S$ and $T_\epsilon \subset T$ such that $g(S_\epsilon) = T_\epsilon$ and $P\{\mathbf{X}\epsilon S_\epsilon\} \geq 1 - \epsilon$.

We now divide S_ϵ into a finite number of cubes whose sides are no more than ϵ in length. (This can always be done. Suppose S_ϵ, which is bounded, can be put into a box whose maximum side has length m. Divide each dimension in 2, leading to 2^n boxes whose maximum length is $m/2$. Continue this process k times, until $m/2k < \epsilon$.) For now, we'll concentrate on what happens inside one particular such box, B_ϵ.

Suppose $x^0 \epsilon B_\epsilon$, and let $y^0 = g(x^0)$, so $x^0 = u(y^0)$. Taylor's Theorem says that

$$y_j - y_j^0 = \sum_{i=1}^{n} \frac{dg_j}{dx_i}(x_i - x_i^0) + HOT$$

where $\mathbf{y} = (y_1, \ldots, y_n) = r(x_1, \ldots, x_n)$ and HOT stands for "higher order terms," which go to zero as ϵ goes to zero. This equation can be expressed in vector notation as

$$\mathbf{y} - \mathbf{y}^0 = \begin{bmatrix} \frac{\partial g_1}{\partial x_1} & \cdots & \frac{\partial g_1}{\partial x_n} \\ \vdots & & \\ \frac{\partial g_n}{\partial x_1} & & \frac{\partial g_n}{\partial x_n} \end{bmatrix} (\mathbf{x} - \mathbf{x}_0) + HOT$$

$$\mathbf{y} - \mathbf{y}^0 = J^*(\mathbf{x} - \mathbf{x}_0) + HOT$$

or

$$\mathbf{y} = J^*\mathbf{x} + \mathbf{b} + HOT \text{ for } x \epsilon B_\epsilon$$

where $\mathbf{b} = \mathbf{y}_0 - J^*\mathbf{x}_0$.

This is exactly the form studied in section 5.8.

Hence

$$f_{\mathbf{y}}(\mathbf{y}) = f_{\mathbf{x}}(u(y)) \cdot \frac{1}{\| J^* \|} + HOT \quad \text{for } x \epsilon B_\epsilon.$$

Putting the pieces together, we have

$$f_{\mathbf{Y}}(\mathbf{y}) = f_{\mathbf{X}}(u(\mathbf{y})) \cdot \frac{1}{\| J^* \|} + HOT \text{ for } x \epsilon T_\epsilon$$

and, letting $\epsilon \to 0$

$$f_{\mathbf{Y}}(\mathbf{y}) = f_{\mathbf{X}}(u(\mathbf{y})) \cdot \frac{1}{\| J^* \|} \text{ for } x \epsilon T.$$

Since

$$\begin{aligned} \mathbf{x} &= u(g(\mathbf{x}) \text{ is an identity in } x, \\ I &= J \cdot J^*, \text{ so} \\ 1 &= \mid I \mid = \mid J \mid \cdot \mid J^* \mid, \text{ so} \\ \mid J \mid &= 1 / \mid J^* \mid \quad \text{so} \\ \| J \| &= 1 / \| J^* \| . \end{aligned}$$

This completes the proof. □

For the one-dimensional case, we obtained

$$f_Y(y) = f_X(g^{-1}(y)) \mid \frac{dg^{-1}(y)}{dy} \mid . \tag{5.46}$$

Once again, $\mid \frac{dg^{-1}(y)}{dy} \mid$ becomes the absolute value of the Jacobian matrix in the n-dimensional case.

There are two difficult parts in using this theorem. The first is checking whether an n-dimensional transformation is 1-1. An excellent way to check this is to compute the inverse function. The second difficult part is to compute the determinant of J. Sometimes it is easier to compute the determinant of J^*, and divide.

As an example, consider the following problem: Let

$$f_{X,Y}(x,y) = \begin{cases} k & \text{if } x^2+y^2 \leq 1 \\ 0 & \text{otherwise} \end{cases}.$$

Find k. From elementary geometry, we know that the area of a circle is πr^2. Here $r = 1$, so $k = 1/\pi$. But we're going to use a transformation to prove this directly, using polar co-ordinates.

Let $x = r\cos\theta$ and $y = r\sin\theta$. These are already inverse transformations. The direct substitutions are $r = \sqrt{x^2+y^2}$ and $\theta = \arctan(y/x)$. Also notice that the point $(0,0)$ has to be excluded, since θ is undefined there. Thus the set $S = \{(x,y) \mid 0 < x^2+y^2 < 1\}$. A single point has probability zero in any continuous distribution, so we still have $P\{S\} = 1$.

The Jacobian matrix is

$$J = \begin{bmatrix} \frac{\partial\ r\cos\theta}{\partial r} & \frac{\partial\ r\sin\theta}{\partial r} \\ \frac{\partial\ r\cos\theta}{\partial\theta} & \frac{\partial\ r\sin\theta}{\partial\theta} \end{bmatrix} = \begin{bmatrix} \cos\theta & \sin\theta \\ r\sin\theta & -r\cos\theta \end{bmatrix}$$

whose determinant is $\mid J \mid = -r\cos^2(\theta) - r\sin^2(\theta) = -r$, thus $\mid\mid J \mid\mid = r$. Hence we have

$$f_{R,\Theta}(r,\theta) = \begin{cases} kr & ,0 < r < 1, \quad 0 < \theta < 2\pi \\ 0 & \text{otherwise} \end{cases}.$$

Therefore

$$\begin{aligned} 1 &= \int_0^1 \int_0^{2\pi} kr d\theta dr = \int_0^1 kr\theta \Big|_0^{2\pi} dr = \int_0^1 2\pi kr dr \\ &= 2\pi k \frac{r^2}{2}\Big|_0^1 = k\pi. \end{aligned}$$

Hence $k = 1/\pi$ as claimed.

5.9.1 *Summary*

This section (finally) shows that the absolute value of the determinant of the Jacobian matrix is the appropriate scaling factor for a general one-to-one multivariate non-linear transformation. This completes the main work of this chapter.

5.9.2 *Exercise*

Let X_1 and X_2 be continuous random variables with joint density $f_{X_1,X_2}(x_1,x_2)$. Let $Y_1 = X_1/(X_1+X_2)$ and $Y_2 = X_1 + X_2$.

(a) Is this transformation one-to-one?

(b) If so, find its Jacobian matrix, and the determinant of that matrix.

(c) Suppose in particular that

$$f_{X_1,X_2}(x_1,x_2) = \begin{cases} 1 & 0 < x_1 < 1\ ,\ 0 < x_2 < 1 \\ 0 & \text{otherwise} \end{cases}.$$

Find the joint density of (Y_1, Y_2).

5.10 The Borel-Kolmogorov Paradox

This paradox is best shown by example, which has the added benefit of giving further practice in computing transformations.

Let $\mathbf{X} = (X_1, X_2)$ be independent and both uniformly distributed on $(0, 1)$. Then their joint density is

$$f_{\mathbf{X}}(\mathbf{x}) = \begin{cases} 1 & 0 < x_1 < 1,\ 0 < x_2 < 1 \\ 0 & \text{otherwise} \end{cases}.$$

Now consider the transformation given by $g(x_1, x_2) = (x_2/x_1, x_1)$, *i.e.*, $y_1 = x_2/x_1, y_2 = x_1$. The inverse transformation is $u(y_1, y_2) = (y_2, y_1 y_2)$, so, because the inverse transformation can be found, g is one-to-one.

The Jacobian matrix is

$$J = \begin{bmatrix} \frac{du_1}{dy_1} & \frac{du_1}{du_2} \\ \frac{du_2}{dy_1} & \frac{du_2}{dy_2} \end{bmatrix} = \begin{bmatrix} 0 & 1 \\ y_2 & y_1 \end{bmatrix}$$

so $\| J \| = y_2$, and

$$f_{\mathbf{Y}}(\mathbf{y}) = \begin{cases} y_2 & 0 < y_2 < 1, \quad 0 < y_1 < 1/y_2 \\ 0 & \text{otherwise} \end{cases}.$$

As a check, it is useful to make sure that the transformed density integrates to 1. If it does not, a mistake has been made, often in finding the limits of integration. In this case

$$\begin{aligned} \int f_{\mathbf{Y}}(\mathbf{y})\mathbf{dy} &= \int_0^1 \int_0^{1/y_2} y_2 dy_1 dy_2 \\ &= \int_0^1 \left[y_2 y_1 \Big|_0^{1/y_2} \right] dy_2 \\ &= \int_0^1 y_2 \left[(1/y_2) - 0 \right] dy_2 \\ &= \int_0^1 1 dy_2 = 1. \end{aligned}$$

We wish to find the conditional distribution of Y_2 given Y_1. To do so, we have to find the marginal distribution of Y_1. And to do that, it is necessary to re-express the limits of integration in the other order.

We have $0 < y_2 < 1$ and $0 < y_1 < 1/y_2$. Clearly y_1 has the limits $0 < y_1 < \infty$, but, for a fixed value of y_1, what are the limits on y_2? We have $0 < y_2 < 1/y_1$, but we also have $0 < y_2 < 1$. Consequently the limits are $0 < y_2 < \min\{1, 1/y_1\}$. Hence $f_Y(\mathbf{y})$ can be re-expressed as

$$f_{\mathbf{Y}}(\mathbf{y}) = \begin{cases} y_2 & 0 < y_1 < \infty,\ 0 < y_2 < \min\{1, 1/y_1\} \\ 0 & \text{otherwise} \end{cases}.$$

Once again, it is wise to check that this density integrates to 1. We have

$$\begin{aligned}
\int f_{\mathbf{Y}}(\mathbf{y})d\mathbf{y} &= \int_0^\infty \int_0^{\min\{1,1/y_1\}} y_2 dy_2 dy_1 \\
&= \int_0^\infty \left. \frac{y_2^2}{2} \right|_0^{\min\{1,1/y_1\}} dy_1 \\
&= \int_0^\infty \frac{(\min\{1,1/y_1\})^2}{2} dy_1 \\
&= \int_0^1 \frac{(\min\{1,1/y_1\})^2}{2} dy_1 + \int_1^\infty \frac{(\min\{1,1/y_1\})^2}{2} dy_1 \\
&= \int_0^1 \frac{1}{2} dy_1 + \int_1^\infty \frac{1}{2y_1^2} dy_1 \\
&= \left. \frac{y_1}{2} \right|_0^1 + (\frac{1}{2})(-1) \cdot \left. \frac{1}{y_1} \right|_1^\infty \\
&= \frac{1}{2} + \frac{1}{2} = 1.
\end{aligned}$$

So our check succeeds.

The marginal distribution of Y_1 is then

$$\begin{aligned}
f_{Y_1}(y_1) = \int f_{\mathbf{Y}}(\mathbf{y}) dy_2 &= \int_0^{\min\{1,1/y_1\}} y_2 dy_2 = \left. \frac{y_2^2}{2} \right|_0^{\min\{1,1/y_1\}} \\
&= (\min\{1,1/y_1\})^2/2 \text{ for } 0 < y_1 < \infty \\
&= \begin{cases} 1/2 & 0 < y_1 < 1 \\ 1/(2y_1^2) & 1 \le y_1 < \infty \\ 0 & \text{otherwise} \end{cases}.
\end{aligned}$$

Then the conditional distribution of Y_2 given Y_1 is

$$f_{Y_2|Y_1}(y_2 \mid y_1) = \frac{f_{Y_2,Y_1}(y_2,y_1)}{f_{Y_1}(y_1)} = \begin{cases} \frac{y_2}{2} & 0 < y_1 < 1 \\ \frac{y_2}{2y_1^2} & 1 \le y_1 < \infty \\ 0 & \text{otherwise} \end{cases}.$$

Now we consider a second transformation of X_1, X_2. (The point of the Borel-Kolmogorov Paradox is to compare the answers derived in these two calculations.)

To distinguish the new variables from the ones just used, we'll let them be $z = (z_1, z_2)$, but the z's play the role of Y in section 5.8. The transformation we now consider is $g(x_1, x_2) = (x_2 - x_1, x_1)$, *i.e.*, $z_1 = x_2 - x_1, z_2 = x_1$. The inverse transformation is $u(z_1, z_2) = (z_2, z_1 + z_2)$. Again, because the inverse transformation has been found, the function g is one-to-one. The Jacobian matrix is

$$J = \begin{bmatrix} \frac{du_1}{dz_1} & \frac{du_1}{dz_2} \\ \frac{du_2}{dz_1} & \frac{du_2}{dz_2} \end{bmatrix} = \begin{bmatrix} 0 & 1 \\ 1 & 1 \end{bmatrix}$$

so $\| J \| = 1$.

Therefore

$$f_{\mathbf{Z}}(\mathbf{z}) = \begin{cases} 1 & 0 < z_2 < 1, \ -z_2 < z_1 < 1 - z_2 \\ 0 & \text{otherwise} \end{cases}.$$

We check, just to be sure, that this integrates to 1:

$$\begin{aligned}\int f_{\mathbf{z}}(\mathbf{z}) &= \int_0^1 \int_{-z_2}^{1-z_2} dz_1 dz_2 \\ &= \int_0^1 z_1 \Big|_{-z_2}^{1-z_2} dz_2 = \int_0^1 [(1-z_2)-(-z_2)]\, dz_2 \\ &= \int_0^1 1 dz_2 = 1.\end{aligned}$$

Now we wish to find the conditional distribution of z_2 given z_1, so we have to find the marginal distribution of z_1. Once again, this requires re-expression of the limits of integration in the other order.

We have $0 < z_2 < 1$ and $-z_2 < z_1 < 1-z_2$. Then z_1 ranges from -1 to 1, *i.e.*, $-1 < z_1 < 1$, and, given z_1, z_2 ranges from z_1 to z_1+1, *i.e.*, $z_1 < z_2 < z_1+1$. Since we already know $0 < z_2 < 1$, we have $\max(0, z_1) < z_2 < \min(1, 1+z_1)$. Hence $f_{\mathbf{z}}(z)$ may be re-expressed as

$$f_{\mathbf{Z}}(\mathbf{z}) = \begin{cases} 1 & -1 < z_2 < 1,\ \max(0, z_1) < z_2 < \min(1, 1+z_1) \\ 0 & \text{otherwise} \end{cases}.$$

Again, we check to make sure that this density integrates to 1, as follows:

$$\begin{aligned}\int f_{\mathbf{Z}}(\mathbf{z})d\mathbf{z} &= \int_{-1}^1 \int_{\max(0,z_1)}^{\min(1,1+z_1)} dz_2 dz_1 \\ &= \int_{-1}^1 \left(z_2 \Big|_{\max(0,z_1)}^{\min(1,1+z_1)} \right) dz_1 \\ &= \int_{-1}^1 (\min(1+z_1) - \max(0, z_1))dz_1 \\ &= \int_{-1}^0 (\min(1, 1+z_1) - \max(0, z_1))dz_1 \\ &+ \int_0^1 (\min(1, 1+z_1) - \max(0, z_1))dz_1 \\ &= \int_{-1}^0 [(1+z_1) - 0]\, dz_1 + \int_0^1 (1-z_1)dz_1 \\ &= (z_1 + z_1^2/2)\Big|_{-1}^0 + (z - z_1^2/2)\Big|_0^1 \\ &= -(-1+1/2) + 1 - 1/2 = 1.\end{aligned}$$

Now we find the marginal distribution of z_1:

$$\begin{aligned}f_{Z_1}(z_1) &= \int f_{\mathbf{Z}}(\mathbf{z})dz_2 = \int_{\max(0,z_1)}^{\min(1,1+2_1)} 1dz_2 \\ &= \begin{cases} \min(1, 1+z_1) - \max(0, z_1) & \text{if } -1 < z_1 < 1 \\ 0 & \text{otherwise} \end{cases}.\end{aligned}$$

$f_{Z_1}(z_1)$ can be conveniently re-expressed as follows:

$$f_{Z_1}(z_1) = \begin{cases} 1+z_1 & -1 < z_1 \leq 0 \\ 1-z_1 & 0 < z_1 < 1 \\ 0 & \text{otherwise} \end{cases}.$$

So now we can write the conditional distribution of Z_2 given Z_1 as

$$f_{Z_2|Z_1}(z_2 \mid z_1) = \begin{cases} \frac{1}{1+z_1} & -1 < z_1 \leq 0 \\ \frac{1}{1-z_1} & 0 < z_1 < 1 \\ 0 & \text{otherwise} \end{cases}.$$

Now (finally!) we are in a position to discuss the Borel-Kolmogorov Paradox. The random variable X_1 is the same as the random variables Y_2 and Z_2. The event $\{Y_1 = 1\}$ is the same as the event $\{Z_1 = 0\}$, yet we observe that

$$f_{Y_2|Y_1}(y_2 \mid y_1 = 1) \neq f_{Z_2|Z_1}(z_2 \mid z_1 = 0).$$

The failure of these two conditional distributions to be equal is what is known as the Borel-Kolmogorov Paradox.

It is certainly the case that X_1, Y_2 and Z_2 are the same random variables, so that's not where the problem lies. Consequently it must lie in the conditioning event. Recall that in section 4.3 we defined the conditional density of Y given X as follows:

$$f_{Y|X}(y \mid x) = \lim_{\Delta \to 0} \frac{d}{dy} P\{Y \leq y \mid X \epsilon N_\Delta(x)\} \tag{4.11}$$

where $N_\Delta(x) = \{x - \frac{\Delta}{2}, x + \frac{\Delta}{2}\}$.

What is going on in the Borel-Kolmogorov Paradox is that $N_\Delta(y_1)$ at $y_1 = 1$ is not the same as $N_\Delta(z_1)$ at $z_1 = 0$. Since limits are a function of the behavior of the function in the neighborhood of, but not at, the limiting point, there is no reason to expect that $f_{Y_2|Y_1}(y_2 \mid y_1 = 1)$ should equal $f_{Z_2|Z_1}(z_2 \mid z_1 = 0)$. Perhaps one can interpret this analysis as a reminder that observing $Y_1 = 1$ is not the same as observing $Z_1 = 0$. However, the fact that they are different is a salutary reminder not to interpret conditional densities too casually.

This example is illustrated by Figure 5.5. In this figure, the dark solid line is the line $x_1 = x_2$. The dotted lines (in the shape of an x) represent a sequence of lines that approach the line $x_1 = x_2$ by lines of the form $x_1/x_2 = b$, where $b \to 1$. This is the sense of closeness (topology, for those readers who know that term) suggested by y_1. The dashed lines (parallel to the line $x_1 = x_2$) represent a sequence of lines that approach the line $x_1 = x_2$ by lines of the form $x_2 = x_1 + a$, where $a \to 0$. This is the sense of closeness suggested by z_1.

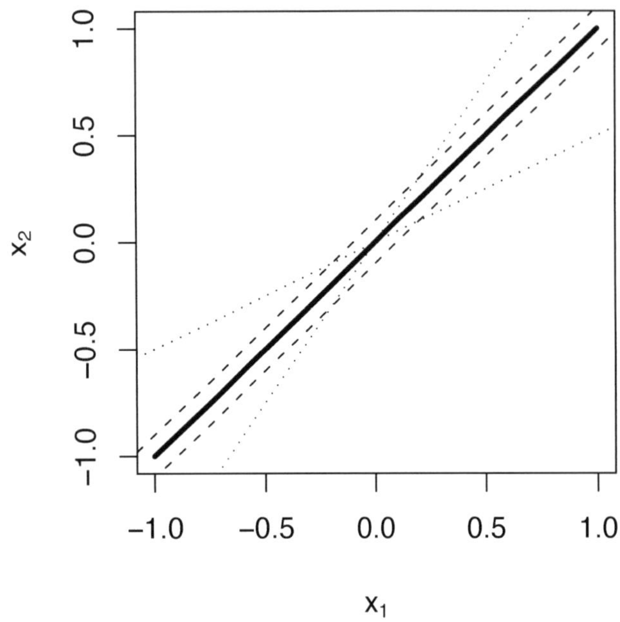

Figure 5.5: Two senses of lines close to the line $x_1 = x_2$.

```
Commands:
x = (-100:100)/100
y = x
plot (x,y, type ="l", xlab = expression (x[1]),
                       ylab = expression (x[2]), lwd = 3)
# expression makes the label with the subscript
abline (-.1, 1, lty=2)
#lty = 2 gives the lightly dotted line
abline (.1,1,lty=2)
abline (0,0.5,lty=3)
#lty = 3 gives the heavily dotted line
abline (0,1.5,lty=3)
```

5.10.1 Summary

When considering conditional densities, the conditional distributions given the same point described in different co-ordinate systems may be distinct. This is called the Borel-Kolmogorov Paradox.

5.10.2 Exercises

1. What is the Borel-Kolmogorov Paradox?
2. Is it a paradox?
3. Is it important? Why or why not?

Chapter 6

The Normal Distribution and the Central Limit Theorem

6.1 Introduction

The purpose of this chapter is to introduce the normal distribution, and to show that, in great generality, the distribution of averages of independent random variables approaches a normal distribution as the number of summands get large (*i.e.*, to prove a central limit theorem). The proof of a weak law of large numbers follows a similar argument.

6.2 Moment generating functions

The probability generating function, introduced in section 3.6, is limited in its application to distributions on the non-negative integers. The function introduced in this section relaxes that constraint, and applies to continuous distributions as well as discrete ones, and to random variables with negative as well as positive values. The expectations in this chapter are to be taken in the McShane (Lebesgue) sense, so that the bounded and dominated convergence theorems apply.

The moment generating function of a random variable X is defined to be

$$M_X(t) = E(e^{tX}). \tag{6.1}$$

For all random variables X,

$$M_X(0) = 1. \tag{6.2}$$

Before exploring the properties of the moment generating function, I first display the moment generating function for some familiar random variables.

First, suppose X takes the value 0 with probability $1-p$ and the value 1 with probability p. Then

$$M_X(t) = E(e^{tX}) = (1-p)e^0 + pe^t = 1 - p + pe^t. \tag{6.3}$$

Now suppose Y has a binomial distribution (see section 2.9), with parameters n and p, that is

$$P\{Y = k\} = \begin{cases} \binom{n}{k,n-k} p^k (1-p)^{n-k} & k = 0, 1, \ldots, n \\ 0 & \text{otherwise.} \end{cases} \tag{6.4}$$

Then

$$\begin{aligned} M_Y(t) &= \sum_{k=0}^{n} \binom{n}{k, n-k} p^k (1-p)^{n-k} e^{tk} \\ &= \sum_{k=0}^{n} \binom{n}{k, n-k} (pe^t)^k (1-p)^{n-k} \\ &= (1 - p + pe^t)^n \end{aligned} \tag{6.5}$$

using the binomial theorem (section 2.9). The last expression in (6.5) is the nth power of (6.3), a matter I'll return to later.

If Z has the Poisson distribution (section 3.9) with parameter λ, then the moment generating function of Z is

$$\begin{aligned} M_Z(t) &= \sum_{j=0}^{\infty} \frac{e^{-\lambda}\lambda^j}{j!} e^{jt} \\ &= e^{-\lambda} \sum_{j=0}^{\infty} \frac{(\lambda e^t)^j}{j!} = e^{-\lambda} e^{\lambda e^t} \\ &= e^{-\lambda(1-e^t)}. \end{aligned} \tag{6.6}$$

Now suppose W has a uniform distribution on (a, b), that is, W was the probability density function

$$f_W(w) = \begin{cases} \frac{1}{b-a} & a < w < b \\ 0 & \text{otherwise} \end{cases}. \tag{6.7}$$

Then the moment generating function of W is

$$\begin{aligned} M_W(t) &= E(e^{tW}) \\ &= \int_a^b \frac{e^{tx}}{b-a} dx \\ &= \left. \frac{e^{tx}}{(b-a)t} \right|_a^b \\ &= \frac{e^{tb} - e^{ta}}{(b-a)t}. \end{aligned} \tag{6.8}$$

To show why moment generating functions are interesting, I first remind you about moments, and then explain how the moment generating function "generates" them.

The kth moment of a random variable X is defined to be

$$\alpha_k = E(X^k) \tag{6.9}$$

when it exists, which means when $E(| X |^k) < \infty$ (see sections 3.3 and 4.4).

I now prove a theorem showing why (6.1) is called the moment generating function:

Theorem 6.2.1. *If the moment generating function $M_X(t)$ of a random variable X exists for all t in a neighborhood of $t = 0$, then all moments of X exist, and*

$$M_X(t) = \sum_{k=0}^{\infty} E(X^k) \frac{t^k}{k!}. \tag{6.10}$$

Proof. Suppose $M_X(t)$ exists for all $t \epsilon (-t_0, t_0)$ where $t_0 > 0$. Then $e^{|tX|} \leq e^{tX} + e^{-tX}$, and the latter function has a finite expectation for all $| t |< t_0$, then $e^{|tX|} = \sum_{k=0}^{\infty} \frac{|tX|^k}{k!}$ also has finite expectation.

Since this is the sum of positive quantities, $E\frac{|tX|^k}{k!}$ is finite, so $E|X^k|$ is finite for all k, so all moments of X exist, and

$$M_X(t) = \sum_{k=0}^{\infty} \frac{E(X^k)t^k}{k!}.$$

□

This is the first proof, among many in this chapter, in which the McShane (Lebesgue) sense of integral is crucial.

Corollary 6.2.2. *If the moment generating function $M_X(t)$ exists for all t in a neighborhood of $t = 0$, then*

$$E(X^k) = M^{(k)}(0) \text{ for } k = 1, 2, \ldots \tag{6.11}$$

Proof. Differentiate (6.10) k times and evaluate the result at $t = 0$. □

One especially attractive feature of the moment generating function is the ease with which it handles sums of independent random variables. Suppose $Z = X + Y$, where X and Y are independent random variables that have moment generating functions. Then

$$\begin{aligned} M_Z(t) &= E(e^{tZ}) = E(e^{t(X+Y)}) = E(e^{tX}e^{tY}) \\ &= E(e^{tX})E(e^{tY}) = M_X(t)M_Y(t). \end{aligned} \tag{6.12}$$

The key step here is that because X and Y are independent, the expectation of a product of a function of X, here $g(X) = e^{tX}$ times a function of Y, here $h(Y) = e^{tY}$, is the product of the expectations (see sections 2.8, 3.4 and 4.4).

This explains why the moment generating function of the binomial distribution, $(1 - p + pe^t)^n$ (see (6.5)) is the nth power of the moment generating function $(1 - p + pe^t)$ of the $0 - 1$ variable (6.3): the binomial random variable is the sum of n independent and identically distributed $0 - 1$ random variables.

Theorem 6.2.3. *Suppose X is a random variable with a moment generating function in a neighborhood of $t = 0$. Then the random variable $Y = aX + b$ also has a moment generating function in a neighborhood of $t = 0$, and*

$$M_Y(t) = e^{bt}M_X(at).$$

Proof.

$$M_Y(t) = E(e^{tY}) = E(e^{t(aX+b)}) = e^{tb}E(e^{atX}) = e^{tb}M_X(at).$$

□

Moment generating functions can be extended to multivariate random variables. Suppose $\mathbf{X} = (X_1, X_2, \ldots, X_k)$ is a k-dimensional random variable, and let $\mathbf{t} = (t_1, \ldots, t_k)$ be a k-dimensional real vector. Then

$$M_{\mathbf{X}}(\mathbf{t}) = E(e^{\mathbf{t}'\mathbf{X}}) = E(e^{\sum_{i=1}^k t_i X_i}).$$

As an example, suppose that $X_i = 1$ (and, if this happens, $X_j = 0$ for $j \neq i$) with probability p_i, where $\sum_{i=1}^k p_i = 1$. Then

$$M_{\mathbf{X}}(\mathbf{t}) = \sum_{i=1}^k p_i e^{t_i}.$$

Suppose $\mathbf{Y}$ is the sum of n such independent random vectors. Then

$$M_{\mathbf{Y}}(\mathbf{t}) = \left(\sum_{i=1}^k p_i e^{t_i}\right)^n,$$

the moment generating function of the multinomial random variable (section 2.9).

Then Theorem 6.2.3 can be extended to the multivariate case as follows:

Theorem 6.2.4. *Suppose $\mathbf{X} = (X_1, \ldots, X_k)$ is a k-dimensional random variable with moment generating function in a neighborhood of $\mathbf{t} = \mathbf{0}$. Then the random vector $\mathbf{Y} = A\mathbf{x} + \mathbf{b}$ also has a moment generating function in a neighborhood of $\mathbf{t} = \mathbf{0}$, and*

$$M_{\mathbf{Y}}(\mathbf{t}) = e^{\mathbf{b}'\mathbf{t}} M_{\mathbf{X}}(\mathbf{t}'A).$$

Proof.

$$\begin{aligned} M_{\mathbf{Y}}(\mathbf{t}) &= E(e^{\mathbf{t}'\mathbf{y}}) \\ &= E\left[e^{\mathbf{t}'}(A\mathbf{x} + \mathbf{b})\right. \\ &= e^{\mathbf{t}'\mathbf{b}} E(e^{t' A\mathbf{x}}) \end{aligned}$$

□

Useful though moment generating functions are, they have limitations. You have already seen an example of random variables for which means do not exist (section 3.3), and another for which, although means do exist, variances do not (section 3.3.4, Exercise 3). Because by Theorem 6.2.1 the moment generating function exists in a neighborhood of zero implies that all moments exist, moment generating functions do not apply to such random variables.

Moment generating functions are known in other parts of mathematics as Laplace Transforms.

6.2.1 *Summary*

The moment generating function defined in (6.2) generates moments, as shown in Theorem 6.2.1. Moment generating functions of a sum of independent random variables is the product of the moment generating functions of the summands. Moment generating functions do not always exist in a neighborhood of $t = 0$.

6.2.2 *Exercises*

1. Find the moment generating function of a geometric random variable (see section 3.7).
2. Find the moment generating function of a negative binomial random variable.
3. Use the moment generating function of the binomial distribution to verify:
   a) $E(Y) = np$
   b) $V(Y) = np(1 - p)$
   where Y has a binomial distribution (6.3).
4. Use the moment generating function of the Poisson distribution to verify:
   a) $E(Z) = \lambda$
   b) $V(Z) = \lambda$

6.3 The normal distribution (also known as the Gaussian distribution)

The standard normal distribution has the following density function:

$$\phi(x) = \frac{1}{\sqrt{2\pi}} e^{-x^2/2} \qquad -\infty < x < \infty. \tag{6.13}$$

This density is shown in Figure 6.1.

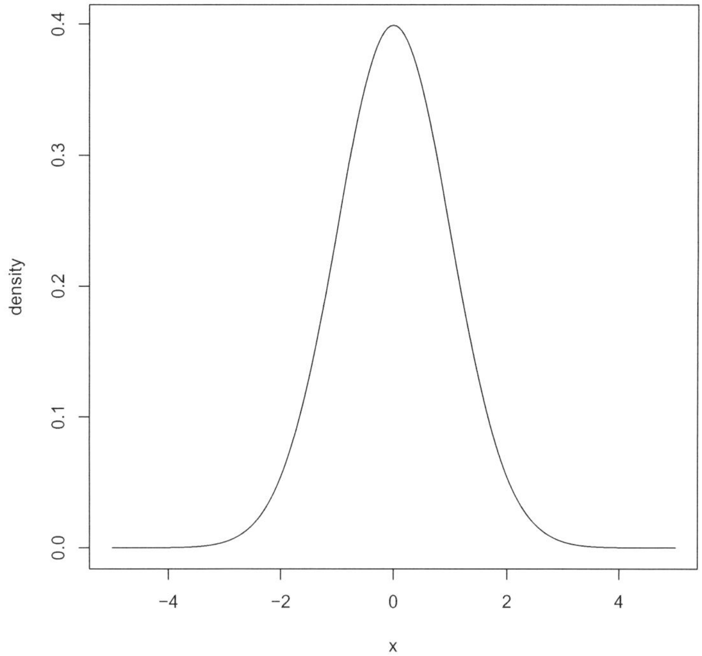

Figure 6.1: Density of the standard density normal distribution.

```
Commands:  x=(-100:100)/20
           y=(1/sqrt(2*pi))*exp (-(x**2)/2)
           plot (x,y,ylab="density", ty = "l")
```

Clearly $\phi \geq 0$ for all real x, but it is not clear whether its integral is 1. This is accomplished with a surprisingly effective trick. Instead of evaluating the integral, I evaluate its square:

$$\begin{aligned} I =& \frac{1}{2\pi} \int_{-\infty}^{\infty} e^{-x^2/2} dx \int_{-\infty}^{\infty} e^{-y^2/2} dx \\ =& \frac{1}{2\pi} \int_{-\infty}^{\infty} \int_{-\infty}^{\infty} e^{-(x^2+y^2)/2} dxdy. \end{aligned}$$

Now I transform to polar co-ordinates: $x = r \sin \theta$, $y = r \cos \theta$, as discussed in section 5.9. The Jacobian found there is r. Then

$$I = \frac{1}{2\pi} \int_{-\pi}^{\pi} \int_{0}^{\infty} e^{-r^2/2} r dr d\theta = \int_{0}^{\infty} re^{-r^2/2} dr.$$

Now let $w = r^2/2$ so $dw = rdr$. Then

$$I = \int_{0}^{\infty} e^{-w} dw = -e^{-w} \mid_{0}^{\infty} = 1 \; .$$

Since the square of the integral in question is 1, and since a non-negative function cannot integrate to a negative number, the integral takes the value 1. Therefore $\phi(x)$ is a legitimate probability density.

Now suppose that a random variable Y is related to a standard normal random variable X by the relation $Y = \sigma X + \mu$. Then Y has the probability distribution

$$f_Y(y) = \frac{1}{\sqrt{2\pi}\sigma} e^{-(y-\mu)^2/2\sigma^2} \qquad -\infty < y < \infty, \tag{6.14}$$

using the theory of transformations developed in Chapter 5.

I now derive the moment generating function of the standard normal random variable:

$$\begin{aligned} M_X(t) =& E(e^{tX}) = \int_{-\infty}^{\infty} e^{tx} \frac{1}{\sqrt{2\pi}} e^{-x^2/2} dx \\ =& \frac{1}{\sqrt{2\pi}} \int_{-\infty}^{\infty} e^{-\frac{1}{2}(x^2-2tx)} dx \\ =& \frac{1}{\sqrt{2\pi}} \int_{-\infty}^{\infty} e^{-\frac{1}{2}(x^2-2tx+t^2)} e^{t^2/2} dx \\ =& \frac{e^{t^2/2}}{\sqrt{2\pi}} \int_{-\infty}^{\infty} e^{-\frac{1}{2}(x-t)^2} dx \\ =& e^{t^2/2}. \end{aligned} \tag{6.15}$$

Expanding $e^{t^2/2}$ in a Taylor series,

$$\begin{aligned} e^{t^2/2} =& \sum_{k=0}^{\infty} \frac{1}{k!} \left(\frac{t^2}{2}\right)^k = \sum_{k=0}^{\infty} \frac{1}{k!2^k} t^{2k} \\ =& \sum_{k=0}^{\infty} \frac{(2k)!}{k!2^k} \frac{t^{2k}}{(2k)!}. \end{aligned}$$

Hence the odd moments of X are 0, and the kth even moments are

$$E(X^{2k}) = \frac{(2k)!}{k!2^k}. \tag{6.16}$$

In particular

$$\begin{aligned} E(X) &= 0, \\ E(X^2) &= 1, \end{aligned}$$

and so $V(X) = E(X^2) - (E(X))^2 = 1 - 0^2 = 1$. Therefore the standard normal distribution has mean 0 and variance 1. Hence also the transformed normal distribution $Y = \sigma X + \mu$ has mean μ and variance σ^2, and is often written $Y \sim N(\mu, \sigma^2)$. In this notation, $X \sim N(0, 1)$. If $Y \sim N(\mu, \sigma^2)$, then $X = \frac{Y-\mu}{\sigma} \sim N(0, 1)$.

It is worthwhile to know that the cdf of a standard normal distribution

$$\Phi(x) = \frac{1}{\sqrt{2\pi}} \int_{-\infty}^{x} e^{-y^2/2} dy \tag{6.17}$$

is not available in closed form. The solution to this issue is typical of mathematical custom, namely to make friends with Φ. There are both tables of Φ (available in many books) and algorithms for computing Φ. Some of its important properties are:

$$\begin{aligned} \Phi(x) =& 1 - \Phi(-x). \\ \Phi(0) =& 0.5 \\ \Phi(1) =& 0.8413 \\ \Phi(2) =& 0.9772. \end{aligned}$$

If $Y \sim N(\mu, \sigma^2)$, then $F_Y(x) = \Phi(\frac{x-\mu}{\sigma})$, since

$$\begin{aligned} F_Y(x) =& P\{Y \leq x\} = P\left\{\frac{Y-\mu}{\sigma} \leq \frac{x-\mu}{\sigma}\right\} = P\left\{X \leq \frac{x-\mu}{\sigma}\right\} \\ =& \Phi\left(\frac{x-\mu}{\sigma}\right). \end{aligned} \tag{6.18}$$

The moment generating function for a random variable $Y \sim N(\mu, \sigma^2)$ is

$$M_Y(t) = e^{\mu t} e^{(\sigma t)^2/2} = e^{\mu t + \sigma^2 t^2/2} \tag{6.19}$$

using Theorem 6.2.3 with $a = \sigma$ and $b = \mu$.

Lemma 6.3.1. *Suppose $X_i \sim N(0, \sigma_i^2), i = 1, 2,$ independent. Then $U = X_1 + X_2 \sim N(0, \sigma^2)$ where $\sigma^2 = \sigma_1^2 + \sigma_2^2$.*

Proof. The strategy I use is to transform the density of (X_1, X_2) to that of (X_1, U) and then to integrate out X_1, yielding the distribution of U.

The Jacobian matrix of the transformation from (X_1, X_2) to (X_1, U) is $\begin{pmatrix} 1 & 0 \\ 1 & 1 \end{pmatrix}$, which has determinant 1. The joint density of (X_1, X_2) is proportional to

$$\exp\left(-\frac{1}{2}\left(X_1^2/\sigma_1^2 + X_2^2/\sigma_2^2\right)\right). \tag{6.20}$$

After transformation, the joint density of X_1 and U is proportional to

$$\exp\left(-\frac{1}{2}\left[X_1^2/\sigma_1^2 + (U - X_1)^2/\sigma_2^2\right]\right). \tag{6.21}$$

Now I will work only with the part in square brackets, which I denote Q.

$$\begin{aligned} Q &= X_1^2/\sigma_1^2 + (U - X_1)^2/\sigma_2^2 \\ &= \frac{1}{\sigma_1^2\sigma_2^2}(X_1^2\sigma_2^2 + (U - X_1)^2\sigma_1^2) \\ &= \frac{1}{\sigma_1^2\sigma_2^2}\left(\sigma_2^2 X_1^2 + \sigma_1^2 X_1^2 - 2\sigma_1^2 U X_1 + \sigma_1^2 U^2\right) \\ &= \frac{1}{\sigma_1^2\sigma_2^2}\left(X_1^2(\sigma_1^2 + \sigma_2^2) - 2\sigma_1^2 U X_1 + \sigma_1^2 U^2\right) \\ &= \frac{\sigma^2}{\sigma_1^2\sigma_2^2}\left(X_1^2 - \frac{2\sigma_1^2 U X_1}{\sigma^2} + \frac{\sigma_1^2 U^2}{\sigma^2}\right) \\ &= \frac{\sigma^2}{\sigma_1^2\sigma_2^2}\left[\left(X_1^2 - \frac{2\sigma_1^2 U X_1}{\sigma^2} + \frac{\sigma_1^4 U^2}{\sigma^4}\right) + \frac{\sigma_1^2 U^2}{\sigma^2} - \frac{\sigma_1^4 U^2}{\sigma^4}\right] \\ &= \frac{\sigma^2}{\sigma_1^2\sigma_2^2}\left(X_1 - \sigma_1^2 U/\sigma^2\right)^2 + \frac{\sigma^2}{\sigma_1^2\sigma_2^2}\left(\frac{\sigma_1^2 U^2}{\sigma^2} - \frac{\sigma_1^4 U^2}{\sigma^4}\right). \end{aligned} \tag{6.22}$$

Now I am in a position to integrate with respect to X_1. The density of X_1 is proportional to

$$\exp\left\{-\frac{1}{2}(X_1 - \sigma_1^2 U/\sigma^2)^2\right\}. \tag{6.23}$$

I recognize this as a normal distribution with mean $\sigma_1^2 U/\sigma^2$ and a variance that does not depend on U. Therefore, its integral only changes the constant of proportionality, which I'm not keeping track of anyway.

What remains of Q after integration with respect to X_1 is

$$\begin{aligned} Q' &= \frac{\sigma^2}{\sigma_1^2\sigma_2^2}\left(\frac{\sigma_1^2 U^2}{\sigma^2} - \frac{\sigma_1^4 U^2}{\sigma^4}\right) \\ &= U^2/\sigma_2^2\left(1 - \frac{\sigma_1^2}{\sigma^2}\right) \\ &= \frac{U^2}{\sigma_2^2}\left(\frac{\sigma^2 - \sigma_1^2}{\sigma^2}\right) \\ &= U^2/\sigma^2. \end{aligned} \tag{6.24}$$

Therefore the density of U is proportional to

$$\exp\left(-\frac{1}{2}U^2/\sigma^2\right), \tag{6.25}$$

so U has a normal distribution with mean 0 and variance σ^2.

It might be supposed that this result could be obtained using moment generating functions, but such a "proof" would not be rigorous because moment generating functions are not necessarily unique.

□

6.3.1 *Remark*

The technique of completing the square is common in working with normal distributions, and is used several times in Chapter 8.

Corollary 6.3.2. *Let $X_i \sim N(\mu_i, \sigma_i^2), i = 1, 2$ independent. Let d_i, $i = 1, 2$ be constants, not both zero. Then $X = \sum_{i=1}^{2} d_i X_i$ has a normal distribution with mean $\mu = \sum_{i=1}^{2} d_i \mu_i$ and variance $\sigma^2 = \sum_{i=1}^{2} d_i^2 \sigma_i^2$.*

Proof. Let $Y_i = X_i - \mu_i$, so $Y_i \sim N(0, \sigma_i^2)$. Then $|d_i|Y_i \sim N(0, d_i^2\sigma_i^2)$ $i = 1, 2$. Since the normal distribution is symmetric around zero, $d_i Y_i \sim N(0, d_i^2\sigma_i^2)$. Applying the Lemma, $d_i Y_i + d_2 Y_2 \sim N(0, \sigma^2)$. Finally, $X = d_1 Y_1 + d_2 Y_2 + \mu = \sum_{i=1} d_i(Y_i + \mu_i) \sim N(\mu, \sigma^2)$. □

Theorem 6.3.3. *Let $X_i \sim N(\mu_i, \sigma_i^2)$, $i = 1, \ldots, n$ independent and let $d_i, i = 1, \ldots, n$ be constants not all zero. Then $X = \sum_{i=1}^{n} d_i X_i$ has a normal distribution with mean $\mu = \sum_{i=1}^{n} d_i \mu_i$ and variance $\sigma^2 = \sum_{i=1}^{n} d_i^2 \sigma_i^2$.*

Proof. By induction on n. □

Corollary 6.3.4. *Let $X_i \sim N(\mu, \sigma^2)$, $i = 1, \ldots, n$ independent, and let $T_n = \sum_{i=1}^{n} X_i/\sqrt{n}$. Then $T_n \sim N(\mu, \sigma^2)$.*

Proof. Let $d_i = 1/\sqrt{n}$ in the theorem. □

6.3.2 *Exercises*

1. What is the missing constant of proportionality in (6.20)?
2. What is the missing constant of proportionality in (6.23)?
3. Use the answers from Exercises 1 and 2 to show that the missing constant of proportionality in (6.24) is $\frac{1}{\sigma\sqrt{2\pi}}$.
4. Why is it important that the variance of X_1 does not depend on U? How would the analysis change if this were not true?

6.4 Multivariate normal distributions

Our treatment of the multivariate normal distribution traces our treatment of the univariate case, as follows: Suppose $\mathbf{X} = (X_1, \ldots, X_k)$ is a vector of k independent standard normal random variables. Then the pdf of $\mathbf{X}$ is

$$\begin{aligned} f_{\mathbf{X}}(x) &= \prod_{j=1}^{k} \frac{1}{\sqrt{2\pi}} e^{-x_j^2/2} \\ &= \frac{1}{(2\pi)^{k/2}} e^{-\sum_{j=1}^{k} x_j^2/2}, -\infty < x_j < \infty \text{ for all } j = 1, \ldots, k. \end{aligned} \tag{6.26}$$

Also, its moment generating function is

$$\begin{aligned} M_X(\mathbf{t}) = \prod_{j=1}^{k} M_{X_j}(t_j) &= \prod_{j=1}^{k} e^{-t_j^2/2} \\ &= e^{-\sum t_j^2/2} = e^{-\mathbf{t}'\mathbf{t}/2}. \end{aligned} \tag{6.27}$$

Such a random vector's distribution is denoted $\mathbf{X} \sim N(\mathbf{0}, I)$, for reasons that will become apparent.

Now let $\sum$ be a symmetric matrix with positive eigenvalues. (I hope that the use of $\sum$ here, to represent a covariance matrix, as is traditional, will not confuse a reader used to thinking of $\sum$ as a sign for summation.) Then by the decomposition (Theorem 1 of 5.8), I may write $\sum$ in the form

$$\sum = PDP'$$

where P is an orthogonal matrix, and D is diagonal with positive numbers on its diagonal. Let Δ be a diagonal matrix with diagonal elements equal to the (positive) square root of those of D. Finally let

$$\sum{}^{1/2} = P\Delta P'.$$

When $\sum^{1/2}$ is defined this way,

$$\begin{aligned} \sum{}^{1/2}\left(\sum{}^{1/2}\right)' &= P\Delta P' P\Delta' P' \\ &= P\Delta\Delta' P' \\ &= PDP' \\ &= \sum. \end{aligned} \tag{6.28}$$

Using this definition of $\sum^{1/2}$, let $\mathbf{Y} = \sum^{1/2}\mathbf{X} + \boldsymbol{\mu}$, where $\mathbf{X} \sim N(0, I)$. Then $E(\mathbf{Y}) = \boldsymbol{\mu}$, and

$$\begin{aligned} \text{Cov}(\mathbf{Y}) &= E[(\mathbf{Y} - \boldsymbol{\mu})'(\mathbf{Y} - \boldsymbol{\mu})] = E\left[\sum{}^{1/2}\mathbf{X}\mathbf{X}'\left(\sum{}^{1/2}\right)'\right] \\ &= \sum{}^{1/2} E(\mathbf{X}\mathbf{X}')\left(\sum{}^{1/2}\right)' = \sum{}^{1/2} I \left(\sum{}^{1/2}\right)' \\ &= \sum{}^{1/2}\left(\sum{}^{1/2}\right)' = \sum. \end{aligned} \tag{6.29}$$

Furthermore, the absolute value of the determinant of the Jacobian of the transformation $\mathbf{y} = \sum^{1/2}\mathbf{x} + \boldsymbol{\mu}$ is

$$\begin{aligned} \left\|\sum{}^{1/2}\right\| &= \left\|P\Delta P'\right\| = \left\|P\right\| \left\|\Delta\right\| \left\|P'\right\| \\ &= \left\|\Delta\right\| = \left|D\right|^{1/2} = \left|\sum\right|^{1/2}, \end{aligned}$$

using the fact that an orthogonal matrix has absolute value of determinant equal to 1 (see 5.7.3).

Hence, $\mathbf{Y}$ has the pdf

$$f_{\mathbf{Y}}(\mathbf{y}) = \frac{1}{(2\pi)^{k/2}|\sum|^{1/2}} e^{-\frac{1}{2}(y-\mu)'\sum^{-1/2}(\sum^{-1/2})'(y-\mu)}$$

$$-\infty < y_i < \infty \text{ for } i = 1, \ldots, k \quad (6.30)$$

where $\sum^{-1/2} = P\Delta^{-1}P'$, so

$$\begin{aligned}\sum^{-1/2}\left(\sum^{-1/2}\right)' =& P\Delta^{-1}P'P\Delta^{-1}P' \\ =& P\Delta^{-2}P \\ =& PD^{-1}P' \\ =& \sum^{-1},\end{aligned}$$

using notation from section 5.8.

Hence,

$$f_{\mathbf{Y}}(\mathbf{y}) = \frac{1}{(2\pi)^{k/2}}\left|\sum\right|^{-1/2} e^{-\frac{1}{2}(\mathbf{y}-\boldsymbol{\mu})'\sum^{-1}(\mathbf{y}-\boldsymbol{\mu})}, \quad -\infty < y_i < \infty, \quad i = 1, \ldots, k. \quad (6.31)$$

Furthermore, the random variable $\mathbf{Y}$ has moment generating function

$$M_Y(t) = e^{\boldsymbol{\mu}'\mathbf{t}} M_{\mathbf{X}}(t'A) = e^{\mu' t} e^{t' \frac{\sum^{-1/2}(\sum^{-1/2})' t}{2}} = e^{\mu' t + \frac{t'\sum^{-1} t}{2}}. \quad (6.32)$$

It comes, then, as no surprise that the distribution of Y is denoted $Y \sim N(\boldsymbol{\mu}, \sum)$, and is said to have a normal distribution with mean $\boldsymbol{\mu}$ and covariance matrix $\sum$.

6.4.1 Exercises

1. In equation (6.31), suppose $k = 1$. Show that (6.31) reduces to (6.14).

6.5 The Central Limit Theorem

The central limit theorem, the main target of this chapter, says roughly that if X_i, $i = 1, \ldots, n$ are identically distributed independent random variables with mean 0 and variance 1, then $T_n = \sum_{i=1}^{n} X_i/\sqrt{n}$ has a limiting normal distribution with mean 0 and variance 1. The exact meaning of the word "limiting" hasn't been specified yet, so treat this paragraph as an advertisement of a coming attraction.

To study the limit theorems, the first task is to be precise about exactly what is meant by $F_n(x)$ approaching $F(x)$. One possible meaning for this is

$$\lim_{n\to\infty} F_n(x) = F(x) \text{ for all } x. \quad (6.33)$$

Consider, however, the following example:

Example 1: Let X_n be a random variable that takes the value $-1/n$ with probability 1/2 and $1/n$ with probability 1/2. Then

$$F_n(x) = \begin{cases} 0 & x < -1/n \\ 1/2 & -1/n \le x < 1/n \\ 1 & x \ge 1/n \end{cases}.$$

With this specification,

$$\lim_{n\to\infty} F_n(x) = G(x) = \begin{cases} 0 & x < 0 \\ 1/2 & x = 0 \\ 1 & x > 0 \end{cases} .$$

This limiting function $G(x)$ is not a distribution function, because at $x = 0$ it is not right-continuous, that is,

$$\lim_{\substack{x\to 0 \\ x>0}} G(x) = 1 \neq G(0) = 1/2.$$

It is reasonable, however, to think that this sequence of random variables should have a limiting distribution, namely one that equals 0 with probability 1. Such a random variable, Y, has distribution function

$$F_Y(x) = \begin{cases} 0 & x < 0 \\ 1 & x \geq 0 \end{cases}$$

which coincides with $G(x)$ except at $x = 0$. For this reason, one may exclude the point $x = 0$ from the requirement stated in (6.33), and say that $F_n(x)$ *converges weakly* to $F(x)$ provided

$$\lim_{n\to\infty} F_n(x) = F(x) \text{ at points } x \text{ of continuity of } F. \tag{6.34}$$

This definition has a second issue, namely that, so defined, $F(x)$ need not be a cumulative distribution function, as the following example shows:

Example 2: Let X_n be random variables that take the value $-n$ with probability 1/2 and n with probability 1/2. Then

$$F_n(x) = \begin{cases} 0 & x < -n \\ 1/2 & -n \leq x < n \\ 1 & x > n \end{cases} .$$

Now for each x, $\lim_{n\to\infty} F_n(x) = 1/2$.

Thus the limiting function fails to satisfy the conditions on a cumulative distribution function that $\lim_{x\to-\infty} F(x) = 0$ and $\lim_{x\to\infty} F(x) = 1$. In this example, the probability has "escaped" toward $-\infty$ and ∞, and there does not appear to be a reasonable sense of a limiting distribution here. Consequently, the focus is on weak convergence as defined in (6.34), with the reminder that the limiting function is not necessarily a distribution function.

6.5.1 *A supplement on a relation between convergence in probability and weak convergence*

Recall from Section 4.11.2, the definition that Y_n converges in probability to Y if

$$P\{|Y_n - Y| > \epsilon\} \to 0 \text{ for all } \epsilon > 0.$$

Slutsky's Theorem has to do with the special case in which Y takes a single value c with probability 1.

Theorem 6.5.1. *(Slutsky) Suppose Y_n converges in probability to c, a constant, and X_n converges weakly to X, a random variable. Then*

a) $X_n + Y_n$ converges weakly to $X + c$.

b) X_nY_n converges weakly to Xc.

c) If $c \neq 0, X_n/Y_n$ converges weakly to X/c.

Proof.

a) Choose x so that $x - c$ is a point of continuity of F_X, and choose $\epsilon_j > 0, j = 1, \ldots,$ to be a sequence of real numbers satisfying

 (i) $\epsilon_j > 0$ for all j

 (ii) $\lim_{j\to\infty} \epsilon_j = 0$

 (iii) $x - c + \epsilon_j$ and $x - c - \epsilon_j$ are points of continuity of F_X. This can always be done because points of discontinuity are at most countable, so points of continuity are uncountable.

$$\begin{aligned} P\{X_n + Y_n \leq x\} &= S_1 + S_2, \text{ where} \\ S_1 &= P\{(X_n + Y_n \leq x) \cap (|Y_n - c| > \epsilon_j)\} \text{ and} \\ S_2 &= P\{(X_n + Y_n \leq x) \cap (|Y_n - c| \leq \epsilon_j)\}. \end{aligned}$$

Consider first S_1.

$$\{X_n + Y_n \leq x) \cap (|Y_n - c| > \epsilon_j) \subseteq \{|Y_n - c| > \epsilon_j\}.$$

Hence,

$$P(S_1) = P\{(X_n + Y_n \leq x) \cap (|Y_n - c| > \epsilon_j)\} \leq P\{|Y_n - c| > \epsilon_j\}.$$

By assumption on Y_n,

$$\lim_{n\to\infty} P\{|Y_n - c| > \epsilon_j\} = 0 \text{ for all } j.$$

Hence, $\lim_{n\to\infty} P(S_1) = 0$ for all j.

Returning to S_2, $|Y_n - c| \leq \epsilon_j$ iff

$$-\epsilon_j \leq Y_n - c \leq \epsilon_j \text{ iff } c - \epsilon_j \leq Y_n \leq c + \epsilon_j \text{ iff } X_n + c - \epsilon_j \leq X_n + Y_n \leq X_n + c + \epsilon_j.$$

Therefore,

$$P\{X_n + c - \epsilon_j \leq x\} \geq P\{X_n + Y_n \leq x\} \geq P\{X_n + c + \epsilon_j \leq x\}. \tag{6.35}$$

Concentrating on the first inequality in (6.35),

$$\lim_{n\to\infty} \sup P\{X_n + c - \epsilon_j \leq x\} \geq \limsup P(X_n + Y_n \leq x).$$

Now $\limsup_{n\to\infty} P\{X_n + c - \epsilon_j \leq x\} = \limsup P\{X_n \leq x - c + \epsilon_j\}$.

Since $x-c+\epsilon_j$ is a point of continuity of F_X, and by assumption on X, $\lim_{n\to\infty} P\{X_n \leq x - c + \epsilon_j\}$ exists and equals $P\{X \leq x - c + \epsilon_j\}$.

Therefore,

$$\limsup P\{X_n + c - \epsilon_j \leq x\} = P\{X \leq x - c + \epsilon_j\} = F_X(x - c + \epsilon_j) \text{ for all } j.$$

Finally, because $x - c$ is a point of continuity of F_X,

$$\lim_{j\to\infty} F_X(x - c + \epsilon_j) = F_X(x - c) = P\{X \leq x - c\}. \tag{6.36}$$

Now turning to the second inequality in (6.35),

$$P\{X_n + Y_n \leq c\} \geq P\{X_n + c + \epsilon_j \leq x\}.$$

Therefore,

$$\lim_{n\to\infty} \inf P\{(X_n + Y_n \leq x)\} \geq \lim_{n\to\infty} \inf P\{X_n + c + \epsilon_j \leq x\}.$$

Now

$$\lim_{n\to\infty} \inf P\{X_n + c + \epsilon_j \leq x\} = \lim_{n\to\infty} \inf P[\{X_n \leq x - c - \epsilon_j\}.$$

Since $x - c - \epsilon_j$ is a point of continuity of F_X, and by assumption on X,

$$\lim_{n\to\infty} P\{X_n \leq x - c - \epsilon_j\} \text{ exists and equals } P\{X \leq x - c - \epsilon_j\} = F_X(x - c - \epsilon_j).$$

Because $x - c - \epsilon_j$ is a point of continuity of F_X,

$$\lim_{j\to\infty} F_X(x - c + \epsilon_j) = F_X(x - c) = P\{X \leq x - c\}. \tag{6.37}$$

Putting (6.36) and (6.37) together,

$$\begin{aligned} P\{X \leq x - c\} &\geq \lim_{n\to\infty} \sup P\{X_n + Y_n \leq x\} \geq \lim_{n\to\infty} \inf P\{X_n + Y_n \leq x\} \\ &\geq P\{X \leq x - c\}. \end{aligned}$$

Then $\lim_{n\to\infty} P\{X_n + Y_n \leq x\}$ exists and equals $P\{X \leq x - c\}$.

□ a)

To prove b), consider first the case $c = 0$.

Assume $\epsilon > 0$ and $\delta > 0$ are such that $\pm\ \epsilon/\delta$ are points of continuity of F_X.

$$\begin{aligned} P\{|X_nY_n| > \epsilon\} &= T_1 + T_2, \text{ where} \\ T_1 &= P\{(|X_nY_n > \epsilon) \cap (Y_n > \delta)\} \text{ and} \\ T_2 &= P\{(|X_nY_n > \epsilon) \cap (Y_n \leq \delta)\}. \end{aligned}$$

Taking T_1 first,

$$\{(|X_nY_n| > \epsilon) \cap (Y_n > \delta)\} \subseteq (Y_n > \delta).$$

Therefore,

$$T_1 \leq P\{Y_n > \delta\} \to 0 \quad \text{as } n \to \infty \text{ for all } \delta > 0.$$

Returning to T_2,

$$|Y_n| \leq \delta \to \frac{1}{|Y_n|} \geq \frac{1}{\delta} \text{ implies } \frac{|X_nY_n|}{|Y_n|} \geq \epsilon/\delta \text{ implies } |X_n| \geq \epsilon/\delta.$$

Therefore,

$$\begin{aligned} T_2 &\leq P\{|X_n| \geq \epsilon/\delta\} \leq P\{X_n > \epsilon/\delta\} + P\{X_n < -\epsilon/\delta\} \\ &= 1 - F_n(\epsilon/\delta) + Fn(-\epsilon/\delta). \end{aligned}$$

Now, as $n \to \infty$, since ϵ/δ and $-\epsilon/\delta$ are points of continuity of F_X,

$$1 - F_n(\epsilon/\delta) + F_n(-\epsilon/\delta) \to 1 - F(\epsilon/\delta) + F(-\epsilon/\delta). \tag{6.38}$$

As $\delta \to 0$, (6.38) approaches

$$1 - F(\infty) + F(-\infty) = 1 - 1 + 0 = 0.$$

Hence, $T_2 \to 0$. Therefore, $P\{|X_nY_n| > \epsilon\} \to 0$ as $n \to \infty$. If $c \neq 0$, substituting $Z_n = Y_n - c$ for Y_n suffices.

□ b)

To prove c), the route is through two lemmas:

Lemma 6.5.2. *Suppose X_n converges weakly to 1, then $1/X_n$ converges weakly to 1.*

To show $P(\{|\frac{1}{X_n} - 1| > \epsilon\}) \to 0$ as $n \to \infty$ for all $\epsilon > 0$,

$$|\frac{1}{X_n} - 1| > \epsilon \text{ if and only if } |X_n - 1| > \epsilon |X_n|.$$

Then

$$\left\{ \left| \frac{1}{X_n} - 1 \right| > \right\} = W_1 \cup W_2,$$

where

$$\begin{aligned} W_1 &= (|X_n - 1| > \epsilon|X_n|) \cup (|X_n| \geq 1/2) \\ W_2 &= (|X_n - 1| > \epsilon|X_n|) \cup (|X_n| < 1/2). \end{aligned}$$

Now,

$$\begin{aligned} P\{W_1\} &\leq P\{|X_n - 1| > \epsilon/2\} \to 0 \text{ as } n \to \infty \text{ by assumption on } X_n. \\ P\{W_2\} &\leq P\{|X_n| < 1/2\} \leq P\{|X_n - 1| > 1/2\} \to 0 \text{ as well.} \end{aligned}$$

Hence,

$P\{|\frac{1}{X_n} - 1| > \epsilon\} \to 0.$ □ Lemma 1

Lemma 6.5.3. *If X_n converges weakly to $c \neq 0$, then $\frac{1}{X_n}$ converges weakly to $1/c$.*

Let $Y_n = X_n/c$. Then Y_n converges weakly to 1 because

$$|Y_n - 1| = |X_n/c - 1| = \frac{|X_n - c|}{|c|}.$$

Therefore,

$$P\{|Y_n - 1| < \epsilon\} = P\{\frac{|X_n - c|}{|c|} < \epsilon\} = P\{|X_n - c| < \epsilon|c|\} \to 0.$$

Hence,

$$X_n = cY_n \text{ and } \frac{1}{X_n} = \frac{1}{c}\frac{1}{Y_n} \to 1/c.$$

□ Lemma 2

Part c) of Slutsky's Theorem now follows, from Lemma 2 and Part b).
This concludes the proof of Slutsky's Theorem. □

The fact that weak convergence requires an exception for points of discontinuity in the limiting distribution function is a hint that trouble, if trouble there is, lies there. The technique for avoiding that trouble is what I call "smudging," adding a small amount of smooth independent noise to both the approximand and the approximation.

The proof to come requires that the noise be limited to a (small) interval $[-\gamma, \gamma]$ around 0, and that its cdf have continuous second derivatives, or equivalently that its pdf have continuous first derivatives. A pdf that does this has the form

$$f(w) = \begin{cases} k(\gamma + w)^2(\gamma - w)^2 & -\gamma \leq w \leq \gamma \\ 0 & \text{otherwise} \end{cases}. \tag{6.39}$$

The symbol W is used for a random variable having this distribution for the rest of this chapter.

6.5.2 A supplement on uniform continuity and points of accumulation

Section 4.7.11 introduced the concept of uniform convergence of a sequence of functions $f_n(x)$ to a limit function $f(x)$. I now introduce a related concept, that of uniform continuity of a function.

To review: a function $f(x)$ is said to be continuous at a point x_0 if for every $\epsilon > 0$, there exists a $\delta(\epsilon, x_o)$ such that, if $\mid x - x_0 \mid < \delta$, then $\mid f(x) - f(x_0) \mid < \epsilon$. Thus, continuity is a local condition at the point x_0. Uniform continuity, by contrast, is a global condition that applies to all points x_0 in an interval. Its definition is the same as that for continuity, except that δ is no longer permitted to depend on x_0, and therefore can be written $\delta(\epsilon)$. So the same δ has to work for every x_0 in the interval. While this distinction may seem technical and unimportant, it has important implications.

Recall the definition of accumulates on point given in Section 4.7. The result that is needed is the following:

Theorem 6.5.4. *Let $f(x)$ be a continuous function on a closed bounded interval, say $[a, b]$. Then $f(x)$ is uniformly continuous on $[a, b]$.*

Proof. This proof is by contradiction. Thus assume that f is continuous, but not uniformly continuous. Then there exists an $\epsilon > 0$ such that, for every sequence $\delta_1, \delta_2, \ldots$ approaching zero, there are values u_n, v_n in the interval $[a, b]$ such that $\mid u_n - v_n \mid < \delta_n$, but $\mid f(u_n) - f(v_n) \mid > \epsilon$. The sequences u_n and v_n have limit points u and v by Lemma 4.7.4. Furthermore, $u = v$ since $\mid u_n - v_n \mid < \delta_n$. Also, $u \epsilon [a, b]$. Therefore the function f is continuous at u. But continuity at u implies that there is some $\delta > 0$ such that, if $\mid u_n - v_n \mid < \delta, \mid f(u_n) - f(v_n) \mid < \epsilon$, which is a contradiction. □

The next investigation is to examine the continuity of the function $f(w)$ defined in (6.39), and its derivatives. Since $f(w) = 0$ if $w < -\gamma$ or $w > \gamma$, f and all its derivatives are continuous there. Similarly, since f is a polynomial in w in the open interval $(-\gamma, \gamma)$, f and all its derivatives are continuous there. Hence, the sensitive points are $w = \gamma$ and $w = -\gamma$.

Since $f(\gamma) = f(-\gamma) = 0$, f is continuous at both points. The first derivative of f is

$$f'(w) = k[2(\gamma + w)(\gamma - w)^2 + 2(\gamma + w)^2(\gamma - w)]. \tag{6.40}$$

Therefore, $f'(\gamma) = f'(-\gamma) = 0$, and f' is continuous at both points.

However

$$\begin{aligned} f''(w) &= 2k[(\gamma - w)^2 - 2(\gamma + w)(\gamma - w) - (\gamma + w)^2 + 2(\gamma + w(\gamma - w)] \\ &= 2k[(\gamma - w)^2 - (\gamma + w)^2]. \end{aligned}$$

Therefore

$$\begin{aligned} f''(\gamma) &= 2k(2\gamma)^2 = 8k\gamma^2 > 0 \\ f''(-\gamma) &= 2k(-2\gamma)^2 = 8k\gamma^2 > 0. \end{aligned}$$

Hence $f''(w)$ is discontinuous at γ and $-\gamma$.

Now $f(w)$ is the derivative of its cumulative distribution function F_W. Hence, F_W, F'_W and F''_W are all continuous functions on $[-\gamma, \gamma]$, and therefore, by Theorem 6.5.4, uniformly continuous on $[-\gamma, \gamma]$. This is important in the proof to come.

6.5.3 Exercises

1 Show that the function $g(x) = 1/x,\ 0 < x \leq 1$ is continuous but not uniformly continuous.

2. Show that the function $h(x) = x^2,\ x \geq 0$ is continuous but not uniformly continuous.

6.5.4 Resuming the proof of the central limit theorem

I now introduce a lemma that justifies the addition of the smudge variable W to both the sequence being approximated and the approximating limit.

Lemma 6.5.5. *Let $S_n, n \geq 1$ and T be random variables. If, for every $\gamma > 0$, there exists a random variable W, independent of S_n and T, satisfying $P\{\mid W \mid \leq \gamma\} = 1$ and such that $F_{S_n+W}(s) \rightarrow F_{T+W}(s)$ as $n \rightarrow \infty$ for all s, then $F_{S_n}(t) \rightarrow F_T(t)$ as $n \rightarrow \infty$ for all t at which F_T is continuous.*

Proof. Let t be a point of continuity of F_T. Given $\epsilon > 0$, find $\gamma > 0$ such that

$$\begin{array}{ll} (i) & F_T(t+2\gamma) \leq F_T(t) + \epsilon \quad \text{and} \\ (ii) & F_T(t-2\gamma) \geq F_T(t) - \epsilon. \end{array}$$

Let W satisfy the hypotheses of the lemma. Since $W \leq \gamma$,

$$\begin{aligned} F_{S_n}(t) &= P\{S_n \leq t\} \\ &\leq P\{S_n + W \leq t + \gamma\} \\ &= F_{S_n+W}(t+\gamma). \end{aligned} \tag{6.41}$$

Similarly, since $W \geq -\gamma$,

$$\begin{aligned} F_{T+W}(t+\gamma) &= P\{T + W \leq t + \gamma\} \\ &= P\{T \leq t + \gamma - W\} \\ &\leq P\{T \leq t + 2\gamma\} \\ &= F_T(t+2\gamma). \end{aligned} \tag{6.42}$$

Since

$$\lim_{n \rightarrow \infty} F_{S_n+W}(t+\gamma) = F_{T+W}(t+\gamma), \tag{6.43}$$

there exists an N such that for all $n \geq N$,

$$\begin{aligned} F_{S_n}(t) &\leq F_{S_n+W}(t+\gamma) & &\text{uses (6.41)} \\ &\leq F_{T+W}(t+\gamma) + \epsilon & &\text{uses (6.43)} \\ &\leq F_T(t+2\gamma) + \epsilon & &\text{uses (6.42)} \\ &\leq F_T(t) + 2\epsilon & &\text{uses } (i)\ . \end{aligned} \tag{6.44}$$

Similarly, since $W \geq -\gamma$,

$$\begin{aligned} F_{S_n}(t) &= P\{S_n \leq t\} \\ &\geq P\{S_{n+W} \leq t - \gamma\} \\ &= F_{S_n+W}(t-\gamma). \end{aligned} \tag{6.45}$$

Since $W \leq \gamma$,

$$\begin{aligned} F_{T+W}(t-\gamma) &= P\{T + W \leq t - \gamma\} \\ &= P\{T \leq t - \gamma - W\} \\ &\geq P\{T \leq t - 2\gamma\} \\ &= F_T(t-2\gamma). \end{aligned} \tag{6.46}$$

Again, since (6.43) holds, there exists an N^* such that for all $n \geq N^*$,

$$\begin{aligned} F_{S_n}(t) &\geq F_{S_n+W}(t-\gamma) & &\text{uses (6.45)} \\ &\geq F_{T+W}(t-\gamma) - \epsilon & &\text{uses (6.43)} \\ &\geq F_T(t-2\gamma) - \epsilon & &\text{uses (6.46)} \\ &\geq F_T(t) - 2\epsilon & &\text{uses } (ii). \end{aligned} \tag{6.47}$$

Putting (6.45) and (6.47) together yields

$$-2\epsilon \leq F_{S_n}(t) - F_T(t) \leq 2\epsilon \tag{6.48}$$

for all $n \geq \max(N, N^*)$. Therefore, $\lim F_{S_n}(t) = F_T(t)$ at all continuity points t of $F_T(\cdot)$.

□

6.5.5 Supplement on the sup-norm

For the next lemma, it is useful to have a distance function between cdf's. The distance function that will be used is called the sup norm. It is written with double lines, and is defined as follows:

$$\|F - G\| = \sup_t \mid F(t) - G(t) \mid . \tag{6.49}$$

Note that this use of double bars is different from its use to mean the determinant of a matrix. Double bars means the sup-norm for the rest of this chapter.

The sup-norm has the following properties:

(i) $\|F - G\| \geq 0$ for all F and G

(ii) $\|F - F\| = 0$

(iii) $\|F - G\| > 0$ if $F \neq G$

(iv) $\|F - G\| = \|G - F\|$

(v) $\|F - H\| \leq \|F - G\| + \|G - H\|$ for all F, G and H.

Properties (i) to (iv) are left to the reader to prove. Here is a proof for property (v), called the triangle inequality.

Proof. Choose $\epsilon > 0$. Then there exists a t^* such that

$$\begin{aligned}
\|F - H\| &\leq \mid F(t^*) - H(t^*) \mid +\epsilon \\
&= \mid F(t^*) - G(t^*) + G(t^*) - H(t^*) \mid +\epsilon \\
&\leq \mid F(t^*) - G(t^*) \mid + \mid G(t^*) - H(t^*) \mid +\epsilon \\
&\leq \sup_t \mid F(t^*) - G(t) \mid + \sup_t \mid G(t) - H(t) \mid +\epsilon \\
&= \|F - G\| + \|G - H\| + \epsilon.
\end{aligned}$$

Since this holds for all $\epsilon > 0$,

$$\|F - H\| \leq \|F - G\| + \|G - H\|. \tag{6.50}$$

□

6.5.6 Resuming the development of the central limit theorem

The proof to follow makes use of the following relationship;

Lemma 6.5.6. $F_{W+X}(w) = E_X F_W(w - X)$.

Proof.

$$\begin{aligned}
F_{W+X}(w) &= P\{W + X \leq w\} \\
&= E_{W,X} I\{W + X \leq w\} \\
&= E_X E_{W|X} I\{W \leq w - X\} \\
&= E_X F_W(w - X).
\end{aligned}$$

□

The idea of the next lemma is to bound the sup-norm of two sums of independent and identically distributed random variables. Specifically, if S_n is the sum of n independent and identically distributed random variables $X_1, \ldots, X_n$ and T_n is the sum of n independent and identically distributed random variables $Y_1, \ldots, Y_n$, then one can sequentially replace each X_i by a Y_i. In n-steps, one goes from S_n to T_n. Adding the smudge variable to both T_n and S_n does not disturb the arguments. More formally,

Lemma 6.5.7. *Let $X, X_1, X_2, \ldots, X_n$ be independent and identically distributed random variables, and let $S_n = \sum_{i=1}^n X_i$. Similarly let $Y, Y_1, Y_2, \ldots, Y_n$ be independent and identically distributed random variables, also independent of $X, X_1, X_2, \ldots, Xn$. Let $T_n = \sum_{i=1}^n Y_i$. Let W be a random variable independent of X, the X_i's, Y, and the Y_i's. Then*

$$\|F_{W+S_n} - F_{W+T_n}\| \leq n\|F_{W+X} - F_{W+Y}\|.$$

Proof. For $0 \leq i \leq n$, let $Q_i = Y_1, +Y_2 + \ldots, Y_i$ and $R_i = X_i + \ldots, +X_n$.

Then

$$\begin{aligned} &F_{W+S_n}(w) - F_{W+T_n}(w) \\ &\quad = \sum_{i=1}^n \left[F_{W+Q_{i-1}+X_i+R_{i+1}}(w) - F_{W+Q_{i-1}+Y_i+R_{i+1}}(w) \right] \\ &\quad = \sum_{i=1}^n E_i \left[F_{W+X_i}(w - Q_{i-1} - R_{i+1}) - F_{W+Y_i}(w - Q_{i-1} - R_{i+1}) \right] \end{aligned} \tag{6.51}$$

using Lemma 6.5.6, where E_i is the expectation over Q_{i-1} and R_{i+1}, i.e. over $Y_1, \ldots, Y_{i-1}$ and $X_{i+1}, \ldots, X_n$.

Then

$$\begin{aligned} &\left|F_{W+S_n}(w) - F_{W+T_n}(w)\right| \\ &\quad = \left| \sum_{i=1}^n E_i \left[F_{W+X_i}(w - Q_{i-1} - R_{i+1}) - F_{W+Y_i}(w - Q_{i-1} - R_{i+1}) \right] \right| \\ &\quad \leq \sum_{i=1}^n \left| E_i \left[F_{W+X_i}(w - Q_{i-1} - R_{i+1}) - F_{W+Y_i}(w - Q_{i-1} - R_{i+1}) \right] \right| \\ &\quad \leq \sum_{i=1}^n E_i \left| F_{W+X_i}(w - Q_{i-1} - R_{i+1}) - F_{W+Yi}(w - Q_{i-1} - R_{i+1}) \right|. \end{aligned} \tag{6.52}$$

Each of the summands is bounded by $\|F_{W+X} - F_{W+Y}\|$, and there are n of them. Consequently,

$$\left| F_{W+S_n}(w) - F_{W+T_n}(w) \right| \leq n\|F_{W+X} - F_{W+Y}\|. \tag{6.53}$$

This inequality holds for all w, so

$$\|F_{W+S_n} - F_{W+T_n}\| \leq n\|F_{W+X} - F_{W+Y}\|.$$

□

Lemma 6.5.8. *Let W and Z be independent random variables with F_W having two bounded and uniformly continuous derivatives. Also, suppose $E(Z) = 0$ and $E(Z^2) = 1$. Then, for every $\epsilon > 0$, there exists an N such that for all $n \geq N$,*

$$\left| F_{W+X}(w) - F_W(w) - \frac{1}{2n} F_W''(w) \right| \leq \epsilon/n$$

for all w, where $X = Z/\sqrt{n}$.

Proof. At the outset, note that N will not be chosen until the end. Hence, references to n are provisional pending that choice. Lemma 6.5.6 yields

$$F_{W+X}(w) = E_X[F_W(w - X)].$$

Since F_W has two continuous derivatives, a Taylor expansion with remainder can be used:

$$F_W(w - X) = F_W(w) - \frac{1}{n^{1/2}}F'(w)Z + \frac{1}{2n}F''(\nu(Z))Z^2, \tag{6.54}$$

where v is a function satisfying $w - Z/\sqrt{n} \leq \nu(Z) \leq w$. Furthermore, $F''_W(v(Z))Z^2$ is integrable because each of the other terms in (6.54) is integrable.

(6.54) can be reexpressed as

$$F_W(w - X) = F_W(w) - \frac{1}{n^{1/2}}F'(w)Z + \frac{1}{2n}F''_W(w)Z^2 + \frac{1}{2n}(F''(v(Z)) - F''_W(w))Z^2.$$

Taking expectations of both sides yields

$$F_{W+X}(w) = F_W(w) + \frac{1}{2n}F''(w) + \frac{1}{2n}E_X\left[F''_W(v(Z)) - F''_W(w)\right]Z^2 \tag{6.55}$$

This is nearly in the form required by the theorem. What remains is to bound the last expectation in (6.55). To do so requires two facts:

(i) Because F_W has a uniformly continuous second derivative, there exists a $\delta > 0$ such that if $\mid y - w \mid \leq \delta$, then $\mid F''_W(y) - F''_W \mid \leq \epsilon$ for all y and w.

(ii) Because the second derivative is bounded, there exists an M such that $\mid F''(y) \mid \leq M$ for all y.

Now I break the expectation in (6.55) into two parts:

$$I_1 = E_X\left[(F''_W(v(Z)) - F''_W(w))Z^2 I(\mid Z \mid \leq \delta n^{1/2})\right)$$

$$I_2 = E_X\left[F''_W(v(Z)) - F''_W(w)Z^2 I(\mid Z \mid > \delta n^{1/2}\right).$$

Taking I_1 first, I have $w - Z/\sqrt{n} \leq v(Z) \leq w$, or $-Z/\sqrt{n} \leq v(Z) - w \leq 0$, so

$$\mid v(Z) - w \mid \leq Z/\sqrt{n} \leq \delta.$$

Using (i), this implies $\mid F''_W(v(Z) - F''_W(w) \mid \leq \epsilon$. Then

$$\begin{aligned} \mid I_1 \mid &\leq \epsilon E Z^2 I(\mid Z \mid \leq \delta n^{1/2}) \\ &\leq \epsilon E Z^2 \\ &\leq \epsilon. \end{aligned} \tag{6.56}$$

Furthermore, note that so far the proof has not required restrictions on N. The bound $\mid I_1 \mid \leq \epsilon$ is proved for all n.

To address I_2, note that

$$\mid F''_W(v(Z)) - F''_W(w) \mid \leq 2M, \quad \text{using } (ii).$$

Because $E(Z^2) = 1$, by truncating Z sufficiently far from 0, $EZ^2 I(\mid Z \mid > \delta\sqrt{n})$ can be made arbitrarily small, and in particular less than $\epsilon/2M$ by making N sufficiently large. Therefore, for all $n \geq N$,

$$\mid I_2 \mid \leq 2M(\epsilon/2M) = \epsilon. \tag{6.57}$$

Putting the pieces together, from (6.55),

$$\mid F_{W+X}(w) - F_W(w) - \frac{1}{2n}F''(w) \mid \leq (1/2n)(\mid I_1 \mid + \mid I_2 \mid) \leq \frac{1}{2n}(\epsilon + \epsilon) = \epsilon/n$$

for all $n \geq N$ and for all w. □

Corollary 6.5.9. *Let W, Z and Z' be independent random variables satisfying $E(Z) = E(Z') = 0$ and $E(Z^2) = E(Z'^2) = 1$. Then there exists an N such that, for all $n \geq N$,*

$$\|F_{W+X} - F_{W+X'}\| < \epsilon/n$$

where $X = Z/\sqrt{n}$ and $X' = Z'/\sqrt{n}$.

Proof. Using Lemma 6.5.8, there is an N such that for all $n \geq N$,

$$| F_{W+X}(w) - F_W(w) - \frac{1}{2n}F''(w) | \leq \epsilon/2n.$$

Similarly, there exists an N' such that for all $n \geq N'$,

$$| F_{W+X'}(w) - F_W(w) - \frac{1}{2n}F''(w) | \leq \epsilon/2n.$$

Let $N^* = \max(N, N')$. Then for all $n \geq N^*$,

$$\begin{aligned} &| F_{W+X}(w) - F_{W+X'}(w) | = \\ &\quad \left| \left[F_{W+X}(w) - F_W(w) - \frac{1}{2n}F''(w)\right] - \left[F_{W+X'}(w) - F_W(w) - \frac{1}{2n}F''(w)\right] \right| \\ &\quad \leq | F_{W+X}(w) - F_W(w) - \frac{1}{2n}F''(w) | + | F_{W+X'}(w) - F_W(w) - \frac{1}{2n}F''(w) | \\ &\quad \leq \epsilon/2n + \epsilon/2n = \epsilon/n. \end{aligned}$$

Since this holds for all w,

$$\|F_{W+X} - F_{W+X'}\| \leq \epsilon/n.$$

Thus N^* satisfies the conditions of the corollary. □

Theorem 6.5.10. *(Central limit theorem, simplest form). Let $Z_1, \ldots, Zn$ be independent and identically distributed random variables with mean 0 and variance 1, and let $S_n = \sum_{i=1}^n Z_i/\sqrt{n}$. Then*

$$\lim_{n\to\infty} F_{S_n}(t) = \Phi(t) \text{ for all real } t.$$

Proof. Assume the notation and conditions of Lemma 6.5.7.

Then

$$\|F_{W+S_n} - F_{W+T_n}\| \leq n\|F_{W+X} - F_{W+Y}\|. \tag{6.58}$$

Corollary 6.5.9 implies that for all $\epsilon > 0$, there exists an N such that for all $n \geq N$,

$$\|F_{W+X} - F_{W+Y}\| \leq \epsilon/n. \tag{6.59}$$

Combining these results yields

$$\|F_{W+S_n} - F_{W+T_n}\| \leq n \cdot \epsilon/n = \epsilon. \tag{6.60}$$

Hence, $W + S_n$ and $W + T_n$ are arbitrarily close in sup-norm. So far, surprisingly, S_n and T_n are sums of independent random variables with mean 0 and variance 1, each divided by $\sqrt{n}$, but are unrestricted. It isn't obvious why the limit is normal.

Consider, however, the impact of Corollary 6.3.4. Thus, if the Y's are normal with mean 0 and variance 1, then T_n has a normal distribution with mean 0 and variance 1, for all n. Therefore, $T_n = T$, not depending on n.

Then

$$\|F_{W+S_n} - F_{W+T}\| \leq \epsilon. \tag{6.61}$$

Now Lemma 6.5.5 may be applied, with the consequence that

$$\lim_{n\to\infty} F_{S_n}(t) = \Phi(t) \tag{6.62}$$

at all points of continuity of $\Phi(t)$. However, this is all real t, since $\Phi(t)$ is continuous. □

6.5.7 The delta method

While it is poor naming practice to include notation in a name, this name is so deeply embedded in the literature that renaming is hopeless. So the "delta method" it is.

The concept is to use Taylor's Theorem to extend the central limit theorem to smooth functions, as follows: Suppose X_n is a sequence of random variables satisfying the following version of the central limit theorem:

$$\sqrt{n}(X_n - \theta) \text{ converges weakly to } N(0, \sigma^2).$$

Can some useful conclusion be found for $g(X_n)$, where g is a smooth function? The result is

Theorem 6.5.11.

$$\sqrt{n}[g(X_n) - g(\theta)] \textit{converges weakly to } N(0, [g'(\theta)]^2\sigma^2),$$

provided $g'(\theta)$ *is continuous and non-zero.*

Proof. A first order Taylor expansion can be written

$$g(X_n) = g(\theta) + g'(\tilde{\theta})(X_n - \theta), \tag{6.63}$$

where $\tilde{\theta}$ lies between X_n and θ. Since X_n converges in probability to θ and $\tilde{\theta}$ lies between X_n and θ, $\tilde{\theta}$ also converges in probability to θ. Since $g'(\theta)$ is continuous, it follows that $g'(\tilde{\theta})$ converges in probability to $g'(\theta)$.

Rearranging (6.63) yields

$$\sqrt{n}[g(X_n) - g(\theta)] = g'(\tilde{\theta})\sqrt{n}[x_n - \theta]. \tag{6.64}$$

Since $\sqrt{n}(X_n - \theta)$ converges in distribution to $N(0, \sigma^2)$ by assumption, an appeal to Slutsky's Theorem 6.5.1 yields the result. □

The delta method can be extended to multivariable X_n's using a multivariate version of Taylor's Theorem. For extension and further results, see van der Vaart (1998, Chapter 3)

6.5.8 A heuristic explanation of smudging

The purpose of this subsection is to give an explanation of why smudging is a reasonable strategy for proving convergence in distribution. There's an irony here, in that the reason I replaced the proof of the central limit theorem in the first edition of this book was to give a simple proof (involving smudging) that doesn't involve characteristic functions. Now I'm going to explain why smudging works by appealing to characteristic functions.

The characteristic function of a random variable X is defined by

$$\psi_X(t) = E_X e^{itx},$$

where t is a real number, and of course $i = \sqrt{-1}$. There are two important facts about characteristic functions that are not proved here (but were in the first edition).

1. Characteristic functions characterize distributions. If $\psi_X(t) = \psi_Y(t)$ for all t, then $F_X(t) = F_Y(t)$ for all t.

   This is the content of Theorem 6.6.1 of the first edition of this book.

2. (Continuity)

   A sequence of random variables X_n converges in distribution to the random variable X if and only if

   $$\lim_{n \to \infty} \psi_{X_n}(t) = \psi_X(t) \quad \text{for all } t.$$

   This is Theorem 6.8.3 in the first edition of this book.

How do these facts bear on smudging? It is important to recall that the smudging variable W is independent of the sequence X_n and of X. The characteristic function of $X_n + W$ is then

$$\psi_{X_n+W}(t) = E_{X_n+W}e^{it(X_n+W)} = E_{x_n}e^{itX_n}E_We^{itW},$$

using independence. Similarly,

$$\psi_{X+W}(t) = E_Xe^{itx}E_We^{itW}.$$

Therefore, $\lim_{n\to\infty}\psi_{x_n+W}(t) = \psi_{X+W}(t)$ for all t if and only if

$$\lim_{n\to\infty}\psi_{X_n}(t) = \psi_X(t) \text{ for all } t.$$

This calculation shows why smudging does not disturb convergence in distribution, but does not yet explain why one might want to use it. The reason to use it is that the smudged cdf is smooth, sufficiently smooth that Taylor's Theorem can be applied. The particular smoothing chosen in equation (6.40) is sufficiently smooth to allow the needed number of derivatives of F. But other calculations could impose yet smoother W's. Smudging may be a technique useful in other calculations.

6.5.9 *Summary*

The central limit theorem is one of the most important in probability theory (which is why it is called "central," and why the normal distribution is called "normal"). The classical proof involves characteristic functions, which are complex-variable generalizations of moment generating functions (and are called Fourier transforms in other parts of applied mathematics).

6.5.10 *Exercises*

1. Prove the statements *(i)*, *(ii)*, *(iii)* and *(iv)* found between (6.49) and (6.50).
2. How would you interpret the central limit theorem in terms of subjective probability and coherence?
3. Let $Z, Z_1, Z_2, \ldots$ be independent and identically distributed random variables satisfying $E(Z) = 0$ and $V(Z) = 1$. Suppose $\sum_{i=1}^n Z_i/\sqrt{n}$ has the same distribution as Z for all n. Prove that Z has a normal distribution.

6.6 The Weak Law of Large Numbers

The purpose of this section is to show that the techniques used to prove the central limit theorem also yield a proof of a second important result, the weak law of large numbers.

The goal is the following:

Theorem 6.6.1. *Let $X, X_1, \ldots, X_n$ be independent and identically distributed random variables with $E(X) = 0$. Let $S_n = \sum_{i=1}^n X_i/n$. Let D be a random variable satisfying $P\{D = 0\} = 1$. Then*

$$\lim_{n\to\infty} F_{S_n}(t) = F_D(t) \textit{ for all real } t \neq 0.$$

To prove this theorem, start first with the following Lemma, which is similar to Lemma 6.5.8:

Lemma 6.6.2. *Let W and Z be independent random variables (recall the density of W is defined previously at (6.39), and suppose Z satisfies $E(Z) = 0$ and $E \mid Z \mid\leq 1$. Let $X = X/n$. Then, for all $\epsilon > 0$, there exists an N such that for all $n \geq N$.*

$$\mid F_{W+X}(w) - F_W(w) \mid\leq \epsilon/n \textit{ for all } w.$$

Proof. Recall from Lemma 6.5.6 that

$$F_{W+X}(w) = E_X F_W(w - X).$$

Since F_W has a continuous derivative, it can be expanded as a Taylor series as follows:

$$\begin{aligned} F_W(w - X) &= F_W(w - X/n) \\ &= F_W(w) - \frac{1}{n} F'_W(v(Z))Z \\ &= F_W(w) + \frac{1}{n} F'_W(w) + \frac{1}{n}(F'_W(v(Z)) - F'_W(w)), \end{aligned} \tag{6.65}$$

where v is a function satisfying $w - Z/n < v(Z) < w$. Also $F'_W(v(Z))$ is integrable because the other terms in (6.65) are integrable. Then, taking the expectation of (6.65) with respect to Z yields

$$F_{W+X}(w) = F_W(w) + \frac{1}{n} E_Z \Big[(F'_W(v(Z)) - F'_W(w))\, Z \Big]. \tag{6.66}$$

Again, the proof requires the expectation to be bounded. But first, record two facts:

(i) Since F_W has a uniformly continuous derivative, there exists a $\delta > 0$ such that $\mid v - w \mid < \delta$ implies $\mid F'_W(v) - F'_W(w) \mid \leq \epsilon/2$.

(ii) Since the derivative is bounded, there exists an M such that $\mid F'_W(v) \mid \leq M$ for all v.

Again, the expectation is broken indexsubjectindexweak law of large numbers (WLLN) into two parts:

$$\begin{aligned} I_1 &= E_X \left[(F'_W(v(Z)) - F'_W(w))\, Z I (\mid Z \mid \leq \delta n) \right] \\ I_2 &= E_X \left[(F'_W(v(X)) - F'_W(w))\, Z I (\mid Z \mid > \delta n) \right]. \end{aligned}$$

Taking I_1 first, from $w - Z/n \leq v(Z) \leq w$, $-Z/n \leq v(Z) - w \leq 0$, so $\mid v(Z) - w \mid < \delta$. This implies, using *(i)*,

$$\mid F'_W(v(Z) - F'_W(w) \mid \leq \epsilon/2.$$

Then

$$I_1 \leq (\epsilon/2) E \mid Z \mid \leq \epsilon/2. \tag{6.67}$$

To address I_2, note that, using *(ii)*

$$\mid F'_W(v(Z) - F'_W(w) \mid \leq 2M.$$

Because $E \mid Z \mid < \infty$, there is an N large enough that, for all $n \geq N$,

$$E \left[\mid Z \mid I(\mid Z \mid > \delta n) \right] \leq \epsilon/4M.$$

Then

$$\mid I_2 \mid \leq 2M(\epsilon/4M) = \epsilon/2. \tag{6.68}$$

To summarize,

$$\mid F_{W+X}(w) - F_W(w) \mid \leq \frac{1}{n}(\mid I_1 \mid + \mid I_2 \mid) \leq \frac{1}{n}(\epsilon/2 + \epsilon/2) = \epsilon/n.$$

□

Corollary 6.6.3. *Let W, Z and Z' be independent random variables, with Z and Z' satisfying $E[Z] = E[Z'] = 0$, $E[\mid Z \mid \leq 1$ and $E(\mid Z' \mid) \leq 1$. Let $X = Z/n$ and $X' = Z'/n$. Then there exists an N such that, for all $n \geq N$,*

$$\|F_{W+X} - F_{W+X'}\| \leq \epsilon/n.$$

Proof. Using Lemma 6.6.2, there is an N such that, for all $n \geq N$

$$\mid F_{W+X}(w) - F_W(w) \mid < \epsilon/2n. \tag{6.69}$$

Similarly, there is an N' such that, for all $n \geq N'$,

$$\mid F_{W+X'} - F_W(w) \mid < \epsilon/2n. \tag{6.70}$$

Let $N^* = \max(N, N')$. Then, for all $n \geq N^*$,

$$\begin{aligned} \mid F_{W+X}(w) - F_{W+X'}(w) \mid &= \mid (F_{W+X}(w) - F_W(w)) - (F_{W+X'}(w) - F_W(w)) \mid \\ &\leq \mid F_{W+X}(w) - F_W(w) \mid + \mid F_{W+X}(w) - F_W(w) \mid \\ &\leq \epsilon/2n + \epsilon/2n = \epsilon/n. \end{aligned} \tag{6.71}$$

Since this holds for all w,

$$\|F_{W+X} - F_{W+X'}\| \leq \epsilon/n.$$

Thus N^* satisfies the conditions of the Corollary. □

Proof. Proof of Theorem 6.6.1. Take $Z' = D$. Then $X' = D$ also, and from Corollary 6.6.3,

$$\|F_{W+X} - F_{W+X'}\| \leq \epsilon/n. \tag{6.72}$$

In combination with Lemma 6.5.7,

$$\begin{aligned} \|F_{W+S_n} - F_{W+T_n}\| &\leq n\|F_{W+X} - F_{W+Y}\| \\ &\leq n \cdot \epsilon/n = \epsilon. \end{aligned} \tag{6.73}$$

But T_n is the sum of n variables indexsubjectindexweak law of large numbers (WLLN) taking the value 0 with probability 1, so $T_n = D$ for all n. Hence,

$$\|F_{W+S_n} - F_{W+T}\| < \epsilon \tag{6.74}$$

for all $n \geq N^*$.

Now Lemma 6.5.5 applies, and yields $F_{S_n}(t) \to F_D(t)$ for all t at which $F_D(t)$ is continous, i.e., for all t except $t = 0$. □

Corollary 6.6.4. *Let $X, X_1, \ldots, X_n$ be independent and identically distributed with $E(X) = 0$. Let $S_n = \sum_{i=1}^n X_i/n$. Then $\lim_{n\to\infty} F_{S_n}(t) = F_D(t)$ for all $t \neq 0$, where $P\{D = 0\} = 1$.*

Proof. Let $Y_i = X_i/E(\mid X \mid)$. Then Theorem 6.6.1 shows that $T_n = \sum_{i=1}^n Y_i/n$ satisfies $\lim_{n\to\infty} F_{S_n}(t) \to F_D(t)$ for all $t \neq 0$.

But

$$\begin{aligned} F_{S_n}(t) &= P\{S_n \leq t\} = P\{\textstyle\sum_{i=1}^n X_i/n \leq t\} = P\{\sum_{i=1}^n Y_i/n \leq tE(\mid X \mid)\} \\ &= P\{T_n \leq tE(\mid X \mid)\} \to F_D(tE(\mid X \mid)) = F_D(t) \text{ for all } t \neq 0. \end{aligned}$$

□

Corollary 6.6.5. *Let $X, X_1, X_2, \ldots$ be independent and identically distributed with $E(X) = \mu$ and $E(X) < \infty$. Let $S_n = \sum_{i=1}^n X_i/n$. Then*

$$\lim_{n\to\infty} F_{S_n}(t) = F_{D+\mu}(t) \textit{ for all } T \neq \mu.$$

Proof. Let $Y_i = X_i - \mu$. Then Corollary 6.6.4 shows that $T_n/n = \sum Y_i/n$ satisfies $\lim_{n\to\infty} F_{T_n}(t) = F_D(t)$.

But

$$\begin{aligned} F_{T_n}(t) &= P\{T_n \leq t\} = P\{\sum Y_i/n \leq t\} = P\{\tfrac{\sum(X_i-\mu)}{n} \leq t\} \\ &= P\{\tfrac{\sum X_i}{n} - \mu \leq t\} = P\{S_n \leq t+\mu\} = F_{S_n}(t+\mu). \end{aligned}$$

Therefore, $\lim_{n\to\infty} F_{S_n}(t) = \lim_{n\to\infty} F_{T_n}(t-\mu) = F_D(t-\mu) = F_{D+\mu}(t)$, i.e., the distribution that takes the value μ with probability 1. □

Corollary 6.6.6. *Let $X_1, X_2, \ldots$ be independent and identically distributed with $E(X) = \mu$. Let $S_n = \sum_{i=1}^n X_i/n$. Then for every $\epsilon > 0$ and $n > 0$, there is an N such that for all $n \geq N$,*

$$P\{| S_n - m |> \eta\} < \epsilon.$$

Proof. Let $\epsilon > 0$ and $\eta > 0$ be given. Then

$$\begin{aligned} P\{| S_n - m |> \eta\} &= P\{(S_n - m) > \eta\} + P\{(S_n - m) < -\eta\} \\ &= 1 - F_{S_n - m}(\eta) + F_{S_n - m}(-\eta) \end{aligned}$$

Corollary 6.6.5 shows that there is an N_1 such that, for all $n \geq N_1$,

$$1 - F_{S_n - m}(-\eta) < \epsilon/2.$$

Similarly, there is an N_2 such that, for all $n \geq N_2$,

$$F_{S_n - m}(-\eta) < \epsilon/2.$$

Let $N = \max(N_1, N_2)$. Then for all $n \geq N$,

$$P\{| S_n - m |> \eta\} < \epsilon/2 + \epsilon/2 = \epsilon.$$

□

The conclusion of Corollary 6.6.6 is the same as the indexsubjectindexweak law of large numbers (WLLN) weak law of large numbers found in section 2.13 using Tchebychav's Inequality. However, unlike the result in section 2.13, Corollary 6.6.6 does not require that the random variables X_i have variances.

6.6.1 *Exercises*

1. Think of an example in which the weak law of Corollary 6.6.6 applies, but that of section 2.13 does not.

6.6.2 *Related literature*

The treatment of the central limit theorem and the weak law of large numbers relies very heavily on Pippenger (2012), which in turn relies on Trotter (1959), who traces his ideas back to Lindeberg (1922). The other main route to these results relies on characteristic functions (essentially Fourier transforms). The advantage of the proof given here is that it relies solely on calculus.

6.6.3 *Remark*

It is striking to me that both proofs are in a sense fixed-point theorems. The role of the normal distribution in the central limit theorem and the degenerate distribution in the weak law of large numbers is only to identify what the fixed point is. I wonder if other important theorems in probability theory can be usefully understood in a similar way.

6.7 Stein's Method

This name refers to many ideas stemming from the seminal paper of Stein (1972), in which he proved a central limit for a class of dependent random variables, and found a rate of convergence as well. The perspective opened by this work applies to many metrics (i.e., senses of convergence) and to other limiting distributions, starting with the Poisson limit (Chen, 1975). Some important expositions and extensions are given in Chen et al. (2011); Ross (2011). This is an exciting area of current research in probability theory.

Chapter 7

Making Decisions

"Did you ever have to make up your mind?
And pick up on one and leave the other behind?
It's not often easy and not often kind.
Did you ever have to make up your mind?"

—John Sebastian, The Lovin' Spoonful

7.1 Introduction

We now shift gears, returning from serious mathematics and probability, to a more philosophical inquiry, the making of good decisions. The sense in which the recommended decisions are good is an important matter to be explained. In addition to explaining utility theory, this chapter explains why the conditional distribution of the parameters $\boldsymbol{\theta}$ after seeing the data $\mathbf{x}$ is a critical goal of Bayesian analyses, as is shown in section 7.7.

7.2 An example

Just as in Chapter 1 there was no suggestion that you should have particular probabilities for certain events, in this chapter there is no suggestion that you should have particular values, that is, that you should prefer certain outcomes to others. This book offers a disciplined language for representing your beliefs and goals, with minimal judgment about whether others share, or should share, either.

Suppose you face a choice. The set of decisions available to you is $\mathcal{D}$, and you are uncertain about the outcome of some random variable θ. For the moment, assume that $\mathcal{D}$ is a finite set. The more general case is taken up later. The set of pairs (d, θ), where $d \in \mathcal{D}$ and $\theta \in \Omega$, is called the set of consequences $\mathcal{C}$. You can think of a consequence as what happens if you choose $d \in \mathcal{D}$ and $\theta \in \Omega$ is the random outcome.

To take a simple example, suppose that you are deciding whether to carry an umbrella today, so $\mathcal{D}$ = {carry, not carry}. Suppose also you are uncertain about whether it will rain, so $\theta = 1$ if it rains, and $\theta = 0$ if it does not. Then you are faced with four possible consequences: $\{c_1 =$ (take, rain), $c_2 =$ (do not take, rain), $c_3 =$ (take, no rain), and $c_4 =$ (do not take, no rain)$\}$.

The possible consequences can be displayed in a matrix as follows:

		uncertain outcome	
		rain	no rain
decision	take umbrella	c_1	c_3
	do not take umbrella	c_2	c_4

Table 7.1: Matrix display of consequences.

A second way of displaying this structure is with a decision tree. Decision trees code decisions with squares and uncertain outcomes with circles. Time is conceived of as moving from left to right. Then a decision tree for the umbrella problem is shown in Figure 7.1:

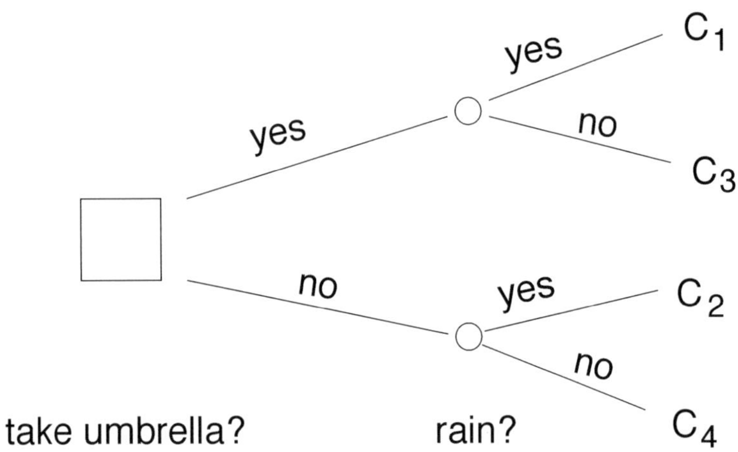

Figure 7.1: Decision tree for the umbrella problem.

I need to understand how you value these consequences relative to one-another, so I need to ask you some structural questions.

We are now going to explore your utilities for the various consequences. You can think of your utility for c, which we will write as $U(c) = U(d, \theta)$ as how you would fare if consequence c occurs, that is, if you make decision $d \epsilon \mathcal{D}$ and $\theta \epsilon \Omega$ is the random outcome.

First, I need you to identify which you consider the best and the worst outcome to be. Suppose you consider $c_4 = c_b$ to be the best consequence. This means that you most prefer the consequence in which you do not bring your umbrella and it does not rain. We assign the consequence c_b to have utility 1, so $U(c_b) = 1$. Suppose also that you consider c_2, where you do not bring your umbrella and it does rain, to be the worst outcome. Then $c_2 = c_w$, and we assign c_w to have utility 0, so $U(c_w) = 0$. The choices of 1 and 0 for the utilities of c_b and c_w, respectively, may seem arbitrary now, but soon you will understand the reason for these choices.

Now consider a new kind of ticket, T_p, that gives you c_b, the best consequence, with probability p, and c_w, the worst consequence, with probability $1 - p$. Clearly, if T_p and $T_{p'}$ are two such tickets, with $p > p'$, you prefer T_p to $T_{p'}$ because T_p gives you a greater chance of the best outcome, c_b, and a smaller chance of the worst outcome, c_w.

Now consider a consequence that is neither the best nor the worst, say c_1, which means that you take an umbrella and it does rain. Now we suppose that there is some $p_1, 0 \leq p_1 \leq 1$

such that you are indifferent between T_{p_1} and c_1. Then we assign to c_1 the utility p_1. Thus we write $U(c_1) = p_1$ where p_1 is chosen so that you are indifferent between T_{p_1} and c_1. You can now appreciate why 1 and 0 are the right utilities for c_b and c_w, respectively. Also it is important to notice that there cannot be two values, say p_1 and p_1', such that you are indifferent between T_{p_1} and c_1 and also indifferent between $T_{p_1'}$ and c_1, since you prefer T_{p_1} to $T_{p_1'}$ if $p_1 > p_1'$. The situation can be illustrated with the following diagram:

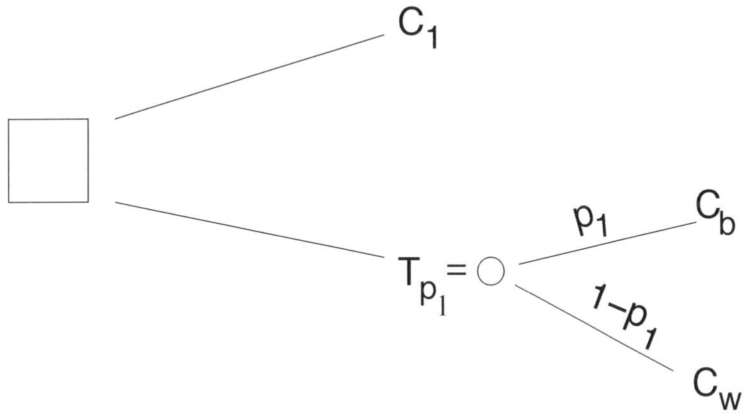

Figure 7.2: The number p_1 is chosen so that you are indifferent between these two choices.

Let's suppose you choose $p_1 = 0.8$, which means that the consequence that you take the umbrella and it rains, is indifferent to you to the ticket $T_{0.8}$, under which, with probability 0.8 you get c_b (no rain, no umbrella) and with probability 0.2 you get c_w (rain, no umbrella).

Similarly we may suppose there is some number p_3 such that you are indifferent between consequence c_3 (no rain, took umbrella) and T_{p_3}. As we did with c_1, we let $U(c_3) = p_3$. We'll suppose you choose $p_3 = 0.4$.

Thus for each consequence $c_i, i = 1, 2, 3, 4$, we take $U(c_i) = p_i$, where you are indifferent between T_{p_i} and c_i. Utility gives a measure of how desirable you find each consequence to be, relative to c_b, the best outcome, and c_w, the worst outcome.

Now how shall we assess the utility of a decision, such as taking the umbrella? There are two possible consequences of taking the umbrella, c_1 and c_3. Suppose your probability of rain is r. Then taking the umbrella is equivalent to you to consequence c_1 with probability r and c_3 with probability $1-r$. Since c_i is indifferent to you to a ticket giving you c_b with probability p_i and c_w with probability $1 - p_i$, taking the umbrella is equivalent to a ticket giving you c_b with probability $p_1r + p_3(1 - r)$ and c_w with probability $(1 - p_1)r + (1 - p_3)(1 - r) = 1 - [p_1r + p_3(1 - r)]$. And, in general, the utility of a decision d is the expected utility of the consequences (d, θ) where the expectation is taken with respect to your opinion about θ, or, put into symbols,

$$U(d) = EU(\theta \mid d).$$

Here d is indifferent to you to a ticket T_u, where $u = EU(\theta \mid d)$.

Suppose your probability of rain is $r = 0.5$. Then, with the chosen numbers, the expected utility of bringing the umbrella is

$$p_1r + p_3(1 - r) = (0.8)(0.5) + (0.4)(0.5) = (0.4) + (0.2) = 0.6.$$

This means that, for you, if the hypothesized numbers were your choices, bringing the umbrella is equivalent to you to $T_{0.6}$, which gives you 0.6 probability of c_b, and 0.4 probability of c_w.

We can also assess the expected utility of not bringing the umbrella. Here the possible outcomes are c_2 and c_4, which happen to be c_w and c_b, respectively, in our scenario, and therefore have utilities 0 and 1, respectively. Then not to bring the umbrella is equivalent to you to a 0.5 probability of $c_4 = c_b$ and a 0.5 probability of $c_2 = c_w$, and therefore you are indifferent between not bringing the umbrella and $T_{0.5}$. The expected utility of not bringing the umbrella is then

$$1(0.5) + 0(0.5) = 0.5.$$

Since $T_{0.6}$ is preferred to $T_{0.5}$, the better decision is to bring the umbrella. The choices, with nodes labeled with probabilities and utilities, are given in Figure 7.3 (in which time goes from left to right, as you make the decision before you find out whether it rains):

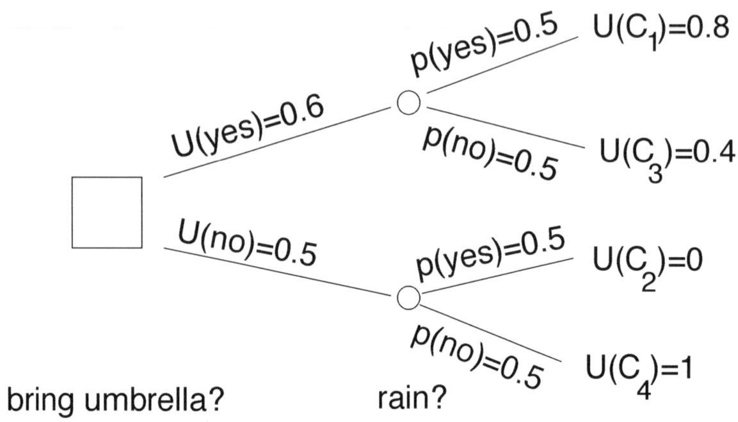

Figure 7.3: Decision tree with probabilities and utilities.

It is now easy to see that choosing $d \epsilon \mathcal{D}$ to maximize $U(d)$ gives you the equivalent of the largest probability of the best outcome, and hence is the best choice for you.

7.2.1 *Remarks on the use of these ideas*

The scheme outlined above starts from a very common-sense perspective. First, it asks you what alternatives $\mathcal{D}$ you are deciding among. Second, it asks you what uncertainties Ω you face. Third, it asks you how you value the consequences C, which consists of pairs, one from $\mathcal{D}$ and one from Ω, against each other, in a technique that articulates well with probability theory. Finally, it asks how likely you regard each of the possible uncertain outcomes. It is hard to see how any sensible organization of the requisite information for making good decisions would avoid asking these questions.

The usefulness of this way of thinking depends critically on the ability of the decision maker to specify the requested information. Often, for example, what appears to be a difficult decision problem is alleviated by the suggestion of a previously uncontemplated alternative decision. Similarly the space of uncertainties is sometimes too narrow. In my experience, the careful structuring of the problem can lead the decision maker to consider the right, pertinent questions, which can be an important contribution in itself.

I should also remind you of the sense in which these are "good" decisions. There should be no suggestion that decisions reached by maximizing expected utility have, ipso facto, any moral superiority. Whether or not they do depends on the connection between moral values and the declared utilities of the decision maker. Thus the decisions made by maximizing expected utility are good only in the sense that they are the best advice we have to achieve the decision maker's goals, whether those are morally good, bad or indifferent.

There is also nothing in the theory of expected utility maximization that bars deciding to let others choose for you. For example, in her wise and insightful book "The Art of Choosing," Sheena Iyengar (2010) relates the story of her parents' arranged marriage. She presents it as accepting a centuries-old tradition, and of wanting to do one's duty within that tradition (see pages 22-45). If "abiding by tradition" is what's most important to you, then that can be expressed in your utility function.

7.2.2 *Summary*

To make the best decisions, given your goals, maximize your expected utility with respect to your probabilities on whatever uncertainties you face.

7.2.3 *Exercises*

1. Vocabulary. State in your own words the meaning of:
   (a) consequence
   (b) utility of a consequence
   (c) utility of a decision
2. Assess your own utilities for the decision problem discussed in this section. Is there a probability for rain, r, above which maximization of expected utility suggests taking an umbrella, and below which not? If so, what is that probability? Would you, in fact, choose to take an umbrella if your probability were above that critical value, and not take an umbrella if it were below? Why or why not?
3. Suppose that in the example of section 7.2, your utilities are as follows:

$$U(c_4) = 1, U(c_3) = 1/3, U(c_2) = 0, U(c_1) = 2/3.$$

   Suppose your probability of rain is 1/2. What is your optimal decision?

7.3 In greater generality

To be more precise, it is important to distinguish $\mathcal{D}$ from Ω. The set of decisions $\mathcal{D}$ that you can make are in your control, but which $\theta \epsilon \Omega$ is, in general, not. To make this distinct salient in the notation, I follow Pearl (2000), and use the function $do(d_i)$ to indicate that you have chosen d_i. Furthermore, it is possible that your probability distribution may depend on which $d_i \epsilon \mathcal{D}$ you choose. Consequently, I should in general ask you for your probabilities $p\{\theta \mid do(d_i)\}$.

In the case of whether or not to carry an umbrella, it is implausible that your probability of rain will depend on whether you carry an umbrella (joking aside). However, suppose that your decisions $\mathcal{D}$ are whether to drive carefully or recklessly, and your uncertainty is about whether you will have an accident. Here it is entirely reasonable that your probability of having an accident depends on your decision about whether to drive carefully or recklessly, i.e., on what you do. (It is a wonder of the English language that reckless driving can cause a wreck).

So start with decisions $\mathcal{D} = \{d_1, \ldots, d_m\}$ and $\Omega = \{\theta_1, \ldots, \theta_n\}$ of uncertain events. Suppose your probabilities are $p\{\theta_j \mid do(d_i)\}$. A consequence C_{ij} is the outcome if you decide to do d_i and θ_j ensues. Let c_b be at least as desirable as any C_{ij}, and let c_w be no more desirable than any C_{ij}. Let $u(C_{ij})$ be the probability of getting c_b, and otherwise getting c_w, such that you are indifferent between getting C_{ij} for sure, and this random prospect. In symbols,

$$u(C_{ij}) = p\{c_b \mid \theta_j, do(d_i)\}. \tag{7.1}$$

Then if you decide on d_i, your probability of getting c_b (and otherwise c_w), is

$$\begin{aligned} p\{c_b \mid do(d_i)\} &= \sum_{j=1}^{n} p\{c_b \mid \theta_j, do(d_i)\} p\{\theta_j \mid do(d_i)\} \\ &= \sum_{j=1}^{n} u(C_{ij}) p\{\theta_j \mid do(d_i)\}. \end{aligned} \tag{7.2}$$

Therefore you maximize your probability of achieving the best outcome for you by choosing d_i to maximize

$$\bar{u}(d_i) = \sum_{j=1}^{n} u(C_{ij}) p\{\theta_j \mid do(d_i)\}. \tag{7.3}$$

When the set of possible decisions $\mathcal{D}$ has more than finitely many choices, there may not exist a maximizing choice. For example, suppose $\mathcal{D}$ consists of the open interval $\mathcal{D} = \{x \mid 0 < x < 1\}$. Suppose also (to keep it very simple) that there is no uncertainty, and that your utility function is $U(x) = x$. There is no choice of x that will maximize $U(x)$. However, for every $\epsilon > 0$, no matter how small, I can find a choice, such as $x = 1 - \epsilon/2$, that gets better than ϵ-close. The casual phrase "maximization of expected utility" will be understood to mean "choose such an ϵ-optimal decision" if an optimal decision is not available. (The word "ϵ-optimal" is pronounced "epsilon-optimal".)

Suppose you are debating between two decisions that, as near as you can calculate, are close in expected utility, and therefore you find this a hard decision. Because these decisions are close in expected utility, it does not matter very much (in prospect, which is the only reasonable way to evaluate decisions you haven't yet made) which you choose. The important point is to avoid really bad decisions. Consequently, "hard" decisions are not hard at all. If necessary, one way of deciding is to flip a coin, and then to think about whether you are disappointed in how the coin came out. If so, ignore the coin and go with what you want. If not, go with the coin.

Decisions can be thought of as tools available to the decision maker to achieve high expected utility. Thus the right metric for whether a decision is nearly optimal is whether it achieves nearly the maximum expected utility possible under the circumstances, and not whether the decision is close, in some other metric, to the optimal decision.

When Ω has more than finitely many elements, the finite sum in (7.3) is replaced by an infinite sum (as in Chapter 3) in the case of a discrete distribution, or by an integral (as in Chapter 4) in the case of a continuous one.

So far the utilities in (7.1), (7.2) and (7.3) depend on the choice of c_b and c_w. The argument I now give shows that if instead other choices were made, the only effect would be a linear transformation of the utility, which has no effect on the ordering of the alternative decisions by maximation of expected utility. Suppose instead that c'_b is at least as desirable as c_b, and that c'_w is no more desirable than c_w. Again, suppose there is some probability P such that you would be indifferent between c_b for sure, and the random prospect that would give you c'_b with probability P and would otherwise give you c'_w. Similarly, suppose there is some probability p such that you would be indifferent between c_w for sure and the random prospect that would give you c'_b with probability p and would otherwise give you c'_w. As in the material before (7.1), let $u'(C_{ij}) = P\{C' \mid \theta_j, d_0(d_i)\}$ be the probability such that you would be indifferent between C_{ij} and the random prospect that gives you c'_b with probability $u'(C_{ij})$ and c'_w with probability $1 - u'(C_{ij})$. What is the relationship between $u(C_{ij})$ and $u'(C_{ij})$?

The consequence C_{ij} is indifferent to you to a random prospect that gives you c_b with probability $u(C_{ij})$ and c_w with probability $1 - u(C_{ij})$. But c_b itself is indifferent to you to a random prospect giving you c'_b with probability P and the c'_w with probability $1 - P$.

Similarly c_w is indifferent to you to a random prospect giving you c'_b with probability p and c'_w with probability $1 - p$. Therefore C_{ij} is indifferent to you to a random prospect giving you c'_b with probability $Pu(C_{ij}) + p(1 - u(C_{ij}))$ and otherwise gives you c'_w. Therefore

$$\begin{aligned} u'(C_{ij}) &= Pu(C_{ij}) + p(1 - u(C_{ij})) \\ &= p + (P - p)u(C_{ij}). \end{aligned} \tag{7.4}$$

In interpreting (7.4) it is important to notice that $P - p > 0$ since c_b is more desirable than c_w to you. Hence, using c'_b and c'_w instead of c_b and c_w, leads to choosing d_i to maximize

$$\begin{aligned} \bar{u}'(d_i) &= \sum_{i=1}^{n} u'(C_{ij})p\{\theta_j \mid do(d_i)\} \\ &= p + (P - p)u(d_i). \end{aligned} \tag{7.5}$$

Therefore the optimal (or ϵ-optimal) choices are the same (note that ϵ has to be rescaled). Also the resulting achieved expected utilities are related by

$$\bar{u}'(d_i) = a + bu(d_i) \tag{7.6}$$

where $b > 0$ [of course, $a = p$ and $b = P - p$]. A transformation of the type (7.6) is always possible for a utility function, and always leads to the same ranking of alternatives as the untransformed utilities. The construction of utility as has been done here amounts to an implicit choice of a and b by using $u(C) = 1$ and $u(c) = 0$, where C is more desirable than c, leading to $b > 0$.

To maximize expected utility is of course the same as to minimize expected loss, if loss is defined as

$$\ell(C_{ij}) = -u(C_{ij}). \tag{7.7}$$

Much of the statistical literature is phrased in terms of losses, possibly reflecting the dour personalities that seem to be attracted to the subject.

As developed here, utilities can be seen as a special case of probability. Conversely, probability, as developed in Chapter 1, can be seen as a special case of utility. There we took c_b = \$1.00 and c_w = \$0.00. As a result, probability and utility are so intertwined as to be, from the perspective of this book, virtually the same subject.

Rubin (1987) points out that from a person's choice of decisions, all that might be discerned is the product of probability and utility. The ramifications of this observation are still being discussed.

7.3.1 *A supplement on regret*

Another transformation of utility is regret, defined as $r(C_{ij}) = \max_i u(C_{ij}) - u(C_{ij})$. Now $g_j = \max_i u(C_{ij})$ does not depend on i. It turns out that there are circumstances under which minimizing expected regret is equivalent to maximizing expected utility, and other circumstances in which it is not. To examine this, write the minimum expected regret as follows:

$$\begin{aligned} \min_i E\ r(C_{ij}) &= \min_i E[g_j - u(C_{ij})] \\ &= \min \left\{ \sum_j g_j p(\theta_j \mid do(d_i)) - \sum_j u(C_{ij})p(\theta_j \mid do(d_i)) \right\}. \end{aligned}$$

The second term is exactly expected utility, thus minimizing expected regret is equivalent to maximizing expected utility provided $\sum_j g_j p(\theta_j \mid do(d_i))$ does not depend on i, which in general is true if $p(\theta_j \mid do(d_i))$ does not depend on i. As previously explained in section 7.3, jokes aside, we do not think the weather is influenced by a decision about whether to carry an umbrella, so in this example, $p(\theta_j \mid do(d_i))$ is reasonably taken not to depend on i. Hence for the decision about whether to take an umbrella, you can either maximize expected utility or minimize expected regret, and the best decision will be the same, as will the achieved expected utility.

However, there are other decision problems in which it is quite reasonable to suppose that $p(\theta_j \mid do(d_i))$ does depend on i, and thus on what you do. In the example given in section 7.3, $\boldsymbol{\Theta}$ is whether or not you have an automobile accident, and $do(d_i)$ is whether or not you drive carefully. In this case, it is very reasonable to suppose that your probability of having an accident depend on your care in driving. For such an example, minimizing expected regret is not the same as maximizing expected utility. It will lead, in general, to suboptimal decisions and loss of expected utility.

For more on expected regret, see Chernoff and Moses (1959, pp. 13, 276).

7.3.2 Notes and other views

There is a lot of literature on this subject, dating back at least to Pascal (born 1623, died 1662). Pascal was a mathematician and a member of the ascetic Port-Royal group of French Catholics. Pascal developed an argument for acting as if one believes in God, which went roughly as follows: If God exists and you ignore His dictates during your life, the result is eternal damnation (minus infinity utility). While if He exists and you follow His dictates, you gain eternal happiness (plus infinity utility). If God does not exist and you follow His dictates, you lose some temporal pleasures you would have enjoyed by not following God's dictates, but so what (difference of some finite number utility). Therefore the utility optimizing policy is to act as if you believe God exists. This is called Pascal's Wager. (See Pascal (1958), pp. 65-96.)

More recent important contributors include Ramsey (1926), Savage (1954), DeGroot (1970) and Fishburn (1970, 1988). Much of the recent work concerns axiom systems. For instance, an Archimedean condition says that c_b and c_w are comparable (to you), in the sense that for each consequence C_{ij}, there is some $P^* < 1$ that you would prefer c_b with probability P^* and c_w otherwise to C_{ij} for sure, and some other $p^* > 0$ such that you would prefer C_{ij} for sure to the random prospect yielding c_b with probability p^* and c_w otherwise. From this assumption it is easy to prove the existence of a p such that you are indifferent between C_{ij} and the random prospect yielding c_b with probability p and otherwise c_w. Pascal's argument violates the Archimedean condition.

A distinction is drawn in some economics writing between "risk" and "uncertainty," the rough idea being that "risk" concerns matters about which there are agreed probabilities, while "uncertainty" deals with the remainder. This distinction is attributed by some to Knight (1921), a view challenged by LeRoy and Singell (1987). Others attribute it to Keynes (1937, pp. 213, 214). The view taken in this book is that from the viewpoint of the individual decision-maker, this distinction is not useful, a point conceded by Keynes (ibid, p. 214).

> The sense in which I am using the term **uncertain** is that in which the prospect of a European war is uncertain, or the price of copper and the rate of interest twenty years hence, or the obsolescence of a new invention, or the position of private wealth-owners in the social system in 1970. About these matters there is no scientific basis on which to form any calculable probability whatever. We simply do not know. Nevertheless, the necessity for action and for decision compels us as practical men to do our best to overlook this awkward fact and to behave exactly as we should if we had behind us a good Benthamite calculation of a series of prospective advantages and disadvantages, each multiplied by its appropriate probability, waiting to be summed.

There is a whole other literature dealing with descriptions of how people actually make decisions. A good summary of this literature can be found in von Winterfeld and Edwards (1986) and Luce (2000). In risk communication, researchers try to find effective ways to combat systematic biases in risk perception. The field of behavioral finance tries to make money by taking advantage of systematic errors people make in decision making.

The development here closely follows that of Lindley (1985), which I highly recommend.

7.3.3 Summary

Utilities are defined in such a way that the optimal decision is to maximize expected utility. When optimal decisions do not exist, ϵ-optimal decisions are nearly as good. Minimizing expected loss is the same as maximizing expected utility, where loss is defined as negative utility.

7.3.4 Exercises

1. Vocabulary. Define in your own words:
   (a) consequence
   (b) utility
   (c) loss
   (d) ϵ-optimality
   (e) Pascal's Wager
2. Prove that, if losses are defined in (7.7), minimizing expected loss is the same as maximizing expected utility.

7.4 Unbounded utility and the St. Petersburg Paradox

The utilities or losses found as suggested in sections 7.2 and 7.3 for finite sets $\mathcal{D}$ of possible decisions, are bounded. Indeed, the bounds are 0 and 1 in the untransformed case. To discuss unbounded utilities, it is useful to distinguish utility functions that are bounded above (i.e., loss functions bounded below), from those that are unbounded in both directions.

To set the stage, it is a good idea to have an example in mind. Suppose a statistician has decided to estimate a parameter $\theta \epsilon \mathcal{R}$, which means to replace the distribution of θ, which we'll denote $p(\theta)$, with a single number $\hat{\theta}$. (The reasons why I regard this as an over-used maneuver in statistics are addressed in Chapter 12.) The most commonly used loss function in statistics for such a circumstance is squared error: $(\theta - \hat{\theta})^2$. Because of the simple relationship

$$\begin{aligned} E(\theta - \hat{\theta})^2 &= E(\theta - \mu + \mu - \hat{\theta})^2 \\ &= E(\theta - \mu)^2 + (\mu - \hat{\theta})^2 \end{aligned} \tag{7.8}$$

where $\mu = E(\theta)$, it is easy to see that expected loss is minimized, or utility maximized, by the choice $\hat{\theta} = \mu = E(\theta)$, and the expected loss resulting from this is $E(\theta - \mu)^2$, which is the variance of θ. (Indeed squared error is so widely used that sometimes $E(\theta)$ is referred to as "the Bayes estimate," as though it were inconceivable that a Bayesian would have any other loss function.)

We have seen examples of random variables θ, starting in Chapter 3, in which the mean and/or variance do not exist. Taking squared error seriously, this would say that any possible choice $\hat{\theta}$ would be as good (or bad) as any other, leading to infinite expected loss, or minus infinity expected utility. What's to be made of this?

To me, what's involved here is taking squared error entirely too seriously. When an integral is infinite, the sum is dominated by large terms in the tails, which is exactly where the utility function is least likely to have been contemplated seriously. Therefore, I prefer to think of utility as inherently bounded, and to use unbounded utility as an approximation only when the tails of the distribution do not contribute substantially to the sums or integrals involved.

The same principle applies to the much less common case in which utility (or loss) is unbounded both above and below.

A second example of this kind was proposed by Daniel Bernoulli in 1738 (see the English translation of 1954). He proposes that a fair coin be flipped until it comes up tails. If the number of flips required is n, the player is rewarded $\$2^n$. If utility is linear in dollars, then

$$EU = \sum_{n=1}^{\infty} \frac{1}{2^n}(2^n) = \sum_{n=1}^{\infty} 1 = \infty, \tag{7.9}$$

so a player should be willing to pay any finite amount to play, which few of us are. This is called the St. Petersburg Paradox.

The first objection to this is that in practice nobody has 2^n dollars for every n, and hence nobody can make this offer. Suppose for example, that a gambling house puts a maximum of 2^k on what it is willing to pay, and if the player obtains k heads in a row, then the game stops at that point. Then the expected earnings to the player are

$$\begin{aligned} EU &= \sum_{n=1}^{k-1} \frac{1}{2^n}(2^n) + \frac{1}{2^k} \cdot 2^k \\ &= k. \end{aligned}$$

Since $2^{10} = 1024$, 2^{20} will be slightly over \$1 million, and 2^{30} will be slightly over a billion. Thus practical limits on the gambling house's resources make the St. Petersburg game much less valuable, even with utility linear in money. While that's true, it should not stop us from thinking about the possibility of unbounded payoffs.

Bernoulli proposed that the trouble lies in the use of utility that's linear in dollars, and proposed utility equal to log dollars instead. But of course prizes of e^{2^n} foil this maneuver.

I think that the difficulty lies instead in unbounded utility. The following result shows that if utility is unbounded, there is a random variable such that expected utility is unbounded as well.

Suppose X is a discrete random variable taking infinitely many values $x_1, x_2, \ldots$. Suppose $U(x)$ is an unbounded utility function.

Lemma 7.4.1. *For every real number B, there are an infinite number of x_i's such that $U(x_i) > B$.*

Proof. Suppose there are only a finite number of x_i's such that $U(x_i) > B$, say $i_1, \ldots, i_k$. Let $B^* = \max_{1 \leq j \leq k} U(x_{i_j})$. Since k is finite, $B^* < \infty$. Then $U(x) \leq B^*$ for all x, so U is bounded. A contradiction. □

Theorem 7.4.2. *If U is unbounded, there is a probability distribution for X such that*

$$EU(X) = \infty.$$

Proof. We construct this probability with the following algorithm by induction:

Take $i = 1$. There are an infinite number of x_i's such that $U(x_i) > 1$. Choose one of them, and let $q_i = 1$. In the inductive step, now for $j < i$, suppose we have chosen $x_j \neq x_1, \ldots, x_{j-1}$ such that $U(x_j) > j^2$. Because there are an infinite number of x_i's with

$U(x_i) > i^2$, excepting $x_1, \ldots, x_{j-1}$ (finite in number) doesn't change this. Choose one of these to be x_i, and let $q_i = 1/i^2$. Now $\sum_{j=1}^{\infty} 1/j^2 = \sum_{j=1}^{\infty} q_j = k < \infty$. Then $p_j = (\frac{1}{k})q_j$ is a probability distribution on $x_1, \ldots,$ and

$$EU(X) \geq \frac{1}{k}\sum_{j=1}^{\infty} q_j U(x_j) = (\frac{1}{k})\sum_{j=1}^{\infty} 1/j^2 U(x_j) > (\frac{1}{k})\sum_{j=1}^{\infty} \frac{j^2}{j^2}$$

$$= (\frac{1}{k})\sum_{j=1}^{\infty} 1 = \infty.$$

□

In light of this result, a St. Petersburg-type paradox may be found for every unbounded utility. This confirms my belief that unbounded utility can be used as an approximation only for some random variables, namely those that do not put too much weight in the tails of a distribution.

One possible way to make infinite expected utility a useful concept is to say that we prefer a random variable with payoff X to one with payoff Y provided $E[U(X) - U(Y)] > 0$, even if $E[U(X)] = E[U(Y)] = \infty$. However, it is possible to have random variables X and Y with the same distribution such that $E[U(X) - U(Y)] > 0$.

For this example, take the space to be $\mathbb{N} \times \{0, 1\}$, so that a typical element is (i, x) where $x = 0$ or 1 and $i\epsilon\{1, 2, \ldots\}$ is a positive integer. The probability of $\{(i, x)\}$ is $1/2^{i+1}$. Define the random variables W, X and Y as follows:

$$\begin{array}{lll} W\{(i,x)\} = 2^i & \text{for } x = 0, 1 & \\ X\{(i,0)\} = 2^{i+1} & ; X\{(i,1)\} = 2 & \text{for } i = 1, 2, \ldots \\ Y\{(i,0)\} = 2 & ; Y\{(i,1)\} = 2^{i+1} & \text{for } i = 1, 2, \ldots \end{array}$$

This specification has the following consequences

$$\begin{array}{lll} P\{W = 2^i\} = & P\{(i,0) \cup (i,1)\} & = 1/2^{i+1} + 1/2^{i+1} = 1/2^i \\ P\{X = 2\} = & P\{\cup_{i=1}^{\infty}(i,1)\} & = \sum_{i=1}^{\infty} 1/2^{i+1} = 1/2 \end{array}$$

and for $i = 1, 2, \ldots$

$$P\{X = 2^{i+1}\} = P\{(i, 0)\} = 1/2^{i+1}.$$

Thus X and W have the same distribution. Similarly Y also has the same distribution. Now consider the random variable $X + Y - 2W$. First

$$X\{(i,0)\} + Y\{(i,0)\} - 2W\{i,0\} = 2^{i+1} + 2 - 2(2^i) = 2.$$

Similarly

$$X\{(i,1)\} + Y\{(i,1)\} - 2W\{i,1\} = 2 + 2^{i+1} - 2(2^i) - 2.$$

Therefore we have

$$X + Y - 2W = 2.$$

Now suppose we have the opportunity to choose among the random variables X, Y and W, and have the utility function $U(R, \{(i, x)\}) = i$ for $R = X, Y$ and W. (All this means is that we rank random variables by their expectations.) Then we have

$$E[U(X) - U(W)] + E[U(Y) - U(W)] = 2$$

so either X is preferred to W or Y is preferred to W, or both, although X, Y and W have the same distribution. However,

$$E[U(X)] = E[U(Y)] = E[U(W)] = \sum_{i=1}^{\infty} 2^i(1/2^i) = \infty.$$

Thus ranking random variables with infinite expected utility according to the difference in their expected utilities leads to ranking identically distributed random variables differently. This example comes from Seidenfeld et al. (2009). □

Another example of anomalies in trying to order decisions with infinite expected utility comes from a version of the "two envelopes paradox." Suppose an integer is chosen, where

$$P\{N = n\} = \frac{1}{3}(2/3)^n \quad n = 0, 1, 2, \ldots \tag{7.10}$$

Two envelopes are prepared, one with 2^N dollars, and the other with 2^{N+1} dollars. Your utility is linear in dollars, so $u(x) = x$. You choose an envelope without knowing its contents, and are asked whether you choose to switch to the other envelope. Your expected utility from choosing the envelope with the smaller amount is

$$\sum_{n=0}^{\infty} \frac{1}{3} 2^n (2/3)^n = \sum_{n=0}^{\infty} (4/3)^n \cdot \frac{1}{3} = \infty \tag{7.11}$$

so you really don't care which envelope you have, and are indifferent between switching and not switching.

Now suppose you open your envelope, and find $\$x$ there. If $x = 1$, then you know $N = 0$, the other envelope has \$2, and it is optimal to switch. Now suppose $x = 2^k > 1$. Then there are two possibilities, $N = k$ and $N = k - 1$. Then we have

$$\begin{aligned} P\{N = k - 1 \mid x\} &= \frac{P\{x \text{ and } N = k-1\}}{P\{x \text{ and } N = k\} + P\{x \text{ and } N = k-1\}} \\ &= \frac{P\{x|N = k-1\}P\{N = k-1\}}{P\{x|N = k-1\}P\{N = k-1\} + P\{x|N = k\}P\{N = k\}} \\ &= \frac{1/2[\frac{1}{3}(2/3)^k - 1]}{\frac{1}{2}[\frac{1}{3}(2/3)^k + \frac{1}{3}(2/3)^{k-1}]} \\ &= \frac{1}{1 + 2/3} = \frac{3}{5}. \end{aligned} \tag{7.12}$$

Therefore $P\{N = k \mid x\} = 2/5$.

Consequently the expected utility of the unseen envelope is

$$\frac{3}{5}\frac{x}{2} + \frac{2}{5}(2x) = \frac{11x}{10} > x. \tag{7.13}$$

Therefore it is to your advantage to switch. Since you would switch whatever the envelope contains, there's no reason to bother looking. It seems that the optimal thing to do is to switch. Your friend, who has the other envelope, reasons the same way, and willingly switches. Now you start over again, and, indeed, switch infinitely many times! This is pretty ridiculous, since there's no reason to think either envelope is better than the other.

Whenever one can go from a reasonable set of hypotheses to an absurd conclusion, there must be a weak step in the argument. In this case, the weak step is going from dominance ("whatever amount x is in your envelope, it is better to switch") to the unconditional conclusion ("Therefore you don't need to know x, it is better to switch"). That step is true if the expected utilities of the options are finite. However, here the expected utilities of both choices are infinite, and so the step is unjustified. Indeed, even though if you knew x it would be in your interest to switch envelopes, in the case where you do not know x, switching and not switching are equally good for you. So beware of hasty analysis of problems with infinite expected utilities!

There are decisions that many people would refuse to make regardless of the consequences to other values they care about. These choices come up especially in discussions of ethics. It is convenient to think of these ultimately distasteful decisions as having minus infinity utility. Thus the theory here, which casts doubt on unbounded utility, contrasts with many discussions in philosophy that go by the general title of utilitarianism.

Such concerns can be accommodated, however, by lexicographic utility which does not satisfy the Archimedean condition. To give a simple example, imagine a bivariate utility function, together with a decision rule that maximizes the expectation of the first component, and, among decisions that are tied on the first component, maximizes the expectation of the second. So perhaps the first component is "satisfies my ethical principles" (and suppose there is no uncertainty about whether a decision does so), and the second component is some, perhaps uncertain, function of wealth. Then provided there is at least one ethically acceptable decision, maximizing this utility function would choose the expected function of wealth maximizing decision subject to the ethical constraint. Hence, I believe the issue with unacceptable choices is more properly focused on the Archimedean condition, and not on unbounded utility. The Archimedean condition might still apply within each component, but not across components. (See Chipman (1960).) For applications of this kind, a natural generalization of the theory presented here would provide a bounded utility function for the first coordinate of a lexicographic utility function, a bounded utility for the second, etc. I do not pursue this theme further in this book.

7.4.1 *Summary*

Unbounded utilities lead to paradoxical behavior if taken too literally, as they can lead to infinite expected utility.

7.4.2 *Notes and references*

The two-envelopes problem is also called the necktie paradox and the exchange paradox. Some articles concerning it are Arntzenius and McCarty (1997) and Chalmers (2002). An excellent website on it is `http://en.wikipedia.org/wiki/two_envelopes_problem`, last visited 11/15/2007.

7.4.3 *Exercises*

1. Vocabulary. Define in your own words:
   (a) St. Petersburg Paradox
   (b) Pascal's Wager
   (c) Archimedean condition
   (d) Lexicographic utility
2. Is Pascal's Wager an example of unbounded utility?
3. What's wrong with infinite expected utility, anyway?
4. Suppose utility is log-dollars. Find a random variable such that expected utility is infinite.
5. Why does lexicographic utility violate the Archimedean condition?

7.5 Risk aversion

People give away parts of their fortunes all the time (it's called charity). Having given away whatever part of their fortunes they wish, we can assume that they make their financial decisions reflecting a desire for a larger fortune rather than a smaller one. Thus it is reasonable to assure that, if f is their current fortune, $u(f)$ is increasing in f. If u is differentiable, this means $u'(f) > 0$. Suppose that there are two decision-makers $(i = 1, 2)$ (think of them as gamblers), each of whom like risk in the sense that

$$\frac{1}{2}u_i(f_i + x) + \frac{1}{2}u_i(f_i - x) > u_i(f_i), \quad i = 1, 2 \tag{7.14}$$

for all x, where f_i is the current fortune of gambler i and u_i is her utility function. Then each prefers a 1/2 probability of winning x, and otherwise losing x, to forgoing such a gamble. Then these gamblers would find it in their interest to flip coins with each other, for stakes x, until one or the other loses his entire fortune. Consequently, risk-lovers will have an incentive to find each other, and, after doing their thing, be rich or broke. The more typical case is risk aversion, where

$$\frac{1}{2}u(f + x) + \frac{1}{2}u(f - x) < u(f). \tag{7.15}$$

7.5.1 *A supplement on finite differences and derivatives*

For this discussion, it is useful to think of a derivative of the function f at the point x in a symmetric way:

$$g'(x) = \lim_{\epsilon \downarrow 0}\left[\frac{g(x + \epsilon) - g(x - \epsilon)}{2\epsilon}\right]. \tag{7.16}$$

Using this idea, what would we make of the second derivative, $f''(x)$? Well,

$$\begin{aligned} f''(x) &= \lim_{\epsilon \downarrow 0}\frac{g'(x + \epsilon) - g'(x - \epsilon)}{2\epsilon} \\ &= \lim_{\epsilon \downarrow 0}\frac{g(x + 2\epsilon) - g(x) - g(x) + g(x - 2\epsilon)}{(2\epsilon)^2} \\ &= \lim_{\epsilon \downarrow 0}\frac{g(x + 2\epsilon) - 2g(x) + g(x - 2\epsilon)}{4\epsilon^2}. \end{aligned} \tag{7.17}$$

Thus, just as the first difference, $g(x + \epsilon) - g(x - \epsilon)$ is the discrete analog of the first derivative, the second difference, $g(x + 2\epsilon) - 2g(x) + g(x - 2\epsilon)$ is the discrete analog of the second derivative. This idea can be applied any number of times.

7.5.2 *Resuming the discussion of risk aversion*

Now the inequality (7.15) can be rewritten as

$$0 > \frac{1}{2}u(f + x) - u(f) + \frac{1}{2}u(f - x) = \frac{1}{2}[u(f + x) - 2u(f) + u(f - x)]. \tag{7.18}$$

The material in square brackets is just a second difference. Thus the condition (7.15) for all f and x is equivalent to

$$u''(f) < 0 \tag{7.19}$$

for all f. A function obeying (7.19) is called concave.

Now for the typical financial decision-maker whose utility satisfies $u'(f) > 0$ and $u''(f) < 0$, we wish to investigate the extent to which this decision-maker is risk averse. Thus we ask what risk premium m makes the decision-maker indifferent between a risk (*i.e.*, uncertain prospect) and the amount $E(Z) - m$. Then m satisfies

$$u(f + E(Z) - m) = E\{u(f + Z)\}, \tag{7.20}$$

and m is a function of f and Z. Now if any constant c is added to f and subtracted from

Z, m is unchanged. It is convenient to take $c = E(Z)$, and, equivalently, consider only Z such that $E(Z) = 0$. Then (7.20) becomes

$$u(f - m) = E\{u(f + Z)\}. \tag{7.21}$$

We consider a small risk Z, that is, one with small variance σ^2. This implies also that the risk premium m is small. These conditions permit expansion of both sides of (7.21) in Taylor series as follows:

$$u(f - m) = u(f) - mu'(f) + \text{ HOT}, \tag{7.22}$$

and

$$\begin{aligned} E\{u(f + Z)\} &= E\{u(f) + Zu'(f) + \frac{1}{2}Z^2u''(f) + \text{ HOT}\} \\ &= u(f) + \frac{1}{2}\sigma^2u''(f) + \text{ HOT}. \end{aligned} \tag{7.23}$$

Equating these expressions, as (7.21) mandates, we find

$$m = -\frac{1}{2}\sigma^2\frac{u''(f)}{u'(f)} = \frac{1}{2}\sigma^2r(f) \tag{7.24}$$

where

$$r(f) = \frac{-u''(f)}{u'(f)}. \tag{7.25}$$

The quantity $r(f)$ is called the decision-maker's local absolute risk aversion.

To be meaningful for utility theory, a quantity like $r(f)$ should not change if instead of u, our decision-maker used the equivalent utility $w(f) = au(f) + b$, where $a > 0$. But $w'(f) = au'(f)$, and $w''(f) = au''(f)$, so

$$-\frac{w''(f)}{w'(f)} = -\frac{au''(f)}{au'(f)} = -\frac{u''(f)}{u'(f)} = r(f). \tag{7.26}$$

Another idea about how risk aversion might be modeled is to think about proportional risk aversion, in which the decision-maker is assumed to be indifferent between fZ and a non-random $E(fZ) - fm$.

If this is the case, then m^* satisfies the following equation:

$$u(f + E(fZ) - m^*) = E\{u(f + fZ)\}. \tag{7.27}$$

Again an arbitrary constant c may be subtracted from Z and compensated for by adding fc to f. Thus again we may take $c = E(Z)$, or, equivalently, take $E(Z) = 0$. Then we have

$$u(f - m^*) = E\{u(f + fZ)\}. \tag{7.28}$$

Again we expand both sides in a Taylor series for small variance σ^2 of Z, as follows:

$$u(f - m^*) = u(f) - m^*u'(f) + \text{ HOT}, \tag{7.29}$$

$$\begin{aligned} E\{u(f + fZ)\} &= E\{u(f) + fZu'(f) + f\frac{Z^2}{2}u''(f) + \text{ HOT}\} \\ &= u(f) + f\frac{\sigma^2}{2}u''(f) + \text{ HOT}. \end{aligned} \tag{7.30}$$

Equating (7.29) and (7.30) yields

$$m^* = -\frac{1}{2} f\sigma^2 \frac{u''(f)}{u'(f)} = \frac{1}{2}\sigma^2 fr(f). \tag{7.31}$$

Therefore we define the quantity $r^* = fr(f)$ to be the decision-maker's local relative risk aversion.

Under the assumptions that $u'(f) > 0$ and $u''(f) < 0$, the absolute risk premium $r(f)$ and the relative risk premium $r^*(f)$ are both positive. Let's see what happens if they happen to be constant in f.

If $r(f)$ is some constant k, we have

$$\frac{u''(f)}{u'(f)} = -k, \tag{7.32}$$

which is an ordinary differential equation. It can be solved as follows: Let $y(f) = u'(f)$. Then (7.32) can be written

$$-k = \frac{u''(f)}{u'(f)} = \frac{y'(f)}{y(f)} = \frac{d}{df} \log y(f). \tag{7.33}$$

Consequently

$$-kx = \int_0^x -kdx = \log y(f)\Big|_0^x = \log y(x) - \log y(0). \tag{7.34}$$

We'll take $-\log y(0)$ to be some constant c_1. Then (7.34) can be written

$$\log y(x) + c_1 = -kx, \tag{7.35}$$

from which

$$u'(x) = y(x) = e^{-kx-c_1}. \tag{7.36}$$

Finally

$$u(x) = -\frac{e^{-kx-c_1}}{k} + c_2. \tag{7.37}$$

In this form, the constants $e^{c_1} > 0$ and c_2 are simply the constants a and b in the equivalent form of the utility $au(x) + b$. Consequently the typical form of the constant absolute risk aversion utility with constant k is

$$u(x) = -\frac{e^{-kx}}{k}. \tag{7.38}$$

For this utility, it is easy to see that $u'(x) = e^{-kx}$ and $u''(x) = -ke^{-kx}$, from which $r(x) = ke^{-kx}/e^{-kx} = k$ as required.

Similarly we might ask what happens with constant relative risk aversion $r^*(f)$. Using the same notation, (7.33) is replaced by

$$-k/f = \frac{u''(f)}{u'(f)} = \frac{y'(f)}{y(f)} = \frac{d}{df} \log y(f). \tag{7.39}$$

Consequently

$$\begin{aligned} \log y(x) &= \int_c^x -k/w\, dw = -k \log x + k \log c_1 \\ &= k \log(c_1/x) = \log(c_1/x)^k. \end{aligned} \tag{7.40}$$

Hence

$$y(x) = (c_1/x)^k, \tag{7.41}$$

so

$$\begin{aligned} u(x) &= c_1^k \int_{c_2}^{x} (\frac{1}{y})^k dy = c_1^k \frac{y^{-k+1}}{-k+1}\Big|_{c_2}^{x} \\ &= c_1^k \left[\frac{x^{1-k}}{1-k} - \frac{c_2^{-k+1}}{1-k} \right]. \end{aligned} \tag{7.42}$$

Again, we may get rid of an additive constant and a positive multiplicative constant, to get the reduced form of the constant relative risk utility:

$$u(x) = x^{1-k}. \tag{7.43}$$

Again it is useful to check that the differential equation is satisfied. But $u'(x) = (1-k)x^{-k}$, and $u''(x) = (1-k)(-k)x^{-k-1}$. Hence

$$r^*(x) = -\frac{xu''(x)}{u'(x)} = \frac{(-x)(1-k)(-k)x^{-k-1}}{(1-k)x^{-k}} = k, \tag{7.44}$$

as required.

7.5.3 *References*

The theory in this section is usually attributed to Pratt (1964) and Arrow (1971), and is usually referred to as Arrow-Pratt risk aversion. The argument here follows Pratt's. However, Pratt and Arrow were preceded by DeFinetti (1952), with respect to absolute risk aversion (see Rubinstein (2006) and Kadane and Bellone (2009)).

7.5.4 *Summary*

This section motivates and derives measures of local absolute risk aversion and local relative risk aversion. It also derives explicit forms of utility for constant local absolute and relative risk aversion.

7.5.5 *Exercises*

1. Vocabulary. Explain in your own words:
   (a) local absolute risk aversion
   (b) local relative risk aversion
   (c) concave function
2. Are you risk averse? If so, does absolute or relative risk aversion describe you better? Are you comfortable with constant risk aversion as describing the way you want to respond to financial risk? What constant k would you choose?
3. Suppose a decision-maker has absolute local risk aversion $r(f)$.
   (a) Show that the risk of gain or loss of h with equal probability ($\pm h$, each with probability $\frac{1}{2}$), is equivalent, asymptotically as $h \to 0$, to the sure loss of $\frac{h}{2}^2 r(f)$.
   (b) Show that the gain of $\pm h$ with respective probabilities $(1 \pm d)/2$ is indifferent to you, asymptotically as $h \to 0$, if $d = \frac{hr(f)}{2}$.
   (c) The price of a gain h with probability p is $\frac{ph(1-qh)\cdot r(f)}{2}$, where $q = 1 - p$.

7.6 Log (fortune) as utility

A person with $\log(f)$ as utility is indifferent between the status quo and a gamble that, with probability $\frac{1}{2}$, increases their fortune by some factor x, and with probability $\frac{1}{2}$, decreases it by the factor $\frac{1}{x}$, as the following algebra shows:

$$\begin{aligned} & \log f - \tfrac{1}{2}(\log xf) - \tfrac{1}{2}\log(f/x) \\ = & \log f - \tfrac{1}{2}\log f - \tfrac{1}{2}\log x - \tfrac{1}{2}\log f + \tfrac{1}{2}\log x = 0. \end{aligned}$$

Thus such a person would be indifferent between the status quo and a flip of a coin that leads to doubling his fortune with probability $\frac{1}{2}$, and halving his fortune otherwise. This is the same as local relative risk aversion equal to one.

In the light of the results of section 7.4, we need first to consider the implications of the fact that the log function is unbounded both from above and from below. The fact that it is unbounded from below, so $\lim_{f \to 0} \log(f) = -\infty$, might be regarded as a good quality for a utility function to have. Its implication is that a person with such a utility function will accept no gambles having positive subjective probability of bankruptcy. A way around having log utility unbounded below, if such were thought desirable, would be to use $\log(1 + f)$, where $f \geq 0$.

That log fortune is unbounded from above, so $\lim_{f \to \infty} \log(f) = \infty$, implies, as found in section 7.4, vulnerability to St. Petersburg paradoxes. Thus we have to recognize that at the high end of possible fortunes, f, there may not be counter-parties able or willing to accept the bets a gambler with this utility function wishes to make.

Consider first an individual who starts with some fortune f, whose utility function is $\log f$ and who has the opportunity to buy an unlimited number of tickets that pay \$1 on an event A, at a price x. He can also buy an unlimited number of tickets on event $\overline{A}$, at price $1 - x = \overline{x}$. How should he respond to these opportunities?

If there is some amount c of his fortune he chooses not to bet, he can achieve the same result by spending cx on tickets for A, and $c\overline{x}$ on tickets for $\overline{A}$, with a total cost of $cx + c\overline{x} = c$. If A occurs, his $\frac{cx}{x} = c$ tickets on A offset exactly his cost c. If $\overline{A}$ occurs, his $\frac{c\overline{x}}{\overline{x}} = c$ tickets on $\overline{A}$ offset exactly his cost c. Consequently without loss of generality, we may suppose that the gambler bets his entire fortune. He needs to know how to divide his fortune f between bets on A and bets on $\overline{A}$. Suppose he chooses to devote a portion ℓ of his fortune to tickets on A, and the rest to $\overline{A}$. He now wants to know the optimal value of ℓ to maximize his expected utility. His answer must satisfy $0 \leq \ell \leq 1$.

Then he spends ℓf on tickets for A. Since they cost x, he buys a total of $\frac{\ell f}{x}$ tickets on A. Similarly he purchases $\frac{\overline{\ell} f}{\overline{x}}$ tickets on $\overline{A}$, where $\overline{\ell} = 1 - \ell$. Since he spends his entire fortune f on tickets, his resulting fortune is $\frac{\ell f}{x}$ if A occurs and $\frac{\overline{\ell} f}{\overline{x}}$ if $\overline{A}$ occurs. Finally suppose that his probability on A is q, so his probability on $\overline{A}$ is $p = 1 - q$. Then his expected utility is

$$\begin{aligned} q \log\left(\frac{\ell f}{x}\right) + p \log\left(\frac{\overline{\ell} f}{\overline{x}}\right) = \qquad (7.45) \\ (q + p)\log f + q \log \ell + p \log(1 - \ell) - q \log x - p \log \overline{x}. \end{aligned}$$

The only part of (7.45) that depends on ℓ is the second and third terms. Taking the derivative with respect to ℓ and setting it equal to zero we obtain

$$\frac{q}{\ell} = \frac{p}{1 - \ell}. \qquad (7.46)$$

Then $q(1 - \ell) = p\ell$, or $q = q\ell + p\ell = \ell$. This solution satisfies the constraint, since $0 \leq \ell \leq 1$. Thus the optimal strategy for this person is to bet on A in proportion to his personal probability, q, on A, and on $\overline{A}$ in proportion to his personal probability, p, on $\overline{A}$.

The achieved utility for doing so is

$$\log f + q \log q + p \log p - q \log x - p \log(1 - x). \tag{7.47}$$

Thus the optimal strategy for this person does not depend on his fortune, f, nor on x. The quantity $-[q \log q + p \log p]$ is known as entropy, or information rate (Shannon (1948)).

7.6.1 *A supplement on optimization*

The analysis given above to maximize (7.45) is just a little too quick. What we have shown is that the choice $\ell = q$ is the unique choice that makes (7.45) have a zero derivative. But zero derivatives of a function with a continuous derivative can occur for maxima, minima, or a third possibility, a saddle point. As an example of a saddle point, consider the function $g(x) = x^3$ at $x = 0$. It has zero derivative at $x = 0$, but is neither a relative maximum nor a relative minimum.

In the case of (7.45), think about the behavior of the function $q \log \ell + p \log(1 - \ell)$ as $\ell \to 0$. Because $\lim_{\ell \to 0} q \log \ell = -\infty$ and $\lim_{\ell \to 0} p \log(1 - \ell) = 0$, we have

$$\lim_{\ell \to 0} [q \log \ell + p \log(1 - \ell)] = -\infty. \tag{7.48}$$

Similarly, we also have

$$\lim_{\ell \to 1} [q \log \ell + p \log(1 - \ell)] = -\infty. \tag{7.49}$$

Thus the function increases as ℓ increases from zero to some point, and decreases again as ℓ increases toward ℓ one. As the derivative of (7.45) is zero only at $\ell = q$, this must be the global maximum of the function.

A second way to check whether a point found by setting a derivative equal to zero is a relative maximum is to compute the second derivative of the function at the point. In this case, the second derivative of (7.45) is

$$-\frac{q}{\ell^2} - \frac{p}{(1 - \ell)^2}. \tag{7.50}$$

Evaluated at the point $\ell = q$, we have

$$-q/q^2 - p/p^2 = -1/q - 1/p < 0. \tag{7.51}$$

Thus the second derivative is negative, so the function rises as ℓ approaches q from below, and then falls afterward. Since there is only one point at which the derivative is zero, this must be the global maximum.

Now suppose that we are asked to find the maximum of a function like (7.45) subject to the constraint $a \leq \ell \leq b$, where $0 \leq a < b \leq 1$. If the unconstrained optimal value $\ell = q$ satisfies the constraint, then it is the optimal value subject to the constraint as well. In this case, we say that the constraint is not binding. But what if the constraint is binding, that is, what if, in the case of (7.45), we have $q < a$ or $q > b$?

Let's take first the case of $0 < q < a$. Then we know that the unconstrained maximum occurs at $\ell = q$, and that throughout the range $a < \ell < b$, the function (7.45) is decreasing. Hence the optimal value of ℓ is $\ell = a$. Similarly, if $q > b$, then throughout the range $a \leq \ell \leq b$, the function (7.45) is rising, and has its maximum at $\ell = b$. Therefore the optimal value of ℓ can be expressed as follows:

$$\ell = \begin{cases} a & \text{if } q < a \\ q & \text{if } a \leq q \leq b \\ b & \text{if } q > b \end{cases}. \tag{7.52}$$

There is a little trick that can express this solution in a more convenient form. The median of a set of numbers is the middle value: half are above and half below. When the number of numbers in the set is even, by convention the average of the two numbers nearest the middle is taken. Consider the median of the numbers a, q and b. When $q < a < b$, the median is a. When $a \leq q \leq b$, the median is q. When $a < b < q$, the median is b. Hence, we may express (7.52) as

$$\ell = \text{ median } \{a, b, q\}. \tag{7.53}$$

We'll use this trick in the next subsection.

When optimizing a function of several variables, the same principles apply. If the point where the partial derivatives are zero is unique, and if the function at the boundary goes to minus infinity, then the point found by setting the partial derivatives to zero is the maximum. The multi-dimensional analog of the second derivative being negative is that the matrix of second partial derivatives is negative-definite. In the multi-dimensional case there isn't an analog of (7.52) and (7.53) that I know of.

Finally, there's a very useful technique for maximizing functions subject to equality constraints known as the method of undetermined multipliers or as Lagrange multipliers. The problem here is to maximize a function $f(\mathbf{x})$, subject to a constraint $g(\mathbf{x}) = 0$, where $\mathbf{x} = (x_1, \ldots, x_k)$ is a vector. One method that works is to solve $g(\mathbf{x})$ for one of the variables $x_1, \ldots, x_k$, substitute the result into $f(\mathbf{x})$, and maximize the resulting function with respect to the remaining $k-1$ variables. This method breaks the symmetry often present among the k variables $x_1, \ldots, x_k$. The method of Lagrange multipliers, by contrast, maximizes, with respect to $\mathbf{x}$ and λ, the new function

$$f(\mathbf{x}) + \lambda g(\mathbf{x}). \tag{7.54}$$

If $\mathbf{x}^0$ maximizes $f(\mathbf{x})$ subject to $g(\mathbf{x}) = 0$, it is obvious that it also maximizes (7.54). To see the converse, notice that the derivative of (7.54) with respect to λ yields the constraint $g(\mathbf{x}) = 0$. The derivatives of (7.54) with respect to the x_i's yield equations of the form

$$\frac{\partial f(\mathbf{x})}{\partial x_i} + \lambda \frac{\partial g(\mathbf{x})}{\partial x_i} = 0 \quad i = 1, \ldots, k. \tag{7.55}$$

On an intuitive basis, if (7.55) failed to hold, it would be possible to shift the point x, while maintaining the constraint $g(\mathbf{x}) = 0$, in a way that would increase f. Lagrange multipliers can be used for more than one constraint. If there are several constraints $g_j(\mathbf{x}) = 0$, $(j = 1, \ldots, J)$, then the maximum of

$$f(\mathbf{x}) + \sum_{j=1}^{J} \lambda_j g_j(\mathbf{x}) \tag{7.56}$$

with respect to $\mathbf{x}$ and $\lambda_1, \ldots, \lambda_J$ yields the maximum of $f(\mathbf{x})$ subject to the constraints $g_j(\mathbf{x}) = 0$, $j = 1, \ldots, J$. A rigorous account of Lagrange multipliers may be found in Courant (1937, Volume 2, pp. 190-199).

We'll use Lagrange multipliers in the next subsection.

7.6.2 Resuming the maximization of log fortune in various circumstances

Now we extend the problem by supposing that the person has a budget $B \leq f$ which cannot be exceeded in his purchases. Suppose he chooses to spend y on tickets for A and $B - y$ on tickets for $\overline{A}$. For notational convenience, let $f^* = f - B$. Then he buys $\frac{y}{x}$ tickets on A, and $\frac{(B-y)}{\overline{x}}$ tickets on $\overline{A}$, resulting in a fortune of $f^* + y/x$ if A occurs, and $f^* + (B - y)/\overline{x}$ if A^c occurs. So his expected utility is

$$q \log(f^* + y/x) + p \log(f^* + (B - y)/\overline{x}). \tag{7.57}$$

Setting the derivative with respect to y equal to zero, we have

$$\frac{q/x}{f^* + y/x} = \frac{p/\overline{x}}{f^* + (B-y)/\overline{x}}, \quad \text{or} \tag{7.58}$$
$$\frac{q}{xf^* + y} = \frac{p}{\overline{x}f^* + (B-y)}.$$

Then

$$q(\overline{x}f^* + (B-y)) = p(xf^* + y),$$
$$q\overline{x}f^* - pxf^* + Bq = qy + py = y.$$

Since the second derivative of (7.57) is negative, the y found by setting the first derivative equal to zero indeed maximizes (7.57).

Since the optimal y must satisfy the bounds $0 \leq y \leq B$, we have that the optimal y is

$$y_{\text{opt}} = \text{median } \{0, B, qB + x\overline{x}f^* \left(\frac{q}{x} - \frac{p}{\overline{x}}\right)\}. \tag{7.59}$$

When $B = f$, so the budget constraint is non-binding, then $y_{\text{opt}} = \text{median } \{0, B, qB\} = qf$, so he optimally spends proportion q of his fortune f on tickets for A, as we found before.

Now suppose that there are n events $A_1, \ldots, A_n$ that are mutually exclusive and exhaustive. Suppose also that $q_i = P\{A_i\}$. There are dollar tickets available on them with respective prices $x_1, \ldots, x_n$ such that $\sum_{i=1}^n x_i = 1$. Again the person has fortune f. The argument given in the third paragraph of this section still applies. Thus we can assume that the person chooses to devote portion ℓ_i to buying tickets on A_i, where $0 \leq \ell_i$ and $\sum_{i=1}^n \ell_i = 1$. Then he buys $\ell_i f/x_i$ tickets on A_i, and his expected fortune is

$$\sum q_i \log(\ell_i f/x_i) = \log f - \sum q_i \log x_i + \sum q_i \log \ell_i. \tag{7.60}$$

Thus we seek ℓ_i, $0 \leq \ell_i$ and $\sum \ell_i = 1$ to maximize $\sum q_i \log \ell_i$. Using the technique of Lagrange multipliers, we maximize

$$\sum_{i=1}^n q_i \log \ell_i - \lambda(\sum_{i=1}^n \ell_i - 1), \tag{7.61}$$

with respect to ℓ_i and λ.

Taking the derivative, we have

$$\frac{q_i}{\ell_i} - \lambda = 0, \text{ or}$$
$$q_i = \lambda \ell_i.$$

Since $\sum q_i = 1 = \sum \ell_i$, we have $\lambda = 1$ and

$$\ell_i = q_i, i = 1, \ldots, n. \tag{7.62}$$

Again, since (7.60) approaches $-\infty$ as any ℓ_i approaches zero, the solution to setting the first derivative of (7.61) equal to zero yields a maximum.

This result suggests a rationale for the investment strategy called re-balancing. Dividing the possible investments into a few categories, such as stocks, bonds and money-market funds, re-balancing means to sell some from the categories that did well, and buy more of those that did poorly, to maintain a predetermined proportion of assets in each category. (This analysis neglects transaction fees.)

7.6.3 Interpretation

The mathematics in section 7.6 are due to Kelly (1956), with some conversion to put them in the framework of this book. While the mathematics are solid, the interpretation of them has been beset with controversy. It began with Kelly's discussion:

> The gambler introduced here follows an essentially different criterion from the classical gambler. At every bet he maximizes the expected value of the logarithm of his capital. The reason has nothing to do with the value function which he attached to his money, but merely with the fact that it is the logarithm which is additive in repeated bets and to which the law of large numbers applies. (pp. 925, 926)

To understand Kelly, he means by "value function" what we mean by utility, and his "classical gambler" has a utility function that is linear in his fortune. His reference to the law of large numbers comes from the fact that if the gambler makes bets on a large number of independent events with some probability, the proportion of success will approach the (from the perspective of this book, subjective) probability the event occurs.

Kelly's argument here is, I think, circular. He basically is saying that if you don't maximize log fortune, your fortune will grow at an exponential rate smaller than the rate you expect to enjoy if you do maximize log fortune. This is obviously true, but isn't relevant to someone whose utility is something other than log fortune.

Kelly then poses the question of what a gambler should do, who is allowed a limited budget (one dollar per week!). He proposes that such a gambler should put the whole dollar on the event yielding the highest expectation. It seems to me that this is correct for a gambler with a utility function linear in his fortune, but not for a budget-limited player with a utility that is log fortune, as shown in (7.59).

Kelly also poses the question of the optimal strategy when there is a "track take," which means when $\sum_{i=1}^{n} x_i > 1$ (in Britain, this is called an "overround"). In this case a gambler using log fortune as utility will not bet his entire fortune. Also there are some offers (maybe all!) so unfavorable that he will not bet on them at all. It turns out, not unreasonably, that in this modified problem, gambles are ranked by q_i/x_i, the gambler's probability of a ticket on A_i succeeding, divided by its cost.

Kelly's work, and the resulting "Kelly criterion," were criticized by a group of economists led by the eminent Paul Samuelson. In an article entitled "The 'Fallacy' of Maximizing the Geometric Mean in Long Sequences of Investing or Gambling," Samuelson (1971) argues essentially that the Kelly strategy leads to large volatility of returns. He concedes that $\log f$ is analytically tractable, "but this will not endear it to anyone whose psychological tastes differ significantly from $\log f$" (Samuelson, 1971, p. 2496). Finally, and famously, Samuelson wrote an article entitled "Why we should not make mean log of wealth big though years to act are long" (Samuelson (1979)); in which he limits himself to words of one syllable.

One has to be careful, though, about arguments based on the volatility of returns. A standard method of portfolio analysis, going back to Markowitz (1959), proposes that one should examine the mean and variance of the return on a portfolio, and choose to minimize some linear functional of them.

To model this, the only way that expected utility can be made to depend on only the mean and variance of the returns X is for utility to be a linear function of X and X^2, so the utility is of the form $U(X) = aX + bX^2$.

The expected utility is then

$$EU(X) = a\mu + b(\mu^2 + \sigma^2),$$

where $\mu = E(X)$ and $\sigma^2 = \text{Var}(X)$, assuming both exist. In order to express the idea that our investor prefers less variance for a given mean, we must have $b < 0$. Then the change in expected utility from changing μ, as measured by the first derivative, is

$$\frac{dE(U(X))}{d\mu} = a + 2b\mu.$$

If $a \leq 0$, $\frac{dE(U(X))}{d\mu} < 0$, which would mean that our investor would always prefer less expected return, which is unacceptable. However, for $a > 0$, we still have $\frac{dE(U(X))}{d\mu} < 0$ if $\mu > -a/2b$, so our investor would dis-prefer large expected returns. Consequently there is no utility function that rationalizes Markovitz's approach. A more modern approach, consistent with expected utility theory, is given in Campbell and Viceira (2002).

Markowitz gets around this by using variance only to compare portfolios with the same mean return. If the returns on an optimal strategy are too volatile for your taste, then perhaps you are using a candidate utility function that does not properly reflect your aversion to risk. I think that's the point Samuelson is making about $\log f$ as a utility. However, it is worth remembering that within the theory of decision-making on the basis of expected utility, there is no place for Var $[U(\theta \mid d)]$.

There is a lot of literature surrounding this debate. Some important contributions include Rotando and Thorp (1992), Samuelson (1973) and Breiman (1961). An entertaining verbal account of Kelly's work, the characters surrounding it and its implications, is in Poundstone (2005).

Markowitz's work on this subject was preceded by DeFinetti (1940) [English translation by Barone (2006)], a point generously conceded by Markowitz (2006) in an article entitled "DeFinetti Scoops Markowitz." See also Rubinstein (2006). Interestingly, DeFinetti justifies the mean-variance approach by appeal to the central limit theorem and asymptotic normality. He does not mention the incompatibility of this approach with the maximization of subjective expected utility, of which he is one of the modern founders.

From the perspective of this book, it is no use to argue what a person's utility function ought to be, any more than it is useful to argue what their probabilities ought to be. Exploring the consequences of various choices is a contribution, and can lead people to change their views upon more informed reflection.

7.6.4 Summary

This section explores some of the consequences of investing (or gambling – is there a difference?) using $\log f$ as a utility function. In the simplest cases one bets one's entire fortune, dividing the proportion bet according to one's subjective probability of the event.

7.6.5 Exercises

1. Vocabulary. Explain in your own words:
   (a) Lagrange multipliers
   (b) Median
2. In your view, what is the significance of Kelly's work?
3. Suppose a person's fortune is f = \$1000, and his utility function is $\log(f)$. Suppose this person can buy tickets on the mutually exclusive events A_1, A_2 and A_3 with prices $x_1 = 1/6$, $x_2 = 1/3$ and $x_3 = 1/2$. Suppose this person's probabilities on these three events are, respectively $q_1 = 1/2$, $q_2 = 1/3$ and $q_3 = 1/6$.
   (a) How much should such a person invest in each kind of ticket to maximize his expected utility?
   (b) How many tickets of each kind should he buy?

(c) Does your optimal strategy propose that he buy tickets on event A_3, even though such tickets are expensive ($x_3 = 1/2$) in relation to the person's probability that event A_3 will occur ($q_3 = 1/6$)? Explain why or why not.

4. Consider the family of utility functions indexed by γ, and of the form,

$$u(f;\gamma) = \frac{f^{1-\gamma}-1}{1-\gamma} \quad 0 < \gamma.$$

These are the constant relative risk aversion utilities, with constant γ.

(a) Use L'Hôpital's Rule (see section 2.7) to show that, as $\gamma \to 1$,

$$\lim_{\gamma \to 1} u(f;\gamma) = \log f \text{ for each } f > 0.$$

(b) Suppose $A_1, \ldots, A_n$ are n mutually and exclusive events. Tickets paying \$1 if event A_i occurs are available at cost x_i, where $x_i > 0$ and $\sum_{i=1}^n x_i = 1$. Also suppose that a person has utility $u(f;\gamma) = \frac{f^{1-\gamma}-1}{1-\gamma}$, for $0 < \gamma$, and wishes to invest this fortune to maximize expected utility. If this person's probabilities are $q_i > 0$ that event A_i will occur, where $\sum_{i=1}^n q_i = 1$, how should such a person divide their fortune among these opportunities?

(c) In part (b), how many tickets of each kind will the person optimally choose to buy?

(d) Find the limiting result, as $\gamma \to 1$, of your answers to (b) and (c). Do they equal the result obtained by using $\log f$ as utility?

5. Suppose your utility is $\log f$ and you are offered the opportunity to buy as many tickets paying \$1 if event A occurs and 0 otherwise. You have probability q that event A will occur. Tickets cost \$ x each. How many tickets would you optimally buy?

6. Reconsider the maximization of (7.60) subject to the constraint $\sum_{i=1}^n \ell_i = 1$. Perform this maximization by substituting $\ell_n = 1 - \ell_1 - \ell_2 - \ldots - \ell_{n-1}$ into (7.60) and maximize with respect to $\ell_1, \ldots, \ell_{n-1}$. Do you get the same result? Which method do you prefer, and why?

7. Suppose that your investment advisor informs you that she believes you face an infinite series of independent favorable bets, where your probability of success is 0.55. Suppose that she proposes that you use log (fortune) as your utility function, and that therefore at each opportunity, she proposes that you bet 0.55 of your fortune on the event in question, and 0.45 of your fortune against.

(a) Run a simulation, assuming that your advisor is correct about your probability of success at each trial and you follow the recommended strategy. Plot your fortune after a (simulated) sequence of 100 such bets.

(b) Now suppose that you are slightly less optimistic than your investment advisor, and believe that your probability of success is only 0.45 at each independent trial. Plot your fortune after 100 trials, again following the recommended strategy.

(c) Now suppose that you have utility which has constant relative risk aversion instead of log (fortune) utility. Suppose that your utility takes the form mentioned in problem 4, and consider the cases $\gamma = 0.5, 0.3$ and 0.1. Rerun your simulations of part (a) and (b) above (your investment advisor's beliefs and your own) for these cases.

(d) In the light of these simulations, which value of $\gamma, 0.5, 0.3, 0.1$, or 0 (which is log (fortune)) best reflects your own utility function? Explain your reasons.

7.7 Decisions after seeing data

We can never know about the days to come
But we think about them anyway.

—Carly Simon

Now suppose that you will have a decision to make after seeing some data. One way to think about how to make such a decision is to wait until you have the data, decide on your (then) current probability $p(\theta)$ for the uncertain θ you then face, and maximize (7.3). This allows for the possibility that you may change your mind after seeing the data, as discussed in section 7.6.3.

A second way to think about such a decision is to use the idea that you now anticipate that, after seeing data $\mathbf{x}$, your opinion about θ will be $p(\theta \mid x)$. Under this assumption, you can calculate now what decision you anticipate to be optimal, as follows.

You will make your decision after seeing the data $\mathbf{x}$, so your decision can be a function of $\mathbf{x}, d(\mathbf{x})$. Since you are now uncertain about both $\mathbf{x}$ and $\boldsymbol{\theta}$, you wish to maximize, over choices $d(\mathbf{x})$, your expected utility, *i.e.*,

$$\begin{aligned}\bar{U} &= \max_{d(\mathbf{x})} \int\int U(d, \boldsymbol{\theta}, \mathbf{x}) p(\boldsymbol{\theta}, \mathbf{x}) d\boldsymbol{\theta} d\mathbf{x} \\ &= \max_{d(\mathbf{x})} \int\int U(d, \boldsymbol{\theta}, \mathbf{x}) p(\boldsymbol{\theta} \mid \mathbf{x}) d\boldsymbol{\theta} p(\mathbf{x}) d\mathbf{x}. \qquad (7.63)\end{aligned}$$

Because $d(\mathbf{x})$ is allowed to be a function of x, we can take it inside the first integral sign, obtaining

$$\bar{U} = \int \left[\max_{d(\mathbf{x})} \int U(d, \boldsymbol{\theta}, \mathbf{x}) p(\boldsymbol{\theta} \mid \mathbf{x}) d\boldsymbol{\theta} \right] p(\mathbf{x}) d\mathbf{x}. \qquad (7.64)$$

Thus you would use your posterior distribution of θ after seeing $\mathbf{x}, p(\boldsymbol{\theta} \mid \mathbf{x})$, and choose $d(\mathbf{x})$ accordingly to maximize posterior expected utility.

This is the reason why Bayesian computation is focused on computing posterior distributions.

7.7.1 *Summary*

A Bayesian makes decisions by maximizing expected utility. When data are to be collected, a Bayesian makes future decisions by maximizing expected utility, where the expectation is taken with respect to the distribution of the uncertain quantity θ after the data are observed. This is anticipated to be the conditional distribution of the θ given the data $\mathbf{x}$.

7.7.2 *Exercise*

1. (a) Suppose that a gambler has fortune f and uses as utility the function $\log f$. Suppose there is a partition $A_1, \ldots, A_n$ of n mutually exclusive and exhaustive events. Suppose that event A_i has probability q_i and that dollar tickets on A_i cost $\$x_i$. Suppose also $\sum q_i = \sum x_i = 1$. Use the results of section 7.6 to find the expected utility of the optimal decision this gambler can make on how to bet.

   (b) Suppose that the gambler receives a signal S such that

   $$P\{S = s \mid A_i\} = p_{s,i}.$$

   Find the gambler's posterior probabilities q'_i that event i will occur. Show that $\sum_{i=1}^{n} q'_i = 1$.

   (c) Now suppose that the gambler receives a signal, from whatever source, that changes his probabilities from q_i to q'_i on event i, where $\sum_{i=1}^{n} q'_i = 1$. What are the gambler's optimal decisions now? What is the resulting expected utility?

7.8 The expected value of sample information

Suppose you have a decision to make. You are uncertain about θ, and are contemplating whether to observe data $\mathbf{x}$ before making the decision. Would you maximize your expected utility by ignoring this opportunity, even if the data were cost-free?

An intuitive argument suggests not. After all, you could ignore the data and do just what you would have done anyway. Alternatively, the data might be helpful to you, allowing you to make a better, more informed, decision. This argument can be made precise, as follows.

Let $U(d, \theta, \mathbf{x})$ be your utility function, depending on your decision d, the unknown θ about which you are uncertain, and the data $\mathbf{x}$ that you may or may not choose to observe. Without the data $\mathbf{x}$, you would maximize

$$\int_X \int_\Theta U(d, \theta, \mathbf{x}) p(\theta, x) d\boldsymbol{\theta} d\mathbf{x}. \tag{7.65}$$

If you learn $\mathbf{x}$, your conditional distribution is $p(\theta \mid \mathbf{x})$, and you would choose d to maximize

$$\int_\Theta U(d, \boldsymbol{\theta}, \mathbf{x}) p(\boldsymbol{\theta} \mid \mathbf{x}) d\boldsymbol{\theta}, \tag{7.66}$$

which has current expectation with respect to the unknown value of X,

$$\int_X \left[\max_d \int_{\boldsymbol{\theta}} U(d, \boldsymbol{\theta}, \mathbf{x}) p(\boldsymbol{\theta} \mid \mathbf{x}) d\boldsymbol{\theta} \right] p(\mathbf{x}) d\mathbf{x}. \tag{7.67}$$

It remains to show that (7.67) is at least as large as (7.65). Suppose d^* maximizes (7.65). Then, for each $\mathbf{x}$,

$$\max_d \int_{\boldsymbol{\theta}} U(d, \boldsymbol{\theta}, \mathbf{x}) p(\boldsymbol{\theta} \mid \mathbf{x}) d\boldsymbol{\theta} \geq \int_{\boldsymbol{\theta}} U(d^*, \boldsymbol{\theta}, \mathbf{x}) p(\boldsymbol{\theta} \mid \mathbf{x}) d\boldsymbol{\theta}. \tag{7.68}$$

Integrating both sides of this equation with respect to the marginal distribution of X, yields

$$\begin{aligned} &\int_X [\max_d \int_{\boldsymbol{\theta}} U(d, \boldsymbol{\theta}, \mathbf{x}) p(\boldsymbol{\theta} \mid \mathbf{x}) d\boldsymbol{\theta}] p(\mathbf{x}) d\mathbf{x} \\ &\geq \int_X \int_{\boldsymbol{\theta}} U(d^*, \boldsymbol{\theta}, \mathbf{x}) p(\boldsymbol{\theta} \mid \mathbf{x}) d\boldsymbol{\theta} p(\mathbf{x}) d\mathbf{x} \\ &= \int_X \int_{\boldsymbol{\theta}} U(d^*, \boldsymbol{\theta}, \mathbf{x}) p(\boldsymbol{\theta}, \mathbf{x}) d\boldsymbol{\theta} d\mathbf{x} \\ &= \max_d \int_X \int_{\boldsymbol{\theta}} U(d, \boldsymbol{\theta}, \mathbf{x}) p(\boldsymbol{\theta}, \mathbf{x}) d\boldsymbol{\theta} d\mathbf{x}, \end{aligned} \tag{7.69}$$

which was to be shown.

Thus a Bayesian would never pay not to see data.

The example in section 3.2 shows that with finite but not countable additivity, you would pay not to see data in certain circumstances. The same is true if you use an improper prior distribution (one that integrates to infinity), even one that is a limit of proper priors (see Kadane et al. (2008)).

7.8.1 *Summary*

A Bayesian with a countably additive proper prior distribution does not pay to avoid seeing data. However, a finitely additive prior, or an improper prior, can lead to such situations.

7.8.2 Exercise

1. Recall the circumstances of exercise 7.7.2. Calculate the expected utility to the gambler of the signal S. Must it always be non-negative? Why or why not?

7.9 An example

Sometimes to figure out how much tax is owed by a taxpayer, an enormous body of records must be examined. A natural response to this is to take a random sample, and to analyze the results. From such a sample, following the ideas expressed in this book, the best that can be obtained is a probability distribution for the amount owed. Suppose θ is the amount owed, and has some (agreed) distribution with density $p(\theta)$. [The idea that the taxpayer and the taxing authority would agree on $p(\theta)$ often does not comport with reality, but that's another story.] The issue here is that the taxpayer can't write a check for a random variable. How much tax t should the taxpayer actually pay?

A natural first reaction to this problem is that the taxpayer should pay some measure of the central tendency of θ, perhaps $E(\theta)$. But there are three reasons why this might be too much. In many situations, the taxpayer has the right to have his records – all of his records - examined. By imposing sampling, the taxing authority is in effect asking the taxpayer to give up this right, and the taxpayer should be compensated for doing so. Second, the taxing authority typically chooses the sample size, imposing risk of overpayment on the taxpayer. The cost of too large a sample should be born by the same party as the cost of too small a sample, namely the taxing authority. Finally, taxation relies for the most part on voluntary compliance. As a result, the state cannot afford to have a reputation as a pirate. For all these reasons, while the state wants its taxes, it has reasons to think that over-collection is worse for it than under-collection.

Suppose that the state's interests are summarized by a loss function $L(t, \theta)$, expressing the idea that to over-collect $(t > \theta)$ its loss is b times the extent of over-collection, while if it under-collects, its loss is a times the extent of under-collection, and the arguments above suggest $b > a > 0$. Such a loss function can be expressed as

$$L(t, \theta) = \begin{cases} a(\theta - t) & \text{if } \theta > t \\ b(t - \theta) & \text{if } \theta < t \end{cases}. \tag{7.70}$$

Then expected loss is

$$\begin{aligned} \bar{L}(t) &= \int_{-\infty}^{\infty} L(t, \theta) p(\theta) d\theta \\ &= \int_{-\infty}^{t} b(t - \theta) p(\theta) d\theta + \int_{t}^{\infty} a(\theta - t) p(\theta) d\theta. \end{aligned} \tag{7.71}$$

We minimize (7.71) by taking its first derivative. Since t occurs in several places in (7.71), this requires use of the chain rule. In this case it also requires remembering the Fundamental Theorem of Calculus, to handle the derivative of a limit of integration, thus:

$$\begin{aligned} \frac{d\bar{L}(t)}{dt} &= b(t - \theta)p(\theta)\Big|^{\theta = t} - a(\theta - t)p(\theta)\Big|^{\theta = t} \\ &+ \int_{-\infty}^{t} bp(\theta) d\theta - \int_{t}^{\infty} ap(\theta) d\theta \\ &= 0 + 0 + bP\{\theta \leq t\} - aP\{\theta > t\}. \end{aligned} \tag{7.72}$$

To justify the differentiation under the integral sign in (7.72) we have implicitly assumed that $E \mid \theta \mid < \infty$, but we needed that assumption anyway to have finite expected loss.

Setting (7.72) to zero and using the fact that $P\{\theta \leq t\} = 1 - P\{\theta > t\}$, we have

$$\begin{aligned} a(1 - P\{\theta \leq t\}) &= bP\{\theta \leq t\}, \text{ or} \\ a &= (a+b)F_\Theta(t), \text{ so} \\ t &= F_\Theta^{-1}\left(\frac{a}{a+b}\right). \end{aligned} \tag{7.73}$$

Since $L(t, \theta) \to \infty$ as $t \to \infty$ and as $t \to -\infty$, the stationary point found in (7.73) is a minimum. Thus (7.73) says that the optimal tax is the $\left(\frac{a}{a+b}\right)^{th}$ quantile of the distribution of θ. In Bright et al. (1988), to which the reader is referred for further details, we argue that b/a should be in the neighborhood of 2 to 4 (*i.e.*, that over-collection might be 2 to 4 times worse than under-collection), which has the consequence under (7.73) that the appropriate quantile of θ for taxation should be between .33 and .2. Current practice at the time we wrote (and still, I believe) uses either .5 (which is equivalent to $a = b$) or .05, which is equivalent to $b/a = 19$.

Of course it is a bit of an exaggeration to think of the state as a rational actor with a utility function, but it is still a useful exercise to model it as if it were.

7.9.1 *Summary*

This example shows how a simple utility function may be used to examine a public policy, and make suggestions for its improvement.

7.9.2 *Exercises*

1. Suppose the result of a taxation audit using sampling is that the amount of tax owed, θ, has a normal distribution with mean \$100,000 and a standard deviation of \$10,000. Using the loss function (7.69), how much tax should be collected if:
   (a) $b/a = 1$
   (b) $b/a = 2$
   (c) $b/a = 4$
   (d) $b/a = 19$?
2. An employer's health plan offers to employees the opportunity to put money, before tax, into a health account the employee can draw upon to pay for health-related expenditures. Any funds not used in the account by the end of the year are forfeited. Suppose the employee's probability distribution for his health-related expenditures over the coming year has density $f(\theta)$. Suppose also that his marginal tax rate is α, $0 < \alpha < 1$, and that he wishes to maximize his expected after-tax income. How much money, d, should he contribute to the health account?

7.9.3 *Further reading*

Optimal sample size calculations are addressed in Kadane (2017).

7.10 Randomized decisions

There are some statistical theories that suggest using randomized decisions. Thus, instead of choosing decision d_1 or decision d_2, such a theory would suggest using a randomization device such as a coin-flip that has probability α of heads, and choosing decision d_1 with probability α and decision d_2 with probability $1 - \alpha$. The outcome of this coin flip is to be regarded as independent of all other uncertainties regarding the problem at hand. Under what conditions would such a policy be optimal?

Suppose decision d_1 has expected utility $U(d_1)$, and decision d_2 has expected utility $U(d_2)$. Then the expected utility of the randomized decision would be

$$U(\alpha d_1 + (1-\alpha)d_2) = \alpha U(d_1) + (1-\alpha)U(d_2). \tag{7.74}$$

There are two important subcases to consider. Suppose first that one decision has greater expected utility than the other. There is no loss of generality in supposing $U(d_1) > U(d_2)$, reversing which decision is d_1 and which is d_2, if necessary. Then the optimal α is $\alpha = 1$, since, for $\alpha < 1$,

$$U(d_1) > \alpha U(d_1) + (1-\alpha)U(d_2). \tag{7.75}$$

Thus in this case, randomized decisions are suboptimal.

Now suppose that $U(d_1) = U(d_2)$. Then any α in the range $0 \leq \alpha \leq 1$ is as good as any other, and each choice achieves utility $U(d_1) = U(d_2)$. Thus a randomized decision is weakly optimal, as utility maximization can be achieved without randomized decisions.

Lest the reader think that randomization is not uniquely optimal to a utility-maximizing Bayesian is so trivial a point as not to be worth discussing, please remember that sampling theory and randomized experimental designs use randomization extensively. I believe that these methods are very useful in statistics. However, I believe that a proper understanding of them belongs to the theory of more than one decision maker. Hence, further discussion of this matter is postponed to Chapter 11, section 4.

An alternative view of the role of randomization from a Bayesian perspective, can be found in Rubin (1978). The core of his argument is that randomization might simplify certain likelihoods, making the findings more robust and hence more persuasive.

7.10.1 Summary

Randomized decisions are not uniquely optimal. In any problem in which randomized decisions are optimal, the non-randomized decisions that are given positive probability under the optimal randomized decision, are also optimal.

7.10.2 Exercise

Recall the circumstances of exercise 3 in section 7.2.3: The decision-maker has to choose whether to take an umbrella, and faces uncertainty about whether it will rain. The four consequences she faces are c_1 = (take, rain), c_2 = (do not take, rain), c_3 = (take, no rain) and c_4 = (do not take, no rain). These have respective utilities $U(c_4) = 1$, $U(c_3) = 1/3$, $U(c_2) = 0$ and $U(c_1) = 2/3$. Suppose the decision maker's probability of rain is p.

(a) For what value p^* of p is the decision-maker indifferent between taking and not taking the umbrella?

(b) Suppose the decision-maker has the probability of rain p^*, and decides to randomize her decision. With probability θ she takes the umbrella and with probability $1-\theta$ she does not. Does she gain expected utility by doing so?

(c) Now suppose her probability of rain is $p > p^*$. What is her optimal decision? Answer the same question as in part (b).

(d) Finally, suppose $p < p^*$. Again, what is her optimal decision? Again answer the same question as in part (b).

7.11 Sequential decisions

So far, we have been studying only a single stage of decision-making. In such a problem, the posterior distribution of the parameters given the data is used as the distribution to compute expected utility, and the decision with maximum expected utility is optimal. However there is no reason to be so restrictive. There can be several stages of information-gathering and decisions. Furthermore, those decisions may affect the information subsequently available, for example by deciding on the nature and extent of information to be collected. The important thing to understand is that the principles of dealing with multiple decision points are exactly those of a single decision point: at each decision point, it is optimal to choose that decision that maximizes expected utility, where the expectation is taken with respect to the distribution of all random variables conditional on the information available at the time of the decision.

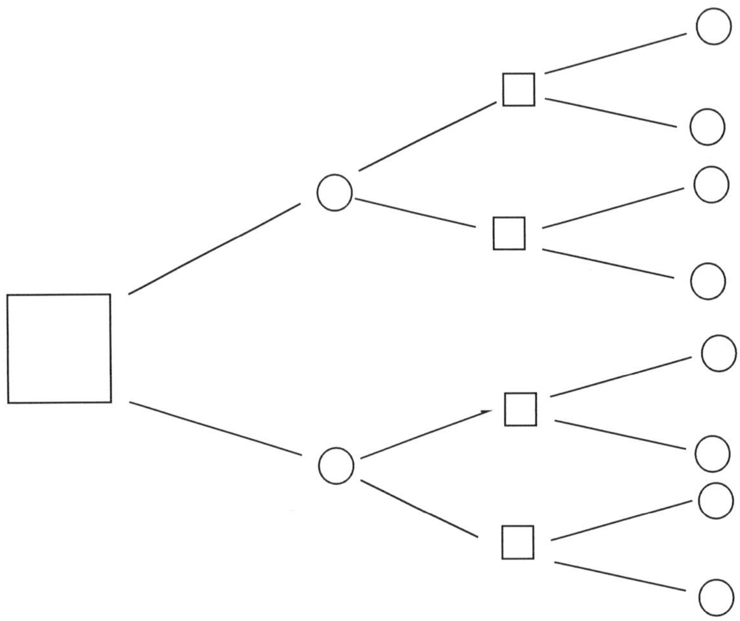

Figure 7.4: Decision tree for a 2-stage sequential decision problem.

Figure 7.4 illustrates a decision tree for a two-stage sequential decision problem. The posterior from the kth decision stage becomes the prior for the $(k+1)^{st}$ decision stage. This suggests that the names "prior" and "posterior" are not very useful, since to make sense they must refer to a particular time point in the decision process. It is probably better practice to keep in mind what is uncertain, and therefore random, and what is known, and therefore to be conditioned upon, at each stage of that process.

Now let's consider some examples. The first example is a class of problems known in other parts of statistics as (static) experimental design. Here there are two decision points: first deciding what data to collect, and then, after the data are available, making whatever terminal decision is required. The first decision requires expected utility of each possible design where the expectation is taken with respect to both the (as yet unobserved) data and the other parameters in the problem. At the second decision point, expected utility is calculated with respect to the conditional distribution of the parameters given the (now observed) data.

In some situations, data are collected in batches, and several decision points can be envisioned. At each decision point, the available decisions are either to stop collecting data and make a terminal decision, or to continue. Sometimes an upper limit on the number of decision points is imposed, so at the last decision point, a terminal decision must be made. These problems are called batch-sequential problems. One application is to the data-monitoring committees of a clinical trial. At each meeting a decision must be made either to stop the trial and make a treatment recommendation, or to continue the trial.

A special case of batch sequential designs are designs in which each batch is of size one. Such designs are called fully sequential.

Because at each stage of a sequential decision process decisions are optimally made by maximizing expected utility, the results of section 7.10 apply to each stage. Hence randomization is never strictly optimal. If a randomized strategy is optimal, so are each of the decisions the randomized strategy puts positive probability on.

7.11.1 *Notes*

The literature on Bayesian sequential decision making is not large; many of the analytically tractable cases are found in DeGroot (1970). An interesting special case is studied in Berry and Fristedt (1985). Computing optimal Bayesian sequential decisions can be difficult because natural methods lead to an exponential explosion in the dimension of the decision space, but Brockwell and Kadane (2003) give some methods to overcome this difficulty.

There is literature on static experimental design in a Bayesian perspective. A review of many of the analytically tractable cases is given by Chaloner and Verdinelli (1995). Other important contributions are those of Verdinelli (2000), DuMouchel and Jones (1994), Joseph (2006) and Lohr (1995).

Bayesian analysis allows the graceful incorporation of new data as it becomes available. This contrasts sharply with sampling theory methods, which are sensitive to how often and when data are analyzed in a sequential setting. This is especially critical in the design of medical experiments, in which early stopping of a clinical trial can save lives or heartache.

7.11.2 *Summary*

At each stage in a sequential decision process, optimal decisions are made by maximizing expected utility. The probability distribution used to take the expectation conditions on all the random variables whose values are known at the time of the decision, and treats as random all those still uncertain at the time of the decision.

7.11.3 *Exercise*

1. Consider the following two-stage decision problem. The investor starts at the first stage with a fortune f_0, and has log fortune as utility. At each stage there are n mutually exclusive and exhaustive events $A_1, \ldots, A_n$ that will be observed after each stage, outcomes after the second stage are independent of those of the first stage. At each stage, there are dollar tickets available for purchase on A_i for a price of $x_i > 0$, where $\sum_{i=1}^{n} x_i = 1$. The investor's probability on A_i in q_i at each stage.
   (a) Suppose the investor's fortune after the first stage is f_1. What proportions ℓ_i should he use for the second stage to purchase tickets on event A_i? What is the amount the investor will optimally spend on tickets on A_i?
   (b) Now consider the investor's problem at the first stage, when his fortune is f_0. What proportions ℓ_i should he use for the first stage to purchase tickets on event A_i? What is the amount the investor will optimally spend on tickets on A_i? If A_i occurs at the first stage, what will the investor's resulting fortune be?
   (c) Now consider both stages together. How does the outcome of the first stage affect the proportions and amounts spent on tickets at the second stage?

(d) What is the expected utility of the two-stage process, with optimal decisions made at each stage?

Chapter 8

Conjugate Analysis

The results of Chapter 7 make it clear that the central computational task in Bayesian analysis is to find the conditional distribution of the unobserved parts of the model (otherwise known as parameters θ) given the observed parts (otherwise known as data x), written in notation as $p(\theta \mid x)$. There are some models for which this computation can be done analytically, and others for which it cannot. This chapter deals with the former.

8.1 A simple normal-normal case

Suppose that you observe data $X_1, X_2, \ldots, X_n$ which you believe are independent and identically distributed with a normal distribution with mean μ (about which you are uncertain) and variance σ_0^2 (about which you are certain). Also suppose that your opinion about μ is described by a normal distribution with mean μ_1 and variance σ_1^2, where μ_1 and σ_1^2 are assumed to be known. Before proceeding, it is useful to reparametrize the normal distribution in terms of the precision $\tau = 1/\sigma^2$. Thus the data are assumed to come from a normal distribution with mean μ and precision $\tau_0 = 1/\sigma_0^2$, and your prior on μ is normal with mean μ_1 and precision $\tau_1 = 1/\sigma_1^2$. Such a reparameterization does not change the meaning of any of your statements of belief, but it does simplify some of the formulae to come.

Our task is to compute the conditional distribution of μ given the observed data $\mathbf{X} = (X_1, X_2, \ldots, X_n)$. We start with the joint distribution of μ and $\mathbf{X}$, and then divide by the marginal distribution of $\mathbf{X}$. This marginal distribution is the integral of the joint distribution, where the integral is with respect to the distribution of μ. Consequently, after integration, the marginal distribution of $X_1, \ldots, X_n$ does not involve μ.

It is a general principle, in the calculations we are about to undertake, that we may neglect factors that do not depend on the parameter whose posterior distribution we are calculating. The result is then proportional to the density in question, so at the end of the calculation, the constant of proportionality must be recovered.

Now the joint distribution of μ and $(X_1, \ldots, X_n) = \mathbf{X}$ comes to us as the conditional distribution of $\mathbf{x}$ given μ times the density of μ. Hence

$$f(\mathbf{X}, \mu) = \left(\frac{1}{\sqrt{2\pi}}\right)^n \tau_0^{n/2} e^{-\frac{\tau_0}{2}\sum(X_i-\mu)^2} \cdot \frac{(\tau_1)^{1/2}}{\sqrt{2\pi}} e^{-\frac{\tau_1}{2}(\mu-\mu_1)^2}. \tag{8.1}$$

Now the factor $\left(\frac{1}{\sqrt{2\pi}}\right)^n \tau_0^{n/2} \frac{(\tau_1)^{1/2}}{\sqrt{2\pi}}$ does not depend on μ, so we may write

$$f(\mathbf{X}, \mu) \propto e^{-Q(\mu)/2} \tag{8.2}$$

where $Q(\mu) = \tau_0 \sum_{i=1}^n (X_i - \mu)^2 + \tau_1(\mu - \mu_1)^2$.

Since $Q(\mu)$ occurs in the exponent in (8.2), to neglect a constant factor in (8.2) is equivalent to neglecting an additive factor in $Q(\mu)$. I write

$$Q(\mu) \,\Delta\, Q'\,(\mu)$$

to mean that $Q(\mu)-Q'(\mu)$ does not depend on μ. Therefore if $Q(\mu)\Delta Q'(\mu)$, then $e^{-Q(\mu)/2} \propto e^{-Q'(\mu)/2}$. I rewrite $Q(\mu)$ as follows:

$$Q(\mu) = \tau_0 \sum_{i=1}^{n} (\mu^2 - 2\mu X_i + X_i^2) + \tau_1(\mu^2 - 2\mu\mu_1 + \mu_1^2).$$

Let $Q'(\mu) = n\tau_0\mu^2 - 2\tau_0\mu \sum X_i + \tau_1\mu^2 - 2\mu\tau_1\mu_1$.

Then $Q(\mu)\Delta Q'(\mu)$ because

$$Q(\mu) - Q'(\mu) = \sum X_i^2 + \tau_1\mu_1^2$$

does not depend on μ.

Hence

$$Q(\mu)\ \Delta\ [\mu^2(n\tau_0 + \tau_1) - 2\mu(n\tau_0\overline{X} + \tau_1\mu_1)].$$

But

$$\mu^2(n\tau_0 + \tau_1) - 2\mu(n\tau_0\overline{X} + \tau_1\mu_1) = (n\tau_0 + \tau_1)\left[\mu^2 - 2\mu\left(\frac{n\tau_0\overline{X} + \tau_1\mu_1}{n\tau_0 + \tau_1}\right)\right].$$

To simplify the notation, let

$$\tau_2 = n\tau_0 + \tau_1 \tag{8.3a}$$

and

$$\mu_2 = \frac{n\tau_0\overline{X} + \tau_1\mu_1}{n\tau_0 + \tau_1}. \tag{8.3b}$$

Then in this notation,

$$Q(\mu)\ \Delta\ \tau_2[\mu^2 - 2\mu\mu_2].$$

The material in square brackets is a perfect square, except that it needs $\tau_2\mu_2^2$, which does not depend on μ. Therefore we may write

$$Q(\mu)\ \Delta\ \tau_2(\mu - \mu_2)^2.$$

Returning to (8.2), we may then write

$$f(\mathbf{X}, \mu) \propto e^{-\tau_2(\mu-\mu_2)^2/2}. \tag{8.4}$$

We can recognize this as the form of a normal distribution for μ, with mean μ_2 and precision τ_2. We therefore know that the missing constant is $\frac{\tau_2^{1/2}}{\sqrt{2\pi}}$.

Now let's return to (8.3) to examine the result found in (8.4). Equation (8.3a) says that the posterior precision τ_2 of μ is the sum of the prior precision τ_1 and the "data precision" $n\tau_0$. Thus if the prior precision τ_1 is small compared to the data precision $n\tau_0$, then the posterior precision is dominated by $n\tau_0$. Conversely, if the prior precision τ_1 is large compared to the data precision $n\tau_0$, then the posterior precision is dominated by τ_1. The result of data collection in this example is always to increase the precision with respect to which μ is known.

Equation (8.3b) can be revealingly re-expressed as

$$\mu_2 = \left(\frac{n\tau_0}{n\tau_0 + \tau_1}\right)\overline{X} + \left(\frac{\tau_1}{n\tau_0 + \tau_1}\right)\mu_1. \tag{8.5}$$

Here μ_2 is a linear combination of $\overline{X}$ and μ_1, where the weights are non-negative and sum to one (such a combination is called a convex combination). Indeed we may say that μ_2 is a precision-weighted average of $\overline{X}$ and μ_1. The intuition is that two information sources are being blended together here, the prior and the sample. The mean of the posterior distribution, μ_2, is a blend of the data information, $\overline{X}$, and the prior mean μ_1, where the weights are proportional to the precisions of the two data sources. Again, if the prior precision τ_1 is small compared to the data precision $n\tau_0$, then the posterior mean μ_2 will be close to $\overline{X}$. Conversely if the prior precision τ_1 is large compared to the data precision $n\tau_0$, then the posterior mean μ_2 will be close to the prior mean μ_1.

Another feature of the calculation is that the data $\mathbf{X}$ enter the result only through the sample size n and the data sum $\sum_{i=1}^{n} X_i$, or equivalently its mean $\overline{X}$. Such a data summary is called a sufficient statistic, because, under the assumptions made, all you need to know about the data is summarized in it.

With respect to the normal likelihood where only the mean is uncertain, the family of normal prior opinions is said to be closed under sampling. This means that whatever the data might be, the posterior distribution is also in the same family. The family of normal distributions is not unique in this respect. The following other families are also closed under sampling:

(i) The family of all prior distributions on μ.

(ii) Each of the opinionated prior distributions that puts probability one on some particular value of μ, say μ_0. In this case, whatever the data turn out to be, the posterior distribution will still put probability one on μ_0. This corresponds to taking τ_0 to be infinity.

(iii) If the normal density for the prior is multiplied by any non-negative function $g(\mu)$ (it has to be positive somewhere), that factor would also be a factor in the posterior. Hence $g(\mu)$ times a normal prior results in $g(\mu)$ times a normal posterior, so it is in the same family. (Indeed (i) and (ii) above can be regarded as special cases of (iii)).

Despite this lack of uniqueness of the family closed under sampling, it is convenient to single out the family of normal prior distributions for μ, and to refer to the pair of likelihood and prior as a conjugate pair.

It should also be emphasized that the calculation depends critically on the distributional assumptions made. Nonetheless, calculations like this one, where they are possible, are useful both for themselves and as an intuitive background for calculations in more complicated cases.

In finding a conjugate pair of likelihood and prior, there should not be implied coercion on you to believe that your data have the form of a particular likelihood (here normal), nor, if they do, that your prior must be of a particular form (here also normal). You are entitled to your opinions, whatever they may be.

8.1.1 Summary

If $X_1, \ldots, X_n$ are believed to be conditionally independent and identically distributed, with a normal distribution with mean μ and precision τ_0, where μ is uncertain but τ_0 is known with certainty, and if μ itself is believed to have a normal distribution with mean μ_1 and precision τ_1 (both known), then the posterior distribution of μ is again normal, with mean μ_2 given in (8.3b) and precision τ_2 given in (8.3a).

8.1.2 Exercises

1. Vocabulary. Explain in your own words the meaning of:
   (a) precision

(b) sufficient statistic

(c) family closed under sampling

(d) conjugate likelihood and prior

2. Suppose your prior on μ is well represented by a normal distribution with mean 2 and precision 1. Also suppose you observe a normal random variable with mean μ and precision 2. Suppose that observation turns out to have the value 3. Compute the posterior distribution that results from these assumptions.

3. Do the same problem, except that the observation now has the value 300.

4. Compare your answers to questions 2 and 3 above. Do you find them equally satisfactory? Why or why not?

8.2 A multivariate normal-normal case with known precision

We now consider a generalization of the calculation in section 8.1 to multivariate normal distributions. In this case, the precision, which in the univariate case was a positive number, now becomes a positive-definite matrix, the inverse of the covariance matrix. Thus we suppose that the data now consist of n vectors, each of length $p, \mathbf{X}_1, \ldots, \mathbf{X}_n$. These vectors are assumed to be conditionally independent and identically distributed with a p-dimensional normal distribution having a p-dimensional mean $\boldsymbol{\mu}$ about which you are uncertain, and a $p \times p$ precision matrix $\boldsymbol{\tau}_0$ which you are certain about. Your prior opinion about $\boldsymbol{\mu}$ is represented by a p-dimensional normal distribution with mean $\boldsymbol{\mu}_1$ and $p \times p$ precision matrix τ_1. Again we wish to find the posterior distribution of μ given the data.

We begin, as before, by writing down the joint density of $\boldsymbol{\mu}$ and the data $\mathbf{X} = (\mathbf{X}_1, \ldots, \mathbf{X}_n)$. This joint density is

$$f(X, \boldsymbol{\mu}) = \left(\frac{1}{\sqrt{2\pi}}\right)^{pn} \mid \tau_0 \mid^{n/2} e^{-\frac{1}{2}\sum_{i=1}^{n}(\mathbf{X}_i - \boldsymbol{\mu})'\tau_0(\mathbf{X}_i - \boldsymbol{\mu})} \cdot \left(\frac{1}{\sqrt{2\pi}}\right)^{p} \mid \tau_1 \mid^{\frac{1}{2}} e^{-\frac{1}{2}(\boldsymbol{\mu} - \boldsymbol{\mu}_1)'\tau_1(\boldsymbol{\mu} - \boldsymbol{\mu}_1)}. \tag{8.6}$$

Expression (8.6) is a straight-forward generalization of (8.1). Again the constant

$$\left(\frac{1}{\sqrt{2\pi}}\right)^{pn} \mid \tau_0 \mid^{n/2} \left(\frac{1}{\sqrt{2\pi}}\right)^{p} \mid \tau_1 \mid^{\frac{1}{2}}$$

does not involve μ, and may be absorbed in a constant of proportionality. Thus we have

$$f(X, \boldsymbol{\mu}) \propto e^{-\frac{1}{2}Q(\boldsymbol{\mu})} \tag{8.7}$$

where $Q(\boldsymbol{\mu}) = \sum_{i=1}^{n}(\mathbf{X}_i - \boldsymbol{\mu})'\tau_0(\mathbf{X}_i - \boldsymbol{\mu}) + (\boldsymbol{\mu} - \boldsymbol{\mu}_1)'\tau_1(\boldsymbol{\mu} - \boldsymbol{\mu}_1)$, which is a generalization of (8.2). Using the same Δ notation as before,

$$\begin{aligned}
Q(\boldsymbol{\mu}) = \quad & \Sigma_{i=1}^{n}(\boldsymbol{\mu} - \mathbf{X}_i)'\tau_0(\boldsymbol{\mu} - \mathbf{X}_i) + (\boldsymbol{\mu} - \boldsymbol{\mu}_1)'\tau_1(\boldsymbol{\mu} - \boldsymbol{\mu}_1) \\
= \quad & n\boldsymbol{\mu}'\tau_0\boldsymbol{\mu} - \boldsymbol{\mu}'\tau_0\Sigma\mathbf{X}_i - \Sigma\mathbf{X}_i'\tau_0\boldsymbol{\mu} \\
+ \quad & \Sigma\mathbf{X}_i'\tau_0\mathbf{X}_i + \ \boldsymbol{\mu}'\tau_1\boldsymbol{\mu} - \boldsymbol{\mu}'\tau_1\boldsymbol{\mu}_1 - \boldsymbol{\mu}_1'\tau_1\boldsymbol{\mu} + \boldsymbol{\mu}_1'\tau_1\boldsymbol{\mu}_1 \\
\Delta \quad & [\boldsymbol{\mu}'(n\tau_0)\boldsymbol{\mu} - \boldsymbol{\mu}'\tau_0\Sigma\mathbf{X}_i - \Sigma\mathbf{X}_i'\tau_0\boldsymbol{\mu} + \boldsymbol{\mu}'\tau_1\boldsymbol{\mu} - \boldsymbol{\mu}'\tau_1\boldsymbol{\mu}_1 - \boldsymbol{\mu}_1'\tau_1\boldsymbol{\mu}] \\
= \quad & \boldsymbol{\mu}'(n\tau_0 + \tau_1)\boldsymbol{\mu} - \boldsymbol{\mu}'\boldsymbol{\gamma} - \boldsymbol{\gamma}'\boldsymbol{\mu} = Q_1(\boldsymbol{\mu}) \\
\text{where } \boldsymbol{\gamma} = \quad & \tau_0\sum_{i=1}^{n}\overline{\mathbf{X}}_i + \tau_1\boldsymbol{\mu}_1 = n\tau_0\overline{\mathbf{X}} + \tau_1\boldsymbol{\mu}_1.
\end{aligned}$$

Let

$$\tau_2 = n\tau_0 + \tau_1 \tag{8.8a}$$

and

$$\boldsymbol{\mu}_2 = \tau_2^{-1}\boldsymbol{\gamma}, \tag{8.8b}$$

and compute

$$\begin{aligned} Q_1(\boldsymbol{\mu}) - (\boldsymbol{\mu} - \boldsymbol{\mu}_2)'\tau_2(\boldsymbol{\mu} - \boldsymbol{\mu}_2) =& [\boldsymbol{\mu}'\tau_2\boldsymbol{\mu} - \boldsymbol{\mu}'\boldsymbol{\gamma} - \boldsymbol{\gamma}'\boldsymbol{\mu}] - \boldsymbol{\mu}'\tau_2\boldsymbol{\mu} + \boldsymbol{\mu}'\tau_2\tau_2^{-1}\boldsymbol{\gamma} \\ &+ \boldsymbol{\gamma}'\tau_2^{-1}\tau_2\boldsymbol{\mu} - \boldsymbol{\mu}_2'\tau_2\boldsymbol{\mu}_2 \\ =& - \boldsymbol{\mu}_2'\tau_2\boldsymbol{\mu}_2 \end{aligned}$$

which does not depend on μ. Therefore (implicitly using transitivity of Δ),

$$Q(\boldsymbol{\mu}) \;\; \Delta \;\; (\boldsymbol{\mu} - \boldsymbol{\mu}_2)'\tau_2(\boldsymbol{\mu} - \boldsymbol{\mu}_2).$$

Returning to (8.7) we may write

$$f(X, \boldsymbol{\mu}) \propto e^{-\frac{1}{2}(\boldsymbol{\mu}-\boldsymbol{\mu}_2)'\tau_2(\boldsymbol{\mu}-\boldsymbol{\mu}_2)} \tag{8.9}$$

which we recognize as a multivariate normal distribution for $\boldsymbol{\mu}$, with mean $\boldsymbol{\mu}_2$ and precision matrix τ_2. So the missing constant is $\left(\frac{|\tau_2|}{(2\pi)^p}\right)^{1/2}$.

I hope that the analogy between this calculation and the univariate one is obvious to the reader. The only difference is that in completing the square for $Q(\boldsymbol{\mu})$, care must be taken to respect the fact that matrix multiplication does not commute. But the basic argument is exactly the same.

Again the precision matrix of the posterior distribution, τ_2, is the sum of the precision matrices of the prior, τ_1, and of the data, $n\tau_0$. Furthermore the posterior mean, $\boldsymbol{\mu}_2$, can be seen to be the matrix convex combination of $\overline{\boldsymbol{X}}$, the data mean, and $\boldsymbol{\mu}$, the prior mean, with weights $(n\tau_0 + \tau_1)^{-1}n\tau_0$ and $(n\tau_0 + \tau_1)^{-1}\tau_1$, respectively.

Again, $\overline{\boldsymbol{X}}$, which is a p-dimensional vector, and is the average, component-wise, of the observations, is a sufficient statistic, when combined with the sample size n.

8.2.1 Shrinkage and Stein's Paradox

To understand Stein's Paradox from Stein's perspective, one has to enter a rather different statistical framework, in which parameters are fixed but unknown, and have no distributions, while data are random, even after they are observed. (My comments on this framework are in Chapter 12).

Suppose the observation of p independent normally distributed random variables is

$$X_i \sim N(\mu_i, 1)\ , i = 1, \ldots, p.$$

The problem is to find good estimates for the μ's, which requires specification of what is meant by "good." What is mean by "good" in this context is the use of the loss function

$$L(\hat{\boldsymbol{\mu}}, \boldsymbol{\mu}) = ||\hat{\boldsymbol{\mu}} - \mu|| \equiv \sum_{i=1}^{p}(\hat{\mu}_i - \mu_i)^2. \tag{8.10}$$

Attention is focused on the risk function

$$R(\hat{\boldsymbol{\mu}}, \boldsymbol{\mu}) = E\{L(\hat{\boldsymbol{\mu}} - \boldsymbol{\mu})\}, \tag{8.11}$$

where the expectation is taken with respect to the data distribution as it impacts $\hat{\boldsymbol{\mu}}$. In this respect, it differs from all the uses of the expectation operator in this book up to this point.

How does (8.11) guide a choice of a "good" estimator? An estimator $\hat{\boldsymbol{\mu}}$ is said to **dominate** another estimator $\tilde{\boldsymbol{\mu}}$ if both

1. $R(\hat{\boldsymbol{\mu}}, \boldsymbol{\mu}) \leq R(\tilde{\boldsymbol{\mu}}, \boldsymbol{\mu})$ for all $\boldsymbol{\mu}$.
2. There is some $\boldsymbol{\mu}$ such that $R(\hat{\boldsymbol{\mu}}, \boldsymbol{\mu}) < R(\tilde{\boldsymbol{\mu}}, \boldsymbol{\mu})$.

If $\tilde{\boldsymbol{\mu}}$ is dominated, it is said to be **inadmissible**. The theory recommends restricting attention to admissible estimators, those that are undominated. However, admissibility is a weak criterion. For example, consider the constant estimator $\hat{\boldsymbol{\mu}}(\mathbf{X}) \equiv 3$, that takes the value 3 whatever the data $\mathbf{X}$ might be. This estimator is admissible, because if $\boldsymbol{\mu} = \mathbf{3}$, one can't do better.

When $p = 1$, it had been shown that $\hat{\boldsymbol{\mu}}_1 = x_1$ is admissible. Stein (1956) showed that when $p = 2, \hat{\boldsymbol{\mu}}_i = x_i, i = 1, 2$ is admissible. However, for $p \geq 3$ the estimator $\hat{\mu}_i = x_i$, $i = 1, \ldots, p$ is inadmissible. James and Stein (1961) showed in particular, that the estimator, for any $\boldsymbol{\mu}_0$,

$$\hat{\boldsymbol{\mu}}_{JS} = \boldsymbol{\mu}_0 \left(\frac{p-2}{||\mathbf{x} - \boldsymbol{\mu}_0||} \right) + \mathbf{X} \left(1 - \frac{p-2}{||\mathbf{X} - \boldsymbol{\mu}_0||} \right) \tag{8.12}$$

dominates $\hat{\boldsymbol{\mu}} = \mathbf{X}$. Hence, $\hat{\boldsymbol{\mu}}$ is inadmissible. This was very surprising at the time, which is why it is called Stein's Paradox.

When $\frac{p-2}{||x - \mu_0||} \leq 1$, (8.12) is in the form of a convex combination of $\boldsymbol{\mu}_0$ and $\mathbf{X}$. In particular, when $\boldsymbol{\mu}_0$ is chosen to be zero, (8.12) shrinks $\mathbf{X}$ toward $\mathbf{0}$. A positive part estimator

$$\hat{\boldsymbol{\mu}}_{pp} = \mu_0 \left(1 - \left(1 - \frac{p-2}{||X - \mu_0||} \right)^+ \right) + \mathbf{X} \left(1 - \frac{p-2}{||\mathbf{X} - \boldsymbol{\mu}_0||} \right)^+, \tag{8.13}$$

where $x^+ = \max(0, x)$ was also developed. The positive part estimator is a convex combination of $\boldsymbol{\mu}_0$ and $\mathbf{x}$.

To put the assumptions used in Stein's work in the notation of Section 8.2, consider the special case in which $n = 1$ and $\tau_0 = I$. Additionally, suppose that the prior on $\boldsymbol{\mu}$ has mean $\boldsymbol{\mu}_1 = (\mu_{1,1}, \ldots \mu_{1,p})$ and $\tau_1 = D$, a diagonal matrix with elements $d_i, i = 1, \ldots, p$. Then, from (8.8a),

$$\tau_2 = I + D, \tag{8.14}$$

a diagonal matrix with elements $d_i + 1, i = 1, \ldots, p$, so τ_2^{-1} is a diagonal matrix with elements $1/(d_i + 1)$, $i = 1, \ldots, p$. Additionally, $\boldsymbol{\gamma}$ is a vector of length p with elements

$$\gamma_i = x_i + d_i \mu_{1,i} \ , i = 1, \ldots p.$$

Finally, then (8.8b) reduces to

$$\mu_{2,i} = \frac{x_i}{1 + d_i} + \frac{d_i \mu_{1,i}}{1 + d_i}, \tag{8.15}$$

a convex combination of x_i and $\mu_{1,i}$. In particular, when $\mu_{1,i} = 0$, (8.15) shows shrinkage of $\mathbf{X}$ toward $\mathbf{0}$.

This analysis delivers a posterior distribution for $\boldsymbol{\mu}_2$, so any symmetric loss function favors the mean, $\boldsymbol{\mu}_2$. It does not rely on admissibility, and has no dimension discontinuity. Shrinkage is thus shown to be a consequence of Bayesian theory.

8.2.2 *Summary*

If $\mathbf{X}_1, \mathbf{X}_2, \ldots, \mathbf{X}_n$ are believed to be conditionally independent and identically distributed, with a p-dimensional normal distribution with mean $\boldsymbol{\mu}$ and precision matrix τ_0, where $\boldsymbol{\mu}$ is uncertain but τ_0 is known with certainty, and if $\boldsymbol{\mu}$ itself is believed to have a normal distribution with mean $\boldsymbol{\mu}_1$ and precision τ_1 (both known), then the posterior distribution of $\boldsymbol{\mu}$ is again normal, with mean $\boldsymbol{\mu}_2$ given in (8.8b), and precision matrix τ_2 given in (8.8a).

8.2.3 *Exercises*

1. Prove that the result derived in section 8.1 is a special case of the result derived in section 8.2.
2. Suppose your prior on $\boldsymbol{\mu}$ (which is two-dimensional) is normal, with mean $(2, 2)$ and precision matrix I, and suppose you observe a normal random variable with mean $\boldsymbol{\mu}$, and precision matrix $\left(\begin{smallmatrix} 2 & 0 \\ 0 & 2 \end{smallmatrix}\right)$. Suppose the observation is (3, 300).
   (a) Compute the posterior distribution on $\boldsymbol{\mu}$ that results from these assumptions.
   (b) Compare the results of this calculation with those you found in section 8.1.2, problems 2 and 3.

8.3 The normal linear model with known precision

The normal linear model is one of the most heavily used and popular models in statistics. The model is given by

$$\mathbf{y} = X\boldsymbol{\beta} + \mathbf{e} \tag{8.16}$$

where $\mathbf{y}$ is an $n \times 1$ vector of observations, X is an $n \times p$ matrix of known constants, $\boldsymbol{\beta}$ is a $p \times 1$ vector of coefficients and $\mathbf{e}$ is an $n \times 1$ vector of error terms. We will suppose for the purpose of this section that $\mathbf{e}$ has a normal distribution with zero mean and known precision matrix τ_0. Additionally, we will assume that $\boldsymbol{\beta}$ has a prior distribution taking the form of a p-dimensional normal distribution with mean $\boldsymbol{\beta}_1$ and precision matrix τ_1, both known.

Before we proceed to the analysis of the model, it is useful to mention some special cases. When the elements of the matrix X are restricted to take the values 0 and 1, the model (8.16) is often called an analysis of variance model. When the X's are more general, (8.16) is often called a linear regression model.

The joint distribution of $\mathbf{y}$ and $\boldsymbol{\beta}$ can be written

$$\begin{aligned} f(\mathbf{y}, \boldsymbol{\beta}) = &\left(\frac{1}{\sqrt{2\pi}}\right)^n \mid \tau_0 \mid^{1/2} e^{-\frac{1}{2}(y - X\beta)'\tau_0(y - X\beta)} \\ &\cdot \left(\frac{1}{\sqrt{2\pi}}\right)^p \mid \tau_1 \mid^{1/2} e^{-\frac{1}{2}(\beta-\beta_1)'\tau_1(\beta-\beta_1)}. \end{aligned} \tag{8.17}$$

Once again we recognize $(\frac{1}{\sqrt{2\pi}})^n \mid \tau_0 \mid^{1/2} (\frac{1}{\sqrt{2\pi}})^p \mid \tau_1 \mid^{1/2}$ as a constant that need not be carried. Thus we can write

$$f(\mathbf{y}, \boldsymbol{\beta}) \propto e^{-\frac{1}{2}Q(\boldsymbol{\beta})} \tag{8.18}$$

where

$$
\begin{aligned}
Q(\boldsymbol{\beta}) =&(\mathbf{y} - X\boldsymbol{\beta})'\tau_0(\mathbf{y} - X\boldsymbol{\beta}) + (\boldsymbol{\beta} - \boldsymbol{\beta}_1)'\tau_1(\boldsymbol{\beta} - \boldsymbol{\beta}_1) \\
=&\big[\boldsymbol{\beta}'X'\tau_0 X\boldsymbol{\beta} - \boldsymbol{\beta}'X'\tau_0\mathbf{y} - \mathbf{y}'\tau_0 X\boldsymbol{\beta} + \mathbf{y}'\tau_0\mathbf{y} \\
&+\boldsymbol{\beta}'\tau_1\boldsymbol{\beta} - \boldsymbol{\beta}'\tau_1\boldsymbol{\beta}_1 - \boldsymbol{\beta}_1'\tau_1\boldsymbol{\beta} + \boldsymbol{\beta}_1'\tau_1\boldsymbol{\beta}_1\big] \\
&\Delta\ \big[\boldsymbol{\beta}'(X'\tau_0 X + \tau_1)\boldsymbol{\beta} - \boldsymbol{\beta}'(X'\tau_0\mathbf{y} + \tau_1\boldsymbol{\beta}_1) \\
&-(\boldsymbol{\beta}_1'\tau_1 + \mathbf{y}'\tau_0 X)\boldsymbol{\beta}\big] \\
=&\boldsymbol{\beta}'\tau_2\boldsymbol{\beta} - \boldsymbol{\beta}'\boldsymbol{\gamma} - \boldsymbol{\gamma}'\boldsymbol{\beta} \\
&\Delta\ \boldsymbol{\beta}'\tau_2\boldsymbol{\beta} - \boldsymbol{\beta}'\boldsymbol{\gamma} - \boldsymbol{\gamma}'\boldsymbol{\beta} + \boldsymbol{\gamma}'\tau_2^{-1}\boldsymbol{\gamma} \\
=&(\boldsymbol{\beta} - \boldsymbol{\beta}_2)'\tau_2(\boldsymbol{\beta} - \boldsymbol{\beta}_2)
\end{aligned}
\tag{8.19}
$$

where

$$\tau_2 = X'\tau_0 X + \tau_1, \tag{8.20a}$$

$$\boldsymbol{\gamma} = X'\tau_0\mathbf{y} + \tau_1\boldsymbol{\beta}_1 \tag{8.20b}$$

and

$$\boldsymbol{\beta}_2 = \tau_2^{-1}\boldsymbol{\gamma}. \tag{8.20c}$$

Therefore the algebra of the last section can be used once again, leading to the conclusion that $\boldsymbol{\beta}$ has a normal posterior distribution with precision matrix τ_2 and mean

$$\boldsymbol{\beta}_2 = \tau_2^{-1}\boldsymbol{\gamma} = (X'\tau_0 X + \tau_1)^{-1}(X'\tau_0\mathbf{y} + \tau_1\boldsymbol{\beta}_1). \tag{8.21}$$

Once again, the posterior precision matrix τ_2 is the sum of the data precision matrix $X'\tau_0 X$ and the prior precision matrix τ_1.

To understand (8.21), consider the special case in which $\tau_0 = \mathbf{I}$ and $\tau_1 = \mathbf{0}$. Then

$$\beta_2 = (X'X)^{-1}X'y,$$

the classical least squares estimate. See Section 2.12.2.

To interpret the mean, let $\hat{\boldsymbol{\beta}} = (X'\tau_0 X)^{-1}X'\tau_0 y$. [In other literature, $\hat{\beta}$ is called the Aitken estimator of β.] Substituting (8.16) yields

$$
\begin{aligned}
\hat{\boldsymbol{\beta}} =&(X'\tau_0 X)^{-1}X'\tau_0(X\boldsymbol{\beta} + \mathbf{e}) \\
=&(X'\tau_0 X)^{-1}X'\tau_0 X\boldsymbol{\beta} + (X'\tau_0 X)^{-1}X'\tau_0\mathbf{e} \\
=&\boldsymbol{\beta} + (X'\tau_0 X)^{-1}X'\tau_0\mathbf{e}.
\end{aligned}
$$

The sampling expectation of $\hat{\beta}$ is then β, and the variance-covariance matrix of $\hat{\beta}$ is

$$
\begin{aligned}
E(\hat{\boldsymbol{\beta}} - \boldsymbol{\beta})(\hat{\boldsymbol{\beta}} - \boldsymbol{\beta})' =&(X'\tau_0 X)^{-1}X'\tau_0 E(\mathbf{e}\mathbf{e}')\tau_0 X(X'\tau_0 X)^{-1} \\
=&(X'\tau_0 X)^{-1}X'\tau_0\tau_0^{-1}\tau_0 X(X'\tau_0 X)^{-1} \\
=&(X'\tau_0 X)^{-1}.
\end{aligned}
$$

Hence the precision-matrix of $\hat{\beta}$ is $X'\tau_0 X$. Thus I may rewrite $\boldsymbol{\beta}_2$ as

$$
\begin{aligned}
\boldsymbol{\beta}_2 =&(X'\tau_0 X + \tau_1)^{-1}\left[(X'\tau_0 X)(X'\tau_0 X)^{-1}X\tau_0\mathbf{y} + \tau_1\boldsymbol{\beta}_1\right] \\
=&(X'\tau_0 X + \tau_1)^{-1}\left[(X'\tau_0 X)\hat{\boldsymbol{\beta}} + \tau_1\boldsymbol{\beta}_1\right],
\end{aligned}
\tag{8.22}
$$

which displays $\boldsymbol{\beta}_2$ as a matrix precision-weighted average of $\hat{\boldsymbol{\beta}}$ and $\boldsymbol{\beta}_1$. For this model $\hat{\boldsymbol{\beta}}$, or equivalently $X'\tau_0\mathbf{y}$, is a vector of sufficient statistics.

One of the issues in linear models is the possibility of lack of identification of the parameters, also known as estimability. To take a simple example, suppose we were to observe $Y_1, \ldots, Y_n$ which are conditionally independent, identically distributed, and have a mean $\beta_1 + \beta_2$ and precision 1. This is a special case of (8.16) in which $p = 2$, the matrix X is $n \times 2$ and has 1 in each entry, and τ_0 is the identity matrix. The problem is that the classical estimate,

$$\hat{\boldsymbol{\beta}} = (X'X)^{-1}X'\mathbf{y} \tag{8.23}$$

cannot be computed, since $X'X$ is singular (multiply it by the vector $(1, -1)'$ to see this). Furthermore, it is clear that while the data are informative about $\beta_1 + \beta_2$, they are not informative for $\beta_1 - \beta_2$. What happens to a Bayesian analysis in such a case? Nothing. Even if $X'X$ does not have an inverse, the matrix $X'\tau_0X + \tau_1$ does have an inverse, because τ_1 is positive definite and $X'\tau_0X$ is positive semi-definite. Thus (8.21) can be computed nonetheless.

In directions such as $\beta_1 - \beta_2$, the posterior is the prior, because the likelihood is flat there. This observation is not special to the normal likelihood (although most classical treatments of identification focus on the normal likelihood). In general, a model is said to lack identification if there are parameter values θ and θ' such that $f(x \mid \theta) = f(x \mid \theta')$ for all possible data points x. In this case, the data cannot tell θ apart from θ'. In the example, $\theta = (\beta_1, \beta_2)$ cannot be distinguished from $\theta' = (\beta_1 + c, \beta_2 - c)$ for any constant c. However, you have a prior distribution and a likelihood. The product of them determines the joint distribution, and hence the conditional distribution of the parameters given the data. Lack of identification does not disturb this chain of reasoning.

I should also mention the issue of multicollinearity, which is a long name for the situation that $X'X$, while not singular, is close to singular. This is not an issue for Bayesians, because again τ_1 in (8.21) creates the needed numerical stability.

8.3.1 Summary

If the likelihood is given by (8.16), with conditionally normally distributed errors with mean 0 and known precision matrix τ_0, and if the prior on β is normal with mean β_1 and precision matrix τ_1, then the posterior on β is again normal with mean given by (8.21) and precision matrix given by (8.20b).

Lack of identification and multicollinearity are not issues in the Bayesian analysis of linear models.

8.3.2 Further reading

There is an enormous literature on the linear model, most of it from a sampling theory perspective. Some Bayesian books dealing with aspects of it include Box and Tiao (1973), O'Hagan and Foster (2004), Raiffa and Schlaifer (1961) and Zellner (1971). For more on identification from a Bayesian perspective, see Kadane (1974), Drèze (1974) and Kaufman (2001).

8.3.3 Exercises

1. Vocabulary. State in your own words the meaning of:
   (a) normal linear model
   (b) identification
   (c) multicollinearity

(d) linear regression
(e) analysis of variance

2. Write down the constant for the posterior distribution for $\boldsymbol{\beta}$, which was found in (8.18) and (8.19) to be proportional to $e^{-\frac{1}{2}(\boldsymbol{\beta}-\boldsymbol{\beta}_2)'\tau_2(\boldsymbol{\beta}-\boldsymbol{\beta}_2)}$.

8.4 The gamma distribution

A typical move in applied mathematics when an intractable problem is found is to give it a name, study its properties, and then redefine "tractable" to include the formerly intractable problem. We have already seen an example of this process in the use of Φ as the cumulative distribution function of the normal distribution in section 6.3. We're about to see a second example.

The gamma function is defined as follows:

$$\Gamma(\alpha) = \int_0^\infty e^{-x} x^{\alpha-1} dx \tag{8.24}$$

defined for all positive real numbers α. Because e^{-x} converges to zero faster than any power of x, this integral converges at infinity. For $\alpha > 0$, it also behaves properly at zero.

To study its properties, we need to use integration by parts. To remind you what that's about, recall that if $u(x)$ and $v(x)$ are both functions of x, then

$$\frac{d}{dx} u(x)v(x) = u(x)\frac{dv(x)}{dx} + v(x)\frac{du(x)}{dx}.$$

Integrating this equation with respect to x, we get

$$u(x)v(x) = \int udv + \int vdu,$$

or, equivalently,

$$\int udv = uv - \int vdu.$$

Applying this to the gamma function, let $u = x^{\alpha-1}$ and $dv = e^{-x}dx$. Then, assuming $\alpha > 1$,

$$\begin{aligned}\Gamma(\alpha) = \int_0^\infty e^{-x} x^{\alpha-1} dx &= -e^{-x} x^{\alpha-1} \Big|_0^\infty + (\alpha - 1) \int_0^\infty e^{-x} x^{\alpha-2} dx. \\ =&(\alpha-1)\Gamma(\alpha-1).\end{aligned} \tag{8.25}$$

Additionally,

$$\Gamma(1) = \int_0^\infty e^{-x} dx = -e^{-x} \Big|_0^\infty = 1.$$

Therefore when $\alpha > 1$ is an integer,

$$\Gamma(\alpha) = (\alpha - 1)! \tag{8.26}$$

Thus the gamma function can be seen as a generalization of the factorial function to all positive real numbers.

In the gamma function, let $y = x/\beta$. Then

$$\Gamma(\alpha) = \int_0^\infty (\beta y)^{\alpha-1} e^{-\beta y} \cdot \beta dy = \beta^\alpha \int_0^\infty y^{\alpha-1} e^{-\beta y} dy. \tag{8.27}$$

Therefore the function

$$f(y \mid \alpha, \beta) = \frac{\beta^\alpha}{\Gamma(\alpha)} y^{\alpha-1} e^{-\beta y} \tag{8.28}$$

is non-negative for $y > 0$ and integrates to 1 for all positive values of α and β. It therefore can be considered a probability density of a continuous random variable, and is called the gamma distribution with parameters α and β.

The moments of the gamma distribution are easily found:

$$\begin{aligned} E(X^k) &= \int_0^\infty x^k \frac{\beta^\alpha}{\Gamma(\alpha)} x^{\alpha-1} e^{-\beta x} dx \\ &= \frac{\beta^\alpha}{\Gamma(\alpha)} \int_0^\infty x^{k+\alpha-1} e^{-\beta x} dx = \frac{\beta^\alpha}{\Gamma(\alpha)} \frac{\Gamma(\alpha+k)}{\beta^{\alpha+k}} = \frac{\Gamma(\alpha+k)}{\Gamma(\alpha)\beta^k}. \end{aligned} \tag{8.29}$$

Therefore

$$E(X) = \alpha/\beta \tag{8.30}$$

$$E(X^2) = \frac{\alpha(\alpha+1)}{\beta^2} \tag{8.31}$$

and

$$V(X) = E(X^2) - (E(X))^2 = \frac{\alpha(\alpha+1)}{\beta^2} - (\alpha/\beta)^2 = \alpha/\beta^2. \tag{8.32}$$

The special case when $\alpha = 1$ is the exponential distribution, often used as a starting place for analyzing life-time distributions. The special case in which $\alpha = n/2$ and $\beta = 1/2$ is called the chi-square distribution with n degrees of freedom.

Now suppose that $\mathbf{X} = (X_1, \ldots, X_n)$ are conditionally independent and identically distributed, and have a normal distribution with known mean μ_0 and precision τ, about which you are uncertain. Also suppose that your opinion about τ is modeled by a gamma distribution with parameters α and β. Then the joint distribution of $\mathbf{X}$ and τ is

$$\begin{aligned} f(X_1, \ldots, X_n, \tau) = \left(\frac{1}{\sqrt{2\pi}}\right)^n \tau^{n/2} e^{-(\tau/2)\sum_{i=1}^n (X_i-\mu_0)^2} \\ \cdot \frac{\beta^\alpha}{\Gamma(\alpha)} \tau^{\alpha-1} e^{-\beta\tau}. \end{aligned} \tag{8.33}$$

Now we recognize $(\frac{1}{\sqrt{2\pi}})^n \frac{\beta^\alpha}{\Gamma(\alpha)}$ as a constant not involving τ. The remainder of (8.33) is

$$f(\mathbf{X}, \tau) \propto \tau^{\alpha+n/2-1} e^{-\tau\left[\beta+\sum_{i=1}^n (X_i-\mu_0)^2/2\right]}. \tag{8.34}$$

$$\text{Let } \alpha_1 = \alpha + n/2 \tag{8.35a}$$

$$\text{and} \quad \beta_1 = \beta + \sum_{i=1}^n (X_i - \mu_0)^2/2. \tag{8.35b}$$

Then (8.34) can be rewritten as

$$f(\mathbf{X}, \tau) \propto \tau^{\alpha_1-1} e^{-\beta_1\tau}, \tag{8.36}$$

and we recognize the distribution as a gamma distribution with parameters α_1 and β_1. Thus the gamma family is conjugate to the normal distribution when the mean is known but the precision is uncertain.

8.4.1 Summary

This section introduces the gamma function in (8.24) and the gamma distribution in (8.28).

If $X_1, \ldots, X_n$ are believed to be conditionally independent and identically distributed, with a normal distribution with mean μ_0 and precision τ, where μ_0 is known with certainty but τ is uncertain, and if τ is believed to have a gamma distribution with parameters α and β (both known), there the posterior distribution of τ is again gamma, with parameters α_1 given by (8.35a) and β_1 given by (8.35b).

8.4.2 Exercises

1. Vocabulary. State in your own words the meaning of:
   (a) Gamma function
   (b) Gamma distribution
   (c) Exponential distribution
   (d) Chi-square distribution
2. Find the constant for the distribution in (8.36).
3. Consider the density $e^{-x}, x > 0$ of the exponential distribution.
   (a) Find its moment generating function.
   (b) Find its nth moment.
   (c) Conclude that $\Gamma(n+1) = n!$
4. Suppose X has a standard normal distribution. Then $E(X) = 0$ and $\text{Var}(X) = 1$, so $E(X^2) = 1$. So

   $$1 = \frac{1}{\sqrt{2\pi}} \int_{-\infty}^{\infty} x^2 e^{-x^2/2} dx = \frac{2}{\sqrt{2\pi}} \int_{0}^{\infty} x^2 e^{-x^2/2} dx.$$

   Substitute $y = x^2/2$ in this integral to evaluate $\Gamma(3/2)$.

8.4.3 Reference

I highly recommend the book by Artin (1964) on the gamma function. It's magic.

8.5 The univariate normal distribution with uncertain mean and precision

Given the result of section 8.1, that when the precision of a normal distribution is known, a normal distribution on μ is conjugate, and the result of section 8.4, that when the mean is known, a gamma distribution on τ is conjugate, one might hope that a joint distribution taking μ and τ to be independent (normal and gamma, respectively) might be conjugate when both μ and τ are uncertain. This would work if the normal likelihood factored into one factor that depends only on μ and another on τ. However, this is not the case, since the exponent has a factor involving the product of μ and τ. However, there is no particular reason to limit the joint prior distribution of μ and τ to be independent. We can, for example, specify a conditional distribution for μ given τ, and a marginal distribution for τ. What we know already, though, is that the conditional distribution for μ given τ must depend on τ for conjugacy to be possible.

The form of prior distribution we choose is as follows: the distribution of μ given τ is normal with mean μ_0 and precision $\lambda_0\tau$, and the distribution on τ is gamma with parameters

α_0 and β_0. This specifies a joint distribution on μ and τ, and, with the normal likelihood, a joint distribution on $\mathbf{X}, \mu$ and τ as follows:

$$\begin{aligned} f(\mathbf{X}, \mu, \tau) = & \left(\frac{1}{\sqrt{2\pi}}\right)^n \tau^{n/2} e^{-\frac{\tau}{2}\sum_{i=1}^n (X_i-\mu)^2} \\ & \cdot \frac{1}{\sqrt{2\pi}} (\lambda_0 \tau)^{1/2} e^{-\frac{\lambda_0 \tau}{2}(\mu-\mu_0)^2} \\ & \cdot \frac{\beta_0^{\alpha_0}}{\Gamma(\alpha_0)} \tau^{\alpha_0 - 1} e^{-\beta_0 \tau}. \end{aligned} \tag{8.37}$$

Again we may eliminate constants not involving the parameters μ and τ. Here the constant is $(\frac{1}{\sqrt{2\pi}})^{n+1} \lambda_0^{1/2} \frac{\beta_0^{\alpha_0}}{\Gamma(\alpha_0)}$. Then we have

$$f(\mathbf{X}, \mu, \tau) \propto \tau^{n/2+1/2+\alpha_0-1} e^{-\tau Q(\mu)}, \tag{8.38}$$

where

$$Q(\mu) = \frac{\sum_{i=1}^n (X_i - \mu)^2}{2} + \frac{\lambda_0}{2}(\mu - \mu_0)^2 + \beta_0.$$

$Q(\mu)$ is a quadratic in μ, which is familiar. However, we cannot eliminate constants from $Q(\mu)$, because in (8.38) it is multiplied by τ, which is one of the parameters in this calculation. Nonetheless, we can re-express $Q(\mu)$ by completing the square, as we have before in analyzing normal posterior distributions.

To simplify the coming algebra a bit, we'll work with

$$Q^*(\mu) = \sum (X_i - \mu)^2 + \lambda_0 (\mu - \mu_0)^2, \tag{8.39}$$

and will substitute our answer into

$$Q(\mu) = \frac{Q^*(\mu)}{2} + \beta_0. \tag{8.40}$$

We begin the analysis of $Q^*(\mu)$ is the usual way, by collecting the quadratic linear and constant terms in μ:

$$\begin{aligned} Q^*(\mu) = & n\mu^2 - 2n\mu\overline{X} + \sum X_i^2 \\ & + \lambda_0\mu^2 - 2\lambda_0\mu\mu_0 + \lambda_0\mu_0^2 \\ = & (n + \lambda_0)\mu^2 - 2\mu\left[n\overline{X} + \lambda_0\mu_0\right] + \sum X_i^2 + \lambda_0\mu_0^2. \end{aligned} \tag{8.41}$$

Completing the square for μ, we have

$$\begin{aligned} Q^*(\mu) = & (n + \lambda_0)\left[\mu - \frac{2\mu(n\overline{X} + \lambda_0\mu_0)}{n + \lambda_0} + \left(\frac{n\overline{X} + \lambda_0\mu_0}{n + \lambda_0}\right)^2\right] \\ & + \sum X_i^2 + \lambda_0\mu_0^2 - \frac{(n\overline{X} + \lambda_0\mu_0)^2}{n + \lambda_0} \\ = & (n + \lambda_0)\left(\mu - \frac{n\overline{X} + \lambda_0\mu_0}{n + \lambda_0}\right)^2 + C \end{aligned} \tag{8.42}$$

where

$$C = \sum X_i^2 + \lambda_0\mu_0^2 - \frac{(n\overline{X} + \lambda_0\mu_0)^2}{n + \lambda_0}.$$

Now we work to simplify the constant C:

$$\begin{aligned}
C =& \sum X_i^2 + \lambda_0\mu_0^2 - \frac{(n\overline{X} + \lambda_0\mu_0)^2}{n+\lambda_0} \\
=& \sum X_i^2 + \frac{(n+\lambda_0)(\lambda_0\mu_0^2) - n^2\overline{X}^2 - 2n\overline{X}\lambda_0\mu_0 - \lambda_0^2\mu_0^2}{n+\lambda_0} \\
=& \sum X_i^2 + \frac{n\lambda_0\mu_0^2 - 2n\overline{X}\lambda_0\mu_0 - n^2\overline{X}^2}{n+\lambda_0} \\
=& \sum X_i^2 - \frac{n^2\overline{X}^2}{n+\lambda_0} + \frac{n\lambda_0}{n+\lambda_0}\left[\mu_0^2 - 2\overline{X}\mu_0\right].
\end{aligned} \tag{8.43}$$

Completing the square for μ_0, (8.43) becomes

$$\begin{aligned}
C =& \sum X_i^2 - \frac{n^2\overline{X}^2}{n+\lambda_0} + \frac{n\lambda_0}{n+\lambda_0}\left[\mu_0^2 - 2\overline{X}\mu_0 + \overline{X}^2\right] - \frac{n\lambda_0}{n+\lambda_0}\overline{X}^2 \\
=& \sum X_i^2 - \frac{(n+\lambda_0)n\overline{X}^2}{n+\lambda_0} + \frac{n\lambda_0}{n+\lambda_0}(\mu_0 - \overline{X})^2 \\
=& \sum_{i=1}^{n} X_i^2 - n\overline{X}^2 + \frac{n\lambda_0}{n+\lambda_0}(\mu_0 - \overline{X})^2 \\
=& \sum_{i=1}^{n} (X_i - \overline{X})^2 + \frac{n\lambda_0}{n+\lambda_0}(\mu_0 - \overline{X})^2.
\end{aligned} \tag{8.44}$$

Now substituting (8.44) into (8.42) and (8.42) into (8.40) we have

$$\begin{aligned}
Q(\mu) =& \beta_0 + \frac{1}{2}\Bigg\{(n+\lambda_0)(\mu - \frac{n\overline{X} + \lambda_0\mu_0}{n+\lambda_0})^2 \\
&+ \sum_{i=1}^{n}(X_i - \overline{X})^2 + \frac{n\lambda_0}{n+\lambda_0}(\mu_0 - \overline{X})^2\Bigg\}.
\end{aligned} \tag{8.45}$$

Let

$$\beta_1 = \beta_0 + \frac{1}{2}\sum_{i=1}^{n}(X_i - \overline{X})^2 + \frac{n\lambda_0}{2(n+\lambda_0)}(\mu - \overline{X})^2, \tag{8.46a}$$

$$\alpha_1 = \alpha_0 + n/2, \tag{8.46b}$$

$$\mu_1 = \frac{\lambda_0\mu_0 + n\overline{X}}{\lambda_0 + n} \tag{8.46c}$$

and

$$\lambda_1 = \lambda_0 + n. \tag{8.46d}$$

Then (8.38) can be re-expressed as

$$f(\mathbf{X}, \mu, \tau) \propto \left[\tau^{1/2} e^{-\frac{1}{2}\lambda_1\tau(\mu-\mu_1)^2}\right] \tau^{\alpha_1 - 1} e^{-\tau\beta_1}, \tag{8.47}$$

which can be recognized (the part in square brackets) as proportional to a normal distribution for μ given τ that has mean μ_1 and precision $\lambda_1\tau$, times (the part not in square brackets) a gamma distribution for τ with parameters α_1 and β_1. Therefore the family specified is conjugate for the univariate normal distribution with uncertainty in both the mean and the precision.

8.5.1 Summary

If $X_1, X_2, \ldots, X_n$ are believed to be conditionally independent and identically distributed with a normal distribution for which both the mean μ and the precision τ are uncertain, and if μ given τ has a normal distribution with mean μ_0 and precision $\lambda_0\tau$, and if τ has a gamma distribution with parameters α_0 and β_0, then the posterior distribution on μ and τ is again in the same family of distributions, with updated parameters given by equations (8.46).

8.5.2 Exercise

1. Find the constant for the posterior distribution of (μ, τ) given in (8.47).

8.6 The normal linear model with uncertain precision

We now consider a generalization of the version of the normal linear model most commonly used. Suppose our data are assembled into an $n \times 1$ vector of observations $\mathbf{y}$, as in (8.16), *i.e.*,

$$\mathbf{y} = X\boldsymbol{\beta} + \mathbf{e} \tag{8.48}$$

where X is an $n \times p$ matrix of known constants, $\boldsymbol{\beta}$ is a $p \times 1$ vector of coefficients and $\mathbf{e}$ is an $n \times 1$ vector of error terms. In distinction to the analysis of section 8.3, we suppose that $\mathbf{e}$ has a normal distribution with zero mean and precision matrix $\tau\boldsymbol{\tau}_0$, where $\boldsymbol{\tau}_0$ is a known $n \times n$ matrix, and τ has a gamma distribution with parameters α_0 and β_0. We also suppose that $\boldsymbol{\beta}$ has a normal distribution, conditional on τ, that is normal with mean $\boldsymbol{\beta}_0$ and precision $\tau\boldsymbol{\tau}_1$, where $\boldsymbol{\tau}_1$ is a known $p \times p$ matrix. (The standard assumptions take $\boldsymbol{\tau}_0$ and $\boldsymbol{\tau}_1$ to be identity matrices, but we can allow the greater generality without added complication.)

Notice that the symbol β_0 is being used here twice to mean different things, which is generally poor notational hygiene. The problem stems from the fact that the traditional notation for the linear model, (8.49) uses the vector $\boldsymbol{\beta}$ for the coefficients, while the traditional notation for the gamma distribution (8.28) uses the scalar β. Furthermore, I have been using the subscript 0 to indicate a prior value. To distinguish them, this section is scrupulous in using $\boldsymbol{\beta}_0$ to mean the vector prior on the normal linear model coefficient, and β_0 to mean the prior on τ.

Once again we write the joint density of the data, $\mathbf{y}$, and the parameters, here τ and $\boldsymbol{\beta}$, as follows:

$$\begin{aligned} f(\mathbf{y}, \tau, \boldsymbol{\beta}) = & \left(\frac{1}{\sqrt{2\pi}}\right)^n \mid \tau\boldsymbol{\tau}_0 \mid^{n/2} e^{-\frac{\tau}{2}(\mathbf{y}-X\boldsymbol{\beta})'\boldsymbol{\tau}_0(\mathbf{y}-X\boldsymbol{\beta})} \\ & \cdot \left(\frac{1}{\sqrt{2\pi}}\right)^p \mid \tau\boldsymbol{\tau}_1 \mid^{p/2} e^{-\frac{\tau}{2}(\boldsymbol{\beta}-\boldsymbol{\beta}_0)'\boldsymbol{\tau}_1(\boldsymbol{\beta}-\boldsymbol{\beta}_0)} \\ & \cdot \frac{(\beta_0)^{\alpha_0}}{\Gamma(\alpha_0)} \tau^{\alpha_0-1} e^{-\beta_0\tau}. \end{aligned} \tag{8.49}$$

Once again we recognize certain constants as being superfluous, namely here

$$\left(\frac{1}{\sqrt{2\pi}}\right)^{n+p} \mid \boldsymbol{\tau}_0 \mid^{n/2} \mid \boldsymbol{\tau}_1 \mid^{p/2} \frac{\beta_0^{\alpha_0}}{\Gamma(\alpha_0)}.$$

So instead of (8.49) we may write

$$
\begin{aligned}
f(\mathbf{y},\tau,\boldsymbol{\beta}) \propto & \tau^{n/2} e^{-\frac{\tau}{2}(\mathbf{y}-X\boldsymbol{\beta})'\boldsymbol{\tau}_0(\mathbf{y}-X\boldsymbol{\beta})} \\
& \tau^{p/2} e^{-\frac{\tau}{2}(\boldsymbol{\beta}-\boldsymbol{\beta}_0)'\boldsymbol{\tau}_1(\boldsymbol{\beta}-\boldsymbol{\beta}_0)} \\
& \tau^{\alpha_0-1} e^{-\beta_0\tau} \\
= & \tau^{n/2+p/2+\alpha_0-1} e^{-\tau Q(\boldsymbol{\beta})}
\end{aligned}
\tag{8.50}
$$

where $Q(\boldsymbol{\beta}) = \frac{1}{2}\left[(\mathbf{y}-X\boldsymbol{\beta})'\boldsymbol{\tau}_0(\mathbf{y}-X\boldsymbol{\beta}) + (\boldsymbol{\beta}-\boldsymbol{\beta}_0)'\boldsymbol{\tau}_1(\boldsymbol{\beta}-\boldsymbol{\beta}_0)\right] + \beta_0$.

Again for simplicity, we work with $Q^*(\beta)$ (the part of $Q(\boldsymbol{\beta})$ in square brackets) and complete the square in $\boldsymbol{\beta}$; again, because here τ is a parameter we are not permitted to discard additive constants from

$$
\begin{aligned}
Q^*(\boldsymbol{\beta}) = & (\mathbf{y}-X\boldsymbol{\beta})'\boldsymbol{\tau}_0(\mathbf{y}-X\boldsymbol{\beta}) + (\boldsymbol{\beta}-\boldsymbol{\beta}_0)'\boldsymbol{\tau}_1(\boldsymbol{\beta}-\boldsymbol{\beta}_0) \\
= & \boldsymbol{\beta}'X'\boldsymbol{\tau}_0X\boldsymbol{\beta} - \boldsymbol{\beta}'X'\boldsymbol{\tau}_0\mathbf{y} - \mathbf{y}'\boldsymbol{\tau}_0X\boldsymbol{\beta} + \mathbf{y}'\boldsymbol{\tau}_0\mathbf{y} \\
& + \boldsymbol{\beta}'\boldsymbol{\tau}_1\boldsymbol{\beta} - \boldsymbol{\beta}'\boldsymbol{\tau}_1\boldsymbol{\beta}_0 - \boldsymbol{\beta}_0'\boldsymbol{\tau}_1\boldsymbol{\beta} + \boldsymbol{\beta}_0'\boldsymbol{\tau}_1\boldsymbol{\beta}_0 \\
= & \boldsymbol{\beta}'(X'\boldsymbol{\tau}_0X + \boldsymbol{\tau}_1)\boldsymbol{\beta} - \boldsymbol{\beta}'(X'\boldsymbol{\tau}_0\mathbf{y} + \boldsymbol{\tau}_1\boldsymbol{\beta}_0) \\
& - (\mathbf{y}'\boldsymbol{\tau}_0X + \boldsymbol{\beta}_0'\boldsymbol{\tau}_1)\boldsymbol{\beta} + \mathbf{y}'\boldsymbol{\tau}_0\mathbf{y} + \boldsymbol{\beta}_0'\boldsymbol{\tau}_1\boldsymbol{\beta}_0.
\end{aligned}
\tag{8.51}
$$

This is a form we have studied before. As in (8.19), let

$$\boldsymbol{\tau}_2 = X'\boldsymbol{\tau}_0X + \boldsymbol{\tau}_1 \tag{8.52a}$$

and

$$\boldsymbol{\gamma} = X'\boldsymbol{\tau}_0\mathbf{y} + \boldsymbol{\tau}_1\boldsymbol{\beta}_0. \tag{8.52b}$$

Then (8.51) becomes

$$Q^*(\boldsymbol{\beta}) = \boldsymbol{\beta}'\boldsymbol{\tau}_2\boldsymbol{\beta} - \boldsymbol{\beta}'\boldsymbol{\gamma} - \boldsymbol{\gamma}'\boldsymbol{\beta} + C_1 \tag{8.53}$$

where $C_1 = \mathbf{y}'\boldsymbol{\tau}_0\mathbf{y} + \boldsymbol{\beta}_0'\boldsymbol{\tau}_1\boldsymbol{\beta}_0$.

Then we complete the square by defining $\boldsymbol{\beta}^* = \tau_2^{-1}\boldsymbol{\gamma}$, and calculating

$$
\begin{aligned}
(\boldsymbol{\beta}-\boldsymbol{\beta}^*)'\boldsymbol{\tau}_2(\boldsymbol{\beta}-\boldsymbol{\beta}^*) = & \boldsymbol{\beta}'\boldsymbol{\tau}_2\boldsymbol{\beta} - \boldsymbol{\beta}'\boldsymbol{\tau}_2\boldsymbol{\beta}^* - \boldsymbol{\beta}^* - \boldsymbol{\tau}_2\boldsymbol{\beta} + \boldsymbol{\beta}^{*\prime}\boldsymbol{\tau}_2\boldsymbol{\beta}^* \\
= & \boldsymbol{\beta}'\boldsymbol{\tau}_2\boldsymbol{\beta} - \boldsymbol{\beta}'\boldsymbol{\gamma} - \boldsymbol{\gamma}'\boldsymbol{\beta} + \boldsymbol{\beta}^{*\prime}\boldsymbol{\tau}_2\boldsymbol{\beta}^*.
\end{aligned}
$$

Therefore

$$
\begin{aligned}
Q^*(\boldsymbol{\beta}) = & (\boldsymbol{\beta}-\boldsymbol{\beta}^*)'\boldsymbol{\tau}_2(\boldsymbol{\beta}-\boldsymbol{\beta}^*) + C_1 - \boldsymbol{\beta}^{*\prime}\boldsymbol{\tau}_2\boldsymbol{\beta}^* \\
= & (\boldsymbol{\beta}-\boldsymbol{\beta}^*)'\boldsymbol{\tau}_2(\boldsymbol{\beta}-\boldsymbol{\beta}^*) + C_2
\end{aligned}
\tag{8.54}
$$

where

$$
\begin{aligned}
C_2 = C_1 - \boldsymbol{\beta}^{*\prime}\boldsymbol{\tau}_2\boldsymbol{\beta}^* = & \mathbf{y}'\boldsymbol{\tau}_0\mathbf{y} + \boldsymbol{\beta}_0'\boldsymbol{\tau}_1\boldsymbol{\beta}_0 \\
& - (\boldsymbol{\beta}_0'\boldsymbol{\tau}_1 + \mathbf{y}'\boldsymbol{\tau}_0X)\tau_2^{-1}(X'\boldsymbol{\tau}_0\mathbf{y} + \boldsymbol{\tau}_1\boldsymbol{\beta}_0).
\end{aligned}
$$

Therefore by substitution, of (8.54) into (8.51) into (8.50), we obtain

$$
\begin{aligned}
f(\mathbf{y},\tau,\boldsymbol{\beta}) \propto & \left\{\tau^{p/2} e^{-\tau(\boldsymbol{\beta}-\boldsymbol{\beta}^*)'\boldsymbol{\tau}_2(\boldsymbol{\beta}-\boldsymbol{\beta}^*)}\right\} \\
& \tau^{n/2+\alpha_0-1} e^{-\tau[\beta_0+(1/2)C_2]}.
\end{aligned}
\tag{8.55}
$$

We recognize the first factor as specifying the posterior distribution of $\boldsymbol{\beta}$ given τ as normal with mean $\boldsymbol{\beta}^*$ and precision matrix $\tau\boldsymbol{\tau}_2$, and the second factor as giving the posterior distribution of τ as a gamma distribution with parameters

$$\alpha_1 = \alpha_0 + n/2 \tag{8.56a}$$

and

$$\begin{aligned}\beta_1 =& \beta_0 + (1/2)C_2 = \beta_0 - (1/2)\left[\mathbf{y}'\boldsymbol{\tau}_0\mathbf{y} + \boldsymbol{\beta}_0'\boldsymbol{\tau}_1\boldsymbol{\beta}_0\right. \\ &\left.-(\boldsymbol{\beta}_0'\boldsymbol{\tau}_1 + \mathbf{y}'\boldsymbol{\tau}_0 X)\tau_2^{-1}(X'\boldsymbol{\tau}_0\mathbf{y} + \boldsymbol{\tau}_1\boldsymbol{\beta}_0)\right].\end{aligned} \tag{8.56b}$$

8.6.1 Summary

Suppose the likelihood is given by the normal linear model in (8.48). We suppose that $\mathbf{e}$ has a normal distribution with mean 0 and precision matrix $\tau\boldsymbol{\tau}_0$, where $\boldsymbol{\tau}_0$ is a known $n \times n$ matrix, and τ has a gamma distribution with parameters α_0 and β_0. Also suppose that $\boldsymbol{\beta}$ has a normal distribution with mean β_0 and precision $\tau\boldsymbol{\tau}_1$, where $\boldsymbol{\tau}_1$ is a known $p \times p$ matrix.

Under these assumptions, the posterior distribution on $\boldsymbol{\beta}$ given τ is again normal, with mean $\boldsymbol{\beta}^*$ defined after (8.53) and precision matrix $\tau\boldsymbol{\tau}_2$, where $\boldsymbol{\tau}_2$ is defined in (8.52a). Also the posterior distribution of τ is a gamma distribution given in (8.56a) and (8.56b).

8.6.2 Exercise

1. What is the constant for the posterior distribution in (8.55)?

8.7 The Wishart distribution

We now seek a convenient family of distributions on precision matrices that is conjugate to the multivariate normal distribution when the value of the precision matrix is uncertain. A $p \times p$ precision matrix is necessarily symmetric, and hence has $p(p+1)/2$ parameters (say all elements on or above the diagonal).

8.7.1 The trace of a square matrix

In order to specify such a distribution, it is necessary to introduce a function of a matrix we have not previously discussed, the trace. If A is an $n \times n$ square matrix, then the trace of A, written $tr(A)$, is defined to be

$$tr(A) = \sum_{i=1}^{n} a_{i,i} \tag{8.57}$$

the sum of the diagonal elements. One of the interesting properties of the trace is that it commutes:

$$tr(AB) = tr\left(\sum_j a_{ij}b_{jk}\right) = \sum_i\sum_j a_{ij}b_{ji} = \sum_j\sum_i b_{ji}a_{ij} = tr(BA). \tag{8.58}$$

Consequently, if A is symmetric, by the Spectral Decomposition (theorem 1 of section 5.8), it can be written in the form $A = PDP'$, where P is orthogonal and D is the diagonal matrix of the eigenvalues of A. Then

$$trA = trPDP' = trDP'P = trDI = trD. \tag{8.59}$$

Therefore the trace of a symmetric matrix is the sum of its eigenvalues.

Also

$$tr(A+B) = \sum_i (a_{ii} + b_{ii}) = \sum_i a_{ii} + \sum_i b_{ii} = trA + trB. \tag{8.60}$$

8.7.2 *The Wishart distribution*

Now that the trace of a symmetric matrix is defined, I can give the form of the Wishart distribution, which is a distribution over the space of $p(p+1)/2$ free elements of a positive definite, symmetric matrix V. That density is proportional to

$$| V |^{(n-p-1)/2} e^{-\frac{1}{2}tr(\tau V)} \tag{8.61}$$

where $n > p-1$ is a number and τ is a symmetric, positive definite $p \times p$ matrix.

When $p = 1$, the Wishart density is proportional to $v^{n-2}e^{-(1/2)\tau v}$, which is (except for a constant) a gamma distribution with $\alpha = n-1$ and $\beta = \tau/2$. Thus the Wishart distribution is a matrix-generalization of the gamma distribution.

In order to evaluate the integral in (8.61), it is necessary to develop the absolute value of the determinants of Jacobians for two important transformations, both of which operate on spaces of positive definite symmetric matrices.

8.7.3 *Jacobian of a linear transformation of a symmetric matrix*

To begin this analysis, we start with a study of elementary operations on matrices, from which the Jacobian is then derivable. In particular we now study the effect on non-singular matrices of two kinds of operations:

(i) the multiplication of a row (column) by a non-zero scalar.

(ii) addition of a multiple of one row (column) to another row (column).

If both of these are available, note that they imply the availability of a third operation:

(iii) interchange of two rows (columns).

To show how this is so, suppose it is desired to interchange rows i and j. We can write the starting position as (r_i, r_j), and the intent is to achieve (r_j, r_i). Consider the following:

$(r_i, r_j) \rightarrow (r_i, r_i + r_j)$	[use (ii) to add r_i to r_j]
$(r_i, r_i + r_j) \rightarrow (-r_j, r_i + r_j)$	[use (ii) to multiply $(r_i + r_j)$ by -1 and add to r_i]
$(-r_j, r_i + r_j) \rightarrow (-r_j, r_i)$	[use (ii) to add $-r_j$ to $r_i + r_j$]
$(-r_j, r_i) \rightarrow (r_j, r_i)$	[use (i) to multiply r_j by -1].

Of course the same can be shown for columns, using the same moves.

Our goal is to use elementary operations to reduce a non-singular $n \times n$ matrix A to the identity by a series of elementary operations E_i on both the rows and columns of A in a way that maintains symmetry. Then we would have

$$A = E_1 E_2 \ldots E_k I.$$

where each E_i is a matrix that performs an elementary operation.

If A is non-singular, there is a non-zero element in the first row. Interchanging two rows, if necessary, brings the non-zero element to the $(1,1)$ position. Subtracting suitable multiples of the first row from the other rows, we obtain a matrix in which all elements in the first column other than the first, are zero. Then, with a move of type (i), multiplying by $1/a$ where a is the element in the first row, reduces the $(1,1)$ element to a 1. Then the resulting matrix is of the form

$$\begin{pmatrix} 1 & c_{12} & \dots & c_{1n} \\ 0 & c_{22} & & c_{2n} \\ \vdots & \vdots & & \\ 0 & c_{n2} & & c_{nn} \end{pmatrix}.$$

Using the same process on the non-singular $(n-1) \times (n-1)$ matrix

$$\begin{pmatrix} c_{22} & \dots & c_{2n} \\ \vdots & & \vdots \\ c_{n2} & \dots & c_{nn} \end{pmatrix}$$

recursively yields the upper triangle matrix

$$\begin{pmatrix} 1 & d_{12} & d_{13} & \dots & & d_{1n} \\ 0 & 1 & d_{23} & \dots & & d_{2n} \\ \vdots & 0 & \ddots & & & \\ \vdots & & & & 1 & d_{n-1,n} \\ 0 & & & & 0 & 1 \end{pmatrix}.$$

Then using only type (ii) row operations reduces the matrix to I.

Each of the operations (i) and (ii) can be represented by matrices premultiplying A (or one of its successors). Thus a move of type (i), which multiplies row i by the scaler c, is accomplished by premultiplying A by a diagonal matrix with c in the ith place on the diagonal and 1's elsewhere. A move of type (ii) that multiples row i by c and adds it to row j is accomplished by premultiplication by a matrix that has 1's on the diagonal, c in the $(i,j)^{th}$ place, and all other off-diagonal elements equal to zero.

We have proved the following.

$$I = F_1 F_2 \dots F_k A$$

where F_i are each matrices of type (i) or type (ii).

Corollary 8.7.1.

$$A = F_k^{-1} F_{k-1}^{-1} \dots F_1^{-1} = E_k E_{k-1} \dots E_1$$

where the E's are matrices of moves of type (i) or (ii).

Proof. The inverse of a matrix of type (i) has $1/c$ in the ith place on the diagonal in place of c; the inverse of a matrix of type (ii) has $-c$ in place of c in the i, j^{th} position. Therefore neither changes type by being inverted. □

Corollary 8.7.2. *Let X be a symmetric non-singular $n \times n$ matrix, and B non-singular. Consider the transformation from X to Y by the operation*

$$Y = BXB'.$$

The Jacobian of this transformation is $\mid B \mid^{n+1}$.

Proof. From Corollary 8.7.1, we may write

$$B = E_k E_{k-1} \dots E_1$$

where each E is of type (i) or type (ii). Then

$$Y = E_k E_{k-1} \dots E_1 X E_1' E_2' \dots E_k'.$$

So the pre-multiplication of X by B and post-multiplication by B' can be considered as a series of k transformations, pre-multiplying by an E of type (i) or (ii) and post-multiplying by its transpose. Formally, let

$$X_0 = X \text{ and } X_h = E_h X_{h-1} E_h' \quad h = 1, \dots, k. \text{ Then } X_k = Y.$$

We now examine the Jacobian of the transformation from X_{i-1} to X_i in the two cases. In doing so, we remember that because the X_i's are symmetric, we take only the differential on or above the diagonal. The elements below the diagonal are determined by symmetry.

Now pre- and post-multiplying by a matrix of a transformation of type (i) yields

$$\begin{aligned} y_{ii} &= a^2 x_{ii} \\ y_{ij} &= a x_{ij} \qquad i \neq j \\ y_{jk} &= x_{jk} \quad j \neq i, k \neq i. \end{aligned}$$

Therefore the Jacobian has $n-1$ factors of a, and one of a^2, with all the others being 1. Therefore the Jacobian is a^{n+1}. But $a^{n+1} = \mid E_h \mid^{n+1}$.

Pre-multiplication by a matrix of type (ii) and post-multiplying by its transpose yields

$$\begin{aligned} y_{ii} &= x_{ii} + 2a x_{ij} + a^2 x_{jj} \\ y_{ki} &= y_{ik} = x_{ik} + a x_{jk} \quad k \neq i \\ y_{jk} &= x_{jk} \quad i \neq j, \ k \neq j. \end{aligned}$$

This yields a Jacobian matrix with 1's down the diagonal and 0's in every place either above or below the diagonal. Hence the Jacobian is 1. Trivially, then, $1 = \mid E_n \mid^{n+1}$.

Then the Jacobian of the transformation from Y to X is

$$\mid E_k \mid^{n+1} \mid E_{k-1} \mid^{n+1} \dots \mid E_1 \mid^{n+1} = \mid E_k E_{k-1} \dots E_1 \mid^{n+1} = \mid B \mid^{n+1}.$$

□

This Jacobian argument comes from Deemer and Olkin (1951) and is apparently due to P.L. Hsu. The analysis of elementary operations is modified from Mirsky (1990).

8.7.4 Determinant of the triangular decomposition

We have $A = TT'$ when T is an $n \times n$ lower triangular matrix and wish to find the Jacobian of this transformation. Because A is symmetric, we need to consider only diagonal and sub-diagonal elements in the differential. That is also true of T. Here we consider the elements of A in the order $a_{11}, a_{12}, \dots, a_{1n}, a_{22}, \dots, a_{2n}$, etc. Similarly we consider $t_{11}, t_{12}, \dots, t_{1n}, t_{22}, \dots, t_{2n}$, etc. There is one major trick to this Jacobian: the Jacobian matrix itself is lower triangular, so its determinant is the product of its diagonal elements. Hence the off-diagonal elements are irrelevant. We'll use the abbreviation NT, standing for negligible terms, for those off-diagonal elements.

Then we have $a_{ik} = \sum_{j=1}^n t_{ij} t_{jk}' = \sum_{j=1}^n t_{ij} t_{kj}$.

Now using the lower triangular nature of T, we need consider only those terms with $j \leq i$ and $j \leq k$, so in summary, $j \leq \min\{i, k\}$. Thus we have

$$a_{ik} = \sum_{j=1}^{\min\{i,k\}} t_{ij} t_{kj}.$$

Writing out these equations, and taking the differentials:

$$\begin{array}{rr}
a_{11} = t_{11}^2 & da_{11} = 2t_{11}dt_{11} \\
a_{12} = t_{11}t_{12} & da_{12} = t_{11}dt_{12} \\
\vdots & \vdots \\
a_{1n} = t_{11}t_{1n} & da_{1n} = t_{11}dt_{1n} \\
a_{22} = t_{11}^2 + t_{22}^2 & da_{22} = 2t_{22}dt_{22} + NT \\
\vdots & \vdots \\
a_{2n} = t_{12}t_{1n} + t_{22}t_{2n} & da_{2n} = t_{22}dt_{2n} + NT \\
\vdots & \vdots \\
a_{nn} = t_1^2 + t_2^2 + \ldots + t_{nn}^2 & da_{nn} = 2t_{nn}dt_{nn} + NT.
\end{array}$$

Therefore the determinant of the Jacobian matrix is the product of the terms on the right, namely

$$2^n t_{11}^n t_{22}^{n-1} \ldots t_{nn} = 2^n \prod_{i=1}^{n} t_{ii}^{n+1-i}.$$

We have proved that the Jacobian of the transform from A to T given by $A = TT'$, where A is $n \times n$ and symmetric positive definite and T is lower-triangular, is

$$2^n \prod_{i=1}^{n} t_{ii}^{n+1-i}.$$

8.7.5 *Integrating the Wishart density*

We now return to integrating the density in (8.61) over the space of positive definite symmetric matrices. We start by putting the trace in a symmetric form:

$$tr(\tau V) = tr\left(\tau^{1/2} V \tau^{1/2'}\right)$$

where $\tau^{1/2} = PD^{1/2}P'$ from Theorem 1 in section 5.8. As V varies over the space of positive-definite matrices, so does $W = \tau^{1/2} V \tau^{1/2'}$. Hence this mapping is one-to-one. Its Jacobian is $| \tau |^{p+1}$, as found in section 8.7.3. Therefore we have

$$\begin{aligned}
C_1 &= \int | V |^{(n-p-1)/2} e^{-\frac{1}{2} tr \tau V} dV \\
&= \int \frac{| W |^{(n-p-1)/2}}{| \tau |^{(n-p-1)/2}} e^{-\frac{1}{2} trW} | \tau |^{(p+1)/2} dW \\
&= \frac{1}{| \tau |^{n/2}} \int | W |^{(n-p+1)/2} e^{-\frac{1}{2} trW} dW.
\end{aligned}$$

Let $C_2 = C_1 | \tau |^{n/2}$.

Then $C_2 = \int | W |^{(n-p-1)/2} e^{-\frac{1}{2} trW} dW$.

Now we apply the triangular decomposition to W, so $W = TT'$, where T is lower triangular with positive diagonal elements. In section 5.8 it was shown that this mapping yields a unique such T. Therefore the mapping is one-to-one. Its Jacobian is computed in section 8.7.4, and is $2^p \prod_{i=1}^{p} \tau_{ii}^{p+1-i}$ in this notation.

Then we have

$$\begin{aligned} C_2 &= \int \mid W \mid^{(n-p-1)/2} e^{-\frac{1}{2}tr(W)} dW \\ &= \int \mid TT' \mid^{(n-p-1)/2} e^{-\frac{1}{2}trTT'} \cdot 2^p \prod_{i=1}^{p} t_{ii}^{p+1-i} dT \\ &= \int \prod_{i=1}^{p} t_{ii}^{n-p-1} e^{-\frac{1}{2}\sum_{i,j}\tau_{ij}^2} \cdot 2^p \prod_{i=1}^{p} t_{ii}^{p+1-i} dT \\ &=2^p \int \prod_{i=1}^{p} t_{ii}^{n-i} e^{-\frac{1}{2}(\sum_{i\neq j} t_{ij}^2 + \sum t_{ii}^2)} dT. \end{aligned}$$

Let $C_3 = C_2/2^p$. The integral now splits into $\frac{p\times(p+1)}{2}$ different independent parts. The off-diagonal elements are each

$$\int_{-\infty}^{\infty} e^{-\frac{1}{2}t_{ij}^2} dt_{ij} = \sqrt{2\pi} \quad (i \neq j)$$

and there are $\frac{p(p-1)}{2}$ of them.

The ith diagonal contributes

$$\int_0^{\infty} t_{ii}^{n-i} e^{-\frac{1}{2}t_{ii}^2} dt_{ii}.$$

Let $y_i = \frac{t_{ii}^2}{2}$. Then $dy = t_{ii}dt_{ii}$, and $t_{ii} = \sqrt{2y_i}$.

Then we have

$$\begin{aligned} \int_0^{\infty} t_{ii}^{n-i} e^{-\frac{1}{2}t_{ii}^2} dt_{ii} &= \int_0^{\infty} e^{-y_i} (\sqrt{2y_i})^{n-i} \cdot \frac{dy_i}{\sqrt{2y_i}} \\ &= \int_0^{\infty} e^{-y_i} (\sqrt{2y_i})^{n-i-1} dy_i \\ &=2^{\frac{n-i-1}{2}} \int_0^{\infty} e^{-y_i} y_i^{\frac{n-i-1}{2}} dy_i \\ &=2^{\frac{n-i-1}{2}} \Gamma\left(\frac{n-i+1}{2}\right). \end{aligned}$$

Hence we have

$$C_3 = (\sqrt{2\pi})^{\frac{p(p-1)}{2}} \prod_{i=1}^{p} \left[2^{\frac{n-i-1}{2}} \Gamma\left(\frac{n-i+1}{2}\right)\right].$$

Let

$$C_4 = \pi \frac{p(p-1)}{4} \prod_{i=1}^{p} \Gamma\left(\frac{n-i+1}{2}\right).$$

Then

$$C_3 = C_4 \left[2^{\frac{p(p-1)}{4} + \sum_{i=1}^{p}\left(\frac{n-i-1}{2}\right)}\right].$$

Now, concentrating on the power of 2 in the last expression, we have

$$
\begin{aligned}
\frac{p(p-1)}{4} + \sum_{i=1}^{p}(\frac{n-i-1}{2}) &= \frac{p(p-1)}{4} + \frac{np}{2} - \frac{p}{2} - \frac{1}{2}\sum_{i=1}^{p} i \\
&= \frac{p(p-1)}{4} + \frac{np}{2} - \frac{p}{2} - \frac{1}{2}\frac{p(p+1)}{2} \\
&= \frac{p^2}{4} - \frac{p}{4} + \frac{np}{2} - \frac{p}{2} - \frac{p^2}{4} - \frac{p}{4} \\
&= \frac{np}{2} - p.
\end{aligned}
$$

Hence $C_3 = C_4[2^{\frac{np}{2}-p}]$.

Putting the results together, we have

$$
\begin{aligned}
C_1 = \frac{C_2}{\mid \tau \mid^{n/2}} &= \frac{2^p C_3}{\mid \tau \mid^{n/2}} = C_4 \frac{2^p[2^{\frac{np}{2}-p}]}{\mid \tau \mid^{n/2}} \\
&= \frac{C_4 2^{\frac{np}{2}}}{\mid \tau \mid^{n/2}} = 2^{\frac{np}{2}} \frac{\pi^{\frac{p(p-1)}{4}} \prod_{i=1}^{p} \Gamma(\frac{n-i+1}{2})}{\mid \tau \mid^{n/2}}.
\end{aligned}
$$

Therefore

$$
f_V(v) = \frac{\mid \tau \mid^{n/2} \mid v \mid^{(n-p-1)/2} \; e^{-\frac{1}{2}tr(\tau v)}}{2^{np/2}\pi^{\frac{p(p-1)}{4}} \prod_{i=1}^{p} \Gamma(\frac{n-i+1}{2})} \tag{8.62}
$$

is a density over all positive definite matrices, and is called the density of the Wishart distribution.

8.7.6 *Multivariate normal distribution with uncertain precision and certain mean*

Suppose that $\mathbf{X} = (\mathbf{X}_1, \mathbf{X}_2, \ldots, \mathbf{X}_n)$ are believed to be conditionally independent and identically distributed p-dimensional vectors from a normal distribution with mean vector $\mathbf{m}$, known with certainty, and precision matrix R. Suppose also that R is believed to have a Wishart distribution with α degrees of freedom and $p \times p$ matrix τ, such that $\alpha > p - 1$ and τ is symmetric and positive definite.

The joint distribution of X and R takes the form

$$
\begin{aligned}
f(\mathbf{X}, R) = &\left(\frac{1}{\sqrt{2\pi}}\right)^{np} \mid R \mid^{n/2} e^{-\frac{1}{2}\sum_{i=1}^{n}(\mathbf{X}_i - \mathbf{m})' R (\mathbf{X}_i - \mathbf{m})} \\
&\cdot c \mid R \mid^{(\alpha-p-1)/2} e^{-\frac{1}{2}tr(\tau R)}.
\end{aligned} \tag{8.63}
$$

We recognize $\left(\frac{1}{\sqrt{2\pi}}\right)^{np} c$ as irrelevant constants, so we can write

$$
f(\mathbf{X}, R) \alpha \mid R \mid^{(n+\alpha-p-1)/2} e^{-\frac{1}{2}[\sum_{i=1}^{n}(\mathbf{X}_i - \mathbf{m})' R (X_i - \mathbf{m}) + tr(\tau R)]}. \tag{8.64}
$$

Now we notice that $\sum_{i=1}^{n}(x_i - m)' R (X_i - m)$ is a number, which can be regarded as a 1×1

matrix, equal to its trace. (I know this sounds like an odd maneuver, but trust me.) Then

$$\begin{aligned}
&\sum_{i=1}^{n}(\mathbf{x}_i - \mathbf{m})'R(\mathbf{x}_i - \mathbf{m}) + tr\ (\tau R) = \\
&tr\left(\sum_{i=1}^{n}(\mathbf{x}_i - \mathbf{m})'R(\mathbf{x}_i - \mathbf{m})\right) + tr\ (\tau R) = \\
&tr\left(\sum_{i=1}^{n}(\mathbf{x}_i - \mathbf{m})(\mathbf{x}_i - \mathbf{m})'R\right) + tr\ (\tau R) = \\
&tr\left[\left(\sum_{i=1}^{n}(\mathbf{x}_i - \mathbf{m})(\mathbf{x}_i - \mathbf{m})' + \tau\right) R\right] \qquad (8.65)
\end{aligned}$$

using (8.58) and (8.60). Therefore (8.64) can be rewritten as

$$f(\mathbf{X}, R) \propto \mid R \mid^{(n^*-p-1)/2} e^{-\frac{1}{2}tr(\tau^* R)} \qquad (8.66)$$

where $\tau^* = \sum_{i=1}^{n}(\mathbf{X}_i - \mathbf{m})(\mathbf{X}_i - \mathbf{m})' + \tau$, which we may recognize as a Wishart distribution with matrix τ^* and $n^* = n + \alpha$ degrees of freedom.

8.7.7 Summary

The Wishart distribution, given in (8.61) is a convenient distribution for positive definite matrices. Section 8.7.6 proves the following result:

Suppose that $X = (\mathbf{X}_1, \mathbf{X}_2, \ldots, \mathbf{X}_n)$ are believed to be conditionally independent and identically distributed p-dimensional vectors from a normal distribution with mean vector $\mathbf{m}$, known with certainty, and precision matrix R. Suppose also that R is believed to have a Wishart distribution with α degrees of freedom and $p \times p$ matrix τ, such that $\alpha > p - 1$ and τ is symmetric and positive definite. Then the posterior distribution on R is again Wishart, with $n + \alpha$ degrees of freedom and matrix τ^* given in (8.66).

8.7.8 Exercise

1. Write out the constant omitted from (8.66). Put another way, what constant makes (8.66) into the posterior density of R given $\mathbf{X}$?

8.8 Multivariate normal data with both mean and precision matrix uncertain

Now, suppose that $X = (\mathbf{X}_1, \mathbf{X}_2, \ldots, \mathbf{X}_n)$ are believed to be conditionally independent and identically distributed p-dimensional random vectors from a normal distribution with mean vector $\mathbf{m}$ and precision matrix R, about both of which you are uncertain. Suppose that your joint distribution over $\mathbf{m}$ and R is given as follows: the distribution of $\mathbf{m}$ given R is p-dimensional multivariate normal with mean $\boldsymbol{\mu}$ and precision matrix vR, and R has a Wishart distribution with $\alpha > p - 1$ degrees of freedom and symmetric positive-definite matrix τ.

Then the joint distribution of $\mathbf{X}, \mathbf{m}$ and R is given by

$$\begin{aligned} f(X, \mathbf{m}, R) =& f(X \mid m, R) f(m \mid R) f(R) \\ =& \left(\frac{1}{\sqrt{2\pi}}\right)^{np} \mid R \mid^{n/2} e^{-\frac{1}{2}\sum_{i=1}^{n}(\mathbf{X}_i - \mathbf{m})' R (\mathbf{X}_i - \mathbf{m})} \\ & \cdot \left(\frac{1}{\sqrt{2\pi}}\right)^{p} \mid vR \mid^{1/2} e^{-\frac{1}{2}(\mathbf{m}-\boldsymbol{\mu})' vR (\mathbf{m}-\boldsymbol{\mu})} \\ & \cdot\ c \mid R \mid^{(\alpha-p-1)/2} e^{-\frac{1}{2} tr(\tau R)}. \end{aligned} \tag{8.67}$$

Again we recognize $\left(\frac{1}{\sqrt{2\pi}}\right)^{(n+1)p} \cdot c \cdot v^{1/2}$ as irrelevant constants that can be absorbed. This yields

$$f(X, \mathbf{m}, R) \propto \mid R \mid^{(n+\alpha-p)/2} e^{-\frac{1}{2}Q(m)} \tag{8.68}$$

where

$$Q(\mathbf{m}) = \sum_{i=1}^{n} (\mathbf{X}_i - \mathbf{m})' R (\mathbf{X}_i - \mathbf{m}) + \nu(\mathbf{m} - \boldsymbol{\mu})' R (\mathbf{m} - \boldsymbol{\mu}) + tr\ \tau R.$$

We now have some algebra to do. We begin by studying the first summand in $Q(\mathbf{m})$:

$$\begin{aligned} \sum_{i=1}^{n} (\mathbf{X}_i - \mathbf{m})' R (\mathbf{X}_i - \mathbf{m}) &= \sum_{i=1}^{n} (\mathbf{X}_i - \overline{\mathbf{X}} + \overline{\mathbf{X}} - \mathbf{m})' R (\mathbf{X}_i - \overline{\mathbf{X}} + \overline{\mathbf{X}} - \mathbf{m}) \\ &= \sum_{i=1}^{n} (\mathbf{X}_i - \overline{X})' R (\mathbf{X}_i - \overline{X}) + n(\overline{\mathbf{X}} - \mathbf{m})' R (\overline{\mathbf{X}} - \mathbf{m}), \end{aligned} \tag{8.69}$$

since

$$\sum_{i=1}^{n} (\mathbf{X}_i - \overline{\mathbf{X}})' R (\overline{\mathbf{X}} - \mathbf{m}) = (n\overline{\mathbf{X}} - n\overline{\mathbf{X}}) R (\overline{\mathbf{X}} - \mathbf{m}) = 0$$

and similarly

$$\sum_{i=1}^{n} (\overline{\mathbf{X}} - \mathbf{m})' R (\mathbf{X}_i - \overline{\mathbf{X}}) = 0.$$

Now

$$\begin{aligned} \sum_{i=1}^{n} (\mathbf{X}_i - \overline{\mathbf{X}})' R (\mathbf{X}_i - \overline{\mathbf{X}}) =& tr \sum_{i=1}^{n} (\mathbf{X}_i - \overline{\mathbf{X}})' R (\mathbf{X}_i - \overline{\mathbf{X}}) \\ =& \sum_{i=1}^{n} tr R (\mathbf{X}_i - \overline{\mathbf{X}})(\mathbf{X}_i - \overline{\mathbf{X}})' \\ =& tr R \sum_{i=1}^{n} (\mathbf{X}_i - \overline{\mathbf{X}})(\mathbf{X}_i - \overline{\mathbf{X}})' \\ =& tr(RS) = tr(SR) \end{aligned} \tag{8.70}$$

where $S = \sum_{i-1}^{n} (\mathbf{X}_i - \overline{\mathbf{X}})(\mathbf{X}_i - \overline{\mathbf{X}})'$.

Our next step is to put together the two quadratic forms in $\mathbf{m}$ and complete the square, as we have done before: taking the second term in $Q(\mathbf{m})$ in (8.68) and the second term in

(8.69) we have

$$\begin{aligned}
& n(\overline{\mathbf{X}}-\mathbf{m})'R(\overline{\mathbf{X}}-\mathbf{m})+\nu(\mathbf{m}-\boldsymbol{\mu})'R(\mathbf{m}-\boldsymbol{\mu}) \\
= \; & n\mathbf{m}'R\mathbf{m}-n\mathbf{m}'R\overline{\mathbf{X}}-n\overline{\mathbf{X}}'R\mathbf{m}+n\overline{\mathbf{X}}'R\overline{\mathbf{X}} \\
& +\nu\mathbf{m}'R\mathbf{m}-\nu\mathbf{m}'R\boldsymbol{\mu}-\nu\boldsymbol{\mu}'R\mathbf{m}+\nu\boldsymbol{\mu}'R\boldsymbol{\mu} \\
= \; & (n+\nu)(\mathbf{m}'R\mathbf{m})-\mathbf{m}'R(\nu\boldsymbol{\mu}+n\overline{\mathbf{X}})-(\nu\boldsymbol{\mu}'+n\overline{\mathbf{X}}')R\mathbf{m} \\
& +\nu\boldsymbol{\mu}'R\boldsymbol{\mu}+n\overline{\mathbf{X}}'R\overline{\mathbf{X}} \\
= \; & (\nu+n)[\mathbf{m}'R\mathbf{m}-\mathbf{m}'R\boldsymbol{\mu}^*-\boldsymbol{\mu}^{*\prime}R\mathbf{m}+\boldsymbol{\mu}^{*\prime}R\boldsymbol{\mu}^*] \\
& +\nu\boldsymbol{\mu}'R\boldsymbol{\mu}+n\overline{\mathbf{X}}'R\overline{\mathbf{X}}-(n+\nu)(\boldsymbol{\mu}^{*\prime}R\boldsymbol{\mu}^*) \\
= \; & (\nu+n)(\mathbf{m}-\boldsymbol{\mu}^*)'R(\mathbf{m}-\boldsymbol{\mu}^*)+\nu\boldsymbol{\mu}'R\boldsymbol{\mu}+n\overline{\mathbf{X}}'R\overline{\mathbf{X}} \\
& -(n+\nu)(\boldsymbol{\mu}^{*\prime}R\boldsymbol{\mu}^*)
\end{aligned}$$

$$\text{where } \boldsymbol{\mu}^* = \frac{\nu\boldsymbol{\mu}+n\overline{\mathbf{X}}}{\nu+n}. \tag{8.71}$$

Now, working with the constant terms from the completion of the square,

$$\begin{aligned}
& \nu\boldsymbol{\mu}'R\boldsymbol{\mu}+n\overline{\mathbf{X}}'R\overline{\mathbf{X}}-(\boldsymbol{\mu}^{*\prime}R\boldsymbol{\mu}^*)(n+\nu) \\
= \; & \nu\boldsymbol{\mu}'R\boldsymbol{\mu}+n\overline{\mathbf{X}}'R\overline{\mathbf{X}}-\frac{1}{n+\nu}(\nu\boldsymbol{\mu}+n\overline{\mathbf{X}})'R(\nu\boldsymbol{\mu}+n\overline{\mathbf{X}}) \\
= \; & \frac{1}{n+\nu}\Big[(n\nu+\nu^2)(\boldsymbol{\mu}'R\boldsymbol{\mu})+(n^2+n\nu)\overline{\mathbf{X}}'R\overline{\mathbf{X}} \\
& -\nu^2\boldsymbol{\mu}'R\boldsymbol{\mu}-n^2\overline{\mathbf{X}}'R\overline{\mathbf{X}} \\
& -\nu n\boldsymbol{\mu}'R\overline{\mathbf{X}}-\nu n\overline{\mathbf{X}}'R\boldsymbol{\mu}\Big] \\
= \; & \frac{n\nu}{n+\nu}\left[\boldsymbol{\mu}'R\boldsymbol{\mu}+\overline{\mathbf{X}}'R\overline{\mathbf{X}}-\boldsymbol{\mu}'R\overline{\mathbf{X}}-\overline{\mathbf{X}}'R\boldsymbol{\mu}\right] \\
= \; & \frac{n\nu}{n+\nu}\left(\boldsymbol{\mu}-\overline{\mathbf{X}}\right)'R(\boldsymbol{\mu}-\overline{\mathbf{X}}) \\
= \; & \frac{n\nu}{n+\nu}tr\left[\left(\boldsymbol{\mu}-\overline{\mathbf{X}}\right)'R\left(\boldsymbol{\mu}-\overline{\mathbf{X}}\right)\right]=\frac{n\nu}{n+\nu}tr\left[\left(\boldsymbol{\mu}-\overline{\mathbf{X}}\right)\left(\boldsymbol{\mu}-\overline{\mathbf{X}}\right)'R\right].
\end{aligned} \tag{8.72}$$

Now putting the pieces together, we have

$$\begin{aligned}
Q(\mathbf{m}) = \; & \sum_{i=1}^{n}(\mathbf{X}_i-\mathbf{m})'R(\mathbf{X}_i-\mathbf{m})+\nu(\mathbf{m}-\boldsymbol{\mu})'R(\mathbf{m}-\boldsymbol{\mu})+tr\ (\tau R) \\
= \; & tr\ [SR+(\nu+n)(\mathbf{m}-\boldsymbol{\mu}^*)'R(\mathbf{m}-\boldsymbol{\mu}^*)] \\
& +\frac{n\nu}{n+\nu}(\boldsymbol{\mu}-\overline{\mathbf{X}})'R(\boldsymbol{\mu}-\overline{\mathbf{X}})+tr\ (\tau R) \\
= \; & tr\left[(\tau+S+\frac{n\nu}{n+\nu}(\boldsymbol{\mu}-\overline{\mathbf{X}})(\boldsymbol{\mu}-\overline{\mathbf{X}})')R\right] \\
& +(\nu+n)(\mathbf{m}-\boldsymbol{\mu}^*)'R(\mathbf{m}-\boldsymbol{\mu}^*).
\end{aligned} \tag{8.73}$$

Substituting (8.73) into (8.68) yields

$$\begin{aligned}
f(X,\mathbf{m},R) \propto \; & \mid R\mid^{p/2} e^{-\frac{1}{2}(v+n)(\mathbf{m}-\boldsymbol{\mu}^*)'R(\mathbf{m}-\boldsymbol{\mu}^*)} \\
& \cdot \mid R \mid^{(\alpha+n-p-1)/2} e^{-\frac{1}{2}tr[(\tau+S)+(\frac{nv}{n+v})(\boldsymbol{\mu}-\overline{\mathbf{X}})(\boldsymbol{\mu}-\overline{\mathbf{X}})']R},
\end{aligned} \tag{8.74}$$

which we recognize as a conditional normal distribution for $\mathbf{m}$ given R, with mean $\boldsymbol{\mu}^*$ and precision matrix $(\nu+n)R$, and a Wishart distribution for R, with $\alpha+n$ degrees of freedom, and matrix

$$\tau^* = \tau + S + \frac{n\nu}{n+\nu}(\boldsymbol{\mu} - \overline{\mathbf{X}})(\boldsymbol{\mu} - \overline{\mathbf{X}})'. \tag{8.75}$$

8.8.1 *Summary*

Suppose that $X = (\mathbf{X}_1, \ldots, \mathbf{X}_n)$ are believed to be conditionally independent and identically distributed p-dimensional random vectors from a normal distribution with mean vector $\mathbf{m}$ and precision matrix R, about both of which you are uncertain. Suppose that your belief about $\mathbf{m}$ conditional on R is a p-dimensional normal distribution with mean $\boldsymbol{\mu}$ and precision matrix νR, and that your belief about R is a Wishart distribution with α degrees of freedom and precision matrix τ.

Then your posterior distribution on m and R is as follows: your distribution on m given R is multivariate normal with mean μ^* given in (8.71) and precision matrix $(\nu+n)R$, and your distribution for R is Wishart with $\alpha+n$ degrees of freedom and precision matrix τ^* given in (8.75).

8.8.2 *Exercise*

1. Write down the constant omitted from (8.74) to make (8.74) the conditional density of m and R given X.

8.9 The Beta and Dirichlet distributions

The Beta distribution is a distribution over unit interval, and turns out to be conjugate to the binomial distribution. Its k-dimensional generalization, the Dirichlet distribution, is conjugate to the k-dimensional generalization of the binomial distribution, namely the multinomial distribution. The purpose of this section is to demonstrate these results.

I start by deriving the constant for the Dirichlet distribution. I have to admit that the proof feels a bit magical to me.

Let S_k be the k-dimensional simplex, so

$$S_k = \{(p_1, \ldots, p_{k-1}) \mid p_i \geq 0, \sum_{i=1}^{k-1} p_i \leq 1\}.$$

(You may be surprised not to find p_k mentioned. The reason is that if p_k is there, with the constraint $\sum_{i=1}^{k} p_i = 1$, the space has k variables of which only $k-1$ are free. Consequently when we take integrals over S_k, it is better to think of S_k as having $k-1$ variables. For other purposes it is more symmetric to include p_k.)

The Dirichlet density is proportional to

$$p_1^{\alpha_1 - 1} p_2^{\alpha_2 - 1} \ldots p_{k-1}^{\alpha_{k-1}-1}(1 - p_1 - p_2 - \ldots - p_{k-1})^{\alpha_k - 1}$$

over the space S_k. The question is the value of the integral.

Theorem 8.9.1.

$$\int_{S_k} p_1^{\alpha_1 - 1} p_2^{\alpha_2 - 1} \ldots p_{k-1}^{\alpha_{k-1}-1}(1 - p_1 - p_2 - \ldots - p_{k-1})^{\alpha_k - 1} dp_1 dp_2, \ldots, dp_{k-1}$$
$$= \frac{\prod_{i=1}^{k} \Gamma(\alpha_i)}{\Gamma(\sum_{i=1}^{k} \alpha_i)}$$

for all positive α_i.

Proof. Let $I = \int_{S_k} p_1^{\alpha_1 - 1} p_2^{\alpha_2 - 1} \ldots p_{k-1}^{\alpha_{k-1}-1} (1 - p_1 - p_2 \ldots - p_{k-1})^{\alpha_k - 1} dp_1 dp_2, \ldots, dp_{k-1}$ and let $I^* = \prod_{i=1}^k \Gamma(\alpha_i)$.

Then

$$I^* = \int_0^\infty \ldots \int_0^\infty \prod_{i=1}^k x_i^{\alpha_i - 1} e^{-\sum_{i=1}^k x_i} dx_1 \ldots dx_k.$$

Now let $y_1, \ldots, y_k$ be defined as follows:

$$\begin{aligned} y_i &= x_i / \textstyle\sum_{j=1}^k x_j \quad i = 1, \ldots, k-1 \\ y_k &= \textstyle\sum_{j=1}^k x_j. \end{aligned}$$

Then $y_i = x_i / y_k \quad i = 1, \ldots, k-1$,

so

$$x_i = y_i y_k \quad i = 1, \ldots, k-1$$

and

$$x_k = y_k - \sum_{j=1}^{k-1} x_j = y_k - \sum_{j=1}^{k-1} y_j y_k = y_k (1 - \sum_{j=1}^{k-1} y_j).$$

Since the inverse function can be found, the transformation is one-to-one.

The Jacobian matrix of this transformation is (see section 5.9)

$$J = \begin{bmatrix} \frac{\partial x_1}{\partial y_1} & \cdots & \frac{\partial x_1}{\partial y_k} \\ \vdots & & \\ \frac{\partial x_k}{\partial y_1} & & \frac{\partial x_k}{\partial y_k} \end{bmatrix} = \begin{bmatrix} y_k & & & \begin{matrix} y_1 \\ y_2 \end{matrix} \\ & \ddots & & \vdots \\ & & y_k & y_{k-1} \\ -y_k & \ldots & -y_k & 1 - \sum_{j=1}^{k-1} y_j \end{bmatrix}$$

where all the entries not written are zero.

To find the determinant of J, recall that rows may be added to each other without changing the value of the determinant (see Theorem 12 in section 5.7). In this case I add each of the first $n-1$ rows to the last row, to obtain

$$\| J \| = \begin{Vmatrix} y_k & & & y_1 \\ & \ddots & y_k & y_{k-1} \\ 0 & \ldots & 0 & 1 \end{Vmatrix}.$$

In each of the $k!$ summands in the determinant, an element of the last row appears only once. Each of the summands not including the (k, k) element is zero. Among those including the (k, k) element, only the product down the diagonal avoids being zero.

Therefore

$$\| J \| = y_k^{k-1}.$$

Now we are in a position to apply the transformation to I^*.

$$
\begin{aligned}
I^* = & \int \prod_{i=1}^{k-1} (y_i y_k)^{\alpha_i - 1} \left[\left(1 - \sum_{j=1}^{k-1} y_j \right) y_k \right]^{\alpha_k - 1} e^{-y_k} y_k^{k-1} dy_1 \ldots dy_{k-1} dy_k \\
= & \int_{S_k} \prod_{i=1}^{k-1} y_i^{\alpha_i - 1} \left(1 - \sum_{j=1}^{k-1} y_j \right)^{\alpha_k - 1} dy_1 \ldots dy_{k-1} \\
& \int_0^\infty y_k^{\sum_{i=1}^{k-1} (\alpha_i - 1) + \alpha_k - 1 + (k-1)} e^{-y_k} dy_k \\
= & I \int_0^\infty y_k^{\sum_{i=1}^{k-1} \alpha_i - (k-1) + \alpha_k - 1 + k - 1} e^{-y_k} dy_k \\
= & I \int_0^\infty y_k^{\sum_{i=1}^{k} \alpha_i - 1} e^{-y_k} dy_k \\
= & I \; \Gamma(\sum_{i=1}^{k} \alpha_i).
\end{aligned}
$$

Therefore $I = I^* / \Gamma(\sum_{i=1}^k \alpha_i)$ as was to be shown. □

Thus the density

$$
p_1^{\alpha_1 - 1} \ldots p_{k-1}^{\alpha_{k-1} - 1} (1 - p_1 - p_2 - \ldots - p_{k-1})^{\alpha_k - 1} \cdot \frac{\Gamma(\sum_{i=1}^k \alpha_i)}{\prod_{i=1}^k \Gamma(\alpha_i)}, \qquad (p_1 \ldots p_{k-1}) \in S_k
$$

and 0 otherwise, is a probability distribution for all $\alpha_i > 0$. This is the Dirichlet distribution with parameters $(\alpha_1, \ldots, \alpha_k)$.

As long as we're not transforming an integral, we can define $p_k = 1 - p_1 - p_2 - \ldots - p_{k-1}$, and write the Dirichlet more compactly (and symmetrically) as

$$
\prod_{i=1}^k p_i^{\alpha_i - 1} \Gamma \left(\sum_{i=1}^k \alpha_i \right) \Big/ \prod_{i=1}^k \Gamma(\alpha_i), \text{ for } (p_1, \ldots, p_{k-1}) \in S_k \tag{8.76}
$$

and 0 otherwise.

The special case when $k = 2$ is called the Beta distribution.

Its density is usually written as

$$
\begin{cases} p^{\alpha-1}(1-p)^{\beta-1} \frac{\Gamma(\alpha+\beta)}{\Gamma(\alpha)\Gamma(\beta)} & 0 < p < 1 \\ 0 & \text{otherwise} \end{cases}. \tag{8.77}
$$

If X has a binomial distribution with parameters n and p, and p has a Beta distribution with parameters α and β, then the joint distribution of X and p is

$$
\binom{n}{j, n-j} p^j (1-p)^{n-j} p^{\alpha-1} (1-p)^{\beta-1} \frac{\Gamma(\alpha+\beta)}{\Gamma(\alpha)\Gamma(\beta)}.
$$

Recognizing $\binom{n}{j,n-j} \frac{\Gamma(\alpha)\Gamma(\beta)}{\Gamma(\alpha+\beta)}$ as an irrelevant constant, the density is proportional to

$$
p^{\alpha+j-1}(1-p)^{\beta+(n-j)-1}
$$

which is recognized as a Beta distribution with parameters $\alpha + j$ and $\beta + (n - j)$.

The name "Beta Distribution," incidentally, comes from the fact that $\frac{\Gamma(\alpha)\Gamma(\beta)}{\Gamma(\alpha+\beta)}$ is called the Beta Function, and is studied in the theory of special functions.

The relationship between the Dirichlet distribution and the multinomial distribution is a straightforward generalization of the relationship between the Beta distribution and the binomial. Their joint distribution is

$$\binom{n}{n_1, n_2, \ldots, n_k} \prod_{j=1}^{k} p_j^{n_j} \cdot \prod p_j^{\alpha_j - 1} \Gamma(\sum_{i=1}^{k} \alpha_i) / \prod_{i=1}^{k} \Gamma(\alpha_i).$$

Recognizing $\binom{n}{n_1,n_2,\ldots,n_k}\Gamma(\sum_{i=1}^{k}\alpha_i)/\prod_{i=1}^{k}\Gamma(\alpha_i)$ as an irrelevant constant, we have the joint density proportional to

$$\prod_{j=1}^{k} p_j^{n_j} \prod_{j=1}^{k} p_j^{\alpha_j - 1} = \prod_{j=1}^{k} p_j^{\alpha_j + n_j - 1} \tag{8.78}$$

which is recognized as a Dirichlet distribution with parameters $(\alpha_1+n_1, \alpha_2+n_2, \ldots, \alpha_k+n_k)$.

The moments of the Dirichlet distribution are found as follows:

$$\begin{aligned} E(p_i^{\ell}) &= \int_{S_k} p_i^{\ell} \prod_{j=1}^{k} p_j^{\alpha_j - 1}\, \Gamma(\sum_{j=1}^{k} \alpha_j) / \prod_{j=1}^{k} \Gamma(\alpha_j) \\ &= \int_{S_k} \prod_{j=1}^{k} p_j^{\alpha_j^* - 1} \Gamma(\sum_{j=1}^{k} \alpha_j) / \prod_{j=1}^{k} \Gamma(\alpha_j), \end{aligned}$$

$$\begin{aligned} \text{where } \alpha_j^* &= \alpha_j \text{ for } j \neq i \\ \text{and } \alpha_i^* &= \alpha_i + \ell. \end{aligned}$$

Then

$$\begin{aligned} E(p_i^{\ell}) &= \frac{\Gamma(\sum_{j=1}^{k} \alpha_j)}{\prod_{j=1}^{k} \Gamma(\alpha_j)} \cdot \frac{\prod_{j=1}^{k} \Gamma(\alpha_j^*)}{\Gamma(\sum_{j=1}^{k} \alpha_j^*)} \\ &= \frac{\Gamma(\alpha_i^*)}{\Gamma(\alpha_i)} \frac{\Gamma(\sum_{j=1}^{k} \alpha_j)}{\Gamma(\sum_{j=1}^{k} \alpha_j^*)} \\ &= \frac{(\alpha_i + \ell - 1)(\alpha_i + \ell - 2) \ldots (\alpha_i)}{(\sum \alpha_j + \ell - 1) \ldots (\sum \alpha_j)}. \end{aligned}$$

In particular,

$$E(p_i) = \frac{\alpha_i}{\sum_{j=1}^{k} \alpha_j}$$

and

$$E(p_i^2) = \frac{(\alpha_i + 1)(\alpha_i)}{(\sum_{j=1}^{k} \alpha_j + 1)(\sum_{j=1}^{k} \alpha_i)}.$$

Therefore

$$
\begin{aligned}
\text{Var}(p_i) &= E(p_i^2) - (E(p_i))^2 \\
&= \frac{(\alpha_i+1)(\alpha_i)}{(\sum_{j=1}^k \alpha_j + 1)\left(\sum_{j=1}^k \alpha_i\right)} - \left(\frac{\alpha_i}{\sum_{j=1}^k \alpha_j}\right)^2 \\
&= \left(\frac{\alpha_i}{\sum_{j=1}^k \alpha_j}\right)\left[\frac{\alpha_i+1}{\sum_{j=1}^k \alpha_j + 1} - \frac{\alpha_i}{\sum_{j=1}^k \alpha_j}\right] \\
&= \left(\frac{\alpha_i}{\sum_{j=1}^k \alpha_j}\right)\left[\frac{(\alpha_i+1)(\sum_{j=1}^k \alpha_j) - \alpha_i(\sum_{j=1}^k \alpha_j + 1)}{(\sum_{j=1}^k \alpha_j)(\sum_{j=1}^k \alpha_j + 1)}\right] \\
&= \left(\frac{\alpha_i}{\sum_{j=1}^k \alpha_j}\right)\left[\frac{\sum_{j=1}^k \alpha_j - \alpha_i}{(\sum_{j=1}^k \alpha_j)(\sum_{j=1}^k \alpha_1 + 1)}\right] \\
&= \frac{(\alpha_i)(\sum_{j\neq i}^k \alpha_j)}{(\sum_{j=1}^k \alpha_j)^2(\sum_{j=1}^k \alpha_j + 1)}.
\end{aligned}
$$

In particular, for the Beta distribution

$$E(p) = \alpha/(\alpha+\beta)$$

and

$$\text{Var}(p) = \frac{\alpha\beta}{(\alpha+\beta)^2(\alpha+\beta+1)}.$$

8.9.1 Refining and coarsening

Just as the multinomial distribution allows graceful coarsening so does the Dirichlet distribution.

Theorem 8.9.2. *Suppose $(p_1, \ldots, p_{k-1})$ has a Dirichlet distribution with parameters $(\alpha_1, \ldots, \alpha_k)$. Then $(p_1 + p_2, p_3, \ldots, p_{k-1})$ has a Dirichlet distribution with parameters $(\alpha_1 + \alpha_2, \alpha_3, \ldots, \alpha_k)$.*

Proof. The density of $(p_1, \ldots, p_{k-1})$ is

$$p_1^{\alpha_1-1} p_2^{\alpha_2-1} p^{\alpha_3-1}, \ldots, p_{k-1}^{\alpha_k-1}(1 - p_1 - p_2, - \ldots, -p_{k-1})^{\alpha_k-1} \cdot \frac{\Gamma(\sum_{i=1}^k \alpha_i)}{\prod_{i=1}^k \Gamma(\alpha_i)} \tag{8.79}$$

for $(p_1, p_2, \ldots, p_{k-1}) \in S_k$, and 0 otherwise.

Let $x = p_1/(p_1+p_2)$ and $y = p_1 + p_2$. My strategy is to integrate x from the density. Now

$$
\begin{aligned}
p_1^{\alpha_1-1} p_2^{\alpha_2-1} &= \left(\tfrac{p_1}{p_1+p_2}\right)^{\alpha_1-1}\left(\tfrac{p_2}{p_1+p_2}\right)^{\alpha_2-1}(p_1+p_2)^{\alpha_1+\alpha_2-2} \\
&= x^{\alpha_1-1}(1-x)^{\alpha_2-1} y^{\alpha_1+\alpha_2-2}.
\end{aligned}
$$

Solving for p_1 and p_2 yields $p_1 = xy$ and $p_2 = y(1-x)$. Then the Jacobian matrix is

$$J = \begin{bmatrix} \frac{\partial p_1}{\partial x} & \frac{\partial p_1}{\partial y} \\ \frac{\partial p_2}{\partial x} & \frac{\partial p_2}{\partial y} \end{bmatrix} = \begin{bmatrix} y & x \\ -y & 1-x \end{bmatrix},$$

so the determinant is $y(1-x) + xy = y$.

Thus the transformed contribution to the density is

$$x^{\alpha_1-1}(1-x)^{\alpha_2-1}y^{\alpha_1+\alpha_2-1}$$

and the density itself is

$$\frac{\Gamma(\sum_{i=1}^k \alpha_i)}{\prod_{i=1}^k \Gamma(\alpha_i)} x^{\alpha_1-1}(1-x)^{\alpha_2-1}y^{\alpha_1+\alpha_2-1}p_3^{\alpha_3-1}, \ldots, p_{k-1}^{\alpha_k-1}(1-y-p_3,\ldots,p_{k-1})^{\alpha_4-1}.$$

The integral with respect to x is a Beta integral:

$$\int_0^1 x^{\alpha_1-1}(1-x)^{\alpha_2-1}dx\frac{\Gamma(\alpha_1)\Gamma(\alpha_2)}{\Gamma(\alpha_1+\alpha_2)}.$$

Then the remaining density is proportional to

$$y^{\alpha_1+\alpha_2-1}p_3^{\alpha_3-1}, \ldots, p_{k-1}^{\alpha_k-1}(1-y-p_3,\ldots,-p_{k-1})^{\alpha_k-1}.$$

Resubstituting $y = p_1 + p_2$ makes things look more familiar:

$$(p_1+p_2)^{\alpha_1+\alpha_2-1}p_3^{\alpha_3-1}, \ldots, p_{k-1}^{\alpha_k-1}(1-(p_1,+p_2)-p_3,\ldots,-p_{k-1}\frac{\Gamma\left(\sum_{i=1}^k \alpha_i\right)}{\Gamma(\alpha_1+\alpha_2)\prod_{i=3}^k \Gamma(\alpha_i)}$$

which is now recognized as a Dirichlet distribution on $(p_1+p_2, p_3, \ldots, p_{k-1})$ with parameter $(\alpha_1+\alpha_2, \alpha_3, \ldots, \alpha_k)$. □

8.9.2 *The marginal distribution of X*

Theorem 8.9.3. *Suppose X has a multinomial distribution with parameters $(n, \mathbf{p})$, and that $\mathbf{p}$ has a Dirichlet distribution with parameter $\boldsymbol{\alpha}$, then marginally X has a multinomial distribution in which*

$$P\{X=j\} = \alpha_j / \sum_{i=1}^k \alpha_i.$$

Proof.

$$\begin{aligned}
P\{X=j|n,p\alpha\} &= pj \\
P\{X=j|n,\alpha\} &= \int p_j \prod p_i^{\alpha_i-1}\left(\Gamma(\textstyle\sum_{i=1}^k \alpha_i)/\prod_{i=1}^k \Gamma(\alpha_i)\right) d\mathbf{p} \\
&= \left(\Gamma(\textstyle\sum_{i=1}^k \alpha_i)/\prod_{i=1}^k \Gamma(\alpha_i)\right) \int p_j \prod_{i=1}^k p_i^{\alpha_i-1} d\mathbf{p} \\
&= \left(\Gamma(\textstyle\sum_{i=1}^k)/\prod_{i=1}^k \Gamma(\alpha_i)\right) \int p_j^{(\alpha_j-1)+1} \prod_{i\neq j=1}^k p_i^{\alpha_i-1} d\mathbf{p} \\
&= \left(\Gamma(\textstyle\sum_{i=1}^k \alpha_i)/\prod_{i=1}^k \Gamma(\alpha_i)\right)\left(\frac{\Gamma(\alpha_j+1)\prod_{i\neq j=1}^k \Gamma(\alpha_i)}{\Gamma(1+\sum_{i=1}^k \alpha_i)}\right) \\
&= \frac{\Gamma(\alpha_j+1)}{\Gamma(\alpha_j)} \cdot \frac{\Gamma(\sum_{i=1}^k \alpha_i)}{\Gamma(1+\sum_{i=1}^k \alpha_i)} \\
&= \frac{\alpha_j}{\sum_{i=1}^k \alpha_i}
\end{aligned}$$

□

When an observation X_1 is observed, X_1 takes the value δ_{X_1}. Then the associated α is increased by 1, and the other α's are unchanged. By induction, after m observations are observed, suppose n_j of them take the value $j, j = 1, \ldots, k$. Note that $m = \sum_{j=1}^{k} n_j$. Then the α's relevant to X_{m+1} are $\alpha'_j = \alpha_j + n_j, j = 1, \ldots, k$. Then

$$P\{X_{m+1} = j|\boldsymbol{\alpha}, X_1, \ldots, X_m\} = \alpha'_j / \sum_{i=1}^{k} \alpha'_i. \tag{8.80}$$

To establish some notation, let $\sum_{i=1}^{k} \alpha_i = W$ and let $G(j) = \alpha_j/W$ be the prior distribution before any observations are taken. Then

$$\begin{aligned} P\{X_{m+1} = j|\boldsymbol{\alpha}, X_1, \ldots, X_m\} &= \tfrac{\alpha_j+n_j}{W+n} = \tfrac{\alpha_j}{W}\left(\tfrac{W}{W+m}\right) + \tfrac{n_j}{m}\left(\tfrac{m}{W+m}\right) \\ &= G(j)\left(\tfrac{W}{W+m}\right) + \tfrac{n_i}{m}\left(\tfrac{m}{W+m}\right). \end{aligned} \tag{8.81}$$

This displays $P\{X_{m+1} = j|\boldsymbol{\alpha}, X_1, \ldots, X_m\}$ as a convex combination of $G(j)$, the prior, and n_j/m, the empirical frequency of j. The result is a generalization of the Polya Urn Scheme (see 2.10.1). The variables $X_1, \ldots, X_m$ are m-exchangeable, because the n_j's depend only on the number of X's that take the value j, and not on their order.

8.9.3 A relationship to the Gamma distribution

Theorem 8.9.4. *Let $Y_i, i = 1, \ldots, K$ have independent Gamma distributions with parameters α_i and β, respectively. Let $V = \sum_{i=1}^{K} Y_i$ and let $\mathbf{X} = (X_1, \ldots, X_K) = (Y_1/V, Y_2/V, \ldots, Y_K/V)$. Then*

a) $\mathbf{X}$ has a Dirichlet distribution with parameters $(\alpha_1, \ldots, \alpha_K)$.

b) V has a Gamma distribution with parameters $\sum_{i=1}^{K} \alpha_i$ and β.

c) $\mathbf{X}$ and V are independent.

Proof. The joint distribution of $Y_1, \ldots, Y_K$ has density

$$f(y, \ldots, y_k) = \prod_{i=1}^{K} \frac{\beta^{\alpha_i} y_i^{\alpha_i - 1} e^{-\beta y_i}}{\Gamma(\alpha_i)} dy_1, \ldots, dy_K.$$

Let $W_K = \beta \sum_{i=1}^{K} y_i$ and $w_i = \beta y_i / W_K \quad i = 1, \ldots, K-1$.

Then the inverse transformation is $y_i = W_K w_i/\beta, i = 1, \ldots, K-1$ and $y_K = W_K/\beta - \sum_{i=1}^{K-1} y_i = W_K/\beta - \sum_{i=1}^{K-1} W_k W_i/\beta = W_K/\beta(1 - \sum_{i=1}^{K-1} w_i)$.

The Jacobian matrix of the transformation is

$$\begin{bmatrix} \frac{\partial y_1}{\partial w_1} & \frac{\partial y_2}{\partial w_1} & \cdots & \frac{\partial y_K}{d w_1} \\ \vdots & & & \\ \frac{\partial y_K}{\partial w_k} & & & \frac{\partial y_K}{\partial W_K} \end{bmatrix} = \begin{bmatrix} W_K/\beta & 0 & \cdots & 0 & W_1/\beta \\ 0 & W_K/\beta & \cdots & 0 & W_2/\beta \\ \vdots & & W_K/\beta & & \\ 0 & & & & \\ -W_K/\beta & -W_K/\beta & & & \frac{1}{\beta}(1 - \sum_{i=1}^{K-1} W_i) \end{bmatrix}.$$

Adding the top $K-1$ rows to the last yields

$$\begin{bmatrix} W_K k/\beta & \cdots & 0 & W_1/\beta \\ 0 & W_K/\beta & & W_2/\beta \\ \vdots & & & \vdots \\ & & & W_{K-1}/\beta \\ 0 & 0 & \cdots\ 0 & 1/\beta \end{bmatrix},$$

so the determinant is W_K^{K-1}/β^K.

Returning to the joint density with the transformed variables,

$$\begin{aligned} f(W_1, \ldots, W_K) &= \prod_{i=1}^K \frac{\beta^{\alpha_i}(W_K W_i/\beta)^{\alpha_i - 1}}{\Gamma(\alpha_i)} e^{-W_K} W_K^{K-1}/\beta^K dw_1, \ldots, dw_K \\ &= \prod_{i=1}^K \frac{w_i^{\alpha_i - 1}}{\Gamma(\alpha_i)} e^{-W_K} w_K^{K-1+\sum \alpha_i - K} dw_i, \ldots, dw_K \\ &= \prod_{i=1}^{K-1} \frac{W_i^{\alpha_i - 1}(1 - \sum_{i=1}^{K-1} w_i)^{\alpha_K - 1}}{\prod_{i=1}^K \Gamma(\alpha_i)} \Gamma(\textstyle\sum_{i=1}^K \alpha_i)^{dw_i, \ldots, dw_K} \frac{e^{-W_K} w_K^{\sum v_i - 1}}{\Gamma(\sum_{i=1}^K \alpha_i)} dw_K \end{aligned}.$$

Therefore $w_1, \ldots, w_{K-1}$ has a Dirichlet distribution with parameters $(\alpha_1, \ldots, \alpha_K)$, W_K has a Gamma distribution with parameters $\sum_{i=1}^K \alpha_i$ and 1, and they are independent.

Furthermore, $X_i = Y_i/V = \frac{W_K w_i/\beta}{W_K/\beta} = W_i, i = 1, \ldots, K-1$, proving a).

Also, $V = W_K/\beta$, so $W_K = \beta V$ and the density of V is

$$f(V) = \frac{(V\beta)^{\sum \alpha_i - 1} e^{-V\beta}}{\Gamma(\sum_{i=1}^K \alpha_i)} \cdot \beta dV = \frac{\beta^{\sum \alpha_i} V^{\sum \alpha_i - 1} e^{-V\beta}}{\Gamma(\sum_{i=1}^K \alpha_i)} \text{proving c).}$$

b) follows from the independence of W_K and $(W_1, \ldots, W_{K-1})$. □

8.9.4 *Stick breaking*

The stick-breaking representation of the Dirichlet distribution is as follows:

Theorem 8.9.5. *Let $\prod \sim$ Dirichlet (αg_0). Then the distribution of $\prod$ is the same as*

$$\sum_{i=1}^{\infty} V_i \prod_{j=1}^{i-1} (1 - V_j) e_{y_i}$$

where $V_i \sim \beta(1, \alpha), i = 1, 2, \ldots$ and are independent and Y_i has a multinomial distribution with probability vector g_o, independent of each other and of the V's.

Remark: The reason for the name is as follows: consider a stick of length 1. Then V_1 "breaks" off some proportion of it. The remainder is of length $(1 - V_1)$. Then V_2 breaks off some additional proportion of the remainder, so there is now $(1 - V_1)(1 - V_2)$ left, etc.

One of the advantages of the stick-breaking representation is that it offers a way of simulating from $\prod$, truncating the process when the stick becomes very short.

The proof of Theorem 8.9.5 proceeds through some lemmas.

Lemma 8.9.6.

a) If Z has the same distribution as $\sum_{k=1}^K g_{ok} Dir(\alpha g_o + e_k)$ then $Z \sim Dir(\alpha g_o)$.

b) If $Z \sim Dir(\alpha g_o)$, then Z has the same distribution as $\sum_{k=1}^K g_{ok} Dir(\alpha g_o + e_k)$.

Proof.
a) Dir $(\alpha g_o + e_K)$ has density

$$\begin{aligned} &\frac{\left(\prod_{\substack{j \neq k \\ j=1}}^K p_j^{\alpha g_{oj}}\right) p_K^{\alpha g_o + 1} \Gamma(1 + \sum_{j=1}^K \alpha g_{oj})}{\left(\prod_{\substack{j \neq k \\ j=1}}^K \Gamma(\alpha g_{oj})\right) \Gamma(\alpha g_{ok} + 1)} \\ &= \frac{(\prod_{j=1}^K p_j^{\alpha g_{oj}}) P_k \Gamma(\sum_{j=1}^K \alpha g_{oj})(\sum_{j=1}^K \alpha g_{oj})}{(\prod_{j=1}^K \Gamma(\alpha g_{oj}))\alpha g_{ok}} \\ &= \left\{ \frac{\prod_{j=1}^K p_j^{\alpha g_{oj}} \Gamma(\sum_{j=1}^K \alpha g_{oj})}{\prod_{j=1}^K \Gamma(\alpha g_{oj})} \right\} \cdot \frac{p_k}{\alpha g_{ok}} \sum_{j=1}^K \alpha g_{oj}. \end{aligned}$$

Let W be the density in curly brackets. Then $Dir(\alpha g_o + e_K)$ has density $W\frac{P_K}{g_{oK}}$, because $\sum_{j=1}^{K} g_{oj} = 1$.

Finally, $\sum_{k=1}^{K} g_{ok}\text{Dir}(\alpha g_o + e_K)$ has density

$$W\sum_{k=1}^{K}\frac{g_{ok}P_k}{g_{ok}} = W\sum_{k=1}^{K}P_k = W.$$

But W is the density of $Dir(\alpha g_o)$, proving a).

To prove b): Let Y be a random variable that takes the value e_k with probability g_{ok}. Then the density of Z can be written as the sum of the density of $Z|Y = e_k$ times the probability of $Y = e_k$, summed over the possible values of e_k. But this is precisely

$$\sum_{k=1}^{K} g_{oK}Dir(\alpha g_o + e_K), \text{ proving } b).$$

□

Lemma 8.9.7. *Let* $W_1 \sim Dir(w_1, \ldots, w_K), W_2 \sim Dir(v_1, v_2, \ldots, v_k)$, *and let* $V \sim Beta(\sum_{i=1}^{K} w_i, \sum_{i=1}^{K} v_i)$ *and suppose* W_1, W_2 *and* V *are independent. Then*

$$Z = VW_1 + (1 - V)W_2 \sim Dir(w_1 + v_1, \ldots, w_K + v_K).$$

Proof. I use a construction using gamma random variables, and show that the construction satisfies the assumptions of the lemma. Then I show, using the construction, that the conclusion follows:

Let $\gamma_i \sim Gam(w_i, \lambda), i = 1, \ldots, K$ independent. Let $V_1 = \sum_{i=1}^{K} \gamma_i$. Then, by Theorem 8.9.4(b), $V_i \sim Gam(\sum_{i=1}^{K} w_i, \lambda)$. Similarly, let $\gamma_i' \sim Gam(v_i, \lambda), i = 1, \ldots, K$, independent of each other and the γ's. Let $V_2 = \sum_{i=1}^{K} \gamma_i$. Then $V_2 \sim Gam(\sum_{i=1}^{K} v_i, \lambda)$. Let $W_1 = \frac{1}{V_1}(\gamma_1, \gamma_2, \ldots, \gamma_K)$ and $W_2 = \frac{1}{V_2}(\gamma_1', \gamma_2', \ldots, \gamma_K')$. Then $W_1 \sim Dir(w_1, \ldots, w_K)$ by Theorem8.9.4(a) and is independent of V_1 by Theorem 8.9.4(c). Similarly, $W_2 \sim Dir(V_1, \ldots, V_K)$, and is independent of V_2. Let $V = V_1/(V_1 + V_2)$.

Now consider the joint distribution of (W_1, V_1, W_2, V_2). Because the γ's are independent of the γ''s, (W_1, V_1) is independent of (W_2, V_2). Furthermore, W_1 is independent of V_1, and W_2 of V_2. Therefore, W_1, V_1, W_2 and V_2 are independent of each other, and specifically W_1 and W_2 are independent of V. Hence, this construction satisfies the assumptions of the lemma.

Then $Z = \frac{V_1W_1 + V_2W_2}{V} = \frac{(\gamma_1+\gamma_1', \gamma_2+\gamma_2', \ldots, \gamma_n+\gamma_n')}{V}$ has a $Dir(v_1 + w_1, v_2 + w_2, \ldots, v_n + w_n)$ distribution, again using Theorem 8.9.4(a). □

Proceeding now to the proof of Theorem 8.9.5 with the help of Lemmas 8.9.6 and 8.9.7, the following results:

Proof.

$$\begin{aligned}
\pi &\sim Dir(\alpha g_o + e_K)\\
\pi &= VW + (1 - V)\pi',\\
\text{where } W &\sim Dir(e_Y), \pi' \sim Dir(\alpha g_o)\\
\text{and } V &\sim Beta(\textstyle\sum_{k=1}^{K} e_{y_k}, \sum_{k=1}^{K} \alpha g_{ok})\\
\text{also } Y &\sim Mult(g_o).
\end{aligned}$$

Now $\sum_{k=1}^{K} e_{y,k} = 1$ and $\sum_{k=1}^{K} \alpha g_{ok} = \alpha \sum_{k=1}^{K} g_{ok} = \alpha$.
W is degenerate: only one of the K parameters is non-zero, and $P\{W = e_y | g_o = e_y\} = 1$.
So specialized, the result is

$$\pi = V_1 e_Y + (1 - V_1)\pi',$$

where π and π' both are $Dir(\alpha g_o)$ distributed random variables, $V_1 \sim Beta(1, \alpha)$ independent of π, and Y is a multinomial with probability vector g_o.

Now recursion on π' yields the result, noting that

$$\prod_{i=1}^{\infty}(1 - V_i) = 0.$$

□

8.9.4.1 *Acknowledgements*

This section is based partly on Paisley (2010), although my proofs differ.

8.9.5 *Summary*

The Dirichlet distribution is conjugate to the multinomial distribution; its special case when $k = 2$, the Beta distribution, is conjugate to the $k = 2$ special case of the multinomial distribution, namely the binomial distribution. The Dirichlet distribution permits graceful coarsening, just as the multinomial distribution does. The Dirichlet distribution admits a stick-breaking interpretation.

8.9.6 *Exercises*

1. Write down the omitted constant in (8.78).
2. Suppose $(p_1, \ldots, p_k)$ have a Dirichlet distribution with parameters $(\alpha_1, \ldots, \alpha_k)$. Find the covariance between p_i and p_j.

8.10 The exponential families

We have now seen many examples of conjugate pairs of distributions, and there's a sense in which they all are similar. The purpose of this section is to display that similarity.

A distribution is a member of a k-dimensional conjugate family if it can be represented as a density as follows:

$$f(x \mid \theta) \propto \exp\left\{\sum_{j=1}^{k} A_j(\theta) B_j(x) + D(\theta)\right\}. \tag{8.82}$$

Suppose the prior on θ can be represented by

$$f(\theta \mid a_1, \ldots, a_k, d) \propto \exp\left\{\sum_{j=1}^{k} a_j A_j(\theta) + dD(\theta)\right\}. \tag{8.83}$$

Then the posterior on θ is proportional to

$$f(\theta \mid a_1 + B_1(x), a_2 + B_2(x), \ldots, a_k + B_k(x), d + 1).$$

Consider first the example of section 8.1, the univariate normal distribution with precision known with certainty but with uncertain mean μ having a normal prior distribution. Then the density of the observations is

$$f(\mathbf{x}, \mu) \propto e^{-Q_1(\mu)/2}$$

where $Q_1(\mu) = \tau_0 \sum_{i=1}^{n} (\mathbf{X}_i - \mu)^2 = \tau_0[n\mu^2 - 2\tau_0\mu n\overline{X} + \tau_0 \sum_{i=1}^{n} X_i]$, so we may take $A_1(\mu) = \frac{-\mu^2}{2}$, $A_2(\mu) = -\mu/2, B_1(\mathbf{x}) = \tau_0 n$, and $B_2(\mathbf{x}) = -2\tau_0 n\overline{X}$.

The prior is then proportional to

$$e^{-Q_2(\mu)/2}$$

where $Q_2(\mu) = \tau_1(\mu - \mu_1)^2 = \mu^2\tau_1 - 2\mu\tau_1\mu_1 + \tau_1\mu_1^2$, so $a_1 = \tau_1$ and $a_2 = -2\tau_1\mu_1$.

Then the posterior is proportional to

$$e^{-(Q_1(\mu)+Q_2(\mu))/2} = e^{-Q(\mu)/2}$$

where $Q(\mu) = \mu^2(\tau_0 n + \tau_1) + \mu(-2\tau_0 n\overline{X} - 2\tau_1\mu)$.

The rest of example 1 consists in reformulating this quadratic in terms of the normal distribution.

Each of the other examples examined so far in this chapter can be viewed as members of an exponential family of distributions, with an associated conjugate family of prior distributions. However, although the exponential family covers many cases, it does not exhaust the examples of conjugate prior distributions. Consider, for example, data that is uniform on $(0, \theta)$ where θ is uncertain. Then

$$f(x \mid \theta) = \frac{1}{\theta} \quad 0 < x < \theta,$$

and 0 otherwise.

If a sample of size n is observed, we have

$$f(\mathbf{x} \mid \theta) = \begin{cases} \frac{1}{\theta^n} & \theta > \max_{i=1,\ldots n} x_i \\ 0 & \text{otherwise} \end{cases}.$$

The conjugate family for this distribution is the Pareto distribution with parameters α and x_0:

$$f(\theta) = \begin{cases} \frac{\alpha x_0^\alpha}{\theta^{\alpha+1}} & \theta \geq x_0 \\ 0 & \text{otherwise} \end{cases}.$$

Then the posterior on θ is

$$\begin{aligned} f(\mathbf{x}, \theta) &= \frac{1}{\theta^n} \cdot \frac{\alpha x_0^\alpha}{\theta^{\alpha+1}} \qquad \theta \geq x_0, \theta \geq \max_{i=1,\ldots n} x_i \\ &\propto \frac{1}{\theta^{n+\alpha+1}} \qquad \theta \geq \max_{i=0,\ldots,n} x_i \end{aligned}$$

which is recognized as a Pareto distribution with parameters $\alpha' = n + \alpha$ and $x_0' = \max_{i=0,\ldots,n} x_i$.

This distribution is not a member of the exponential family.

Suppose ω is the whole sample, and y is a statistic, i.e., a function of ω. If the joint distribution of ω and y can be factored as follows

$$f_\theta(\omega, y) = f(\omega|y) f_\theta(y)$$

then Fisher (1922) defined y as a sufficient statistic. Note that $f(\omega|y)$ does not depend on θ. In a Bayesian analysis, the factor $f(\omega|y)$ is irrelevant, since

$$\pi(\theta|\omega) = \frac{f(\omega|y)f_\theta(y)\pi(\theta)}{\int f(\omega|y)f_\theta(y)\pi(\omega)d\theta} = \frac{f_\theta(y)\pi(\theta)}{\int f_\theta(y)\pi(\theta)d\theta}.$$

The theorem of Koopman (1936), Pitman (1936) and Darmois (1935) [they discovered it independently at about the same time] shows that if the support of the density does not depend on the parameter, then only exponential families have sufficient statistics of fixed dimension. (The latter condition is necessary since ω (all the data) is always a sufficient statistic, but of course its dimension is the sample size.)

Also in 1922, Fisher (1925) began a discussion of ancillary statistics, which have been defined to mean statistics that don't depend on the parameters. An example would be $\omega|y$ in the discussion of sufficiency. Whether to condition on an ancillary statistic is irrelevant to a Bayesian (see above), but is a matter of controversy among frequentists.

8.10.1 *Summary*

Most examples of conjugate families of likelihoods are members of exponential families. However the uniform distribution on $(0, \theta)$ is an example to show that not all conjugate families are exponential. Bayesian analysis automatically makes use of sufficient statistics when they are available, and is invariant to conditioning on ancillary statistics.

8.10.2 *Exercises*

1. For each of the following, display the likelihood in the form of (8.82) and the conjugate prior in the form of (8.83):
   (a) the multivariate normal case with known precision (section 8.2).
   (b) the normal linear model (section 8.3) with known precision.
   (c) the univariate normal with known mean and unknown precision (section 8.4).
   (d) the univariate normal with both mean and precision uncertain (section 8.5).
   (e) the normal linear model with uncertain scale (section 8.6).
   (f) the multivariate normal distribution with uncertain precision and certain mean (section 8.7.6).
   (g) the multivariate normal distribution with both mean and precision matrix uncertain (section 8.8).
   (h) the multinomial distribution (section 8.9) – hint: you might want to start with the binomial distribution.
2. Show that the density $f(x|\theta) = 1/\theta, 0 < x < \theta$, and 0 otherwise, is not a member of the exponential family.

8.10.3 *Utility*

In an interesting paper, Lindley (1976) explores the possibility of using conjugate forms for utility as well. These have the advantage of making the calculation of expected utility simpler, just as using a conjugate prior makes the calculation of the posterior distribution simpler.

8.11 Large sample theory for Bayesians

While Bayesian analysis usually occurs for a fixed sample size n, it may be useful to see what happens as the sample size gets large. We'll concentrate on the conditionally independent and identically distributed case. The arguments here are only heuristic, intended to give a flavor of the results. To make them rigorous would require controlling the order of the error terms.

The posterior after n observations from a likelihood $g(x \mid \theta)$ and prior $\pi(\theta)$ can be written

$$\begin{aligned} f_n(\theta \mid \mathbf{x}) &\propto \pi(\theta) \prod_{i=1}^{n} g(x_i \mid \theta) \\ &= \pi(\theta) e^{n\left[\sum_{i=1}^{n} \frac{\log g(X_i \mid \theta)}{n}\right]}. \end{aligned}$$

Now $\frac{\sum_{i=1}^{n} \log g(x_i|\theta)}{n}$ is the average of a function of n independent random variables, which, by the law of large numbers, approaches its expectation. However, we must discuss the nature of this expectation. The Bayesian believes there is some "true" θ_0, but doesn't know what it is (if θ_0 were known it would not be necessary to compute the posterior). With respect to this true but unknown θ_0, the distribution of observations x, in the opinion of this Bayesian, is $g(X \mid \theta_0)$. Therefore the Bayesian believes that

$$\frac{\sum_{i=1}^{n} \log g(X_i \mid \theta)}{n} \rightarrow \int [\log g(x \mid \theta)]\, g(x \mid \theta_0) dx.$$

Provided $\pi(\theta_0) > 0$, we then have, for large n

$$f_n(\theta \mid x) \propto \pi(\theta_0) e^{n \int [\log g(x|\theta)] g(x|\theta_0) dx}.$$

8.11.1 A supplement on convex functions and Jensen's Inequality

A function $h(x)$ is strictly convex on an interval $I = [a, b]$ if

$$h(tx + (1-t)y) < th(x) + (1-t)h(y)$$

for all $x, y \epsilon I$ and for all $t, 0 < t < 1$.

By induction, this implies

$$h\left(\sum_{i=1}^{n} p_i x_i\right) < \sum_{i=1}^{n} p_i h(X_i)$$

provided $p_i > 0$ and $\sum_{i=1}^{n} p_i = 1$. Consequently, if h is strictly convex

$$h(E(X)) < Eh(X)$$

provided X is non-trivial. This is known as Jensen's Inequality.

Lemma 8.11.1. *If h'' exists and is positive, then h is strictly convex.*

Proof. Let x and y be given, $x, y \epsilon I$. Without loss of generality, we may suppose $x < y$. Let $0 < t < 1$ be given. Then $x < tx + (1-t)y < y$.

Now $h'' > 0$ implies that h' is an increasing function. Thus if $\xi \epsilon (x, tx + (1-t)y)$ and $\eta \epsilon (tx + (1-t)y, y)$, then $h'(\xi) < h'(\eta)$ because $\xi < \eta$. Then

$$\frac{\int_x^{tx+(1-t)y} h'(\xi) d\xi}{tx + (1-t)y - x} < \frac{\int_{tx+(1-t)y}^{y} h'(\eta) d\eta}{y - tx + (1-t)y},$$

so

$$\frac{h(tx + (1-t)y) - h(x)}{(1-t)(y-x)} < \frac{h(y) - h(tx + (1-t)y)}{t(y-x)}.$$

But this implies

$$t[h(tx + (1-t)y) - h(x)] < (1-t)[h(y) - h(tx + (1-t)y)]$$

or

$$h[tx + (1-t)y] < th(x) + (1-t)h(y)$$

so h is strictly convex. □

8.11.2 *Resuming the main argument*

We now observe that the function $h(x) = x \log x$ is convex, because $h'(x) = \log x + 1$, and $h''(x) = \frac{1}{x} > 0$ for $x > 0$.

Now consider applying Jensen's Inequality to the random variable $Y = \frac{g(X|\theta)}{g(X|\theta_0)}$ with respect to the probability distribution $g(X \mid \theta_0)$ and the convex function $h(x) = x \log x$. Then

$$EY = \int \frac{g(x \mid \theta)}{g(x \mid \theta_0)} g(x \mid \theta_0) dx = \int g(x \mid \theta) dx = 1.$$

Thus $h(E(Y)) = 1(\log 1) = 0$.

Hence we have

$$\int \frac{g(x \mid \theta)}{g(x \mid \theta_0)} \log \left(\frac{g(x \mid \theta)}{g(x \mid \theta_0)} \right) \cdot g(x \mid \theta_0) dx < 0$$

or $\int g(x \mid \theta) \log g(x \mid \theta) dx < \int g(x \mid \theta) \log g(x \mid \theta_0)$ with equality only when

$$g(x \mid \theta) = g(x \mid \theta_0) \text{ for all } x.$$

If there is only one value of θ_0 satisfying this equation, then this argument shows that the probability will all pile up at that point as n gets large. Thus, for large n,

$$f_n(\theta \mid x) \propto \pi(\theta_0) e^{n \int \log[g(x|\theta_0)] g(x|\theta_0)} dx$$

which means that the Bayesian believes that, as the sample size gets large, all the probability will pile up at θ_0.

Now suppose there is more than one value of θ for which $g(x \mid \theta) = g(x \mid \theta_0)$. This is the case of non-identification. Then no amount of data will distinguish θ from θ_0, and so, no matter how large n may be, the relative weight given to such θ and θ_0 will depend on the prior alone. This is a feature, but not a fault, of Bayesian analysis, since it gives a straight-forward consequence of the assumptions made (*i.e.*, beliefs of the Bayesian).

We now extend the argument to examine the posterior distribution around the maximum posterior point; assuming that to be unique:

We already know that

$$f_n(\theta \mid X) \propto e^{L_n(\theta)}$$

where $L_n(\theta) = \log \pi(\theta) + \sum_{i=1}^{n} \log g(X_i \mid \theta)$.

Expand $L_n(\theta)$ in a Taylor series around its maximum, $\hat{\theta}$.

$$L_n(\theta) = L_n(\hat{\theta}) + (\theta - \hat{\theta}) L_n'(\hat{\theta}) + \frac{(\theta - \hat{\theta})^2}{2} L_n''(\hat{\theta}) + HOT.$$

Now $L'_n(\hat{\theta}) = 0$ because $\hat{\theta}$ is chosen to maximize $L_n(\hat{\theta})$. Also $e^{L_n(\hat{\theta})}$ is a constant, that can be absorbed by the constant of proportionality. Therefore

$$f_n(\theta \mid X) \propto e^{\frac{(\theta-\hat{\theta})^2}{2} L''_n(\hat{\theta}) + HOT}.$$

Remembering that $L''_n(\hat{\theta}) < 0$ because $\hat{\theta}$ maximizes $L_n(\hat{\theta})$, we have that the posterior of θ is approximately normal, with mean $\hat{\theta}$ and precision $-L''_n(\hat{\theta})$.

When $\boldsymbol{\theta}$ is a vector, the Taylor expansion looks slightly different:

$$L_n(\boldsymbol{\theta}) = L_n(\hat{\boldsymbol{\theta}}) + (\boldsymbol{\theta} - \hat{\boldsymbol{\theta}})'\delta L^*_n(\hat{\boldsymbol{\theta}}) + (1/2)(\boldsymbol{\theta} - \hat{\boldsymbol{\theta}})'\delta^2 L_n(\hat{\boldsymbol{\theta}})(\boldsymbol{\theta} - \hat{\boldsymbol{\theta}}) + HOT$$

where $\delta L^*_n(\hat{\boldsymbol{\theta}}) = \left(\frac{dL(\hat{\theta})}{\delta\theta_1}, \frac{dL(\hat{\theta})}{\delta\theta_2}, \ldots \frac{dL(\hat{\theta})}{\delta\theta_k}\right)$ and $\delta^2 L_n(\hat{\boldsymbol{\theta}})$ is a $k \times k$ matrix whose i, j^{th} element is $\frac{\delta^2 L_n(\hat{\boldsymbol{\theta}})}{\delta\theta_i \delta\theta_j}$.

Using the same argument, we now see that $f_n(\boldsymbol{\theta} \mid X)$ has an asymptotic k-dimensional normal distribution with mean $\hat{\boldsymbol{\theta}}$, and precision matrix $-\delta^2 L_n(\hat{\theta})$.

The same technique can be used to approximate moments of posterior distributions. Suppose $g(\theta)$ is a positive function of θ. Then

$$\begin{aligned} Eg(\theta) &= \frac{\int g(\theta) \prod_{i=1}^n f(x_i \mid \theta)\pi(\theta)d\theta}{\int \prod_{i=1}^n f(x_i \mid \theta)\pi(\theta)d\theta} \\ &= \frac{\int e^{nL^*_n(\theta)}d\theta}{\int e^{nL_n(\theta)}} \end{aligned}$$

where $L_n(\theta) = \frac{\log \pi(\theta) + \sum_{i=1}^n \log f(X_i|\theta)}{n}$ and $L^*_n(\theta) = L_n(\theta) + \frac{\log g(\theta)}{n}$.

Let $\hat{\theta}$ maximize $L_n(\theta)$ and $\hat{\theta}^*$ maximize $L^*_n(\theta)$.

Then we have

$$\begin{aligned} Eg(\theta) &\simeq \frac{e^{nL^*_n(\hat{\theta}^*)} \int e^{-\frac{1}{2}(\theta-\hat{\theta}^*)^2 L''_n(\hat{\theta}^*) + HOT} d\theta}{e^{nL_n(\hat{\theta})} \int e^{-\frac{1}{2}(\theta-\hat{\theta}^*)^2 L''_n(\hat{\theta}) + HOT} d\theta} \\ &= \frac{e^{nL^*_n(\hat{\theta}^*)} L^{*\prime\prime}_n(\hat{\theta}^*)^{1/2}}{e^{nL_n(\hat{\theta})} L''_n(\hat{\theta})^{1/2}}, \end{aligned}$$

which is the univariate form of the Laplace Approximation. The multivariate version, not surprisingly, is

$$E(g(\boldsymbol{\theta})) = \frac{e^{nL_n(\hat{\boldsymbol{\theta}}^*)} \mid \delta^2 L^*_n(\hat{\boldsymbol{\theta}}^*) \mid^{1/2}}{e^{nL_n(\hat{\boldsymbol{\theta}})} \mid \delta^2 L_n(\hat{\boldsymbol{\theta}}) \mid^{1/2}}.$$

When g might be negative, one approach is to use the above approximation on the moment generating function, and then to take the first derivative at $t = 0$.

8.11.3 Exercises

1. Vocabulary: What is the Laplace Approximation?
2. Consider the integral representation of $n!$, namely

$$n! = \Gamma(n+1) = \int_0^\infty x^n e^{-x} dx = \int_0^\infty e^{L(x)} dx,$$

   where $L(x) = -x + n \log x$.

(a) Expand $L(x)$ in a Taylor series, retaining the constant, linear and quadratic terms.

(b) Evaluate the Taylor series at the point $x = \hat{x}$ that satisfies $L'(\hat{x}) = 0$.

(c) Derive Stirling's Approximation,

$$n! \doteq \sqrt{2\pi}\, n^{n+1/2} e^{-n}.$$

8.11.4 References

For consistency and asymptotic normality, see Johnson (1967, 1970), Walker (1969), Heyde and Johnstone (1979), Poskitt (1987), and Barron et al. (1999). For Laplace's method, see Tierney and Kadane (1986), Kass et al. (1988), Kass et al. (1989a), Tierney et al. (1989). For Stirling's Approximation, see Feller (1957). Laplace's method is also known in applied mathematics as a saddle-point approximation.

8.12 Some general perspective

Conjugate analysis is neat mathematically when it works. However, the slightest deviation in the specification of the likelihood or prior would destroy the property of conjugacy. Consequently, these results are interesting but far from a usable platform from which to do analyses.

Similarly, large sample theory is nice, but gives little guidance on how large a sample is required for large sample theory to yield good approximations. Since Bayesian analyses can and do deal with small samples as well as large ones (indeed Bayesians can gracefully make decisions with no data at all, relying on their prior), large sample theory is also quite limited in scope.

Because of these limitations, Bayesians now rely heavily on computational methods to find posterior distributions, as outlined in Chapter 10.

Chapter 9

Hierarchical Structuring of a Model

9.1 Introduction

Bayesian analysis requires a joint distribution of all the uncertain quantities deemed relevant to a problem, both data (before they are observed) and parameters. After the data are observed, of course, the relevant distribution is that of the parameters conditioned on the observed data. Hierarchical models have proven to be a particularly useful way of structuring that joint distribution.

Suppose the parameters $\boldsymbol{\theta}$ can be divided into sets, so that $\boldsymbol{\theta} = (\boldsymbol{\alpha}, \boldsymbol{\beta}, \boldsymbol{\gamma}, \boldsymbol{\delta}, \ldots)$, and suppose x represents the data. Then the desired joint distribution can be written, without loss of generality, as

$$\begin{aligned} f(x, \boldsymbol{\theta}) &= f(x, \boldsymbol{\alpha}, \boldsymbol{\beta}, \boldsymbol{\gamma}, \boldsymbol{\delta}, \ldots) \\ &= f_1(x \mid \boldsymbol{\alpha}, \boldsymbol{\beta}, \boldsymbol{\gamma}, \boldsymbol{\delta}, \ldots) f_2(\boldsymbol{\alpha} \mid \boldsymbol{\beta}, \boldsymbol{\gamma}, \boldsymbol{\delta}, \ldots) \\ & \quad f_3(\boldsymbol{\beta} \mid \boldsymbol{\gamma}, \boldsymbol{\delta}, \ldots) f_4(\boldsymbol{\gamma} \mid \boldsymbol{\delta}, \ldots) \text{ etc.} \end{aligned} \tag{9.1}$$

In certain circumstances (and this is the special trick of a hierarchical model), the conditional distributions in (9.1) can be simplified as follows.

$$\begin{aligned} f_1(x \mid \boldsymbol{\alpha}, \boldsymbol{\beta}, \boldsymbol{\gamma}, \boldsymbol{\delta}, \ldots) &= g_1(x \mid \boldsymbol{\alpha}) \\ f_2(\boldsymbol{\alpha} \mid \boldsymbol{\beta}, \boldsymbol{\gamma}, \boldsymbol{\delta}, \ldots) &= g_2(\boldsymbol{\alpha} \mid \boldsymbol{\beta}) \\ f_3(\boldsymbol{\beta} \mid \boldsymbol{\gamma}, \boldsymbol{\delta}, \ldots) &= g_3(\boldsymbol{\beta} \mid \boldsymbol{\gamma}) \\ f_4(\boldsymbol{\gamma} \mid \boldsymbol{\delta}, \ldots) &= g_4(\boldsymbol{\gamma} \mid \boldsymbol{\delta}) \text{ etc.} \end{aligned} \tag{9.2}$$

I think an example would be useful at this point. Suppose a standardized mathematics test is given to children in school. The data, $\mathbf{x}$, are the scores of each child. The parameters α might be the "true" ability of the child. Thus we might expect x to be centered on α, with some variance because performance on a test can vary from testing to testing for all sorts of reasons. The children in a single class are taught by the same teacher using the same materials, and thus the abilities, α, of children in the class might reasonably be thought to be related. Thus each α relating to a student in a class might be regarded as coming from a distribution of true abilities of children in that class, characterized by parameters β. Similarly, classes in a school may be related to each other with a distribution characterized by parameters γ, the school district by δ, the state, the nation, etc.

The hierarchical idea applies to this example with the thought that to predict how a particular child will do on the exam, all you need is the parameter α of the ability of that child. The α's for the other children, and the β's, γ's and δ's are irrelevant. Hence, it is reasonable to suppose that

$$f_1(x \mid \alpha, \beta, \gamma, \delta, \ldots) = g_1(x \mid \alpha) \tag{9.3}$$

for some distribution g_1. Similarly, if one wishes to understand the individual effects α, all that matters are the class parameters β. Thus we might write

$$f_2(\alpha \mid \beta, \gamma, \delta, \ldots) = g_2(\alpha \mid \beta) \tag{9.4}$$

for some (possibly different) distribution g_2. The same kind of argument applies to classes in the school, schools in a district, etc.

The benefit of hierarchical structuring is that it permits the modeling of each level in the hierarchy with a model suitable to that level. Additionally it correctly propagates uncertainty at each level, so that the posterior distributions reflect those uncertainties. Experience with hierarchical models suggests that this is a natural way of thinking about many problems, and permits decomposing a complex issue into subproblems, each of which can be understood and modeled.

This idea has old historical roots. Because these roots still play out in the current literature, it is useful to retrace a bit of them. The received wisdom in the early 1960's (see for example Scheffé (1959, 1999)) was to draw a distinction in linear models between "fixed effects" and "random effects." "Mixed effect models" had both random effects and fixed effects. And what was the difference between random effects and fixed effects? It had to do with what you were interested in. If you were interested in the ability of each child, you would treat the α's as fixed-effect parameters. If you were interested in the classes, but not the ability of each child, you would treat the α's as random effects and the βs as fixed effects in the example. There are several peculiarities in this from a Bayesian point of view. First, "random effects" are parameters with priors. The classical analysis integrates those parameters out of the likelihood. But, classically, parameters are not supposed to have distributions, and integrating with respect to a parameter is supposedly an illegitimate move. Second, the distinction between "random" and "fixed" is essentially about what one wishes to estimate, and thus is a matter of the utility function. How can it be that the utility function can affect what the likelihood is, particularly in a classical context in which the likelihood is imagined to be the objective truth about how the data were generated? Third, what if I care both about the children individually and about how classes of children compare? I can't treat the same parameter as both fixed and random in the same analysis! I can remember confused social scientists wanting advice about which parameters to treat as random and which as fixed, and being surprised at the response that all parameters are random (*i.e.*, are uncertain quantities that have distributions). From a Bayesian perspective there is no distinction, and no issue. With a probability model for the data and all the parameters, such as (9.1), the posterior distribution, conditioned on the data $\mathbf{x}$, gives a distribution for each child, each class, each school, etc. These distributions are correlated, in general, but that correlation causes no essential difficulty.

Another variant is called "empirical Bayes." The idea here is that at the highest level of the hierarchy (say at the international level in the example of the standardized mathematics test), no prior is imposed, but instead some classical estimation scheme, such as maximum likelihood, is used. Conditioning on those estimates, the rest of the model is treated in a Bayesian fashion. There is a systematic issue with this program, however. By treating the estimates of the parameters at the highest level of the hierarchy as if they were known to be the parameter value with no uncertainty, one is exaggerating the certainty with which all the other parameters are known as well. This can be seen from the formula

$$V(X) = EV(X \mid Y) + VE(X \mid Y). \tag{9.5}$$

(See section 2.12.7, exercise 3.) Here Y is the symbol for the highest level parameters, and X represents some other parameter in the model. What is desired is the variance of X. However, the empirical Bayes method sets the second term above to zero. Since it is non-negative, use of only the first term leads to systematic under-estimation of $V(X)$.

The solution to this difficulty, like the solution to the quandary of which parameters to treat as random and which as fixed, is instead to state a full hierarchical model in which all parameters are treated as random quantities.

9.1.1 *Summary*

A hierarchical model divides the parameters into groups that permit the imposition of assumptions of conditional independence. Historically they arose from discussions of random effects and mixed models, and of empirical Bayes methods.

9.1.2 *Exercises*

1. Vocabulary. State in your own words the meaning of:
   (a) fixed vs. random effects
   (b) mixed effect model
   (c) empirical Bayes
   (d) fully hierarchical model
2. Think of your own example of a hierarchical structure to model some phenomenon of interest to you.

9.1.3 *More history and related literature*

The impact of von Neumann and Morgenstern (1944)'s work on game theory was immense. (We'll study a bit of the details later, in Chapter 11.) Partly the influence was due to von Neumann's preeminence as a mathematician, and partly it had to do with the many ideas put forward in their book. Among the most important of those ideas was the use of utility functions. Another was the minimax approach to making decisions, which suggests choosing that decision that makes as good as possible the worst outcome that might happen.

These ideas were imported into statistics by Wald (1950), who advocated limiting attention to admissible procedures: those such that no other procedure does at least as well for all values of the parameter space and strictly better for at least one such value. It turns out that the admissible procedures are those supported by a proper prior distribution in the parameter space, together with certain limits of them. The set of admissible procedures is thus vast. For example, the estimate $\hat{\theta}(\mathbf{x}) = 3$ for all possible data sets $\mathbf{x}$, is admissible, because it is supported by the opinionated prior that puts probability 1 on the event $\Theta = 3$. The reason why $\hat{\theta}(\mathbf{x}) = 3$ is generally unacceptable as an estimate is that in most estimation problems, we have more uncertainty about Θ [indeed, why estimate it if you already know the answer?]. However, this subjective line of reasoning was unacceptable to Wald and most of his contemporaries. Various ad hoc methods were then proposed to choose among admissible estimators.

The next important result was due to Stein (1956, 1962) and James and Stein (1961). Using squared error loss, and the model

$$x_i \sim N(\theta_i, 1) \quad i = 1, \ldots, n, \quad \text{(independent)} \tag{9.6}$$

Stein showed that the maximum likelihood estimate $\hat{\theta}_i = x_i$ is admissible if $n = 1$ or 2, but not if $n > 2$. One does better drawing the $\hat{\theta}$'s toward an arbitrary origin. Lindley's discussion of Stein's paper (1962) shows that this shrinkage toward an origin is a simple consequence of a prior on the θ's, for example, that the θ's themselves are independently drawn from a normal distribution. (Chapter 8 of this book shows details of the Bayesian calculations.) Kempthorne (1971), commenting on Lindley (1971), gives references to earlier work in animal genetics where shrinkage was used. Novick (1972) gives a reference to earlier work in educational testing that also uses shrinkage. Lindley and Smith (1972) give a general theory for hierarchical models that have normal distributions at each stage.

Stein's result and Lindley's interpretation gave rise to many applied efforts. An expository paper by Efron and Morris (1977) studies several data sets. Looking at batting averages of baseball players half-way through a season, they show that the players with the highest averages tend to bat less impressively in the second half of the season, while those with the worst batting averages in the first half tend to do better. Thus, drawing in the batting averages toward a common mean seems to lead to better estimates. (While clever and plausible, I was always a bit uncomfortable with this argument for batters with low batting averages, because a manager might bench such a player.)

A second notable example is the paper of DuMouchel and Harris (1983). They use a hierarchical model to study the carcinogenicity of various chemicals (diesel engine emissions, cigarette smoke, coke oven emissions, etc.) on various species (*i.e.*, humans and mice) using various biological indicators. The goal, obviously, was to see to what extent experimental results in animals could be extrapolated to humans. Although the thinking is hierarchical Bayesian, the computations are empirical Bayesian, as the parameters of the highest level in the hierarchy were estimated using maximum likelihood methods, and these were then conditioned upon. (At the time, Bayesian computing did not have available the algorithms to be described in the next chapter.)

The idea of empirical Bayes was championed by Robbins (1956). Kass and Steffey (1989) pointed out that it systematically underestimates variances, using (9.5). Deeley and Lindley (1981) highlight the difference between empirical Bayes and the fully Bayesian methods suggested in this volume.

A modern treatment of hierarchical models is in Gelman and Hill (2007).

9.1.4 Minimax as a decision criterion

The decision theory of Chapter 7 is oriented toward maximizing expected utility. However, there is another way that statistical decision theorists propose that decisions be made, following Wald. They propose that a decision-maker conceive of a zero-sum game against Nature. Thus Nature is endowed with the superpower of knowing (and caring) what the decision-maker's utility and probabilities are, and choosing the parameter value least favorable to the decision-maker. There's a clinical diagnosis for such beliefs, namely paranoia.

As many who suffer from paranoia have discovered, one can always dream-up an even worse possibility to guard against. Thus, the minimax framework is unstable. Indeed, it is hard to understand why a true minimaxer would ever get out of bed in the morning, considering the worst things that might happen.

9.2 Missing data

My intent is to interpret missing data very broadly. The name suggests items that might have been observed but were not. The difficulty with this concept is that one can imagine many different possible worlds in which various unobserved items might have been observed. There seems to be no limit to what might have been observed but was not.

Consequently I take the view that missing data are simply parameters. This fits in with the general view taken here that proper statistical modeling requires a joint distribution of all the quantities of interest. When data becomes available, they are conditioned upon. This avoids all consideration of hypothetical worlds in which some, but not all, sources of uncertainty might have been revealed, but were not.

9.2.1 Examples

a. Speeding on the New Jersey Turnpike

While there are many kinds of examples, a few will suffice to show the scope of missing data. The first example is about a sample survey. To keep things as simple as possible, we'll suppose that there are N items, and a random (equally likely) sample of n is drawn. If all n can be reached and their response obtained, standard sampling theory (*i.e.*, Cochran (1977)) applies to find the uncertainty engendered by the fact that typically n is much less than N. For more on how random sampling fits in with Bayesian ideas, see Chapter 12, section 12.4.

In modern surveys, however, typically of the n sampled items, only m, many fewer than n, actually respond. There are two standard responses to this development, both extreme. One is to ignore the response rate, and treat the m items as if it were the selected random sample. The other is to decline to analyze the results of such a survey, on the grounds that the response rate is so low as to make the data meaningless. As a pragmatic matter, the first response is not too bad if m is close to n, but the second seems either unimaginative or lazy. The methods developed here suggest a third way, one that permits an analysis but that does not ignore the fact that desired data are unavailable.

I was involved as an expert witness in a lawsuit alleging racial bias in the enforcement of the traffic laws at the southern end of the New Jersey turnpike (see State of New Jersey vs. Pedro Soto et al. (1996)).

Together with my colleagues John Lamberth and Norma Terrin, we found the following:

(a) In a stationary survey, with observers on a bridge over the turnpike, about 13.5% of the cars observed on random days and times had an African-American occupant.

(b) In a rolling survey, with a car whose cruise-control was set for 60 miles per hour (the speed limit was 55), a count was made of the number of cars passing this car, the number he passed, and the race of the drivers. Of the cars encountered, over 98% passed the test car, and about 15% had an African-American occupant.

(c) In a study of those stopped for traffic violations, on randomly selected days, 46.2% were African Americans.

From (b) we could conclude that nearly everyone on the New Jersey turnpike was speeding, and hence vulnerable to being stopped. Legally this meant that everyone on the turnpike was "similarly situated." However, the statistical issue was that 69.1% of the race data on stops were missing, some because race data were omitted by the police officer, contrary to police regulations, and some because some data were destroyed pursuant to a police documentation retention policy.

If you ignore the issue of missing data, a simple application of Bayes Theorem yields

$$\begin{aligned}\theta =& \frac{P(\text{stop}|\text{black})}{P(\text{stop}|\text{white})} = \frac{P(\text{black}|\text{stop})P(\text{stop})/P(\text{black})}{P(\text{white}|\text{stop})P(\text{stop})/P(\text{white})} \\ =& \frac{.462/0.15}{.538/0.85} = 4.86. \end{aligned} \tag{9.7}$$

Hence, your odds θ of being stopped if you are black are nearly five times those of being stopped if you are white.

To analyze the situation further, and take into account the possibility of race-biased reporting, we considered the following notation (taken from Kadane and Terrin (1997).)

$$\begin{aligned}
r_1 &= \text{P(race reported} \mid \text{black and stopped)}\\
r_2 &= \text{P(race reported} \mid \text{white and stopped)}\\
t &= \text{P(black} \mid \text{stopped)}\\
1-t &= \text{P(white} \mid \text{stopped)}\\
n_1 &= \text{number of blacks reported as stopped}\\
n_2 &= \text{number of whites reported as stopped}\\
n_3 &= \text{number of people stopped whose race is not reported}
\end{aligned}$$

Three events may occur with a stop: the person stopped is black and the race is reported, the person stopped is white and the race is reported, or the person who is stopped does not have their race reported. These events have respective probabilities $r_1 t$, $r_2(1-t)$, and $(1-r_1)t + (1-r_2)(1-t)$. Since, given these parameters, the stops are regarded as independent and identically distributed, the likelihood function is trinomial:

$$(r_1 t)^{n_1}\{r_2(1-t)\}^{n_2}\{(1-r_1)t + (1-r_2)(1-t)\}^{n_3}. \tag{9.8}$$

Treating the parameters as t, r_1 and r_2, the goal is a distribution for Θ, as in equation (9.7), which in this notation is

$$\Theta = \frac{t/0.15}{(1-t)/0.85} = \frac{0.85t}{0.15(1-t)}. \tag{9.9}$$

Although there are three parameters about which information is sought, r_1, r_2 and t, there are only two free parameters in the trinomial. Hence, this system lacks identification, which is not a problem for Bayesians (see section 8.3).

Using a variety of priors in (r_1, r_2, t) space, and in particular different assumptions on

$$r = \frac{r_1/(1-r_1)}{r_2/(1-r_2)}, \tag{9.10}$$

the odds of having a stopped driver's race reported if black to that of white, we show that even if $r = 3$, the probability that $\theta > 1$, which would mean that blacks are more likely to be stopped than whites, is over 99%. (For more details, see Kadane and Terrin (1997).) This case had important consequences for New Jersey.

In other surveys, there may be useful auxiliary information about the conduct of the survey that may be brought to bear. In a study of Canadians' attitudes toward smoking in the workplace, sampled telephone numbers were called up to 12 times in an effort to get answers to attitude questions. The different responses of those who answered late in the survey compared to those who answered early were used to sharpen the prediction of what persons who were not contacted would have said (see Mariano and Kadane (2001)).

b. Informative missingness

In some circumstances, the fact that data are missing is somewhat informative about what the data would have been had it been observed. For example, it is well known that weak students tend not to be available to take high-stakes, especially multi-school, examinations. This could stem from decisions made by the students themselves, or from pressure from school authorities. Thus, the very fact that a student did not take a particular exam is somewhat informative about the score a student would have gotten had he or she taken the examination. A study that explicitly models this effect is Dunn et al. (2003).

c. Undetectable precision

It is common that environmental and other physical data fall below the level that can be reliably detected. In such circumstances, some analysts use a fixed number, such as zero, the detection limit, or half the detection limit. To do so exaggerates the certainty of the observation, and could even be regarded as fabricating data.

A sounder approach is to regard such missing observations as random variables having support between zero and the detection limit. While it might seem that there is no particular justification for taking one distribution over another for such missing data, theory and/or the shape of the distribution above the detection limit may offer guidance. If the conclusions drawn from the study depend importantly on what distribution is assumed for the missing data, this is an important consideration to make available to readers.

d. Lifetimes

In biostatistics, an important area, that goes under the title of survival analysis, has to do with how long people with a particular disease will survive after various treatments, perhaps as a function of covariates. Typically one does not want to wait until the last patient dies to draw conclusions. Thus, the unknown time of death of patients still alive is a kind of missing data.

In engineering, studies consider how long a machine will last before it breaks, or before it is unrepairable, perhaps as a function of the level of preventive maintenance it is given, and the conditions under which it is used. Again, it is usually inconvenient to wait until the last machine wears out.

In actuarial and demographic studies, the question is the distribution of lifetimes of a cohort of people in a population. Again, it is not useful to wait until the last of them dies to draw conclusions.

Statistically, these are very similar problems. In all three cases, imagining fixed, known, times of death (or machine failure) exaggerates the information in the data. In all three cases, some reasonable distribution over the times of failure (or death) offers an appropriate tool to model the uncertainty inherent in the situation.

e. Design of experiments when one treatment is dangerous

In biostatistics, there are situations in which verification of disease status is expensive and/or dangerous. As a consequence, less intrusive tests are used as proxies. Estimates of the sensitivity and specificity of such tests are influenced by the selection of patients to have the "gold standard," but highly intrusive diagnostic test. Here the issue is what the result of such a test would have been, had it been administered to all patients. For an example and references, see Kosinski and Barnhard (2003) and Buzoianu and Kadane (2008).

f. Regime switching

In many problems, it is useful to think of several possible underlying processes, and a mechanism that switches between processes. Sometimes the most important parameters are those that determine the current regime (*i.e.*, is there now a denial-of-service attack on a computer network, or not), sometimes it is the parameters within the regime that are most important. In both cases, the regime is unobserved, and hence can be regarded as missing data.

g. Measurement error

When important discrepancies are believed between what was measured and what was wished for, it is sound practice to model the discrepancy. This requires notation for the "true, underlying" variable measured with error. These additional variables should be thought of as parameters, and take their place in a hierarchical model. In a sense, they can be thought of as missing data.

h. Selection effects

A statistician should always be thinking about how the data before him or her came to be there. There is an old story, which may have never happened, that illustrates this point. According to the story, in World War II a statistician was asked by an Air Force general to study where the bullet holes were on the fighter planes. The general explained that he wanted to armor the planes, and wanted to do so where the planes were being shot. The statistician's response was, "I'll do the study for you, but I would point out that those are precisely the places not to armor."

Why would the statistician make that recommendation? The planes available for study were the planes that managed to return to the base despite being shot at. The desired inference about armoring has to do with the planes that were shot down, and hence unavailable for study. Hence, the statistician is thinking, "if there are bullet holes in the tail of the airplanes, but they returned to base, don't worry about holes in the tail. But if there are no holes in the fuel tank of the planes that returned, armor the fuel tank!" It doesn't really matter whether this actually happened; the point is to be wary about the relationship between the available data and the desired inference.

i. Finding excellent students

There is another example which I can personally attest to. A study was done of the quantitative and verbal scores of graduate students in statistics at Carnegie Mellon University. The result showed that the strongest students were those with high verbal scores; the quantitative scores were not very predictive. Shortly after that, I visited the Kennedy School at Harvard, where a parallel study had been done. It showed that the strongest students there had high quantitative scores, and that verbal scores were not very predictive. Should we conclude from this that the Kennedy School's program is more quantitative than the program of the Statistics Department at Carnegie Mellon? Not at all. What is happening is that no student with a weak quantitative background would dream of applying to be a student in Statistics at Carnegie Mellon; conversely a student with weak verbal skills would not apply nor be admitted to the Kennedy School. Thus, what distinguishes students in each case is the other skill. In both schools, the best students are both quantitatively and verbally able.

Selection effects are particularly important to think about in analyses of admission to various programs of education and training. If the policy excludes a certain type of student, data on students in the program will not be very informative about how students excluded by policy would have fared if they had been admitted.

9.2.2 Bayesian analysis of missing data

Missing data fits in comfortably with the general scheme of hierarchical models. In missing data problems, how an observation comes to be missing is important to model. The joint distribution of the process that leads observations to be missing and the missing values themselves have to be modeled, in one of the two obvious alternate factorizations. In either case, there is no essentially new problem in computing posterior distributions for problems with missing data. The advantage in doing so is that the resulting posterior distributions correctly reflect the uncertainty due to the fact of missing data, and hence produce a more realistic reflection of the consequences of the analyst's beliefs.

9.2.3 Summary

Missing data are parameters. As such they are to be modeled jointly with all the parameters in a problem. The resulting posterior distributions of the structural parameters appropriately reflect the uncertainty occasioned by the missing data. The resulting posterior distribution of the missing data is itself important in some problems.

9.2.4 Remarks and further reading

The seminal work on missing data is due to Rubin (1976), see also Little and Rubin (2003). While the initial emphasis was on the assumptions needed to justify sampling theory and likelihood based methods, this work also led to the development of the "non-ignorable" case, which today dominates the Bayesian literature.

9.2.5 Exercises

1. Vocabulary. Explain in your own words what missing data are.
2. Choose one of the examples in section 9.2.1. Choose a simple preliminary model for the problem.

9.3 Meta-analysis

Another kind of application of hierarchical models is meta-analysis. The context here is that there may be many studies of the same phenomenon, for example the comparative efficacy of two or more treatments for the same disease. These studies may differ in many details, for example the population studied, the way the treatments were administered, the dosages if drugs were involved, etc. Often the amount of detail reported varies among published studies, and not infrequently the original data are unavailable. Meta-analysis seeks to put these disparate studies together to see what can fairly be concluded about the fundamental question that each of the studies sought to address: the comparative efficacy of the treatments. Often many judgments have to be made about how much weight to give to the various studies. These are natural for Bayesians to make and declare, but somewhat less natural for adherents of other schools of statistics.

9.3.1 Summary

Meta-analysis is an example of a hierarchical model, easily understood in a Bayesian context.

9.4 Model uncertainty/model choice

Often it is not clear to statistical modelers what model to use. In one simple form the question might be what explanatory variables to include in a regression. Alternatively the models might be rather different views of the mechanism that produced the data.

Suppose there are K possible models for the data $\mathbf{x}$, and that the likelihood for each model involves its own parameters $\boldsymbol{\theta}_k$, $k = 1, \ldots, K$. With a prior $\pi_k(\theta_k)$ on each parameter space conditional on k, the probability of the data, conditional on k, can be written as

$$f_k(\mathbf{x} \mid \boldsymbol{\theta}_k)\pi_k(\boldsymbol{\theta}_k). \tag{9.11}$$

In order to specify the model completely, let κ be a random variable indexing model choice, and let $p_k = P\{\kappa = k\} \geq 0$, with $\sum p_k = 1$. Then the joint distribution of the data and the parameters can be written

$$p(\boldsymbol{\theta}, K|x) \propto \sum_{k=1}^{K} p_k I\{\kappa = k\} f_k(\mathbf{x} \mid \boldsymbol{\theta}_k)\pi_k(\boldsymbol{\theta}_k). \tag{9.12}$$

This is a hierarchical model with a discrete parameter specifying a model at the top, and then parameters $\boldsymbol{\theta}_k$ at the next level, and finally the data $\mathbf{x}$.

The data then require the computation of the posterior distributions of all the parameters, including the model choice parameter. Sometimes nearly all the posterior probability concentrates on a single submodel, and in this case little harm is done in concentrating attention on that single submodel. However, when more than one submodel retains substantial posterior probability, there is no reason to choose a single preferred submodel, and substantial reason not to. The strategy of keeping several submodels in play, especially for prediction, is called "model averaging" (Draper (1995)). For a review, see Hoeting et al. (1999).

There are several details of this general picture worth noticing. The parameters in each of the submodels may or may not be *a priori* conditionally independent of one-another. Thus, one cannot necessarily put together priors for each submodel and casually assume independence.

The special case of the choice of explanatory variables in regressions is much discussed in the literature. One way of thinking about it is to consider the (huge) model that incorporates all of the contemplated variables. This move may feel uncomfortable, because it can involve more variables than data. However, in principle, the material of Chapter 8 shows that Bayesian analysis can be conducted when the number of variables exceeds the sample size, so this is not an objection in principle. (Finding a prior to behave in such a high-dimensional space can be a difficult matter of application, admittedly.)

One way suggested by some to deal with these issues is to impose a very simple prior, for example a uniform distribution on all of the 2^k possible subsets of k regressions, and to impose (improper!) flat priors on each of the regressors in each of the submodels. The result is called a "spike and slab" prior. Such a prior is highly discontinuous near the origin, as it puts high probability on zero for each of the coefficients. But in many problems there is no reason to single out zero as a special value deserving more credence than, for example, values close to zero. Consequently, I regard such priors as attempts to avoid the responsibility of stating and defending one's true beliefs. I take the demand to justify to readers one's modeling choices to be the strength of the subjective Bayesian position.

Consider, for example, the simple normal linear regression model

$$y_i = \alpha + \beta x_i + \gamma Z_i + \epsilon_i \quad \epsilon_i \sim N(0, \sigma^2), \epsilon\text{'s independent.} \tag{9.13}$$

Suppose the question of interest is whether this model can be simplified as follows:

$$y_i = \alpha + \beta x_i + \epsilon_i \quad \epsilon_i \sim N(0, \sigma^2), \epsilon\text{'s independent.} \tag{9.14}$$

There are many ways to address this issue; here I'll compare two Bayesian ways. One method is to put priors on the parameters of each model, and create a hierarchy in which, with some probability p, equation (9.13) pertains, and with probability $1-p$, (9.14) pertains. The priors on α, β and σ^2 in (9.13) need not be the same as the priors on them in (9.14) and indeed may not be independent. Together with a prior on p, this creates a full probability model, from which posterior distributions for all the parameters can be calculated. The posterior distribution on p then can be interpreted as offering the best current view of the relative plausibility of the two models (9.13) and (9.14). This method essentially creates a supermodel comprising the two submodels (9.13) and (9.14).

Another way to think about the issue is to take the more general model (9.13) as basic, and then to ask whether the data support the conclusion $\gamma = 0$, which then specializes the model to (9.14). In order for this question to be non-trivial, the prior put on γ must have a discrete lump of probability on $\gamma = 0$. In my experience it is very unusual to have such a belief, because it says that $\gamma = 0$ is special, very different from $\gamma = 10^{-3}$ or $\gamma = -10^{-3}$, for example. Every continuous prior on γ has the consequence that $P\{\gamma = 0\} = 0$ (see equation (4.4)). If any such continuous prior on γ represents your beliefs, then your posterior must also have $P\{\gamma = 0\} = 0$, and, without needing any data or computations, you know that $p = 1$, so you disbelieve (9.14). Again, your prior on $(\alpha, \beta, \sigma^2 \mid \gamma)$ need not be continuous at $\gamma = 0$, which corresponds to the remark above in the hierarchical setting that your prior on $(\alpha, \beta, \sigma^2)$ in (9.13) need not be the same as your prior on $(\alpha, \beta, \sigma^2)$ in (9.14).

These two ways of thinking about the issue of $\gamma = 0$ are in fact equivalent, in that any belief in one setting corresponds to a particular belief in the other. Understanding the equivalence, however, leads one to question more deeply what is meant by the question of whether $\gamma = 0$.

So far, the entire issue has been framed around the question of whether it is reasonable to believe, in any given application, that γ takes exactly the value 0. As explained above, in virtually every applied problem I have seen, the answer to that question is "no." But surely I want to be able to simplify models. I certainly do. How, then, can I explain my wish to simplify models given that to do so apparently is contrary to my belief that the larger model is nearly always closer to the truth?

My answer to this apparent conundrum is that I find it *useful* to simplify models. In other words, the road to simplification of models, in my mind, has to do with the utility function being used, and not with what is believed to be "true." In its most elementary form, one can imagine a trade-off between parsimony (the wish for fewer variables, as in (9.14)) and accuracy (better predictive power, as in (9.13)). Being explicit about how one views that trade-off can be a basis for explaining the choices made in model choice and simplification.

There is literature offering such choices, notably Akaike information criterion (AIC) (Akaike (1973, 1974))

$$AIC = 2k - 2\ l_n\ (L) \tag{9.15}$$

where k is the number of parameters and L is the maximized likelihood. A second measure is BIC (Schwarz (1978))

$$BIC = k\ l_n\ (n) - 2\ l_n\ (L) \tag{9.16}$$

where n is the sample size. Yet another effort in this direction is the DIC (deviance information criterion) of Spiegelhalter et al. (2002). The spirit of each of these is to propose some automatic choice of the trade-off between parsimony and accuracy. Just as I question the idea of canonical prior distributions to be used without explicit consideration of the particular applied context, so too I question the use of these automatic utilities (or, equivalently, losses). In both cases, they offer an apparently cheap way to avoid having to take responsibility for choices being made, and, at the same time, destroy the meaning of tho quantities being computed.

In addressing a complicated model, sometimes one is asked how you know whether the model fits. As a general question, this question has no answer. Practically, however, a better question is "what aspect of this model do you find most questionable?" This focuses attention on the (subjectively chosen) most sensitive matter. Of course, as explained above, the larger model will always "fit" better; whether it fits *usefully* better involves, whether explicitly or implicitly, utility considerations.

9.4.1 Summary

Choosing models and, as a special case, variables in a regression, is easily understood as applications of Bayesian hierarchical modeling.

9.4.2 Further reading

Much of this section relies on the review of Kadane and Lazar (2004). The idea of keeping all plausible models in play is also known as model averaging (Draper (1995), Hoeting et al. (1999)). For another view, Box (1980) advocates using significance testing to choose among models, and the Bayesian analysis of the resulting chosen model. Yet another view is given in Gelman et al. (1995).

Figure 9.1: Representing the relationship between variables in the standardized examination example.

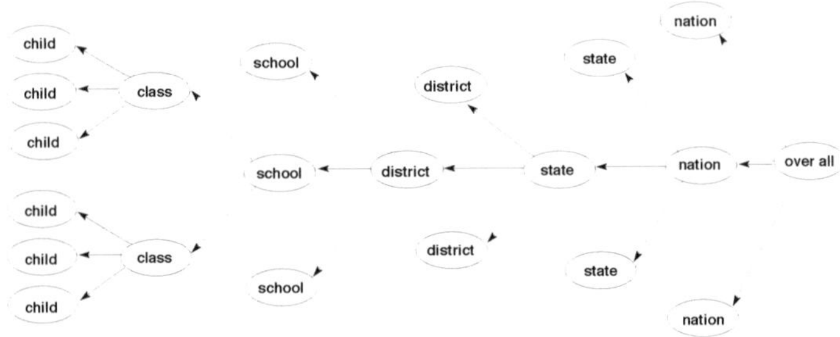

Figure 9.2: A more detailed representation of the relationship between variables in the standardized examination example.

9.5 Graphical representations of hierarchical models

Return now to the example of section 9.1, of children taking a standardized examination. One way to give a graphical picture of the hypothesized structure is given in Figure 9.1.

Figure 9.1 represents equations like (9.2), in that it expresses the idea that to explain (or predict) the scores of children in a particular class, if you know the class parameters, it would be irrelevant to know the school, district, etc., parameters. Similarly, to explain or predict the class variables, only the school and the children in that class are relevant, not the district, state, etc. Figures like 9.1 are a convenient and parsimonious way of displaying conditional independence relationships such as (9.2).

Useful as a figure like Figure 9.1 is, it does not display all of the information implicit in the structure of the hierarchical model for children's performance in the standardized examination. In particular, it does not express the idea that children's performances in a class are conditionally independent of one another, given the class parameters; that classes are conditionally independent of one another given the school parameters, etc. Thus, we might express these relationships with a graph like Figure 9.2.

Graphical representations like Figures 9.1 and 9.2 are called "directed acyclic graphs," or DAGs for short. They are "directed" because each arrow has a direction, and "acyclic" because they do not have cycles, as exemplified by graphs that look like Figure 9.3.

The models represented by DAGs are also called "Bayesian networks" or "Bayes nets" in some literature (a further example of the observation that there are many more names than objects or facts).

DAGs can represent more complicated models than this example suggests. For example, the extent of mathematics education of the teacher might be relevant. Then, (for simplicity reverting to the style of Figure 9.1), we might have

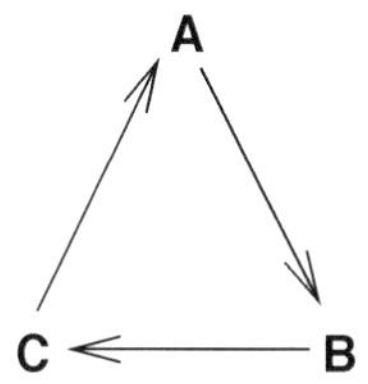

Figure 9.3: A graph with a cycle.

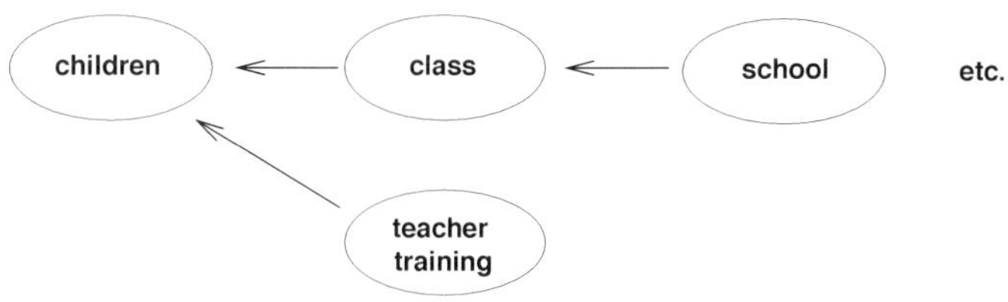

Figure 9.4: Figure 9.1 with teacher training added.

The structure of Figure 9.4 implicitly changes what is meant by "class" to mean those aspects of a class not differentiated by differences in teacher training.

To complicate matters further, it may be the policy of some school districts to make greater efforts to attract especially well-trained mathematics teachers, which would lead to a modified graph as follows:

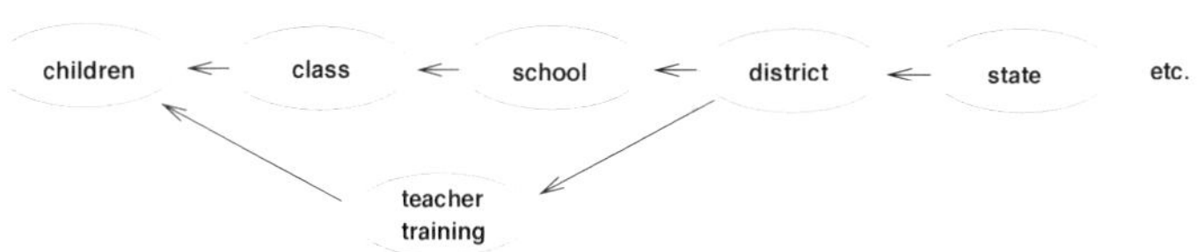

Figure 9.5: District policy influences the extent of teacher training.

In general, if there is an arrow from A to B, then A is a "parent" of B, and B is a "child" of A. Thus, for example in Figure 9.5, "teacher training" is a child of the "district" and a parent of "children." For each variable X_i, we may define parents (X_i) to be the set of variables X_j with arrows from X_j to X_i. Then we have

$$P(x_1, \ldots, x_n) = \prod_{i=1}^{n} P(x_i \mid \text{ parents } (X_i)) \tag{9.17}$$

where $X_1, \ldots, X_n$ are the variables explained by the model.

9.5.1 Summary

Graphical representation, and specifically DAGs are a useful way to visualize a hierarchical model.

9.5.2 Exercises

1. Vocabulary. State in your own words the meaning of:
   (a) directed
   (b) acyclic
   (c) DAG
   (d) Bayesian network
2. Choose one of the examples in section 9.2. Draw a DAG for it. Explain the assumptions implicit in the DAG you drew.

9.5.3 Additional references

The standard work on graphical models is Lauritzen (1996). Heckerman (1999) is a nicely written introduction.

9.6 Causation

> Charlie goes out to his front porch every night at exactly 10 p.m., claps his hands three times, and goes back into his house. His neighbor sees him doing this, and asks him why he does it. "I'm keeping the elephants away," says Charlie. "But Charlie, there are no elephants around here," responds his neighbor. "You see, it works," says Charlie.

The issue of how to discern if x causes y has been the subject of discussion and debate for many centuries, and that debate is not over. My goal here is to explain why causation is a sensitive matter to statisticians, and to give an introduction to the currently active positions about causation.

First, many readers will recognize the slogan "correlation does not imply causation." For example, consider two jointly normal uncertain quantities with correlation ρ. If it were the case that correlation implied causation, should we conclude that X causes Y or that Y causes X?

But the real issue lies deeper. Imagine a study, conducted in London in 1900, of women in London and whether they have tuberculosis. The finding is that women who wear fur coats have less tuberculosis than women who do not wear fur coats. Should we conclude that the wearing of fur prevents tuberculosis? From what we now understand about tuberculosis, the answer is "no." Women who wore fur coats were richer, had better diets, lived in better heated houses, and had better access to medicine. All these would affect their tuberculosis rates.

The general issue this raises is that it is very difficult to measure all of the covariates that might be important in a study. And we saw in the discussion of section 2.3 on Simpson's Paradox that another covariate can reverse the recommendation of a study. It is no wonder that the theme-song of statistics is "It ain't necessarily so."

Progress on the tuberculosis question might have been made by designing a clinical trial among women who did not currently have tuberculosis and who did not have fur coats. Randomly choose half to get fur coats, and see if the rates of tuberculosis are different in the two groups. The results would be a disappointment to the fur industry. For more on why a Bayesian might favor random selection, see section 11.5.

The past few decades have seen very lively discussions among statisticians and others about causation. The observations I give here concerning this debate are intended to put the discussion in the framework suggested in this volume, and to point interested readers to the relevant literature.

One important idea in this discussion is that of "potential outcomes." To introduce some notation, suppose there is a population U of units u. For example, U might be the women of London without fur coats and with no current tuberculosis. Suppose there is a function Y on U of scientific interest. To continue the example, Y might equal 1 if the woman u has tuberculosis a year later, and $Y = 0$ otherwise. Suppose also that there is a decision variable D having two values: $D(u) = t$ indicating that unit u is assigned treatment t, and $D(u) = c$ indicating that unit u is assigned control treatment c. For example, $D(u) = t$ might mean to give woman $u \epsilon U$ a fur coat. The potential outcomes $Y_t(u)$ and $Y_c(u)$ are respectively the value of $Y(u)$ if $D(u) = t$ or $D(u) = c$. Once $D(u)$ has been determined, only one of $Y_t(u)$ and $Y_c(u)$ will be observed. Thus in retrospect, one of $Y_t(u)$ and $Y_c(u)$ is counter-factual – it didn't happen. The causal effect of $D(u) = t$ relative to $D(u) = c$ can then be defined to be

$$Y_t(u) - Y_c(u).$$

Much of the discussion about causal effects centers on the unobserved character of one of the two terms, $Y_t(u)$ and $Y_c(u)$. Various models and assumptions are proposed to deal with this, depending, for example, on whether one has a randomized experiment with complete compliance, a randomized experiment with incomplete compliance, an observational study, etc. Some of the discussion has to do with circumstances under which various assumptions in such models are testable, and whether certain parameters in such models are identified.

An important additional part of the potential outcomes framework is a model for how the treatment assignment was done, a point emphasized by Rubin (2004). The story of the fighter planes in section 9.2.1 (h) makes it clear why this is a crucial consideration for understanding the import of the data available for analysis.

From the perspective of this book, there is nothing wrong with defining and dealing with potential outcomes. They are simply parameters, names for uncertain quantities that one wishes to discuss. There is also nothing wrong with untested (or untestable) assumptions, nor with lack of identification (see section 8.3). Every inference depends in principle on both, so there is nothing novel in causal inference that leads it to be different in kind in these respects. As always, a thorough discussion of the assumptions (models and priors) should accompany inferences, and the sensitivity of the conclusions to the assumptions should be explored. The extent to which a reader will find the conclusions acceptable (whether causal or not) will depend on the plausibility to that reader of the assumptions made. And this in turn will depend on the quality of the arguments adduced to support those assumptions.

The potential outcomes framework goes back to Neyman (1923), Cornford (1965) and Lewis (1973) and has been championed by Rubin (1974, 1978, 1980, 1986), Holland (1986), Robins (1986, 1987) and Robins and Greenland (1989), among others. It has also been criticized, especially by Dawid (2000).

There is a distinction drawn in this literature between discerning the effects of causes on the one hand, and the causes of effects on the other. According to both Holland (1986) and Dawid (2000), the former is simpler than the latter. The former is amenable to direct experimentation (administer one of the treatments and see what happens); the latter would require thinking about each of the possible causes to ascertain your probability of the effect if you or someone or something else took each action regarded as a possible cause, and then invoking Bayes Theorem. This kind of reasoning is exemplified by Sir Arthur Conan Doyle's Sherlock Holmes (Doyle, 1981, pp. 83, 84) in writing about synthetic reasoning:

> "Most people, if you describe a train of events to them, will tell you what the result would be. They can put those events together in their minds, and argue from

> them that something will come to pass. There are few people, however, who, if you told them a result, would be able to evolve from their own inner consciousness what the steps were that led to that result. This power is what I mean when I talk of reasoning backward, or analytically."

Shafer (2000) criticizes the exercise of finding the causes of effects as follows: Suppose I am required to bet $1 on the outcome of the flip of a coin. I bet on heads, and lose. He asks whether my choice of heads "caused" me to lose $1. If I believe that the outcome of the flip is independent of my choice, then the answer to this question is "yes." However, if I assume a different counter-factual world, in which the coin must land opposite to the way I bet, then the answer would be "no." This may sound peculiar, so I pause to give an example. Suppose there is a statistician, we'll call him "Persi," who by dint of much practice, is able to flip a coin and reliably make it come out "heads" or "tails" as he chooses. If Persi is flipping the coin, and wants me to lose the dollar, then I'm going to lose. There is nothing incoherent in believing that Persi can do this, nor that he would. I can think of this causally as that Persi caused me to lose, as I would have lost no matter which way I bet. This is an illuminating example, I think, because it highlights the importance of one's prior beliefs in the making of causal attributions.

Another important perspective on causation is that provided by Spirtes et al. (1993, 2000) and Pearl (2000, 2003). Pearl introduces a "do" operator, to distinguish the case in which the random variable X happens to take the value x_0 from the case in which the decision variable X is set to the value x_0. He accounts for the effect of this by modifying equation (9.17) as follows:

$$P(x_1, \ldots, x_n) = \prod_{i \mid X_i \notin X} P(x_i \mid p(x_i| \text{ parents } (X_i)), \text{ do } (X = x_o)). \tag{9.18}$$

Lindley (2002) gives an interesting review of Pearl (2000). He remarks that it is coherent to have different beliefs about $p(y \mid \text{ see } (x))$ and $p(y \mid \text{ do } (x))$. For example, if y is an indicator function for the presence of tuberculosis and x is the presence of a fur coat, the assumption that

$$p(y \mid \text{ see } (x)) = p(y \mid \text{ do } (x)) \tag{9.19}$$

is doubtful. However no doubt there are other situations in which (9.19) would be acceptable.

Rubin's potential outcomes can be translated into a graphical causal model, and conversely a graphical causal model can be translated into a potential outcomes model. However, the potential outcomes model is restrictive, in that what is to be regarded as an outcome is to be specified in advance. By contrast, Spirtes et al. (1993, 2000) stress "search" (known here as model uncertainty, see section 9.4), which does not specify outcomes in advance. Running through the discussions of these ways of speaking about causation are various matters of style. Pearl (2003) and Lauritzen (2004) like causal diagrams, while Rubin (2004) distrusts them. Rubin (2004) espouses potential outcomes as a framework; Pearl (2003) finds the assumptions awkward to understand and Dawid (2000) and Lauritzen (2004) distrust potential outcomes. I find myself in sympathy with the following remark by Lauritzen (2004):

> I have no difficulty accepting that potential responses, structural equations, and graphical models coexist as languages expressing causal concepts each with their virtues and vices. It is hardly possible to imagine a language that completely prevents users from expressing stupid things.

The issue as I see it is that the proponents of each way of thinking give some examples in which the method favored by that author is used, and then implicitly make the claim that theirs is the only, or best, way to understand causation in general.

It is also useful to recognize limitations of each of the approaches. In the potential outcomes literature there is doubt that one can speak of discrimination "caused by" age, race or sex, since these are not conditions that can be changed in an individual. However, as Fienberg and Haviland (2003) point out, the perceptions of age, race and sex have been altered experimentally, and these experiments do shed light on the issue of discrimination.

Similarly, I would like to be able to say that I believe that the eruption of Mt. St. Helens caused a large mud-slide. No-one can "do" such an eruption, and I trust people would not do so if they could. Nonetheless such a sentence makes sense to me. Thus, while I find the current literature on causation to be helpful and insightful, I believe there is still more to understand about causation.

9.7 Non-parametric Bayes

Roughly, non-parametric Bayes is about processes on infinite dimensional spaces. There's something very audacious about it, as even low dimensional spaces present their challenges. Nonetheless, this area has attracted a lot of attention. New methods and applications are making what might have seemed impossible into something familiar.

My goal in this section is heuristic. I intend to introduce some of the most important infinite-dimensional objects. However, I will not prove the existence of these processes, as that would take the discussion into measure theory beyond the scope of this book.

The first issue is the name. The methods that are referred to as "Non-Parametric Bayes" are the opposite of non-parametric; they have many more parameters than do other methods referred to as parametric. Since there's no such thing as a free lunch in statistical modeling, this choice requires greater reliance on other assumptions. Notably in non-parametric Bayes, the methods generally assume the availability of large data sets and a conditional independence assumption applying to this large data set. That structural assumption allows for greater flexibility in other parts of the model. Another aspect of the assumptions made in non-parametric Bayes is heavy use of conjugacy (see Chapter 8). These are modeling choices that are appropriate in some circumstances and not in others.

9.7.1 Gaussian Process

A Gaussian process is a distribution on function f mapping a space $\mathcal{X}$ to the real line $\mathbb{R}$, i.e., $f : \mathcal{X} \rightarrow \mathbb{R}$. If $\mathcal{X}$ is infinite dimensional, then so is f. Let $\mathbf{x}_n = (x_1, \ldots, x_n)$, so $f(\mathbf{x}_n) = (f(x_1), f(x_2), \ldots, f(x_n))$, an n-dimensional vector of function values evaluated at $\mathbf{x}_n$. Because f is stochastic, it has a probability distribution $P(f)$.

Then $P(f)$ is a Gaussian process if for any finite subset $\mathbf{x}_n$ of $\mathcal{X}$, the marginal distribution of $P(f(\mathbf{x}_n))$ is multivariate Gaussian, i.e., multivariate Normal.

Because they have Gaussian (normal) distributions, they are characterized by their means $\mu(\mathbf{x})$ and covariance functions $\sum(\mathbf{x}, \mathbf{x})$. These may be modeled with arbitrary functions themselves, provided $\sum$ is positive definite.

To relate Gaussian processes to more familiar models, consider first a simple regression model:

$$y_i = \beta_0 + \beta_1 x_i + \epsilon_i, \tag{9.20}$$

where the observations are pairs (y_i, x_i), and ϵ_i is assumed to have a Gaussian (normal) distribution with mean 0 and variance σ^2, and ϵ_i and ϵ_j are assumed to be independent for all $i \neq j$. This is a special case of the model discussed in Section 8.6.

A more general model might be

$$y_i = \sum_{k=1}^{K} \beta_k \phi_k(x_i) + \epsilon_i \tag{9.21}$$

where the (specified) functions ϕ_k are called basis functions. Suppose the prior on $\boldsymbol{\beta} = (\beta_1, \ldots, \beta_k)$ is given by

$$\boldsymbol{\beta} \sim N(0, \wedge) \tag{9.22}$$

where $\wedge$ is diagonal with elements $\boldsymbol{\lambda} = (\lambda_1, \ldots, \lambda_K)$, all positive.

Then $\boldsymbol{\beta}$ can be integrated out, yielding

$$\begin{aligned} E(y_i) &= 0 \\ \text{and } Cov(y_i, y_j) &= \sum_{k=1}^{K} \lambda_k \phi_k(x_i)\phi_k(x_j) + \delta_{ij}\sigma^2 = C_{i,j}, \end{aligned} \tag{9.23}$$

where δ_{ij} is the Dirac delta: $\delta_{ij} = 1$ if $i = j$ and zero otherwise.

So then y_i is a Gaussian process with covariance function $c(x_i, x_j) = C_{i,j}$.

9.7.2 Exercise

1. Prove that $C_{i,j}$ in equation (9.23) is positive definite.

9.7.3 Further reading

Rasmussen and Williams (2006) and Murphy (2012).

9.7.4 Dirichlet Process

A Dirichlet process is a distribution over distributions, written

$$G \sim DP(\cdot | G_o, \alpha) \tag{9.24}$$

where α can be thought of as a scaling parameter, and G_o is called the base measure. The defining property of a Dirichlet process is as follows: Let G_o be a probability distribution on a space Θ, and let $\alpha > 0$ be a real number. Then for all finite partitions $(A_1, \ldots, A_K)$ of Θ,

$$G(A_1), \ldots, G(A_K) \sim Dir(\alpha G_o(A_1), \ldots, \alpha G_o(A_K)). \tag{9.25}$$

The reason why it is plausible that such a process exists is given in Theorem 8.9.2, about refining and coarsening of the Dirichlet distribution. Because all the coarsenings of a very refined Dirichlet distribution are compatible (their α's add), there will be no partitions found from coarsening for which (9.25) fails. Ferguson (1973) proves the existence of the Dirichlet process.

The Dirichlet process can be thought of as an infinite-dimensional Dirichlet distribution, in the same sense that a Gaussian process can be thought of as an infinite dimensional Gaussian (Normal) distribution. To understand the analogy between the Dirichlet distribution and the Dirichlet process, I first have to apologize for the multiple use of the symbol "α," used to mean one thing in the description of the Dirichlet distribution, a different thing in the description of the Dirichlet process, and yet a third thing in the Beta distribution, the two-dimensional special case of the Dirichlet distribution. Unfortunately, all these uses are so deeply embedded in the literature surrounding these objects that I would be doing my readers a disservice if I failed to use the traditional symbols.

To avoid compounding the natural confusion that this triple use naturally occasions, for this discussion I will superscript "*alpha*" as follows: "α^B" means "α" as used in the Beta distribution (see (8.71)), "α^D" means "α" as used in the Dirichlet distribution (see (8.70)) and "α^{DP}" means alpha in the Dirichlet process as discussed in this section. To remind

	dimension	location	scale
Beta Distribution	2 (1 constraint)	$\alpha^B/(\alpha^B+\beta)$	$\alpha^B+\beta$
Dirichlet Distribution	k (1 constraint)	$\alpha_i^D/\sum_{j=1}^k \alpha_j^D$	$\sum_{j=1}^k \alpha_j^D$
Dirichlet Process	∞	$G_o(\cdot)$	α^{DP}

Table 9.1: Relationship of notation for increasingly general Dirichlets.

you, α^B and α^{DP} are positive real numbers, while α^D is a real vector of length k. Table 9.1 indicates the relationships:

Note that in 9.25 the finite partition $(A_1, \ldots, A_K)$ can be scrambled up, and the same Dirichlet distribution arises. Thus, any sense of adjacency among the A's is lost in a Dirichlet process, as is the case for the Dirichlet distribution.

The mean and variance of the Dirichlet process are easily found from (9.25). Let A be any set for which $G_o(A)$ is defined. Then $(A, \bar{A})$ are a partition of Θ. Then (9.25), applied to this partition, says that

$$G(A) \sim Beta(\alpha G_o(A), \alpha G_o(\bar{A})). \tag{9.26}$$

Consequently

$$EG(A) = \frac{\alpha G_o(A)}{\alpha G_o(A) + \alpha G_o(\bar{A})} = G_o(A), \tag{9.27}$$

which can be written

$$EG(\cdot) = G_o(\cdot). \tag{9.28}$$

Also

$$\begin{aligned} Var(G(A)) &= \frac{\alpha G_o(A)}{\alpha G_o(A) + \alpha G_o \bar{A}} \cdot \frac{\alpha G_o(\bar{A})}{\alpha G_o(A) + \alpha G_o(\bar{A})} \cdot \frac{1}{\alpha G_o(A) + \alpha G_o(\bar{A}) + 1} \\ &= \frac{G_o(A) \cdot G_o(\bar{A})}{\alpha + 1}. \end{aligned} \tag{9.29}$$

Note that the Dirichlet process can be thought of as conjugate to an infinite dimensional multinomial distribution, in the same way that the Dirichlet distribution is conjugate to the (finite-dimensional) multinomial distribution.

9.7.5 *The Generalized Polya Urn Scheme and the Chinese Restaurant Process*

To explain the marginal posterior distribution of the Dirichlet process, I begin with analogous facts for the Dirichlet distribution, as displayed in section 8.9.2. To remind you, let $X_1, \ldots, X_m$ be observations from a Dirichlet distribution with parameter vector $\boldsymbol{\alpha}$. Then the posterior distribution is given by

$$P\{X_{m+1} = j | \boldsymbol{\alpha}, X_1, \ldots, X_m\} = G(j)\left(\frac{W}{W+m}\right) + \frac{n_j}{m}\left(\frac{m}{W,+m}\right) \tag{9.30}$$

(see equation (8.81)). For reasons discussed there, (9.30) can be thought of as a Generalized Polya Urn Scheme. Note that G is the prior distribution before any observations are taken, the analog of G_o for the Dirichlet process. Similarly, W is the sum of the α^D's, whose analog is α^P (see Table 9.1 in Section 9.7.4). Thus translated, I have

$$P\{X_{m+1} = j | G_o, \alpha, X_1, \ldots, X_m\} = G_o\left(\frac{\alpha}{\alpha+m}\right) + \frac{n_j}{m}\left(\frac{m}{\alpha+m}\right). \tag{9.31}$$

That this is a valid equation for the Dirichlet process is shown by Blackwell and MacQueen (1973).

Equation (9.31) has been given a story, called the Chinese Restaurant Process. Imagine a restaurant with infinitely many, infinitely large tables. The first customer sits at a table, governed by G_o. The second customer may sit at the table already chosen by the first customer (with probability $1/(\alpha + 1)$) [note $n_j = 1$ and $m = 1$] or at a new table with probability $\alpha/(\alpha + 1)$ (which table again is governed by G_o). As the process continues according to (9.31), large tables tend to attract yet more new customers. This effect (popular tables get even more popular) is used in machine learning to model clustering of similar observations.

9.7.6 *Stick breaking representation*

Once again I turn to analogous results for the Dirichlet distribution. Consider Theorem 8.9.5:

Let π have a Dirichlet distribution with parameter αg_o (here $\alpha > 0$ is a scalor and g_o is a probability distribution). Then the distribution of π is the same as

$$\sum_{i=1}^{\infty} V_i \prod_{j=1}^{i-1} (1 - V_j) e_{y_i} \tag{9.32}$$

where $V_i \sim B(1, \alpha), i = 1, 2, \ldots$ and are independent, and Y_i has a multinomial distribution with probability vector g_o, independent of each other and of the V's.

To translate (9.32) to the Dirichlet process, all that must change is that now Y_i's are independent draws from G_o. That this representation is true of the Dirichlet process is the result of Sethuraman (1994).

Once again, the V_i's have the same stick breaking interpretation, and once again α plays a role only in the distribution of the V's, and G_o only in the distribution of the Y's.

An implication of (9.32) is that samples from a Dirichlet process are discrete. There's no reason to be disturbed by that, since all the integration methods of Chapter 4 are discrete approximations to integrals.

9.7.7 *Acknowledgment*

I benefited especially from Ghahramani (2005) in writing about Bayesian non-parametrics.

9.7.8 *Exercise*

1. Referring to (9.29), find the covariance of $G(A_1)$ and $G(A_2)$, where A_1 and A_2 are disjoint.

Chapter 10

Bayesian Computation: Markov Chain Monte Carlo

10.1 Introduction

Chapter 8 showed many of the most common conjugate analyses used in Bayesian computation, and also introduced Laplace's Method and large sample theory. While those methods are useful, they are limited. Conjugate analysis applies only for particular forms of likelihood and prior; large sample theory applies only when the sample size is "large," and there is little guidance about just how large that is. Consequently, attention is drawn to numerical methods, which are the subject of this chapter.

10.2 Simulation

Generally Bayesian computations are aimed at an integral of some kind, for example

$$I = \int_{[0,1]} f(x)dx. \tag{10.1}$$

One natural way to approximate such an integral is to evaluate the function f on a grid of points $\{\frac{i}{n}, i = 0, 1, \ldots, n\}$, and approximate I by

$$\hat{I} = \sum_{i=0}^{n} f(i/n)/(n+1) \tag{10.2}$$

which is called the **trapezoid rule**. In a sense, the trapezoid rule is closely related to the theory of Riemann integration (see Chapter 4).

An alternative method is to choose $n+1$ points $\{x_0, \ldots, x_n\}$ independently from a uniform distribution on $[0, 1]$, and approximate I with

$$\hat{\hat{I}} = \sum_{i=0}^{n} f(x_i)/(n+1), \tag{10.3}$$

which is called a **Monte Carlo approximation**. Since a different draw of uniformly independent points would lead to a different approximation (10.3), the Monte Carlo approximation is stochastic. However, because the x_i's are independent, the strong law of large numbers applies, provided

$$\int_0^1 \mid f(x) \mid dx < \infty. \tag{10.4}$$

Additionally, the central limit theorem applies, provided

$$\int_0^1 f^2(x)dx < \infty. \tag{10.5}$$

The central limit theorem shows that the rate of convergence of $\hat{\hat{I}}$ to I is a constant times $n^{-1/2}$.

While both methods work satisfactorily for a one-dimensional integral such as (10.1), the situation is different for a multi-dimensional integral. Suppose for example that (10.1) is replaced by

$$I^* = \int_{[0,1]^k} f(y_1, y_2, \ldots, y_k) dy_1 dy_2 \ldots dy_k. \tag{10.6}$$

The trapezoid rule would now require a k-dimensional grid, and $(n+1)^k$ evaluations of the function f. It is easy to imagine that this could be computationally expensive if k is large and f is complicated.

However, the Monte Carlo method scales more gracefully. Let $W_1 = f(U_1, \ldots, U_k), W_2 = f(U_{k+1}, \ldots, U_{2k})$ etc., where $U_1, U_2, \ldots$ are independent draws from a uniform distribution on $[0, 1]$. Then I^* can be approximated by

$$\hat{\hat{I}}^* = \sum_{i=1}^{n} W_i/n. \tag{10.7}$$

Again, because the W's are independent and identically distributed, both the strong law of large numbers and the central limit theorem apply, and again the rate of convergence is the standard deviation times $n^{-1/2}$.

Computerized methods for generating samples from uniform distributions generally rely on pseudo-random number generators, which are deterministic algorithms designed to mimic independent draws from a uniform distribution on $[0, 1]$. Because the algorithms are deterministic, in principle their use could lead to false conclusions about stochastic phenomena. In practice they work quite well.

The Monte Carlo method can be extended to more general integrals, for example of the type

$$\int_{\mathcal{R}^k} g(\mathbf{x}) f(\mathbf{x}) d\mathbf{x} = \int_{\mathcal{R}^k} g(x_1, \ldots, x_k) f(x_1, \ldots, x_k) dx_1, \ldots, dx_k. \tag{10.8}$$

When $f(\cdot)$ is a probability density, this integral can be expressed as $E(g(X))$, where X has density $f(x_1, \ldots, x_k)$.

Again, for a strong law of large numbers to apply, it is necessary to assure

$$E(\mid g(X) \mid) < \infty,$$

and for a central limit theorem, $E(g^2(X)) < \infty$.

There are special tricks to simulate draws from various standard distributions, starting from a pseudo-random number generator producing uniform $(0, 1)$ random variables (and fudging independence). One general method relies on knowing the cumulative distribution function F of a continuous random variable X. Let

$$F^{-1}(t) = \inf_x \{F(x) > t\}. \tag{10.9}$$

If U has a uniform distribution on $[0, 1]$, then $F^{-1}(U)$ has the same distribution as X, since

$$P\{F^{-1}(U) \leq x\} = P\{U \leq F(x)\} = F(x). \tag{10.10}$$

A second general method is called rejection sampling, or accept-reject sampling. Suppose we wish to generate samples from a continuous target density $\pi(x) = f(x)/K$, where $f(x)$ is known but the constant K is not necessarily known (but it might be). Let $h(x)$ be a density that is easy to simulate from, and suppose there is a constant c such that $f(x) \leq ch(x)$ for all x. Then the following algorithm generates independent samples from π:

1. Generate W from $h(x)$ and independently u from a uniform $(0,1)$.
2. If $u \leq f(W)/ch(W)$
   return W [acceptance]
   Else return to 1. [rejection]

The smaller c, the fewer rejections there will be. The smallest c can be and still satisfy the constraint that $f(x) \leq ch(x)$ for all x is to choose

$$c = \inf_x \frac{f(x)}{h(x)}.$$

However, larger c's can be used if they are more convenient, at some loss of algorithmic efficiency.

Why does rejection sampling work?

Theorem 10.2.1. *W generated by the above algorithm has the density π.*

Proof. Let $\alpha(x) = f(x)/h(x)c$ and let N be the index of the first acceptance. Also let $U_1, U_2, \ldots$ be the sequence of generated uniform random variables and $W_1, W_2, \ldots$ be the sequence of generated W's. Then the probability of acceptance at the first step is

$$\begin{aligned} p_1 = P\{U_1 \leq \alpha(W_1)\} &= \int P\{U_1 \leq \alpha(w)\} P_{W_1}(dw) \\ &= \int \alpha(w)h(w)dw = \int f(w)/c\, dw = \frac{1}{c}\int K\pi(x)dx = K/c. \end{aligned}$$

Since the steps are independent, this shows that N has a geometric distribution (see section 3.7) with parameter K/c. Then if A is a set for which $P\{W \epsilon A\}$ is defined,

$$\begin{aligned} P\{W\epsilon A\} &= \textstyle\sum_{n\geq 1} P\{N=n, W\epsilon A\} \\ &= \textstyle\sum_{n\geq 1} P\{\cap_{k\leq n-1}[U_k > \alpha(W_n)] \cap [U_n \leq \alpha(W_n)] \cap [W_n \epsilon A]\} \\ &= \textstyle\sum_{n\geq 1}(1-p_1)^{n-1} P\{[U_1 \leq \alpha(W_1)] \cap [W_1 \epsilon A]\} \\ &= \tfrac{1}{p_1} P\{[U_1 \leq \alpha(W_1)] \cap [W_1 \epsilon A]\} \\ &= \tfrac{1}{p_1} P\{U_1 \leq \alpha(w_1) \mid w_1 \epsilon A\} P\{w_1 \epsilon A\} \\ &= \tfrac{1}{p_1} \textstyle\int_A P\{U_1 \leq \alpha(w_1)\} P_{W_1}(dw_1) \\ &= \tfrac{1}{p_1} \textstyle\int_A \alpha(w_1)h(w_1)dw_1 \\ &= Kc \textstyle\int_A f(w)/c\, dw = K \int_A \pi(w)/K dw \\ &= \textstyle\int_A \pi(w)dw. \end{aligned}$$

Thus W has the density π, as required. $\square$

Because the central limit theorem shows that the convergence of (10.3) to (10.10), or more generally of (10.7) to (10.6), occurs at the rate $\sigma/\sqrt{n}$, techniques have been developed to reduce the variance σ^2. Some of the most important are:

1. Importance Sampling.
   The idea of importance sampling is to reduce the variability of the integrand by choice of the density with respect to which the integral is taken. It is convenient to divide the range of the integral into two parts: those where g is positive and where it is negative. This is accomplished using the following decomposition.
   Let $g^+(x) = \max\{g(x), 0\}$ and $g^-(x) = \min\{g(x), 0\}$. Then $g(x) = g^+(x) + g^-(x)$, so

$$\int g(x)f(x)dx = \int g^+(x)f(x)dx - \int(-g^-(x))f(x)dx. \tag{10.11}$$

Both $g^+(x)$ and $-g^-(x)$ are non-negative. Thus without loss of generality, we may consider integrals of non-negative function g. Now

$$Eg(X) = \int g(x)f(x)dx = \int \frac{g(x)f(x)}{\tilde{f}(x)}\tilde{f}(x)dx \qquad (10.12)$$

where $\tilde{f}(x)$ is a positive density (needed to avoid dividing by 0 in (10.12)). If $\tilde{f}(x)$ can be chosen to be roughly proportional to $g(x)f(x)$ and to be easily simulated from, the resulting Monte Carlo estimate, from $Y(x) = g(x)f(x)/\tilde{f}(x)$ with respect to a random variable with density $\tilde{f}(x)$, will have small variance. The name "importance sampling" comes from the fact that the method will lead to more heavily sampling points where the original integrand $g(x)f(x)$ is large, thus at the points where that contributes most to the integral (10.12).

2. Control Variate
   Here the idea is to find a function $h(x)$ whose expectation is easy to compute and such that the estimate of $E(f(x) - h(x))$ has smaller variance than the estimate of $E(x)$. Since
   $$Ef(x) = E(f(x) - h(x)) + Eh(x), \qquad (10.13)$$
   this results in a simulation with a smaller variance.
3. Antithetic Variables
   In some integration problems, there are transformations that have negative correlations that can be exploited. Consider again the integral in (10.1). Since the transformation $x \to 1 - x$ leaves dx invariant, (10.1) can be rewritten
   $$I = \frac{1}{2}\int_0^1 (f(x) + f(1-x))dx, \qquad (10.14)$$
   so I can be approximated by
   $$\frac{1}{n}\frac{1}{2}\left[f(U_1) + f(1-U_1)] + \ldots + \frac{1}{2}[f(U_n) + f(1-U_n)\right]. \qquad (10.15)$$
   When $f(U)$ and $f(1-U)$ have negative correlation, the result is a simulation with smaller variance.
4. Stratification
   This is a method borrowed from the classical theory of sample surveys. There are simulations in which it is known that in some parts of the domain the function being integrated is much more variable than in other parts of the domain and it is also known which parts of the domain those are. (We have already seen such an example in section 4.9.) Stratification can exploit such knowledge to concentrate sampling in the more variable parts, thus reducing the resulting uncertainty.
   In particular, suppose the goal is to approximate
   $$I = E(g(X)) = \int_D g(x)f(x)dx, \qquad (10.16)$$
   where X has the density $f(x)$.
   Then
   $$I = \sum_{i=1}^{m} E(I_{[X \epsilon D_i]}g(X)) \qquad (10.17)$$
   where $D_1, \ldots, D_m$ are disjoint sets whose union is D (in the language introduced in Chapter 1, $\{D_1, \ldots, D_m\}$ is a partition of D). Let $\sigma_i^2 = V(I_{[X \epsilon D_i]}g(X))$, and suppose

n_i observations are devoted to sampling from D_i. The resulting variance of the Monte Carlo approximation is $\sum_{i=1}^m \sigma_i^2/n_i$. Minimizing this subject to $\sum_{i=1}^m n_i = n$, yields the optimal $n_i = n\sigma_i / \sum_{i=1}^m \sigma_i$. (OK, these may not be integers, but you can use the nearest integer.) The resulting minimized variance is $\sum_{i=1}^m \sigma_i^2/n$. Fortunately even rough guesses of the σ_i^2's can lead to gains (reductions in variance), as is the case in stratification in survey sampling.

5. Conditional Means

   Suppose one wishes to approximate

$$Eg(X,Y) = \int g(x,y)f(x,y)dx\ dy. \tag{10.18}$$

   There are such integrals in which one of the variables (say Y) can be integrated analytically, conditional on the others, X.

   Since

$$Eg(X,Y) = E\{Eg(X,Y) \mid X\} \tag{10.19}$$

   (see section 2.8), it is possible to reduce the dimension of the integral. Furthermore, using the conditional variance formula (see section 2.12.7, exercise 3),

$$V(g(X,Y)) = V\{Eg(X,Y \mid X)\} + E\{Vg(X,Y) \mid X\}, \tag{10.20}$$

   the first term is zero, leading to a reduction in variance by doing so. The general principle is to reduce an integration problem analytically as much as possible, resorting to simulation only when analytic methods are intractable.

10.2.1 *Summary*

Simulation methods are a useful supplement to analytic methods in that they permit the approximation of integrals, particularly multivariate integrals, that are unavailable by analytic methods. Because they are based on independent and identically distributed random draws, they support both a law of large numbers and a central limit theorem. A variety of variance reduction techniques can help make such simulations more efficient.

10.2.2 *Exercises*

1. State in your own words a definition of
   (a) trapezoid rule
   (b) simulation
   (c) Monte Carlo method
   (d) pseudo-random number generator
   (e) rejection sampling
   (f) importance sampling
   (g) control variate
   (h) antithetic variate
   (i) stratification
   (j) conditional means
2. Consider $\int_0^1 x^2 dx$
   (a) Compute it analytically.
   (b) Using evaluation at 10 points, approximate it using the trapezoid rule, in `R`.

(c) Using evaluation at 10 points, approximate it using Monte Carlo simulation, in R.

(d) Do both b and c again with 100 points.

(e) Compare the four approximations computed in b, c and d with the analytic result in a. Which turned out to be most accurate? Why?

3. Suppose that the stratification variance $\sum_{i=1}^{m} \sigma_i^2/n_i$ is to be minimized subject to the constraint $\sum_{i=1}^{m} n_i = n$.

   (a) Let the minimization be taken over all real positive numbers n_i, not just the integers. Show that the optimal n_i's satisfy $n_i = n\sigma_i / \sum_{i=1}^{m} \sigma_i$, $i = 1, \ldots, m$. [Hint: Use a Lagrange multiplier, see section 7.6.1.]

   (b) Show that the resulting minimum value of the variance is $\sum_{i=1}^{m} \sigma_i^2/n$.

   (c) Let $\sigma_1^2 = 4, \sigma_2^2 = 9, \sigma_3^2 = 25$, and $n = 50$. Find the optimal sample sizes n_1, n_2 and n_3. What is the resulting variance?

   (d) Continuing (c), suppose that by mistake a person used $\sigma_1^2 = 4, \sigma_2^2 = 16$ and $\sigma_3^2 = 16$ instead. How would such a person allocate the sample of size 50? If those allocations were used instead of the optimal ones calculated in c, how much higher would the resulting variance be?

10.2.3 References

For more on (pseudo) random number generators, see L'Ecuyer (2002); for stratification in its sampling context, see Cochran (1977); for methods of generating samples from various distributions other than uniform, see Devroye (1985). A good discussion of acceptance sampling and its extensions can be found in Casella and Robert (2004, pp. 47–62). For variance reduction generally, see Dagpunar (2007, Chapter 5) and Rubinstein and Kroese (2008, Chapter 5). A good review of importance sampling in this context is given by Liesenfeld and Richard (2001).

10.3 Markov Chain Monte Carlo and the Metropolis-Hastings algorithm

While sampling independent random variables can be effective in particular cases, many statistical models, especially hierarchical models such as those discussed in Chapter 9, require a general method not dependent on special cases. A natural generalization of independent samples are Markov Chain samplers, in which the next random variable sampled depends only on the most recent value, and not on the history of the sampled values before the most recent one.

A **stochastic process** is a set of uncertain quantities, which is to say, of random variables. A **discrete-time stochastic process** is a stochastic process indexed by the non-negative integers, in notation, $(X_0, X_1, X_2, \ldots)$. A **Markov Chain** is a discrete-time stochastic process satisfying the following **Markov condition**:

$$\begin{aligned} P\{X_n \epsilon A \mid X_0 &= x_0, X_1 = x_1, \ldots, X_{n-1} = x_{n-1}\} \\ &= P\{X_n \epsilon A \mid X_{n-1} = x_{n-1}\} \end{aligned} \tag{10.21}$$

for all $n \geq 1$ and all sets A for which it is defined. Thus the Markov condition says that the probability distribution of where the chain goes next $(X_n \epsilon A)$ depends only on where it is now $(X_{n-1} = x_{n-1})$ and not on the history of how it came to be at x_{n-1}.

A Markov Chain is therefore characterized by the probability distribution of its starting state X_0, and by its transition probabilities $P\{X_n \epsilon A \mid X_{n-1} = x_{n-1}\}$. When the transition probabilities do not depend on n, the Markov Chain is called **time-homogeneous**. Our attention will focus on time-homogeneous Markov Chains, or HMC's.

Markov Chains are also distinguished by their domain E. The three leading cases are when E is finite, when E is countable and when E is $\mathbb{R}^k$ or a subset of $\mathbb{R}^k$.

While there are many examples of Markov Chains, one is already familiar to readers of this book. Recall the gambler's ruin problem, discussed in section 2.7. Gambler A starts with \$$i$, and gambler B with \$$(m-i)$. (For convenience there is a change in notation. What in section 2.7 is denoted "n" is here "m.") At each play, Gambler A wins \$1 with probability p and loses \$1 with probability $q = 1-p$. Let X_n be gambler A's fortune after n plays of this game, and consider discrete-time the stochastic process $\{X_0, X_1, \ldots, \}$, This process is a Markov Chain, because

$$P\{X_n = x_n \mid X_0 = x_0, \ldots, X_{n-1} = x_{n-1}\}$$
$$= \begin{cases} p & \text{if } x_n - x_{n-1} = 1 \text{ and } x_{n-1} \neq 0, m \\ q & \text{if } x_n - x_{n-1} = -1 \text{ and } x_{n-1} \neq 0, m \\ 1 & \text{if } x_{n-1} = x_n = 0 \text{ or } m. \end{cases}$$

Thus the player's next fortune depends only on his current fortune x_{n-1} and not his path to that fortune, which is the Markov condition. Finally, this Markov Chain is time-homogeneous, because the same transition probabilities obtain regardless of the value of n. Of course in this example, E is finite, since $E = \{0, 1, \ldots, m\}$.

Up to now in this book, Roman letters have been used for data and Greek letters for parameters. In this chapter we're going to change that convention. The posterior distribution we would like to simulate from is proportional to the likelihood times the prior, which ordinarily would be written as $\ell(\theta \mid x)p(\theta)$. The algorithms to be discussed move in the parameter space of θ; x is the data, which stays fixed. Nonetheless, it is convenient to write the elements of E, the domain of the Markov Chain, with lower case Roman letters, x, y, etc. Thus $\ell(\theta \mid x)p(\theta)$ will be written in this chapter as a constant times $\pi(x)$.

Thus let $\pi(x)$ be the likelihood times prior, divided by its integral so that it is a pdf.

In what follows below, expectations are written with an integral sign. However, if S is discrete, the same quantities can be interpreted as sums, taking the probability density function to be a probability mass function. In fact, in the mixed case, part discrete and part continuous, the integrals may be understood in the McShane-Stieltjes sense (see section 4.8). For a given set A, the notation $\mid A \mid$ means the volume of A in the continuous case, the number of elements of A in the discrete case, and the sum of these in the mixed case.

At each time point $n \geq 0$, the Markov Chain either stays where it is or makes a jump. Thus, conditional on $X_0 = x_0, X_1 = x_1, \ldots, X_n = x$, where $x_i \epsilon E, i = 1, \ldots, n-1, x \epsilon E$, the next state X_{n+1} is either

(a) equal to X_n, so $X_{n+1} = x$ with some probability $r(x), 0 \leq r(x) < 1$ or

(b) moves to some new state y according to some density $p(x, y)$, where $\int_E p(x, y)dy = 1 - r(x)$.

The quantity $p(x, y)$ is then non-negative, but integrates to a number less than or equal to one. Such a quantity is called a subprobability. To avoid indeterminacy in the discrete case, we limit $p(x, y)$ so that $p(x, x) = 0$, as otherwise the constraint $r(x) < 1$ would not have meaning.

The motion of the Markov Chain is then governed by its transition probabilities. The probability that a Markov Chain at $X_n = x$ moves to some set $A \subset E$ is then

$$\begin{aligned} P(x, A) &= P\{X_{n+1} \epsilon A \mid X_n = x\} \\ &= \int_A p(x, y)dy + r(x)\delta_x(A), \end{aligned} \tag{10.22}$$

where $\delta_x(A) = 1$ if $x \epsilon A$ and 0 otherwise.

Then if X_n has the density function $\lambda(x)$, one can write

$$P\{X_n \epsilon A) = \int_A \lambda(x)dx, \tag{10.23}$$

then the next state X_{n+1} has density function as follows:

$$\begin{aligned} P\{X_{n+1} \epsilon A\} &= \int_E \lambda(x)P(x,A) \\ &= \int_E \lambda(x)\left[\int_A p(x,y)dy + r(x)\delta_x(A)\right]dx. \end{aligned} \tag{10.24}$$

Now

$$\int_E \lambda(x) \int_A p(x,y)dy\ dx = \int_A \int_E \lambda(x)p(x,y)dx\ dy$$

and

$$\int_E \lambda(x)r(x)\delta_x(A)dx = \int_A \lambda(x)r(x)dx = \int_A \lambda(y)r(y)dy.$$

Therefore,

$$P\{X_n \epsilon A\} = \int_A \left[\int_E \lambda(x)p(x,y)dx + \lambda(y)r(y)\right]dy, \tag{10.25}$$

so the density of X_n is

$$\int_E \lambda(x)p(x,y)dx + \lambda(y)r(y), \tag{10.26}$$

which is written as $\lambda P(y)$.

Thus P maps λ to λP, and the nth iterate can be defined recursively as

$$\lambda P^n = (\lambda P^{n-1})P, \tag{10.27}$$

where, by convention, $\lambda P^0 = \lambda$.

A key role in Markov Chain theory is played by an invariant probability density π. A probability density π is called **invariant** (or **stationary**) if $\pi P = \pi$. This means that if X_n has density π, so does X_{n+1}. This is equivalent to

$$\int \pi(x)p(x,y)dx = (1 - r(y))\pi(y). \tag{10.28}$$

There is a huge and growing literature on Markov Chains. One of the concerns of that literature is whether a stationary distribution exists. The nature of the application of Markov Chains discussed here allows us to sidestep this question: it turns out that without any further assumptions the chains generated have a stationary distribution, as will be shown.

Another very important concept for Markov Chains is a **reversible** chain (also known as satisfying detailed balance). A chain is reversible if there is a pdf λ such that

$$\lambda(x)p(x,y) = \lambda(y)p(y,x). \tag{10.29}$$

Lemma 10.3.1. *A reversible chain has an invariant pdf.*

Proof.

$$\begin{aligned} \textstyle\int \lambda(x)p(x,y)dx = \int \lambda(y)p(y,x)dx = \lambda(y) \int p(y,x)dx \\ = \lambda(y)(1 - r(y)). \end{aligned}$$

Therefore, λ is an invariant pdf. □

Indeed, to go further, we know (up to a very important, but unknown constant), what we would like the stationary distribution to be, namely the posterior distribution. Then unlike much of the probability literature, we start with the intended stationary distribution and construct a Markov Chain having the intended stationary distribution, rather than exploring the properties of a given chain.

The particular algorithm considered here is the Metropolis-Hastings algorithm, and works as follows: imagine that the chain has arrived at the state $x \epsilon E$ at some stage n. A proposal is made, according to some distribution $q(x, y)$ to move to $y \epsilon E$. With some probability $\alpha(x, y)$ this proposal is accepted. If the proposal is not accepted, the chain stays at x. Thus we have

$$X_{n+1} = \begin{cases} y \text{ with probability } \alpha(x, y) \\ x \text{ with probability } 1 - \alpha(x, y) \end{cases} .$$

The particular form of $\alpha(x, y)$ that is used in the algorithm is

$$\alpha(x, y) = \begin{cases} \min[\frac{\pi(y)q(y,x)}{\pi(x)q(x,y)}, 1] & \text{if } \pi(x)q(x, y) > 0 \\ 1 & \text{otherwise} \end{cases} \tag{10.30}$$

where π is the posterior distribution, up to an unknown constant. Because the form of α involves the ratio $\pi(y)/\pi(x)$ the algorithm does not require knowledge of the unknown constant. This is quite important, since one of the purposes of using this technique is to deal with ignorance of that constant.

It is useful to give some intuition behind (10.30). The function $\alpha(x, y)$, the acceptance probability for a move from x to y, is larger when y is *a priori* relatively more likely than x, ($\pi(y)/\pi(x)$ large), and when a proposal of a move from y to x is more likely than a proposal of a move from x to y ($q(y, x)/q(x, y)$ large).

I now show that, due to the construction of the Metropolis-Hastings algorithm, it is reversible.

Lemma 10.3.2. *The Metropolis Hastings algorithm is reversible with respect to the density π.*

Proof. We need to show that $\pi(x)q(x, y)\alpha(x, y) = \pi(y)q(y, x)\alpha(y, x)$. Suppose $\pi(y)q(y, x) \geq \pi(x)q(x, y)$. If $\pi(y)q(y, x) = 0$ then $\pi(x)q(x, y) = 0$ and reversibility holds. Assume, then, that $\pi(y)q(y, x) > 0$. Then $\alpha(x, y) = 1$ and $\alpha(y, x) = \pi(x)q(x, y)/\pi(y)q(y, x)$ using the assumption that $\pi(y)q(y, x) > 0$. In this case,

$$\pi(y)q(y, x)\alpha(y, x) = \pi(y)q(y, x)\pi(x)q(x, y)/\pi(y)q(y, x) = \pi(x)q(x, y) = \pi(x)q(x, y)\alpha(x, y).$$

If $\pi(y)q(y, x) \leq \pi(x)q(x, y)$ reverse the roles of x and y above. □

By virtue of Lemmas 10.3.1 and 10.3.2, we know that $\pi(x)$ is an invariant distribution for the Metropolis-Hastings algorithm, for all proposal densities q. Thus the Metropolis-Hastings algorithm satisfies its design criterion: it has the posterior distribution as an invariant distribution.

Let $S = \{x \epsilon E \mid \pi(x) > 0\}$. The next lemma shows that without loss of generality, the space on which the Metropolis-Hastings algorithm operates can be taken to be $S \subseteq E$.

Lemma 10.3.3. *For a Metropolis-Hastings chain, if $x \epsilon S, P(x, S) = 1$.*

Proof. Suppose $x \epsilon S$. Then $\pi(x) > 0$. The first step of a Metropolis-Hastings algorithm may do one of two things. It may reject a proposal, in which case $x \epsilon S$ is the value for X_1. Or it may propose and accept a new point $y \neq x$. If the candidate y is proposed, then $q(x, y) > 0$. Hence $\pi(x)q(x, y) > 0$. Then the candidate y is accepted with probability

$$\alpha(x, y) = \min \left\{ \frac{\pi(y)q(y, x)}{\pi(x)q(x, y)}, 1 \right\}.$$

Now for $\alpha(x, y) > 0$, we must have $\pi(y)q(y, x) > 0$, and hence $\pi(y) > 0$, so $y \epsilon S$. Thus we have $P(x, S) = 1$. A simple induction then shows $X_n \in S$ for all n. □

In view of Lemma 10.3.3, the Metropolis-Hastings algorithm may be conceived of as moving on the space S. Hence all integrals (sums) below in which the range of integration is unspecified is to be taken over the space S.

The next goals are to show that π is the only invariant distribution for a Metropolis-Hastings chain, and then to show that the chain is ergodic, which means that averages of a function of sample paths almost surely approach the expectation of the function with respect to π.

Thus far, no restrictions have been imposed on the proposals $q(x, y)$. It will next be shown that some such conditions must be imposed. To start, consider a Markov Chain with four states, so $E = \{1, 2, 3, 4\}$. Let the transitions between these states governed by the matrix

$$P = \begin{pmatrix} 1/2 & 1/2 & 0 & 0 \\ 1/2 & 1/2 & 0 & 0 \\ 0 & 0 & 1/2 & 1/2 \\ 0 & 0 & 1/2 & 1/2 \end{pmatrix}.$$

If the chain starts in states 1 or 2, it stays in states 1 or 2. Similarly, if the chain starts in 3 or 4, it stays in states 3 or 4. Can such a chain be the result of a Metropolis-Hastings algorithm? Yes, it can, if the proposal distribution satisfies $q(1, 2) = q(2, 1) = q(3, 4) = q(4, 3) = 1$ $\pi(1) = \pi(2)$, and $\pi(3) = \pi(4)$. It is easy to see that both $\begin{pmatrix} 1/2 \\ 1/2 \\ 0 \\ 0 \end{pmatrix}$ and $\begin{pmatrix} 0 \\ 0 \\ 1/2 \\ 1/2 \end{pmatrix}$ are stationary distributions for this chain, so uniqueness will not hold. Furthermore, the long-run averages of a function f will be either $(f(1) + f(2))/2$ or $(f(3) + f(4))/2$, depending on whether the chain starts in states $S_1 = \{1, 2\}$ or states $S_2 = \{3, 4\}$. So it is necessary to have an assumption to prevent this kind of behavior.

More generally, this simple example illustrates the issue that the original Markov Chain decomposes into two subchains that operate on the disjoint sets S_1 and S_2, and it is impossible to go from S_1 to S_2 or S_2 to S_1. A Markov Chain that can be decomposed in this way is called "reducible"; we seek an assumption that guarantees irreducibility of the chain resulting from a Metropolis-Hastings algorithm.

The assumption that we will make is as follows:

Hypothesis: There is a subset $I \subseteq S$ satisfying

(i) For each initial state $x \epsilon S$, there is an integer $n(x) \geq 1$ such that

$$P^{n(x)}(x, I) = P\{X_{n(x)} \epsilon I \mid X_0 = x\} > 0.$$

(ii) There exists a subset $J \subset S$ such that $\mid J \mid > 0$ and a constant $\beta > 0$ such that

$$p(y, z) \geq \beta \text{ for all } y \epsilon I, z \epsilon J.$$

(A subset I satisfying (ii) is called "small.")

Assumptions (i) and (ii) "tug" in opposite directions in the following sense. If a set I satisfies (i), then any set $I' \supseteq I$ also does. However, if a set I satisfies (ii), then any set $I' \subseteq I$ also does. The force of the hypothesis is that there is a set I that is simultaneously small enough to satisfy (ii) and large enough to satisfy (i).

To see how this hypothesis works in practice, reconsider the example of the chain introduced above, and suppose that $\pi(1) = \pi(2) > 0$ and $\pi(3) = \pi(4) > 0$. Then $S = \{1, 2, 3, 4\}$. What shall we choose for I? If $I \subseteq S_1$ or $I \subseteq S_2$, condition (i) fails, since it is not possible to move from one state to the other. Thus these choices for I are too small. On the other hand, if I contains elements of both S_1 and S_2, then condition (ii) fails because there are no choices of J that satisfy the condition.

Now suppose instead that $\pi(1) = \pi(2) = 0$ (the case $\pi(3) = \pi(4) = 0$ is the same, reversing S_1 and S_2). Then $S = \{3, 4\}$, and the choices $I = \{3\}$ and $J = \{4\}$ satisfy the hypothesis.

Now consider an alternating chain, characterized by the transition matrix

$$P = \begin{pmatrix} 0 & 1 \\ 1 & 0 \end{pmatrix}.$$

With this transition matrix, the state is sure to change with every transition. Again, can such a Markov Chain be the result of a Metropolis-Hastings algorithm? Again, yes it can, if the proposal distribution q satisfies $q(1, 2) = q(2, 1) = 1$, and $\pi(1) = \pi(2) = 1/2$. Then it is easy to see that a move will always be proposed and accepted. (A chain of this type is called "periodic," here with period 2.) How does this chain fare with the hypothesis? Clearly we have $S = \{1, 2\}$. Suppose we take $I = \{1\}$ and $J = \{2\}$. Then (ii) is satisfied, with $\beta = 1$. Also (i) is satisfied, since $n(2) = 2$ and $n(1) = 1$ suffices. Thus, this periodic chain satisfies the hypothesis.

The assumptions of the hypothesis say, in order, that: (i) it is possible, in $n(x)$ steps, to go from any arbitrary starting point $x \epsilon S$ to the set I, and (ii) having gotten to the set I, there is some other set J such that the probability (density) is at least β for all moves from points $y \epsilon I$ to points $z \epsilon J$. In the discrete case, I can be taken to be a single point y. Then (i) guarantees that the chain can eventually go from x to y, and, by reversibility, back to x, thus preventing the frozen chain behavior. Finally (ii) is automatically satisfied.

In the continuous case, it is sufficient to assume that there are points y and z in S such that $p(y, z) > 0$, $p(\cdot, \cdot)$ is continuous at (y, z), and the Metropolis-Hastings algorithm visits arbitrarily small neighborhoods of y with positive probability, eventually, from an arbitrary initial state $x \epsilon S$.

A consequence of (i) is that $\lambda(I) > 0$, for every invariant λ. Because λ is invariant, we have $\lambda = \lambda P = \lambda P^2 = \ldots$.

Then

$$\lambda(I) = \int_S \lambda(x) \sum_{n=1}^{\infty} 2^{-n} P^n(x, I) > 0 \tag{10.31}$$

from (i). Because π has a density and is invariant, it also follows that $\mid I \mid > 0$.

Another consequence of these assumptions is that, for $x \epsilon I$,

$$1 - \int_S p(x, y)dy \geq \int_J p(x, y)dy \geq \beta \mid J \mid . \tag{10.32}$$

The heart of the construction to follow is the idea of regeneration or recurrence. Consider a discrete chain that starts at y, wanders around S, and then comes back to y and then does it again, etc. Each tour that starts and ends at y is independent of each other tour, and is identically distributed. For regeneration to be useful it must be shown that the chain will return infinitely often, and in finite expected but stochastically-varying time. Since the law of large numbers then applies to each tour, this opens the way for a law of large numbers to be proved for Markov Chains, specifically, in the cases considered here, to the output of the Metropolis-Hastings algorithm. The same idea applies in the continuous case, but is slightly more delicate since the return is to the set I rather than to a single point y.

In the analysis that follows, assumption (ii) is used immediately and heavily. Assumption (i) also comes up, but in only two (crucial) places in the development and then only through (10.31). Starting with assumption (ii), let ν be a uniform distribution on the set J:

$$\nu(z) = \begin{cases} \mid J \mid^{-1} & \text{for } z \epsilon J \\ 0 & \text{elsewhere} \end{cases}. \tag{10.33}$$

Also define $s(y)$ as follows:

$$s(y) = \begin{cases} \beta \mid J \mid & \text{for } y \epsilon I \\ 0 & \text{elsewhere} \end{cases}. \tag{10.34}$$

Then

$$s(y)\nu(z) = \begin{cases} \beta & \text{if } y \in I,\ z \in J \\ 0 & \text{otherwise} \end{cases}. \tag{10.35}$$

From the definition of a small set, we have

$$p(y, z) \geq s(y)\nu(z) \tag{10.36}$$

for all $y, z \epsilon S$. This is called a minorization condition in the literature.

Let

$$Q(y, A) = P(y, A) - s(y) \int_A \nu(z)dz. \tag{10.37}$$

If $y \in I$ and $A \subseteq S$,

$$Q(y, A) = P(y, A) - \beta \mid J \mid \frac{\mid A \cap J \mid}{\mid J \mid} = P(y, A) - \beta \mid A \cap J \mid \geq 0 \tag{10.38}$$

in view of (10.36).

If $y \notin I$, and $A \subseteq S$,

$$Q(y, A) = P(y, A) \geq 0. \tag{10.39}$$

Q can be regarded as an operator mapping a subprobability λ to the following subprobability:

$$\lambda Q(z) = \lambda P(z) - \left(\int \lambda(y)s(y)dy \right) \nu(z), \tag{10.40}$$

for $z \in S$.

Next, define a bivariate Markov Chain (U_n, Y_n) as follows. Let $U_0, U_1, \ldots$ be a sequence of S-valued random variables, and let $Y_0, Y_1, \ldots$ be a sequence of $\{0, 1\}$-valued random variables. The transition probabilities for this chain are as follows:

$$P\{U_n \in A, Y_n = 1 \mid U_{n-1} = y; Y_{n-1}\} = s(y) \int_A \nu(z)dz \tag{10.41}$$

$$P\{U_n \in A, Y_n = 0 \mid U_{n-1} = y, Y_{n-1}\} = Q(y, A) \tag{10.42}$$

for all $n \geq 1$, $A \subseteq S$, independently of Y_{n-1}.

The law of motion of the random variables U is the same as those of the random variables X, since

$$\begin{aligned} P\{U_n \in A \mid U_0, \ldots, U_{n-2}, U_{n-1} = y\} = \\ P\{U_n \in A, Y_n = 0 \mid U_{n-1} = y\} + P\{U_n \in A; Y_n = 1 \mid U_{n-1} = y\} \\ = Q(y, A) + s(y) \int_A \nu(z)dz = P(y, A), \end{aligned} \tag{10.43}$$

for $n \geq 1$.

One way to think about U_n and X_n is to imagine them as realizations, that is, to imagine starting the process X at X_0 according to some distribution, and then developing according to P. One could also imagine the process (U, Y) realized according to (10.41) and (10.42). Then there is no reason to think that the realized X and U will be equal. However, our purpose is to study the probability distributions of X and U, which by (10.43) are identical if their starting distributions are identical. Consequently, it is not an abuse of notation to use the letter X for U, which we will do.

Also

$$P\{Y_n = 1 \mid X_{n-1} = y, Y_{n-1}\} = s(y), \text{ so} \tag{10.44}$$

$$P\{Y_n = 0 \mid X_{n-1} = y, Y_{n-1}\} = 1 - s(y). \tag{10.45}$$

Thus, when the bivariate chain reaches $(X_{n-1} = y, Y_{n-1})$, at the next stage, $Y_n = 1$ with probability $s(y)$ and $Y_n = 0$ otherwise.

The reason the bivariate chain (X, Y) is a powerful tool analytically is then when $Y_n = 1, X_n$ has the same distribution each time, namely ν. This is shown as follows:

$$\begin{aligned} P\{X_n \in A \mid X_{n-1} = y, Y_{n-1}, Y_n = 1\} = \\ (P\{Y_n = 1 \mid X_{n-1} = y, Y_{n-1}\})^{-1} P\{X_n \in A, Y_n = 1 \mid X_{n-1} = y, Y_{n-1}\} \\ = (s(y))^{-1} s(y) \int_A \nu(z)dz = \int_A \nu(z)dz. \end{aligned} \tag{10.46}$$

In this case, the bivariate chain (X, Y) is said to regenerate at the (random) epochs at which $Y_n = 1$.

From the time-homogeneous Markov Property of the (X, Y) chain,

$$\begin{aligned} & P\{X_n \in A_0, X_{n+1} \in A_1, \ldots; Y_{n+1} = y_1, Y_{n+2} = y_2 \mid \\ & X_0, X_1, \ldots, X_{n-1}, Y_0, \ldots, Y_{n-1}, Y_n = 1\} \\ = \; & P\{X_0 \in A_0, X_1 \in A_1, \ldots; Y_1 = y_2, Y_2 = w_2, \ldots \mid Y_0 = 1\} \\ = \; & P_\nu\{X_0 \in A_0, X_1 \in A_1, \ldots; Y_1 = y_1, Y_2 = y_2, \ldots\} \end{aligned} \tag{10.47}$$

where the subscript ν indicates that X_0 has the initial distribution ν.

The next concept to introduce is a random variable T taking values in $\mathcal{N} \cup \{\infty\}$. In particular, T is the first regeneration epoch time, so

$$T = \min\{n > 0 : Y_n = 1\}. \tag{10.48}$$

More generally, let $1 \leq T_1 \leq T_2 \leq T_3 \ldots$ denote the successive regeneration epochs, where

$$T_1 = T$$

and

$$T_i = \min\{n > T_{i-1} : Y_n = 1\} \text{ for } i = 2, 3, \ldots \tag{10.49}$$

The hard work in the proof to come is showing that the random variables T_i are finite, and have finite expectations. To begin the analysis of T, we have

$$\begin{aligned}
& P\{X_n \in A, T > n \mid X_{n-1} = y, T > n-1\} \\
= \quad & P\{X_n \in A, Y_1 = 0, \ldots, Y_n = 0 \mid X_{n-1} = y, Y_1 = 0, \ldots, Y_{n-1} = 0\} \\
& \text{(by definition of T)} \\
= \quad & P\{X_n \in A, Y_n = 0 \mid X_{n-1} = y, Y_1 = 0, \ldots, Y_{n-1} = 0\} \\
& (P\{CD \mid CF\} = P\{D \mid CF\} \text{ for all events C, D and F}). \\
= \quad & P\{X_n \in A, Y_n = 0 \mid X_{n-1} = y, Y_{n-1} = 0\} \\
& ((X,Y) \text{is a Markov Chain}) \\
= \quad & Q(y, A) \\
& \text{(see (10.42))}. \qquad (10.50)
\end{aligned}$$

Now we can state an important result that will help to control the distribution of T:

Lemma 10.3.4. *Suppose X_0 has the arbitrary initial distribution λ. Then*

$$P_\lambda\{X_n \in A, T > n\} = \int_A \lambda Q^n(x) dx.$$

Proof. By induction on n. At $n = 0$, the lemma is just the definition of λ. Suppose then, the lemma is true at $n-1$ for $n = 1, 2, \ldots$. Then

$$\begin{aligned}
& P_\lambda\{X_n \in A, T > n\} \\
= \quad & \int_S P\{X_n \in A, Y_n = 0 \mid X_{n-1} = x, T > n-1\} \lambda Q^{n-1}(x) dx \\
& \text{(uses inductive hypothesis)} \\
= \quad & \int_S \lambda Q^{n-1}(x) Q(x, A) dx \\
= \quad & \int_A \lambda Q^n(x) dx \qquad (10.51)
\end{aligned}$$

by definition of the nth iterate of the kernel Q. □

Using Lemma 10.3.4,

$$\begin{aligned}
P_\lambda\{T \geq n\} = P_\lambda\{T > n-1\} &= P_\lambda\{X_n \in S, T > n-1\} \\
&= \int \lambda Q^{n-1}(x) dx.
\end{aligned} \qquad (10.52)$$

Also

$$\begin{aligned}
P\{T = n\} &= P\{X_{n-1} \in S, T > n-1, Y_n = 1\} \\
&= \int P\{Y_n = 1 \mid X_{n-1} = x, T > n-1\} \lambda Q^{n-1}(x) dx \\
&= \int s(x) \lambda Q^{n-1}(x) dx.
\end{aligned} \qquad (10.53)$$

Let

$$\mu(x) = \sum_{n=0}^{\infty} \nu Q^n(x). \qquad (10.54)$$

The function $\mu(x)$ is called the potential function. If the starting distribution is ν, the expected number of visits to the set A before T is given by the integral of potential function over A:

$$E_\nu \sum_{n=0}^{T-1} \delta_{X_n}(A) = \sum_{n=0}^{\infty} P_\nu\{X_n \in A, T > n\}$$
$$= \sum_{n=0}^{\infty} \int_A \nu Q^n(x)dx$$
$$= \int_A \mu(x)dx. \tag{10.55}$$

In particular, if $A = S$, the expected regeneration time is

$$M = E_\nu(T) = \int \mu(x)dx. \tag{10.56}$$

The key to further progress is examining when $M < \infty$. If $f(x)$ is any non-negative measurable function $f(x), x \in S$,

$$E_\nu \sum_{n=0}^{T-1} f(X_n) = \int \mu(x)f(x)dx. \tag{10.57}$$

Also, setting $\lambda = \nu$ and summing (10.53) over n, we have

$$P_\nu(T < \infty) = \int \mu(x)s(x)dx. \tag{10.58}$$

For each $n \geq 1$, let L_n be the time elapsed since the last regeneration before n. Then

$$\{T \leq n\} = \cup_{k=0}^{n-1}\{L_n = k\} \tag{10.59}$$

where $L_n = \min\{0 \leq k \leq n-1 : Y_{n-k} = 1\}$.

Let λ be an arbitrary starting density. Then for all $n \geq 1$ and $A \subseteq S$,

$$P\{X_n \in A\} = P\{X_n \in A, T > n\} + \sum_{k=0}^{n-1} P\{L_n = k, X_n \in A\}. \tag{10.60}$$

Now

$$\begin{aligned}
&\sum_{k=0}^{n-1} P\{L_n = k, X_n \in A\} \\
= \quad &\sum_{k=0}^{n-1} P\{Y_{n-k} = 1, Y_{n-k+1} = 0, \ldots, Y_n = 0; X_n \in A\} && (10.61)\\
&&& \text{(uses definition of } L\text{)}\\
= \quad &\sum_{k=0}^{n-1} P\{Y_{n-k} = 1\} P\{Y_{n-k+1} = 0, \ldots, Y_n = 0, X_n \in A \mid Y_{n-k} = 1\} && (10.62)\\
&&& \text{(conditional probability)}\\
= \quad &\sum_{k=0}^{n-1} P\{Y_{n-k} = 1\} P\{Y_1 = 0, \ldots, Y_k = 0, X_k \in A \mid Y_0 = 1\} && (10.63)\\
&&& \text{(time homogeneity)}\\
= \quad &\sum_{k=0}^{n-1} P\{Y_{n-k} = 1\} P_\nu\{Y_1 = 0, \ldots, Y_k = 0, X_k \in A\} && (10.64)\\
&&& \text{(uses (10.47))}
\end{aligned}$$

$$= \sum_{k=0}^{n-1} \int \lambda P^{n-k-1}(y) s(y) dy \int_A \nu Q^k(x) dx.$$

(uses Lemma 10.3.4 and (10.44)) (10.65)

We now suppose that λ is invariant. (We know that at least one invariant distribution exists, namely π. We are getting ready to prove, but have not yet proved, that under our assumptions, π is the *only* invariant distribution.) With this assumption, we have two results:

$$P\{X_n \in A\} = \int_A \lambda(y) dy \tag{10.66}$$

and

$$\int \lambda P^{n-k-1}(y) s(y) dy = \int \lambda(y) s(y) dy. \tag{10.67}$$

Substituting these results into (10.60) and (10.65), we have

$$\int_A \lambda(y) dy = P_\lambda\{X_n \in A, T > n\} + \int \lambda(y) s(y) dy \sum_{k=0}^{n-1} \int_A \nu Q^k(x) dx. \tag{10.68}$$

Now let $A = S$, to obtain

$$1 = \int \lambda(y) dy = P_\lambda(T > n) + \int \lambda(y) s(y) dy \sum_{k=0}^{n-1} \int \nu Q^k(x) dx. \tag{10.69}$$

Letting $n \to \infty$ yields

$$1 = P_\lambda\{T = \infty\} + M \int \lambda(y) s(y) dy. \tag{10.70}$$

Now

$$\int \lambda(y)s(y)dy = \beta \mid J \mid \lambda(I) > 0 \tag{10.71}$$

using (10.31).

Therefore, $M < \infty$. Since $M = E_\nu T$, we also have

$$P_\nu(T < \infty) = \int \mu(x)s(x)dx = 1. \tag{10.72}$$

These results are important, because they say that if the chain started with $Y_0 = 1$, the expected time T until the next time some $Y_n = 1$ is finite. This in turn allows us to return to the random variables T_i, defined at (10.49). Using (10.47),

$$\begin{aligned} P\{X_{T_i} &\in A_0, X_{T_i+1} \in A_1, \dots, X_{T_i+m-1} \in A_{m-1}; T_{i+1} - T_i = m \mid \\ &X_0, X_1, \dots, X_{T_i-1}; T_1, \dots, T_{i-1}; T_i = n\} \\ &= P\{X_0 \in A_0, X_1 \in A_1, \dots, X_{m-1} \in A_{m-1}; T = m \mid Y_0 = 1\} \\ &= P_\nu\{X_0 \in A_0, \dots, X_{n-1} \in A_{m-1}; T = m\}. \end{aligned} \tag{10.73}$$

This has the following implication: Consider the random blocks

$$\xi_0 = (X_0, \dots, X_{T-1}; T)$$

$$\xi_i = (X_{T_i}, \dots, X_{T_{i+1}-1}; T_{i+1} - T_i) \text{ for } 1 = 1, 2, \dots$$

These blocks are independent. Also the blocks $\xi_i, i \geq 1$ have the same distribution, and have the same distribution as the block ξ_0 under the initial distribution ν.

Hence

$$P\{T_{i+1} - T_i = m \mid X_0, X_1, \dots, X_{n-1}; T_1, \dots, T_{i-1}, T_i = n\} = P_\nu(T = m). \tag{10.74}$$

Furthermore, for a given function $f(x), x \in S$, we can define the random sums over the blocks ξ_i as follows

$$\begin{aligned} \xi_0(f) &= \sum_{m=0}^{T-1} f(X_m) \\ \xi_i(f) &= \sum_{m=T_i}^{T_{i+1}-1} f(X_m) \quad i \geq 1. \end{aligned} \tag{10.75}$$

These sums are independent. The random variables $\xi_i(f)(i \geq 1)$ are identically distributed, and have the same distribution as the random variable $\xi_0(f)$ under the initial pdf ν.

Lemma 10.3.5.

$$P\{T_i < \infty \mid X_0 = x\} = P\{T < \infty \mid X_0 = x\}$$

for all $x \in S$ and $i \geq 1$.

Proof. By induction on i. When $i = 1, T_1 = T$ so there is nothing to prove. Suppose then, the lemma is true for i.

Then

$$\begin{aligned}
P\{T_{i+1} < \infty \mid X_0 = x\} &= \sum_{n=1}^{\infty}\sum_{m=1}^{\infty} P\{T_{i+1} - T_i = m, T_i = n \mid X_0 = x\} \\
&= \sum_{n=1}^{\infty}\sum_{m=1}^{\infty} P\{T_{i+1} - T_i = m \mid T_i = n, X_0 = x\} P\{T_i = n \mid X_0 = x\} \\
&= \sum_{n=1}^{\infty}\sum_{m=1}^{\infty} P_\nu\{T = m\} P\{T_i = n \mid X_0 = x\} \\
&= \sum_{m=1}^{\infty} P_\nu\{T = m\} \sum_{n=1}^{\infty} P\{T_i = n \mid X_0 = x\}.
\end{aligned}$$

But $\sum_{m=1}^{\infty} P_\nu\{T = m\} = P_\nu\{T < \infty\} = 1$ using (10.72) and

$$\sum_{n=1}^{\infty} P\{T_i = n \mid X_0 = x\} = P\{T_i < \infty \mid X_0 = x\}.$$

Hence $P\{T_{i+1} < \infty \mid X_0 = x\} = P\{T_i < \infty \mid X_0 = x\} = P\{T < \infty \mid X_0 = x\}$ completing the inductive step. □

We can now address the uniqueness of the invariant distribution. To begin, observe that μ is invariant, as follows:

$$\begin{aligned}
\mu(y) &= \nu(y) + \sum_{n=1}^{\infty} \nu Q^n(y) && \text{(uses (10.54))} \\
&= \nu(y) + \sum_{n=0}^{\infty} (\nu Q^n) Q(y) && \text{(just algebra)} \\
&= \left(\int \mu(x) s(x) dx\right) \nu(y) + \mu Q(y) && \text{(uses (10.72) and (10.54))} \\
&= \mu P(y). && \text{(uses (10.40))}
\end{aligned} \tag{10.76}$$

Now let $n \to \infty$ in (10.68), yielding

$$\int_A \lambda(y) dy \geq \left(\int \lambda(y) s(y) dy\right) \int_A \mu(y) dy \tag{10.77}$$

for all A, and every invariant distribution λ.

Now consider the function

$$\beta(y) = \lambda(y) - \left(\int \lambda(x) s(x) dx\right) \mu(y). \tag{10.78}$$

Because both λ and μ are invariant, so is β. By (10.77), $\beta \geq 0$. I claim now that $\int \beta(y) dy = 0$. Suppose the contrary. Then the function $\beta^*(y) = \beta(y) / \int \beta(y) dy$ would be an invariant pdf, and would satisfy

$$\begin{aligned}
&\int \beta^*(y) s(y) dy \\
&\qquad = \frac{1}{\int \beta(y) dy} \left[\int \lambda(y) s(y) dy - \left(\int \lambda(y) s(y) dy\right) \int s(x) \mu(x) dx\right] \\
&\qquad\qquad = 0
\end{aligned} \tag{10.79}$$

using (10.72).

But this contradicts (10.31). Therefore $\int \beta(y)dy = 0$, and $\beta(y) = 0$ almost everywhere. Integrating (10.78) then yields

$$1 = \left(\int \lambda(x)s(x)dx\right)\int \mu(y)dy \tag{10.80}$$

$$= \left(\int \lambda(x)s(x)dx\right)M \qquad \text{(using (10.56))}$$

so

$$\lambda(y) = \left(\int \lambda(x)s(x)dx\right)\mu(y) = \mu(y)/M. \tag{10.81}$$

and is therefore unique. Since we already know that π is invariant (see Lemmas 10.3.1 and 10.3.2) it is therefore the only invariant pdf, so we have

$$\pi(y) = \mu(y)/M. \tag{10.82}$$

Now that the invariant distribution has been shown to be unique, the next goal is to show that the regeneration times are finite no matter what starting point is used. Thus we seek to prove

Lemma 10.3.6.

$$P\{T_i < \infty \mid X_0 = x\} = 1$$

for all $i = 1, 2, \ldots$ *and all* $x \in S$.

Proof. In view of Lemma 10.3.5, it is sufficient to show

$$P\{T < \infty \mid X_0 = x\} = 1. \tag{10.83}$$

Using (10.70) and (10.80), we have

$$1 = P_\pi\{T = \infty\} + MM^{-1}. \tag{10.84}$$

Thus

$$1 = P_\pi\{T < \infty\} = \int P\{T < \infty \mid X_0 = x\}\pi(x)dx \tag{10.85}$$

which implies

$$P\{T < \infty \mid X_0 = x\} = 1 \tag{10.86}$$

for all $x \in S$ except possibly a set of measure 0. We now prove that (10.86) holds for all $x \in S$.

Let $h_\infty(x) = P\{T = \infty \mid U_0 = x\} = \lim_{n\to\infty} P\{T > n \mid X_0 = x\}$.

Using Lemma 10.3.4, we have

$$h_\infty(x) = \lim_{n\to\infty} Q^n(x, S). \tag{10.87}$$

Because $P\{T > n \mid X_0 = x\}$ is monotone non-increasing in n and hence dominated (see Theorem 4.7.11), we may exchange integrals and limits in the following calculation:

$$\begin{aligned}
h_\infty(x) &= \lim_{n\to\infty} Q^n(x,S) \\
&= \lim_{n\to\infty} \int Q^n(x,dz) \\
&= \int \lim_{n\to\infty} Q^n(x,dz) \\
&= \int \lim_{n\to\infty} \int Q(x,dy)Q^{n-1}(y,dz) \\
&= \int Q(x,dy) \lim_{n\to\infty} \int Q^{n-1}(y,dz) \\
&= \int Q(x,dy) \lim_{n\to\infty} Q^{n-1}(y,S) \\
&= \int Q(x,dy)h_\infty(y) \text{ for all } x \in S.
\end{aligned} \tag{10.88}$$

Now (10.72) implies that

$$\int h_\infty(y)\nu(y)dy = 0, \tag{10.89}$$

so it follows that

$$\int P(x,dy)h_\infty(y) = h_\infty(x) \quad \text{for all } x \in S \tag{10.90}$$

from (10.37). In view of (10.22) this is equivalent to

$$\int p(x,y)h_\infty(y) = (1 - r(x))h_\infty(x). \tag{10.91}$$

Now suppose, contrary to hypothesis, that there is some $x_0 \in S$ such that $h_\infty(x_0) > 0$. Since $r(x) < 1$ for all $x \in S$ (see (a) above (10.22)), we would then have

$$\int p(x_0,y)h_\infty(y)dy > 0. \tag{10.92}$$

But this implies

$$\int h_\infty(y)dy > 0 \tag{10.93}$$

contradicting (10.86). Therefore $h_\infty(x) = 0$ for all $x \in S$, which proves the lemma. □

The property proved in Lemma 10.3.6 is known in the literature as Harris recurrence.

We now turn to the statement and proof of the Strong Law of Large Numbers for the Metropolis-Hastings algorithm.

Theorem 10.3.7. *Let $f(x), x \in S$ be a π-integrable function, so $f(\cdot)$ satisfies*

$$\int_S | f(x) | \pi(x)dx < \infty. \tag{10.94}$$

Let $X_0 = x \in S$ be an arbitrary starting point for the Metropolis-Hastings algorithm. Let

$$S_n = \sum_{i=0}^{n} f(x_i). \tag{10.95}$$

Then, with probability 1,

$$\lim_{n\to\infty} S_n/n = \int_S f(x)\pi(x)dx. \tag{10.96}$$

Proof. Since the random variables $\xi_0(f), \dots, \xi_i(f), \dots$ defined in (10.75) are independent and $\xi_1(f), \xi_2(f) \dots$ are identically distributed (with the same distribution as X_0 under the initial pdf ν), the Strong Law of Large Numbers for independent random variables (see section 4.11) and from $P\{T < \infty \mid X_0 = x\} = 1$ it follows that

$$\lim_{i=\infty} i^{-1} \sum_{j=0}^{i} \xi_j(f) = E\xi_1(f) = E_\nu \xi_0(f) = \int f(x)\mu(x)dx \tag{10.97}$$

$$= M \int_S f(x)\pi(x)dx \tag{10.98}$$

with probability 1 (using (10.57) and (10.82)).

Also

$$\lim_{i\to\infty} i^{-1} T_i = E(T_2 - T_1) = E_\nu T = M \tag{10.99}$$

with probability 1, since $M < \infty$.

It remains to account for the part of S_n that is after the last regeneration time. To that end, let $N(n), n = 1, 2, \dots$ be the (random) number of regeneration epochs T_i up to time n. Then

$$T_{N(n)} \le n < T_{N(n)+1}. \tag{10.100}$$

Since $N(n) \to \infty$ with probability 1 (from Lemma 10.3.6),

$$\lim_{n\to\infty} n^{-1} N(n) = \lim_{n\to\infty} (T_{N(n)})^{-1} N(n) = M^{-1} \tag{10.101}$$

with probability 1. Now

$$S_n = \sum_{m=0}^{n-1} f(X_m) = \sum_{j=0}^{N(n)-1} \xi_j(f) + \xi'_{N(n)} \tag{10.102}$$

where

$$\xi'_{N(n)} = \begin{cases} \sum_{m=T_{N(n)}}^{n-1} f(X_m) & \text{if } T_{N(n)} \le n-1 \\ 0 & \text{if } T_{N(n)} = n \end{cases}.$$

Now $\mid \xi'_{N(n)} \mid$ is bounded as follows:

$$\mid \xi'_{N(n)} \mid \le \sum_{m=T_{N(n)}}^{T_{m(n)+1}-1} \mid f(X_m) \mid . \tag{10.103}$$

The random variable on the right-hand side has the same distribution (with probability 1) as that of $\sum_{m=0}^{T-1} \mid f(X_m) \mid$ (under the initial distribution ν), so it follows that

$$\lim_{n\to\infty} n^{-1} \xi'_{N(n)} = 0 \text{ with probability 1.} \tag{10.104}$$

Then

$$\begin{aligned}
\lim_{n\to\infty} n^{-1}S_n &= \lim_{n\to\infty} n^{-1}\left(\sum_{j=0}^{N(n)-1} \xi_j(f) + \xi'_{N(n)}\right) && \text{(uses (10.102))}\\
&= \lim_{n\to\infty} n^{-1} \sum_{j=0}^{N(n)-1} \xi_j(f) && \text{(uses (10.104))}\\
&= \lim_{n\to\infty}\left(\frac{N(n)}{n}\right) \lim_{n\to\infty}\left(\sum_{j=0}^{N(n)-1} \xi_j\right)\\
&= M^{-1}\left(M\int_S f(x)\pi(x)dx\right) && \text{(uses (10.94) and (10.101))}\\
&= \int_S f(x)\pi(x)dx.
\end{aligned}$$

□

10.3.1 *Literature*

This treatment relies very heavily on Nummelin (2002). There is a vast literature on Markov Chain theory generally. The classic works are Nummelin (1984) and Meyn et al. (2009). Additional results, with additional assumptions, give a central limit theorem and a geometric rate of convergence. An important paper linking the general theory with Markov Chain Monte Carlo is Tierney (1994).

10.3.2 *Summary*

A very general strong law holds for the output of the Metropolis-Hastings algorithm.

10.3.3 *Exercises*

1. State in your own words the meaning of
   (a) stochastic process
   (b) Markov Chain
   (c) time homogeneous Markov Chain
   (d) stationary distribution
   (e) reversible chain
   (f) Metropolis-Hastings algorithm
   (g) minorization condition
   (h) potential function
2. Suppose you have data $X_1, \ldots, X_n$ which you believe come from a normal distribution with mean θ and variance 1. Suppose also that you are uncertain about θ, in fact, for you, θ has the following Cauchy distribution:

$$f(\theta) = \frac{1}{\pi(1+\theta^2)}, -\infty < \theta < \infty.$$

   (a) Show that f is a pdf, by showing that it integrates to 1.
   (b) Can you find the posterior distribution of θ analytically? Why or why not.

(c) If not, write a Metropolis-Hastings algorithm whose limiting distribution is that posterior distribution.

3. Consider the transition matrix

$$P = \begin{pmatrix} 1 & 0 \\ 0 & 1 \end{pmatrix}.$$

(a) What proposal distribution $q(x, y)$ for a Metropolis-Hastings algorithm leads to this transition matrix?

(b) Does this specification satisfy the hypothesis after Lemma 10.3.3? Why or why not?

(c) Show that both $\pi_1 = \binom{1}{0}$ and $\pi_2 = \binom{0}{1}$ are stationary probability vectors for this transition matrix.

(d) What other assumption of the theorem does this transition matrix fail to satisfy?

10.4 Extensions and special cases

This section considers several extensions and special cases of the Metropolis-Hastings algorithm. The first issue has to do with what happens when several such algorithms are used in succession. To be precise, suppose that $P_1, \ldots, P_k$ are Metropolis-Hastings algorithms. Each P_i is assumed to obey the following:

(i) There is a distribution π (not depending on i) with respect to which each P_i is invariant, that is,

$$P_i\pi = \pi \quad i = 1, \ldots, k.$$

(ii) Each P_i satisfies $r_i(x) < 1$ for all $x \in S = \{x \mid \pi(x) > 0\}$.

We now consider the algorithm

$$P = P_k P_{k-1} \ldots P_1$$

which consists of applying P_1 to $X_0 = x_0$, then P_2, etc. Although each of the P_i's may not satisfy the hypothesis of the almost-sure convergence result, P may well. In this case the theorem applies to P. (Although each P_i is reversible, the product need not be.) Also note that P has π as an invariant distribution, because

$$P\pi = P_k P_{k-1} \ldots P_1 \pi = P_k P_{k-1} \ldots P_2 \pi = \ldots = \pi. \tag{10.105}$$

The fact that one can use several Metropolis-Hastings algorithms in succession opens the way for block updates, in which a part, but not all of the parameter space is moved by one of the P_i's. In particular, suppose $\mathbf{x} = (x_1, \ldots, x_p)$ is the parameter space. Let $\mathbf{x}_K$ be a subset of the components of $\mathbf{x}$, and $\mathbf{x}_{\notin K}$ denote the components not in $\mathbf{x}_K$. Then, rearranging the order of the components if necessary, we may write

$$\mathbf{x} = (\mathbf{x}_K, \mathbf{x}_{\notin K}).$$

Now a Metropolis-Hastings algorithm could propose to update only the components of $\mathbf{x}_K$, leaving the components of $\mathbf{x}_{\notin K}$ unchanged. Such a sampler cannot by itself satisfy the hypothesis, since it leaves the elements of $\mathbf{x}_{\notin K}$ unchanged, but several such Metropolis-Hastings algorithms in succession could. Block updating is very useful in designing Metropolis-Hastings samplers. For example, it is natural to use block updating in problems that involve missing data, and more generally in hierarchical models (see Chapter 9). While updating each parameter individually is a valid special case of block updating, it is often more advantageous to update several parameters together, especially if they have a linear regression structure (see Chapter 8).

In certain problems it is possible to derive analytically the conditional posterior distribution of a block of parameters given the others, by deriving $\pi(\mathbf{x}_K \mid \mathbf{x}_{\notin K})$. In this case, one choice of proposal function for a block-sampler sampling $\mathbf{x}_K$ is

$$q(\mathbf{x}_K \mid \mathbf{x}_{\notin K}) = \pi(\mathbf{x}_K \mid \mathbf{x}_{\notin K}).$$

Thus the proposal is to move from the point $\mathbf{x} = (\mathbf{x}_K, \mathbf{x}_{\notin K})$ to a new point $\mathbf{y} = (\mathbf{y}_K, \mathbf{x}_{\notin K})$. Under the choice of proposal function above,

$$\frac{\pi(\mathbf{x})}{q(\mathbf{x},\mathbf{y})} = \frac{\pi(\mathbf{x}_K, \mathbf{x}_{\notin K})}{\pi(\mathbf{x}_K \mid \mathbf{x}_{\notin K})} = \pi(\mathbf{x}_{\notin K}) = \frac{\pi(\mathbf{y}_K, \mathbf{x}_{\notin K})}{\pi(\mathbf{y}_K \mid \mathbf{x}_{\notin K})} = \frac{\pi(\mathbf{y})}{q(\mathbf{y},\mathbf{x})}. \tag{10.106}$$

Consequently, under this choice, every such proposal is accepted, according to 10.30. This is called a Gibbs Step; a Metropolis-Hastings algorithm consisting only of Gibbs Steps is called a Gibbs Sampler.

Some other special cases of note are:

(a) If q is symmetric, so $q(x,y) = q(y,x)$,

$$\alpha(x,y) = \min\left\{\frac{\pi(x)}{\pi(y)}, 1\right\}. \tag{10.107}$$

This is the original Metropolis version (Metropolis et al. (1953)).

(b) A random walk

$$\mathbf{y} = \mathbf{x} + \boldsymbol{\epsilon} \tag{10.108}$$

where $\boldsymbol{\epsilon}$ is independent of $\mathbf{x}$. Often $\boldsymbol{\epsilon}$ is chosen to be symmetric around 0, in which case *(a)* applies.

(c) Independence, where $q(x,y) = s(x)$ for some density $s(x)$. Then

$$\alpha(x,y) = \min\left\{\frac{\pi(x)s(y)}{\pi(y)s(x)}, 1\right\}. \tag{10.109}$$

A joint chain can also be composed of a mixture of chains $P_1 \ldots, P_k$, *i.e.*,

$$P = \sum_{i=1}^{k} \alpha_i P_i \tag{10.110}$$

where $\alpha_i > 0$ and $\sum_{i=1}^{k} \alpha_i = 1$. If each P_i satisfies conditions (i) and (ii) of the hypothesis of section 10.3, then so will P. Furthermore, unlike the case of using the P_i's in succession, a mixture of Metropolis-Hastings chains is reversible. Algorithms of this type are often called "random scans."

10.4.1 Summary

You are introduced to several of the most important special cases of the Metropolis-Hastings algorithm, including especially the Gibbs Sampler.

10.4.2 Exercises

1. State in your own words the meaning of
   (a) Gibbs Step
   (b) Gibbs Sampler
   (c) random walk sampler

(d) independence sampler

(e) blocks of parameters

2. Make a sampler in pseudo-code exemplifying each of the three special cases mentioned in problem 1. Give examples of when it would be useful and efficient to use each, and explain why.

10.5 Practical considerations

The practice of the Metropolis-Hastings algorithm is shadowed by two related considerations. The first is the dependent nature of the resulting chain. A chain that is less dependent will have more information, for a given sample size, about the target posterior distribution. The second important consideration is the sample size. Almost sure convergence is nice, but it is an asymptotic property. How large must the sample size be for the resulting averages to be a good approximation? Since every computer run is of finite duration, this issue is unavoidable.

To overcome these problems, the Metropolis-Hastings algorithm offers great design flexibility in deciding what blocks of parameters to use in each step, what proposal distribution to use and how much of the initial part of the sample to ignore as "burn in." The purpose of this section is to give some practical guidance on how to make these choices wisely.

To emphasize why these considerations are important, imagine a two-state chain whose transition probabilities are given by

$$P = \begin{pmatrix} 1-\epsilon & \epsilon \\ \epsilon & 1-\epsilon \end{pmatrix}.$$

For every ϵ, $0 < \epsilon \leq 1$, such a chain can result from a Metropolis-Hastings algorithm, where $\pi(1) = \pi(2) = 1/2$, $r(x) = 1 - \epsilon$ for $x = 1, 2$, and $p(1, 2) = p(2, 1) = 1$. This algorithm proposes to move with probability ϵ, and always accepts the proposed move. Again, for every $\epsilon, 0 < \epsilon \leq 1$, this algorithm satisfies the hypotheses of the almost-sure-convergence theorem. The sample paths of this algorithm will have identical observations for chunks whose length is governed by a geometric distribution with parameter ϵ and expectation $1/\epsilon$, followed by another such chunk of the other parameter value of length governed by the same distribution. For small $\epsilon > 0$, the chain mixes arbitrarily poorly, and would require arbitrarily large samples for almost sure convergence to set in. However, at $\epsilon = 1/2$, the sample path is that of independent observations. Of course at $\epsilon = 0$, the chain is reducible, and violates the assumptions of the almost-sure-convergence result.

This example illustrates an important point, namely that trouble, in the sense of poor mixing and large required sample sizes, can result from being too close to the boundary of algorithms that satisfy the required conditions for convergence. Another example is proximity to violations of the assumption that the posterior distribution is proper, that is, that it integrates to 1. When the posterior is not proper, the chain resulting from the Metropolis-Hastings algorithm can be run, but the consequence will be at best recurrence that is expected to be infinitely far off in the future (see Bremaud (1999, Theorem 2.3, p. 103)). Such a posterior distribution can be the result of the use of an improper prior distribution used to express ignorance (see discussion in section 1.1.2 about why I think this is misguided as a matter of principle). I have seen such improper posterior distributions come up in practice, in particular in the imposition of improper "ignorance" priors on variances high in a hierarchical model. Some of the default priors in the popular Winbugs program (Spiegelhalter et al. (2003)) are proper but only barely so. These also present the danger that if the likelihood is not sufficiently informative, the posterior density may be so spread out as to be effectively improper. The paper of Natarajan and McCulloch (1998) gives a detailed study of diffuse proper prior distributions in the setting of a normal probit hierarchical model, and shows the damage that can result.

How should blocks be chosen? When the model structure is hierarchical, often it is useful to consider the parameters at a given level of the hierarchy (or a subset of them) as a block. This permits use of the conditional independence conditions frequently found in such models. Another important consideration is that parameters that are highly correlated (positively or negatively) should be considered together. To take an unrealistic extreme example, suppose a model includes two parameters γ and β (together with possibly other parameters as well). Suppose that the posterior distribution requires that $\gamma = \beta$ (realistically in this case, one of the two would be substituted for the other so there would be one fewer parameter in the model). If γ and β were in different blocks, the constraint would not permit either to be moved, leading to no mixing at all. Now suppose instead that γ and β are highly correlated. Then only very small moves in either would be permitted, leading to very slow mixing.

The second design issue is the choice of a proposal distribution q for a block. If the Gibbs Sampler is available, which requires that the required conditional distributions can be found analytically, that is an obvious choice. When the Gibbs Sampler is not available, a key indicator for q is the average acceptance probability $\overline{\alpha}$. If $\overline{\alpha}$ is low, this suggests that q is proposing steps that are too big. Conversely, if $\overline{\alpha}$ is high, then this suggests that q's proposed steps are too small, leading to poor mixing. How should "too high" and "too low" be judged? Some work by Roberts et al. (1997) suggests that $\overline{\alpha}$ in the range of .25 to .5 is good for a random walk chain, and this seems to be good advice more generally. Another consideration is that it is wise to have a proposal distribution that has heavier tails than the posterior distribution being approximated. There are reasons other than ensuring good approximation to the posterior why this is good advice, a matter we'll return to later.

It is not possible to know in advance what the average acceptance rate $\overline{\alpha}$ will be. Consequently common practice is to run a chain with an initial choice of q, examine the results to see which blocks are not mixing well, and then adjust those proposal distributions accordingly. There are proposals to automate this process, leading to adaptive Markov Chain Monte Carlo (MCMC). However, if the proposal distribution depends on the past draws of the chain, the chain may no longer be Markovian. How to design adaptive chains with good properties is a subject of current research.

There has been some debate about whether to start a chain with different starting values (to see if they converge to the same area of the parameter space) (see Gelman and Rubin (1992)) or to run one longer chain (see Geyer (1992) and Tierney (1992)), on the argument that once two separate chains reach the same value, their distributions from then on are identical. Both of these arguments have some force; the choice seems more pressing if computational resources are scarce given the complexity of the model and algorithm.

Often there is a desire to check the sensitivity of the model to various aspects of it. If the motivation for this is personal uncertainty on the part of the person doing the analysis, this can suggest that the model does not yet fully reflect the uncertainty of that person. On the other hand, sensitivity analysis can also be used as a way to communicate to others that variations of a certain size in some aspect of the model may or may not change the posterior conclusions in important ways. The output of a Markov Chain Monte Carlo may be used for such a sensitivity analysis by reweighting the output. Thus, if $\pi(x)$ is the posterior distribution the MCMC was run with, and $\pi^*(x)$ is the newly desired posterior, the trivial calculation

$$\int f(x)\pi^*(x)dx = \int f(x)\pi(x)\left(\frac{\pi^*(x)}{\pi(x)}\right)dx \tag{10.111}$$

suggests that reweighting the output with weights $(\pi^*(x)/\pi(x))$ will yield the needed approximation. (This amounts to importance sampling applied to MCMC output.) It requires that $\pi(x)$ not be zero (or very small, relative to $\pi^*(x)$). The availability of this technique suggests that a prior distribution might be chosen to mix easily in the whole parameter space. Then the prior representing the honest belief of the analyst could be a factor in the $\pi^*(x)$ used in reweighting.

The reweighting idea can be used in the unfortunate situation of two rather disparate high density areas. Suppose for example that the posterior distribution is a weighted average of two densities, say one is $N(-10, 1)$ and the other $N(10, 1)$, where the weight on each is unknown. A chain might take a long time to move from one area of high posterior to the other, so information about the relative weights would be slow in coming. By using a prior that upweights (artificially) the interval $(-9, 9)$, the chain can easily move back and forth, giving information about how much weight belongs in each component. Reweighting will then downweight the $(-9, 9)$ interval appropriately. Reweighting is known in the literature as "Sampling Importance Resampling," or SIR (Rubin (1988)).

There are many techniques that have been proposed for checking how much of the sample output from an MCMC should be disregarded as "burn-in," and whether equilibrium has been achieved. Of course none of these methods is definitive, but each is useful. The package BOA (Bayesian Output Analysis) (Smith (2005)) and CODA (Plummer et al. (2006)) are standardly used for such checking.

Algorithms, called perfect sampling, have been developed in some special cases that sample from the posterior distribution directly, without relying on asymptotics (Propp and Wilson (1996)). It remains to be seen whether these methods can be developed into a practical tool.

10.5.1 Summary

This section gives practical hints for dealing with burn-in, convergence and reweighting.

10.5.2 Exercises

1. State in your own words the meaning of
   (a) mixing of a Markov Chain
   (b) burn-in
   (c) equilibrium
   (d) adaptive algorithms
   (e) importance sampling reweighting of chains
2. Reconsider the algorithm you wrote to answer question 2(*c*) in section 10.3.3. Make up some data, and run your algorithm on a computer. How much burn-in do you allow for, and why? How do you decide whether the output of your algorithm has converged?

10.6 Variable dimensions: Reversible jumps

As discussed so far, the Metropolis-Hastings algorithm is constrained to moves of the same dimension; typically S is a subset of $\mathbb{R}^d$ for some d. However, this can be overly constraining. For example, when there is uncertainty about how many independent variables to include in a regression (see section 9.4), it is natural to want a chain that explores regressions with several such choices.

Fortunately an extension of the Metropolis-Hastings algorithm provides a solution. For this purpose, suppose the parameter space is augmented with a variable indicating its dimension. Thus let $\mathbf{x} = (m, \theta_m)$ where θ_m is a parameter of dimension m. It is proposed to move to the value $\mathbf{y} = (n, \theta_n)$. The question is how to make such a move consonant with the Metropolis-Hastings algorithm.

One idea that doesn't work is to update θ_m to θ_n directly, since θ_m has an interpretation only under the model indexed by m. Thus all of $\mathbf{x}$ has to be updated to $\mathbf{y}$ in a single move. If $m < n$, the idea of a reversible jump is to simulate $n - m$ random variables $\mathbf{u}$ from some density $g(\mathbf{u})$, and to consider the proposed move from (m, θ_m, u) to (n, θ_n). To implement this, a one-to-one (*i.e.*, invertible) differentiable function T maps (m, θ_m, u) to (n, θ_n). This move has acceptance probability

$$\alpha(\mathbf{x}, \mathbf{y}) = \min\left\{1, \frac{\pi(x)}{\pi(y)p(u)} \cdot J\right\} \tag{10.112}$$

where J is the absolute value of the determinant of the Jacobian matrix of the transformation T. The Jacobian is the local ratio between the densities of $\pi(x)$ and $\pi(y)$, which is why it appears. Moving from $\mathbf{y}$ to $\mathbf{x}$ is the same in reverse. Thus what the reversible jump technique does is (artificially) make the dimensions of the two spaces equal, and is therefore a special case (or extension, depending on how you want to think about it) of the Metropolis-Hastings algorithm.

A special warning is needed about the ratio $\pi(x)/\pi(y)$. While constant multipliers need not be accounted for explicitly, those that depend on the dimension of the space cannot be ignored. (This is comparable to the issue of which constants can and cannot be ignored in deriving conjugate distributions, for which see Chapter 8.)

10.6.1 *Summary*

This section introduces the important reversible jump algorithm.

10.6.2 *Exercises*

1. State in your own words the meaning of
   (a) reversible jump algorithm
   (b) variable dimensions in the parameter space
2. Give some examples of when variable dimensions would be important.
3. Explain why the Jacobian appears in the reversible jump algorithm.

10.7 Hamiltonian Monte Carlo

While almost sure convergence is nice, it can involve too long a wait. This is especially true if the parameter space is high-dimensional, resulting in a low acceptance rate for typical MCMC proposals. It makes sense, then, to look for proposal distributions that make better use of the structure to make attractive proposals.

The method studied in this section borrows language and insight from physics, which is not surprising since MCMC in general, and the Metropolis-Hastings algorithm in particular, originated in physics. Suppose, then, that the parameter space has dimensional d, and that $\mathbf{q}$ is a point in that space. If the momentum $\mathbf{p}$ (also d-dimensional) were known, then the trajectory of the particle p would be deterministic, controlled by the equations

$$\begin{aligned} \frac{\delta q_i}{\delta t} &= \frac{\delta \mathcal{H}}{\delta p_i} \\ \frac{\delta p_i}{\delta t} &= \frac{-\delta \mathcal{H}}{\delta q_i} \end{aligned} \tag{10.113}$$

where H is the Hamiltonian function, taken here to be of the form

$$H(q, p) = U(q) + p'm^{-1}p/2, \tag{10.114}$$

where m is positive definite. Also,

$$U(q) = -\log[\pi(q)L(q(D))], \tag{10.115}$$

is minus the log posterior distribution, up to a constant.

Such trajectories are deterministic. In Hamiltonian Monte Carlo, randomness is introduced by drawing moments p from a distribution (here normal with mean $\mathbf{0}$ and covariance matrix m).

Choosing a proposal by replacing (q, p) with new (q', p') in this way has some advantages. First, it is reversible, because $(q', -p')$ leads back to (q, p). Second, the Hamiltonian is preserved, since

$$\frac{\delta H}{\delta t} = \sum_{i=1}^{d} \left[\frac{\delta q_i}{\delta t_i} \frac{\delta H}{\delta q_i} + \frac{\delta p_i}{\delta t_i} \frac{\delta H}{\delta p_i} \right] = \sum_{i=1}^{d} \left[\frac{\delta H}{\delta p_i} \frac{\delta H}{\delta q_i} - \frac{\delta H}{\delta q_i} \frac{\delta H}{\delta p_i} \right] = 0. \tag{10.116}$$

It is also volume-preserving in $(\mathbf{q}, \mathbf{p})$-space (not proved here).

The Hamiltonian equations (10.113) cannot in general be solved exactly. Consequently, what is done is to take a number s of size $\epsilon > 0$ in the direction indicated by momentum $\mathbf{p}$. The most successful way to do this has turned out to be the "Leapfrog method," in which $\mathbf{p}$ is incremented by an $\epsilon/2$ step, followed by incrementing $\mathbf{q}$ by a full ϵ step, and finally followed by a second $\epsilon/2$ step for $\mathbf{p}$.

Hamiltonian Markov Chain Monte Carlo is implemented in STAN (`https:mc-stan.org`), which is free on the web.

10.7.1 Further reading

Neal (2011, Chapter 5) and Bentencourt (2018).

10.8 Variational Inference

The main difficulties with MCMC are that it is slow, and knowing when sufficient sampling has occurred is not easy. Variation Inference (VI) is an alternative worth considering. VI is used for large datasets and high dimensional problems in which MCMC and its variants are either too slow or infeasible. Accordingly, it relies on specific assumptions convenient for large-scale computing. The idea is to replace the sampling done in MCMC with an optimization. It finds the member of a specified family of distributions, Q, closest, in a particular sense, to the posterior distribution.

The particular sense generally chosen is Kullback-Liebler divergence (Kullback and Liebler, 1951), as follows:

Let $\mathbf{z}$ be a vector of the parameters in a problem, and let $p(\mathbf{z}|\mathbf{x})$ represent the target posterior distribution, having observed the data $\mathbf{x}$. Let Q be the specified set of approximating densities. Then VI seeks

$$q^*(\mathbf{z}) = \underset{q(\mathbf{z}) \in Q}{\arg\min} KL(q(\mathbf{z})||p(\mathbf{z}|\mathbf{x})) \tag{10.117}$$

where $KL(q(\mathbf{z})||p(\mathbf{z}|\mathbf{x})) = E_q \log\left(\frac{q(\mathbf{z})}{p(\mathbf{z}|\mathbf{x})}\right)$. Note that KL divergence is not a distance metric, because it is not symmetric in its arguments. Nonetheless, it is non-negative, and zero only when $q(\mathbf{z}) = p(\mathbf{z}|\mathbf{x})$. Therefore, it is not an unreasonable functional to minimize.

Working from the definition of KL divergence,

$$\begin{aligned} KL(q(\mathbf{z})||p(\mathbf{z}|\mathbf{x})) &= E_q \log\left(\frac{q(\mathbf{z})}{p(\mathbf{z}|\mathbf{x})}\right) \\ &= E_q(\log q(\mathbf{z})) - E_q \log p(\mathbf{z}|\mathbf{x}) \\ &= E_q(\log q(\mathbf{z})) - E_q(\log p(\mathbf{z}, \mathbf{x})) + \log p(\mathbf{x}). \end{aligned} \tag{10.118}$$

One reason that KL divergence is computationally attractive is that $\log p(x)$ [which is typically hard to compute] is a constant with respect to q, and hence can be ignored. Thus, to minimize (10.118), one can equivalently maximize, with respect to $q \in Q$,

$$E_q \log p(\mathbf{z}, \mathbf{x}) - E_q \log q(\mathbf{z}). \tag{10.119}$$

Because the KL divergence is non-negative, this quantity is a lower bound for $\log(\mathbf{x})$, and is called in this literature the ELBO (evidence lower bound), where $\log p(\mathbf{x})$ is called the "evidence."

Having decided to maximize (10.119) with respect to the family Q, the next issue is to specify the family Q. A computationally convenient choice goes by the daunting name of "mean field variational family." Despite its name, what it amounts to is choosing Q to have one-dimensional independent factors.

This family Q can be written as

$$q(\mathbf{z}) = \prod_{j=1}^{m} q_j(z_j). \tag{10.120}$$

The point of the analysis to follow is that the maximization in (10.119), where Q is specified as in (10.120), can be reduced from an m-dimensional maximization (which can be daunting) to a series of one-dimensional maximizations (much easier). Each one-dimensional maximization is accomplished by holding the other $m-1$ distributions q fixed. The analysis shows that doing so increases (10.119) at each step. Thus, repeating cycles of the maximization makes (10.119) larger, so it converges. The point at which it converges may be only a local maximum, so it may be necessary to repeat the calculations with different starting points. Having explained the strategy, the next step is to demonstrate the claimed reduction.

In the calculation that follows, q_j is to be maximized, holding each other q fixed at its current value. So the expression in (10.119) is analyzed by separating out the terms involving q_j, as follows:

$$\begin{aligned} L(q_j) &= E_q \log p(\mathbf{z}, \mathbf{x}) - E_q \log q(z) \\ &= E_{q_j} E_{q_{i \neq j}} \left[\log p(\mathbf{z}, \mathbf{x}) - \log q_j - \textstyle\sum_{k \neq j} \log q_k \right] \\ &= S_1 - S_2 - S_3, \\ \text{where} & \\ S_1 &= E_{q_j} E_{q_{i \neq j}} \log p(\mathbf{z}, \mathbf{x}) \\ S_2 &= E_{q_j} E_{q_{i \neq j}} \log q_j \\ \text{and} & \\ S_3 &= E_{q_j} E_{q_{i \neq j}} \textstyle\sum_{k \neq j} \log q_k. \end{aligned}$$

Taking each in turn, let $\log f_j(z_j) = E_{q_{i \neq j}} \log p(\mathbf{z}, \mathbf{x})$, so

$$S_1 = E_{q_j} \log f_j(z_j).$$

For S_2, $E_{q_{i \neq j}} \log q_j = \log q_j$, so $S_2 = E_{q_j} \log q_j$.

Finally, $E_{q_{i \neq j}} \sum_{k \neq j} \log q_k$ does not involve q_j, and hence is a constant for this purpose, so its expectation is a constant.

Hence

$$\begin{aligned} L(q_j) &= E_{q_j} \log f_j(z_j) - E_{q_j} \log q_j + K, \text{ where } K \text{ is a constant.} \\ &= -KL(q_j || f_j) + K. \end{aligned}$$

So maximizing (10.119) is done by minimizing $KL(q_j || f_j)$, which is accomplished by setting $q_j = f_j$ if q_j is unconstrained. This implies

$$q_j(z_j) \propto e^{E_{q_{i \neq j}} \log(\mathbf{z}, \mathbf{x})},$$

where the constant K becomes the multiplicative constant needed to normalize q_j.

Often the factors q_j of Q are taken to be convenient distributions, in which case the optimization is taken with respect to the parameters of the distribution q_j. How this is done in practice depends on the structure of the model $p(\mathbf{z}, \mathbf{x})$ and the approximating product of independent factors q. For examples, see Murphy (2012, Chapter 21) and Blei et al. (2017).

There are many extensions of these methods. For example, the same ideas could be used to divide the global maximization, without insisting that the smaller ones must be of dimension 1. Also, VI is not invariant to reparameterization, so several different parameterizations could be tried.

10.8.1 Summary

This section gives a very brief exposition of an area still under active research.

10.8.2 Exercises

1. State in your own words the principle ideas of variational inference.
2. What are the critical assumptions of variational inference? When are they more or less likely to be a practical issue?
3. Consider the model in Section 8.5 (univariate normal distribution with uncertain mean and precision). In this case the posterior is available analytically, and the posterior on the two parameters are not independent. Apply variational inference, and explore how much error is introduced through the assumption of independence between the posterior mean and the posterior precision.

10.8.3 Further reading:

Murphy (2012) gives excellent explanations of many methods on the border between machine learning and statistics.

Chapter 11

Multiparty Problems

Shlomo the fool was known far and wide for his strange behavior: offered a choice between two coins, he would always choose the less valuable one. People who did not believe this would seek him out and offer him two coins, and he always chose the less valuable.

One day his best friend said to him: "Shlomo, I know you can tell which coin is more valuable, because you always choose the other one. Why do you do this?" "I think," said Shlomo, "that if I chose the more valuable coin, people would stop offering me coins."

11.1 More than one decision maker

The decision theory presented so far in this book, particularly in Chapter 7, is limited to a single person, who maximizes his or her expected utility. The distribution used to compute the expectation reflects that person's beliefs at the time of the decision, and the utility function reflects the desires and values of that person. Thus the decision theory of Chapter 7 focuses on an individual decision maker.

It must be acknowledged, however, that many important decisions involve many decision makers. This chapter explores various facets of multi-party decision making, viewed Bayesianly. To give structure to such problems, specifications must be made about how the various parties relate to the decision process. There are two leading cases: (a) sequential decision-making, in which first one party makes a decision, and then another, and (b) simultaneous decision-making, in which the parties make decisions without knowledge of the others' decisions. Simultaneous decision-making is often called game theory, although many classic games, such as chess, bridge, backgammon and poker, involve sequential decision-making, not simultaneous decision-making.

There isn't a satisfactory over-all theory of optimal decision-making involving many parties. If there were, the social sciences, particularly economics and political science, would be much simpler and better developed than they are. As a result, this chapter should be regarded as exploratory, discussing interesting special cases.

11.2 A simple three-stage game

The case of sequential decision-making is ostensibly less complicated, since the decision maker knows what his or her predecessor has decided. For this reason, we begin with such a case. There are several important themes that emerge from this simple example. One is the usefulness of doing the analysis backward in time. A second is that, even in a case in which everything is known to both parties, there is uncertainty about the action that will be taken by the other party.

This is a game between two parties. They take turns moving an object on the line. Jane moves the object at the first and third stage, and Dick moves the object in the second stage.

Jane's target for the final location of the object is x, while Dick's is y. Each is penalized by an amount proportional to the squared distance of the final location from the target, plus the square of the distance the player moves the object.

To establish notation, suppose the object starts at s_0. At the first stage, Jane moves the object a distance u (positive or negative). The result of this move is that after the first stage, the object is in location $s_1 = s_0 + u$. At the second stage, Dick moves the object by distance v, so that, after the second stage, the object is in location $s_2 = s_1 + v$. Finally, at the third stage, Jane moves the object distance w, and after the third stage the object is in location $s_3 = s_2 + w$.

Figure 11.1 displays the structure of the moves in the game.

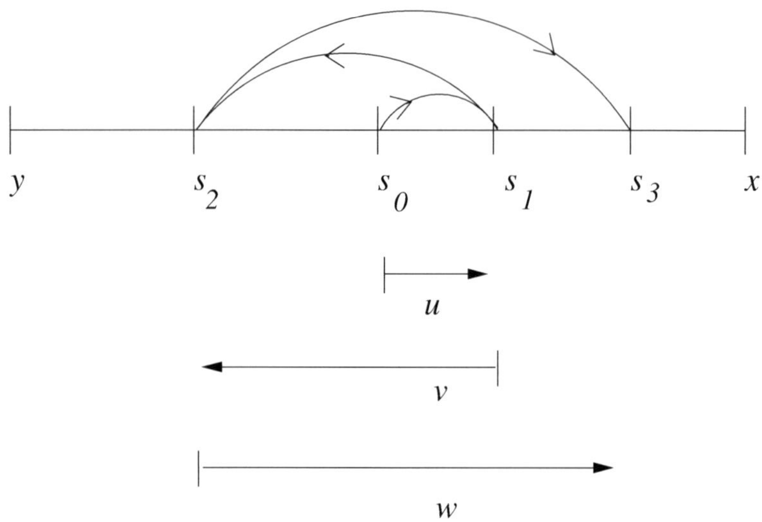

Figure 11.1: Moves in the three-stage sequential game.

Now we suppose that players are charged for playing this game, by the following amounts: for Jane, her charge is

$$L_J = q(s_3 - x)^2 + u^2 + w^2 = q(s_0 + u + v + w - x)^2 + u^2 + w^2 \tag{11.1}$$

and for Dick,

$$L_D = r(s_3 - y)^2 + v^2 = r(s_0 + u + v + w - y)^2 + v^2, \tag{11.2}$$

where q and r are positive. Thus each player is charged quadratically for the distance he or she chooses to move the object, and proportionately to the squared distance of the object's final location (s_3) to that player's target. How might the players play such a game?

It turns out that the principles of this game are better appreciated with a more general loss structure. Thus we can imagine loss functions $L_J(u, v, w)$ and $L_D(u, v, w)$ for Jane and Dick, respectively.

So far, nothing has been specified about the knowledge and beliefs of the players, nor about their willingness and ability to respond to the incentives given them in equations (11.1) and (11.2). Each such specification represents a special case of the game, of greater or lesser plausibility in a particular applied setting. (Yes, of course, the whole setting is quite contrived, and it is hard to imagine an applied setting for it; however, its very simplicity allows us to discuss some important principles.)

For this section, suppose that x, y, q, r and s_0 are known (with certainty) to both players. In section 11.3, we consider a more general scenario in which Jane's target, x, is not known with certainty by Dick, and similarly Dick's target, y, is not known with certainty by Jane.

It is important to keep track of what is known to a given player at a particular stage. For example, s_2 is known to Jane at stage 3, but is not known to Dick at the beginning of stage 2 before Dick decides on v, because s_2 involves v, Dick's move at stage 2. For this reason it is convenient to consider the moves backwards in time. Therefore, let's consider first the problem faced by Jane at stage 3. We suppose that she knows at this stage the current location of the object, $s_2 = s_0 + u + v$, because the choices u and v have already been made.

Suppose Jane wishes to choose w to minimize $L_J(u, v, w)$, and suppose this minimum occurs at $w^*(s_2)$. If L_J motivates Jane, and if she can calculate w^* and execute it, then this is what Jane should do. When L_J takes the form (11.1), the resulting $w^*(s_2)$ satisfies

$$w^*(s_2) = q(x - s_2)/(q + 1). \tag{11.3}$$

Now let's consider Dick's problem in choosing v. In order to choose wisely, Dick must predict Jane's behavior. In doing so, Dick may rationally hold whatever belief he may have about Jane's choice of w. Specifically, he is not obligated to believe, with probability 1, that Jane will choose w^*. Dick is also not excluded, by Bayesian principles, from believing that Jane is likely to, or sure to, behave in accordance with w^*. Hence the assumption that w^* characterizes Jane's behavior at stage 3 is a special case among many possibilities for Dick's beliefs. Dick does well in this game not by casually adopting an idealized version of Jane, but rather by accurately forecasting the behavior of Jane at stage 3.

With all of that as background, how should Dick choose v? At this point in the game, Dick knows s_1, the location of the object after stage 1. Hence, if L_D motivates Dick, the optimal choice minimizes, over choices of v,

$$\int L_D(u, v, w) P_D(w \mid u, do(v)) dw, \tag{11.4}$$

where P_D is Dick's probability density for Jane's choice w, given Jane chooses u, and Dick chooses v.

In the special case in which Dick is sure that Jane will choose w^*, (11.3) specializes to

$$L_D(u, v, w) = L_D(u, v, w^*(s_2)) = L_D(u, v, w^*(s_1 + v)). \tag{11.5}$$

Again, if Dick is motivated by L_D, and can calculate the optimal v, namely v^*, and execute it, then this is what Dick should do. If Dick is sure that Jane will choose according to (11.3), and if L_D takes the form (11.2), then the resulting optimal v takes the form

$$v^*(s_1) = (1 - k)(m - s_1) \tag{11.6}$$

where $k = (q + 1)^2/[r + (q + 1)^2]$ and $m = (q + 1)y - qx$.

Finally, we consider Jane's first stage move. Jane will minimize over choice of u,

$$\int L_J(u, v(s_0 + u), w(s_0 + u + v)) P_J(v \mid do(u)) dv, \tag{11.7}$$

where $P_J(v \mid do(u))$ is Jane's probability density for Dick's choice of v at the second stage, if Jane chooses u at this stage. Again, we can consider the special case in which Jane is sure that Dick will play according to v^*. In this case, (11.7) simplifies to the choice of u to minimize

$$L_J(u, v^*(s_0 + u), w(s_0 + u + v^*)). \tag{11.8}$$

In the special case that L_J takes the form of (11.1) the optimal u, u^* takes the form

$$u^* = qk(x - (1 - k)m - ks_0)/(qk^2 + q + 1). \tag{11.9}$$

Thus w^*, v^* and u^* are the optimal moves, under the assumptions made, for the players.

In a non-cooperative game, such as this one, players are assumed to be motivated only by their own respective loss functions L_J and L_D. Specifically, they are assumed not to have available to them the possibility of making enforceable agreements (contracts) between them. Such a contract, if available, would have the effect of changing the player's losses by including a term for penalties if the contract were violated.

Might such a contract be desirable to the parties if it were available? There are at least two situations in which such contracts would be desirable: consider situation 1, in which $y < s_0 < x$. If there are values of r and q for which $u^* > x - s_0$, then Jane is paying to move the object beyond x (her target), to her apparent detriment and the detriment of Dick. Figure 11.2 displays this situation:

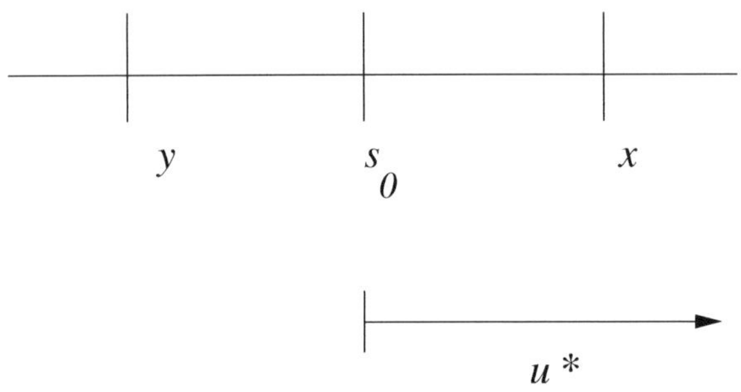

Figure 11.2: Situation 1. Jane's first move, u^*, moves the object further than x, imposing costs on both herself and Dick.

A contract in this case might specify that Jane agrees to restrict the choices of u available at stage 1 to $u \leq x - s_0$ in return for suitable compensation from Dick. The contract would be enforceable if there is an outside party able to fine violations of the contract sufficiently heavily to deter violations. More generally, they might choose to minimize $L_J + L_D$, with such side-payments as might be needed to make this acceptable to both.

Another case in which an enforceable contract between the players would be desirable is situation 2, in which $s_0 < x < y$, and $u^* < 0$. This situation is displayed in Figure 11.3.

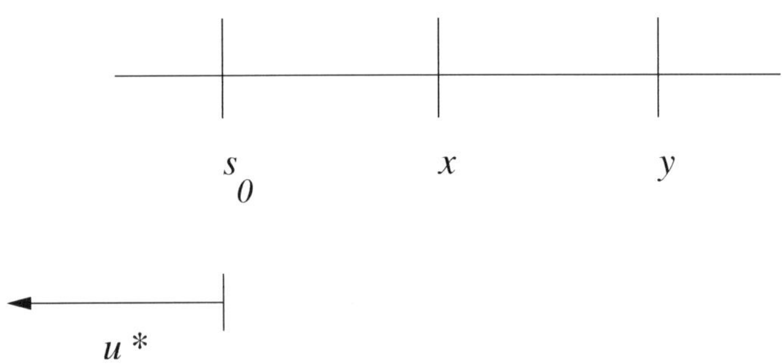

Figure 11.3: Situation 2. Jane's first move, u^*, moves the object further away from both x and y, to both players' detriment.

Specifically, we pose two questions:

(i) If $y < s_0 < x$, are there values of r and q under which the optimal $u^* > x - s_0$?

(ii) If $s_0 < x < y$, are there values of r and q under which the optimal $u^* < 0$?

To address these questions, revert to loss functions L_J and L_D as specified in (11.1) and (11.2), and re-express (11.9) in a more convenient form:

Let $C = qk/(qk^2 + q + 1)$. Then

$$\begin{aligned} u^* =& C(x - (1-k)m - ks_0) \\ =& C(x - (1-k)[(q+1)y - qx] - ks_0) \\ =& C(k(x - s_0) + (1-k)(q+1)(x-y)). \end{aligned} \tag{11.10}$$

In addressing question (i), I use the notation "iff" to mean "if and only if." Then

$$\begin{aligned} & u^* > x - s_0 \\ \text{iff } & C(k(x - s_0) + (1-k)(q+1)(x-y)) > x - s_0 \\ \text{iff } & C(1-k)(q+1)(x-y) > (1 - Ck)(x - s_0) \\ \text{iff } & \frac{C(1-k)(q+1)}{1 - Ck} > \frac{x - s_0}{x - y}. \end{aligned} \tag{11.11}$$

Now

$$\begin{aligned} \frac{C(1-k)(q+1)}{1-Ck} =& \left(\frac{qk}{qk^2 + q + 1}\right)\frac{(1-k)(q+1)}{1 - \frac{qk^2}{qk^2+q+1}} \\ =& \frac{(qk)(1-k)(q+1)}{(q+1)} \\ =& qk(1-k). \end{aligned} \tag{11.12}$$

Therefore, in answer to question (i), if $y < s_0 < x$, $u^* > x - s_0$ if and only if $\frac{x-s_0}{x-y} > qk(1-k)$.

Similarly, to address question (ii),

$$\begin{aligned} & u^* > 0 \\ \text{iff } & k(x - s_0) + (1-k)(q+1)(x-y) > 0 \\ \text{iff } & k(x - s_0) > (1-k)(q+1)(y-x) \\ \text{iff } & \frac{x - s_0}{y - x} > \frac{(1-k)(q+1)}{k}. \end{aligned} \tag{11.13}$$

But

$$\frac{1-k}{k} = \frac{1 - (q+1)^2 \Big/ [r + (q+1)^2]}{(q+1)^2 \Big/ [r + (q+1)^2]} = \frac{r}{(q+1)^2}. \tag{11.14}$$

Hence

$$\frac{(1-k)(q+1)}{k} = \frac{r(q+1)}{(q+1)^2} = r/(q+1). \tag{11.15}$$

Therefore, we find, in answer to question (ii), that if $s_0 < x < y$, then $u^* < 0$ if and only if $\frac{x-s_0}{y-x} < \frac{r}{1+q}$. Hence in these circumstances it would be in the interests of both parties to make an enforceable contract. The solutions u^*, v^* and w^* are inherently non-cooperative.

Now we examine what would happen if Jane's penalty for missing her target, x, is much higher than her cost of moving the object. This can be expressed mathematically by letting $q \to \infty$. Applying this limit to (11.3), we find

$$\lim_{q \to \infty} w^*(q) = x - s_2,$$

which yields the unsurprising insight that no matter where the object is after stage 2, s_2, Jane will move it by the amount $x - s_2$ so that it finally gets to x, and she avoids an arbitrarily large penalty. Next we look at (11.6). As $q \to \infty$, $k \to 1$ so $v^* \to 0$. This means that Dick makes no move in this limiting case. Finally, examining (11.9) we find $u^* \to (x - s_0)/2$. Hence $s_2 = s_0 + u = s_0 + \frac{(x-s_0)}{2}$, so

$$w^* = x - s_2 = x - \left(s_0 + \left(\frac{(x - s_0)}{2} \right) \right) = \frac{x}{2} - \frac{s_0}{2}.$$

Thus Jane has a simple strategy: her first move, u^*, moves the object half of the distance from s_0 to x; her second move, w^*, moves it the rest of the way. This strategy has cost $2(\frac{x-s_0}{2})^2 = \frac{(x-s_0)^2}{2}$, half the cost of making the move from s_0 to x in a single leap, which would cost $(x - s_0)^2$.

What would Dick think if Jane chose $u = x - s_0$ as her first move? This is obviously suboptimal (Dick knows (11.9), and all of the constants in (11.9)). Why would Jane make such a move? Possibly by being aggressive and making an initially costly move, Jane is signaling that she is irrational. If Dick moves the object, perhaps Jane will move it back to her target x. If Dick believes this, his best strategy is not to move. Perhaps Jane wants to establish with Dick (or with an audience) her willingness to accept seemingly irrational costs to establish her dominance.

Reputations are part of our everyday life. People, corporations and governments go to extraordinary lengths to establish and maintain reputations. Brand management and advertising can be understood in terms of reputation. In the context of the three-stage game here, perhaps Jane is trying to establish a reputation for irrationality, which can sometimes be useful (see Schelling (1960), p. 17). One can model the phenomenon of reputation as embedding the initial game in a larger one, perhaps by repetition. Shlomo the fool has embedded his choice of coins in a larger game in which his reputation for choosing the less valuable coin has utility to him in bringing him a steady flow of coins.

This line of thinking suggests that, when we are confronted with behavior that appears not to coincide with notions of rationality we have imposed on it, perhaps the reason for the behavior is that we do not understand the situation in the same way that the players do. If we ignore this possibility, we may find ourselves in the position of those who offer Shlomo his choice of coins.

With those comments as background, I now address the issue of the extent to which the behavior described in (11.3), (11.6) and (11.9) comports with the personal or subjective Bayesian philosophy. With respect to Jane's last move, if L_J given in (11.1) represents her losses, then w^* and only w^* is the optimal move. The situation is more complicated for Dick at stage 2. In the derivation of v^*, we assumed not only that L_D in (11.2) represents Dick's losses at stage 2, but also that Dick is certain that Jane will use w^* at stage 3. Is there some law of nature requiring Dick to have such a belief about Jane? I would argue that the answer to this question is "no." Indeed, I would argue that Dick is entitled to whatever belief he may have concerning Jane's choice at stage 3. Surely it is interesting and useful to Dick to know that w^* minimizes (11.1), but this knowledge does not, in my view, render Dick a sure loser if he does not put full credence into w^*. What serves Dick best is as accurate a descriptive theory of Jane's likely behavior at stage 3 as Dick can devise, which is a matter of opinion for him. Assuming that (11.2) represents Dick's losses, Bayesian principles would argue that the best strategy is to minimize the expectation of (11.2), where the random variable with respect to which the expectation is taken reflects Dick's uncertainty about Jane's choice of w. Thus the assumption that Jane will choose in accordance with w^* is a special case of the possible beliefs of Dick. And it is that special case of belief that supports the choice of v^*.

Finally, we examine Jane's choice of u at stage 1. The derivation of u^* assumed not only the relevance of the loss function (11.1), but also that w^* and v^*, given respectively by (11.3) and (11.6), are accurate predictors of behavior. With respect to w^*, Jane is in a knowledgeable position to predict her own future behavior. There may be some circumstances under which it is useful to model Jane as being uncertain about her own future behavior, essentially treating the future Jane as a new Janelle. But for the moment let us leave this consideration aside, and concentrate on the assumptions embedded in Jane's use of v^* as a prediction of Dick's choice of v. Here Jane is led to consider what she believes about Dick's beliefs about how Jane will choose w at stage 3, as well as the question about how Dick will choose v even if he is sure that Jane will choose w^*. Again the principles of subjective Bayesianism permit Jane a wide range of beliefs about Dick's choice of v. Given whatever that belief may be, Bayesian considerations then recommend minimizing expected loss L_J, with respect to the uncertainty about v reflected in the beliefs of Jane, as given in (11.7).

11.2.1 Summary

This is an example in which all the parameters are known with certainty, and yet uncertainty remains about the strategy of the other player. Consequently it is coherent for the players to depart from the strategies given by v^* and u^*. Jane will optimally depart from her strategy w^* only if L_J in (11.3) does not appropriately reflect all of her losses or gains.

11.2.2 References and notes

The reasoning used in section 11.2 is called backward induction, because time is considered in the reverse direction, from the latest decision, to the next latest, etc. Backward induction is often used in problems of this kind.

The game considered here is from DeGroot and Kadane (1983), and has precursors in Cyert and DeGroot (1970, 1977).

A related sequence of papers examines the (somewhat) practical situation of the use of the peremptory challenges in the selection of jurors in US law. See Roth et al. (1977), DeGroot and Kadane (1980), DeGroot (1987) and Kadane et al. (1999).

11.2.3 Exercises

1. Explain backward induction.
2. Try to find optimal strategies by considering first Jane's first move, then Dick's move, and finally Jane's second move, the third move in the game. Is this simpler or more difficult? Why? You may assume that the loss functions L_J and L_D represent the players' losses, that they are both optimizers, and that Jane knows that Dick knows this.
3. Suppose $x = 1, y = -1, s_0 = 0, r = 2$ and $q = 3$. Find the optimal strategies, again under the assumptions specified in problem 2.
4. Investigate the behavior of u^*, v^* and w^* as $r \to \infty$.
5. Prove (11.3).
6. Prove (11.6).
7. Prove (11.9).
8. Choose what you consider to be a reasonable choice for $P_D(w \mid u, do(v))$ other than the choice of w^* with probability 1, and minimize (11.4) with respect to your choice.
9. Construct a contract that is better for both parties than they can do for themselves by playing the game. Make whatever assumptions you need, for example losses (11.1) and (11.2), and special values of q and r. Is a side-payment necessary to make your proposed contract better for both? If so, what size of side-payment is needed, and which player pays it to the other?

11.3 Private information

> A scorpion asks a frog to take him across the Jordan River. "That would be a foolish thing for me to do," says the frog. "We'd get out to the middle, and you would probably sting me and I would die."
>
> "That would be foolish of me," responds the scorpion, "since I would drown and die if I did sting you."
>
> "You have a good point," says the frog. "OK, climb aboard."
>
> So the scorpion gets on the frog's back, and the frog starts to swim across the river. When they get to the middle of the river, the scorpion stings the frog. As paralysis starts to set in, the frog says "Why did you do that?" As the scorpion is about to sink beneath the water, he says, "Well, that's life for you."

We now suppose that each player knows his or her own target, but is uncertain about the other player's target. Thus Jane knows x, but not y, and Dick knows y, but not x. Both players are assumed to know q, r and s_0, as they did in section 11.2. The important point here is that each player may learn about the other's target by observing the moves of the other player.

Private information, that is, information that one person has and another does not, is ubiquitous in our society. The enormous resources devoted to education, the media, scientific publication, libraries of all sorts, etc. are all evidence of how important the distinction is between private and public information. There are governmental, commercial and personal secrets as well.

Again, we proceed by backwards induction. At stage 3, Jane knows her own target x, the location of the object s_2, and her value of q. Her uncertainty about the value of Dick's target, y, is irrelevant to her choice. Hence she continues to minimize L_J, and chooses $w^*(s_2)$. In the special case of loss L_J satisfying (11.1) her choice is (11.3) as before.

Next, we examine Dick's choice of v. Here Dick's uncertainty about Jane's target, x, matters to him. In general, he minimizes

$$\int L_D(u, v, w) P_D(w, x \mid u, do(v)) dw \; dx \tag{11.16}$$

where $P_D(w, x \mid u, do(v))$ is Dick's joint uncertainty about what Jane will do at stage 3, w, and about her target, x, given Jane's first move u and Dick's decision v. The special case in which Dick is sure that Jane will use w^* simplifies (11.16) to be

$$\int L_D(u, v, w^*) P_D(x \mid u, do(v)) dx. \tag{11.17}$$

However, even this assumption does not help all that much, because w^* is a function of the (unknown to Dick) target x for Jane, even when (11.1) is taken as Jane's loss, as is shown by (11.3). Suppose, then, that (11.1) is Jane's loss and Dick knows this, and is certain that she will implement w^* in (11.3) at stage 3. Also suppose Dick's loss is (11.2). Then

$$v^* = (1 - k)[M(u) - s_1] \tag{11.18}$$

where $M(u) = E_D(m|u) = E_D[(q + 1)y - qx|u] = (q + 1)y - qE_D(x|u)$.

Dick at stage 2 has a cognitively difficult task, to evaluate $M(u)$, or, equivalently, $E_D(x|u)$, his expectation of Jane's target, after seeing her first move u. As in the material studied in Chapters 1 to 10 of this book, there is nothing inherent in the structure of the problem requiring a decision-maker to have a particular likelihood function or prior distribution. So too, here, there is nothing in the structure of the problem requiring Dick to have a particular value of $E_D(x \mid u)$, We can proceed, however, by imagining him to have some specific choice of $E_D(x \mid u)$.

Recall that in the situation of section 11.2, with x and y known to both parties, Jane's choice, u^*, is given by (11.9) [under the assumptions made about how Dick will act at stage 2, which in turn makes an assumption about how Jane will act at stage 3]. The advantage of (11.9) is that it gives an explicit relationship between x and u^*, as follows:

$$u^* = qk(x - (1-k)m - ks_0)/(qk^2 + q + 1).$$

Let $f = (qk^2 + q + 1)/qk$. Then

$$\begin{aligned} fu^* =& x - (1-k)m - ks_0 \\ =& x - (1-k)[(q+1)y - qx] - ks_0 \\ =& x[1 + q(1-k)] - (1-k)(q+1)y - ks_0. \end{aligned}$$

Solving for x yields

$$x = (fu^* + (1-k)(q+1)y + ks_0)/(1 + q(1-k)). \tag{11.19}$$

This relationship might be used by Dick to choose

$$E_D(x \mid u) = (fu + (1-k)(q+1)y + ks_0)/(1 + q(1-k)). \tag{11.20}$$

Dick can implement this choice, as he knows u, y and s_0, even though he knows that Jane cannot implement (11.9) since she does not know y.

We now move back in time again, and consider Jane's choice of u at stage 1. Since under the scenario of this section, Jane is uncertain about Dick's goal y, so (11.7) must be modified to reflect this uncertainty. Thus Jane chooses u to minimize

$$\int\int L_J(u, v(s_0 + w)), w(s_0 + u + v))P_J(y, v \mid do(u))dvdy, \tag{11.21}$$

where $P_J(y, v \mid do(u))$ reflects Jane's uncertainty both about Dick's goal, y, and his action, v, at stage 2.

This minimization is sufficiently complicated that I move immediately to the assumptions that losses are given by (11.1) and (11.2), that (11.3) is Jane's choice at stage 3, and that (11.18) is Dick's choice at stage 2 implemented by (11.20), and that Jane knows this.

We have, from (11.18),

$$v = (1-k)[M(u) - s_1] = (1-k)[M(u) - s_0 - u]. \tag{11.22}$$

Then

$$\begin{aligned} x - s_0 - u - v =& x - s_0 - u - (1-k)[M(u) - s_0 - u] \\ =& x - ks_0 - ku - (1-k)M(u) \\ =& K(u), \end{aligned} \tag{11.23}$$

where $K(u) = x - k(s_0 + u) - (1-k)M(u)$.

Now using (11.3) and (11.23),

$$\begin{aligned} w =& (q/(q+1))(x - s_2) = (q/(q+1))(x - s_0 - u - v) \\ =& (q/(q+1))K(u). \end{aligned} \tag{11.24}$$

Now

$$\begin{aligned}
L_J =& q(s_0 + u + v + w - x)^2 + u^2 + w^2 \\
=& q(-K(u) + (q/(q+1))K(u))^2 + u^2 + (q/(q+1))^2 K^2(u) \\
=& qK^2(u)(1 - q/(q+1))^2 + u^2 + (q^2/(q+1)^2)K^2(u) \\
=& u^2 + K^2(u)\left[\frac{q}{(q+1)^2} + \frac{q^2}{(q+1)^2}\right] \\
=& u^2 + K^2(u)(q/(q+1)).
\end{aligned} \tag{11.25}$$

Then Jane's expected loss is $E_J L_J$, and, differentiating under the integral sign, the optimal u^* satisfies the implicit equation

$$0 = \frac{\partial E_J L_J}{\partial u} = 2u + (q/(q+1))\frac{d}{du}E_J[K^2(u)]. \tag{11.26}$$

With the choice of (11.20) a value for $M(u)$ follows:

$$\begin{aligned}
M(u) =& (q+1)y - qE_2(x \mid u) \\
=& (q+1)y - q[fu + (1-k)(q+1)y + ks_0]/(1 + q(1-k)) \\
=& (q+1)y[1 - q(1-k)/(1 + q(1-k))] \\
& - (qfu + qks_0)/(1 + q(1-k)) \\
=& (q+1)y/(1 + q(1-k)) - (qfu + qks_0)/(1 + q(1-k)) \\
=& [(q+1)y - qfu - qks_0]/(1 + q(1-k)).
\end{aligned} \tag{11.27}$$

Substituting (11.27) into (11.18) yields a value for v^*, as follows:

$$\begin{aligned}
M(u) - s_1 =& [(q+1)y - qfu - qks_0]/[1 + q(1-k)] - s_0 - u \\
=& \frac{1}{1 + q(1-k)}[(q+1)y - qfu - qks_0 - (1 + q(1-k))(s_0 + u)] \\
=& \frac{1}{1 + q(1-k)}\{(q+1)y - u[qf + 1 + q(1-k)] \\
& -s_0[qk + (1 + q(1-k))]\}.
\end{aligned} \tag{11.28}$$

Now

$$qk + 1 + q(1-k) = qk + 1 + q - qk = 1 + q. \tag{11.29}$$

Substituting for f,

$$\begin{aligned}
qf + 1 + q(1-k) =& (qk^2 + q + 1)/k + 1 + q(1-k) \\
=& \frac{1}{k}\{qk^2 + q + 1 + k + kq - qk^2\} \\
=& \frac{1}{k}\{(q+1) + k(q+1)\} \\
=& \frac{1}{k}(q+1)(k+1).
\end{aligned} \tag{11.30}$$

Hence

$$M(u) - s_1 = \left(\frac{q+1}{1 + q(1-k)}\right)\{y - s_0 - [(k+1)/k]u\}. \tag{11.31}$$

Then

$$\begin{aligned}
v^* =& (1-k)[M(u) - s_1] \\
=& \left[\frac{(1-k)(q+1)}{1 + q(1-k)}\right]\{y - s_0 - [(k+1)/k]u\}.
\end{aligned} \tag{11.32}$$

How might Jane think about $\frac{d}{du}E_J[K^2(u)]$, where

$$\begin{aligned}K(u) =& x - k(s_0+u) - (1-k)M(u)\\ =& x - k(s_0+u) - (1-k)[(q+1)y - qE_D(x \mid u)]\\ =& x - ks_0 - ku - (1-k)(q+1)y + (1-k)qE_D(x \mid u)?\end{aligned}$$

Jane knows k, q, x and s_0. Additionally u is her decision variable. The quantities uncertain to Jane are y, Dick's target, and $E_D(x \mid u)$, Dick's expectation of x, Jane's target, after seeing u. Because of the (convenient) squared-error nature of L_J, Jane needs to specify, in principle, five quantities, $E_J(y), E_J(y^2), E_JE_D(x \mid u), E_J\{E_D(x \mid u)\}^2$ and $E_J\{yE_D(x \mid u)\}$. The first two reflect simply Jane's uncertainty about Dick's target. The last three terms are more interesting, as they are moments of Jane's beliefs about what Dick may conclude about Jane's target x, after seeing her first move, u. [It is typical of n-stage games that they require elicitations $n-1$ steps back. Here $n=3$, so we have 2-step elicitations, what Jane thinks Dick will conclude after seeing his first move u. As n increases, these elicitations become dizzyingly difficult to think about.]

That Jane has to think about how her decision u may influence Dick's opinion about her target x is an important point. In bargaining situations, it may or may not be to your advantage to reveal your goals to your counterpart. For more on this, see Schelling (1960); Fisher and Ury (1981) and Raiffa (1985, 2007). A huge literature on negotiations has ensued.

One way to make a tractable special case is to suppose that Jane will believe that Dick will use (11.20) as a guide to $E_D(x \mid u)$. While this helps with Jane's elicitations, it does not resolve everything, as (11.20) involves y, which Jane does not know. However, it does permit simplification, as follows:

$$\begin{aligned}K(u) - x =& [(1-k)(q+1)y + ks_0]\left[\frac{(1-k)q}{1+q(1-k)} - 1\right]\\ &+u\left[\frac{(1-k)gf}{1+q(1-k)} - k\right].\end{aligned} \tag{11.33}$$

Now

$$\frac{(1-k)q}{1+q(1-k)} - 1 = \frac{(1-k)q - 1 - q(1-k)}{1+q(1-k)} = \frac{-1}{1+q(1-k)}.$$

Also

$$\begin{aligned}\frac{(1-k)qf}{1+q(1-k)} - k =& \frac{(1-k)qf - k - kq(1-k)}{1+q(1-k)}\\ =& \frac{(1-k)q(f-k) - k}{1+q(1-k)}.\end{aligned}$$

Recalling $f = (qk^2+q+1)/qk$,

$$f - k = \frac{qk^2+q+1}{qk} - k = \frac{qk^2+q+1-qk^2}{qk} = \frac{q+1}{qk}.$$

Then

$$\frac{(1-k)qf}{1+q(1-k)} - k = \frac{(1-k)(\frac{q+1}{k})}{1+q(1-k)} - k = \frac{(1-k)(q+1)-k^2}{k[1+q(1-k)]}.$$

Summarizing,

$$K(u) - x = u\left[\frac{(1-k)(q+1)-k^2}{k[1+q(1-k)]}\right] - \frac{1}{1+q(1-k)}[(1-k)(q+1)y + ks_0]. \tag{11.34}$$

For the next calculation, we may rewrite the result as follows:

$$K(u) = x + au + by + cs_0 \tag{11.35}$$

where $a = \frac{(1-k)(q+1)-k^2}{k[1+q(1-k)]}$, $b = -\frac{(1-k)(q+1)}{1+q(1-k)}$ and $c = \frac{-k}{1+q(1-k)}$. Both x and s_0 are known to Jane, y is uncertain and u is to be decided. For this reason, we may treat $x + cs_0 = d$ as a single known unit. Thus

$$K(u) = d + au + by,$$

so

$$K^2(u) = d^2 + a^2u^2 + b^2y^2 + adu + bdy + abuy$$

and

$$E_J K^2(u) = d^2 + a^2u^2 + b^2E_J(y^2) + adu + bdE_J(y) + abuE_J(y). \tag{11.36}$$

Therefore

$$\frac{d}{du}E_J K^2(u) = 2a^2u + ad + abE_J(y). \tag{11.37}$$

Hence the only additional elicitation that must be done is $E_J(y)$, which is Jane's expectation of Dick's target y, at stage 1. A not unreasonable choice for $E_J(y)$ is x, Jane's target.

Substituting this result into (11.26) yields

$$\begin{aligned} 0 =& 2u + (q/q+1)[2a^2u + ad + abx] \\ =& 2(q+1)u + q[2a^2u + ad + abx] \\ =& 2u[(q+1) + qa^2] + qa[d + bx] \\ \text{or } u^* =& \frac{-qa[d+bx]}{2[q+1+qa^2]}. \end{aligned} \tag{11.38}$$

The point of this example is to illustrate the kind of reasoning required to implement the optimal strategies found in this very special game. As noted above, as n, the number of stages, grows, the elicitations become increasingly difficult to contemplate. Nonetheless, I believe there is value in having a method that poses the relevant questions, even if they are difficult.

11.3.1 Other views

The issue of what constraints Bayesian rationality implies in situations involving more than one decision maker has been a subject of discussion and debate for some time. Some of the contributors to this literature include Luce and Raiffa (1957), Nash (1951), Bernheim (1984) and Pearce (1984).

An important contribution is that of Aumann (1987). He proposes a model in which each player i has a probability measure p^i on S, the set of all possible states, ω, of the world. He emphasizes the generality he intends for the set S as follows:

> The term 'state of the world' implies a definite specification for all parameters that may be the object of uncertainty on the part of any player... In particular, each ω includes a specification of which action is chosen by each player at that state ω. (p. 6)

Applied to the game under discussion, Aumann's assumption would require each player to have a probability distribution on an Ω that would include a specification of $\{x, y, M(u), K(u), u, v, \text{ and } w\}$. This strikes me as peculiar, because it requires each player to have a probability distribution with respect to his own behavior. Distinguishing decision variables, under the current control of the agent, from quantities uncertain to the agent at the time of the decision, seems essential to me to an understanding of optimal decision-making. Furthermore, to bet with someone about that person's current actions seems to me to be a recipe for immediate sure loss if the stakes are high enough. (In other contexts, such an offer might be construed as a bribe.) Making bets with an agent with respect to his future choices does not seem as problematic, because the agent cannot now make that choice. Furthermore, making bets with an agent about his past actions might make sense, as he might have forgotten what he did.

Aumann defends this feature of his model (pp. 8, 9) by proposing that it is a indexsubjectindexAumann's model model of the beliefs of an outside observer, not one of the players. Of course, it is legitimate for an outsider to be uncertain about what each of the players may do. But it raises another question: why should player i accept this outside observer's opinions as his own?

There is a second issue raised by Aumann's article, namely his assumption that the players share a common prior distribution. This assumption is especially restrictive when added to the previous expansive interpretation of Ω. After conceding that his model could be adapted to incorporate subjective priors, he rejects that route. He justifies the common prior assumptions on two grounds: first, a pragmatic argument that economists want to concentrate on differences among people in their "information," and allowing subjective priors interferes with this program. To some extent this argument is purely linguistic, in that one could extend the notion of "information" to include differences among priors. Aumann's second argument is that incorporating subjective priors "yields results that are far less sharp than those obtained with common priors" (p. 14). I find this argument unappealing. One can get very sharp results by assuming that everybody agrees on what strategies they will play. But the unaddressed question is whether such an assumption has anything to do with the real world in which people face uncertainty in situations involving other decision makers. Sharp results are nice when the assumptions made to get them are plausible in practice, but only then.

The effect of these assumptions together is that each player is assumed to be as uncertain about his own behavior as he is about his opponents'. It is hard for me to imagine situations in which that is a reasonable assumption.

11.3.2 References and notes

The stochastic version of the three-move game is from DeGroot and Kadane (1983). Commentary on the Aumann paper is also found in Kadane and Seidenfeld (1992).

11.3.3 Summary

The stochastic version of the three-move game shows that Jane's last move w is the same as it is in the non-stochastic version. If Dick assumes that Jane will use that strategy in her last move, he still has an inference problem about how to interpret Player 1's first move u. Under our simplified quadratic loss, the conditional expectation $M(u)$ is all that is required. Finally Jane, in choosing u, has to assess $K(u)$, which means thinking about what she believes Dick will infer about her beliefs about his target y from each move u he might make.

Aumann proposes a way through this thicket, but it has some drawbacks, which are discussed.

11.3.4 Exercises

1. Suppose someone offers to buy from you or sell to you for 40 cents a ticket that pays $1 if you snap your fingers in the next minute. Describe two ways in which you could make that person a sure loser.
2. Examine the behavior of the strategies (11.32), (11.38) and (11.43) as $k \to \infty$.
3. Prove (11.18).

11.4 Design for another's analysis

Joshua takes a job digging in a diamond mine near Johannesburg. At quitting time on his first day, he arrives with a wheelbarrow full of dirt. The security guards suspect he is stealing diamonds. They inspect him, and carefully sieve the dirt in the wheelbarrow, but find nothing, so they let him pass. The same thing happens each day for six months, with the same result.

Finally, Joshua decides to quit his job, and join his girlfriend in Cape Town. After his last checkout (with full wheelbarrow), his friends take him for a farewell beer. One of his friends says, "We all want to know how you have been stealing diamonds." "Oh," says Joshua, "I never stole a diamond. But I have been doing a great business in wheelbarrows."

Chapter 7 discusses experimental design as a sequential problem in which the same person both decides what design to use, and then, after the data are available, analyzes the results. This section discusses the case in which those functions are performed by different individuals. Thus the results of this section are a generalization of those in Chapter 7. Why is the general case of interest?

In many practical settings, an experiment is conducted to inform many persons beyond the person designing the experiment. When a pharmaceutical company does an experiment to show the efficacy of a new drug, the audience is not just the company, but also the Food and Drug Administration, and, more generally, the medical community and potential customers. While the company may be convinced that the drug is wonderful (otherwise it would not invest the resources needed to test the drug), the FDA is likely to take a more skeptical attitude. Thus the company needs to design the trial not to convince itself, but to convince the FDA and others.

Similarly in the setting of a criminal investigation, it is generally conceded that the investigator may use his beliefs and hunches in deciding what evidence to collect. He does that collection with the knowledge that the results of the investigation must convince prosecutors, judges and juries likely not to share his beliefs.

Designed experiments are often expensive, and frequently are social undertakings, often publicly funded. The experimenter hopes to use the results to persuade a profession that includes persons with varying levels of prior agreement with the experimenter. For these reasons, I believe that the framework for experimental design explored in this section is far more commonly applicable than is the special case in which the designers' and analysts' priors, likelihoods and utility functions are taken to be identical.

To give a flavor of the kind of analysis that results, I report here on a very simplified special case. Suppose that Dan is the designer and Edward is the estimator, and that they are both uncertain about the parameter θ. Dan's prior density on θ is $\pi_d(\theta)$ and Edward's is $\pi_e(\theta)$. We'll suppose that Dan knows Edward's prior. This is a special case of the more general case in which Dan has a probability distribution on Edward's prior. Also Dan and Edward will be imagined to share a likelihood function. Their posterior distributions are denoted by $\pi_d(\theta \mid \mathbf{x})$ and $\pi_e(\theta \mid \mathbf{x})$, respectively, where $\mathbf{x}$ represents the experimental result of a sample of size n. The goal of this experiment is to find an estimate a of θ. Then

Edward chooses the estimator a to minimize $E^{\pi_e(\theta|x)}L_e(\theta, a)$, where $L_e(\theta, a) = (\theta - a)^2$ is Edward's loss function. Now Dan has some joint distribution for the data $\mathbf{x}$ and the parameter θ, $\pi_e(\theta, \mathbf{x})$. Dan chooses a sample size n to minimize $E^{\pi_e(\theta,\mathbf{x})}L_d(\theta, a, \mathbf{x})$, where a is chosen by Edward, $L_d(\theta, a, \mathbf{x}) = (\theta - a)^2 + cn$, and c is a cost per observation. To be specific, we assume that the likelihood for each of n independent and identically distributed observations is the same for both players, and is normal with mean θ and precision 1. Dan's prior is assumed to be normal with mean μ_d and precision τ_d; similarly Edward's prior is assumed to be normal with mean μ_e and precision τ_e. These choices of likelihood and prior permit the use of conjugate analysis, as explained in Chapter 8.

Then Edward's posterior distribution on θ after seeing the data $\mathbf{x}$ is normal with mean $\frac{n}{n+\tau_e}\overline{X}_n + \frac{\tau_e}{n+\tau_e}\mu_e$, and precision $n + \tau_e$, when $\overline{X}_n$ is the mean of the observations x. Under Edward's squared error loss function, he chooses as his action a, his posterior mean,

$$a = \frac{n}{n + \tau_e}\overline{X}_n + \frac{\tau_e}{n + \tau_e}\mu_e.$$

Now what should Dan do?

Let $f = n/(n + \tau_e)$ and $b = \tau_e\mu_e/(n + \tau_e)$. Then Edward's choice is $a = f\overline{X} + b$. Dan chooses n to minimize his expectation of

$$\begin{aligned} cn + &E_d(a - \theta)^2 = \\ cn + &E_d(f\overline{X}_n + b - \theta)^2 \end{aligned} \tag{11.39}$$

where the expectation is over $\overline{X}_n$ and θ, both unknown to Edward at the time he chooses the sample size. Taking the expectation of the second term in (11.39) with respect to $\overline{X}_n$ first, where $\overline{X}_n \mid \theta \sim N(\theta, 1/n)$,

$$\begin{aligned} &E_d\{(f\overline{X} + b - \theta)^2 \mid \theta, n\} \\ =&E_d\{[f(\overline{X} - \theta) + b + (f - 1)\theta]^2 \mid \theta, n\} \\ =&f^2/n + [b + (f - 1)\theta]^2, \end{aligned} \tag{11.40}$$

Now the expectation of the second term in (11.40) with respect to θ, where $\theta \sim N(\mu_d, 1/\tau_d)$, is

$$\begin{aligned} E_d\{[b + (f - 1)\theta]^2\} =&E_d[(f - 1)(\theta - \mu_d) + b + (f - 1)\mu_d]^2 \\ =&(f - 1)^2/\tau_d + (b + (f - 1)\mu_d)^2. \end{aligned}$$

Then Dan's loss, as a function of n, is

$$\begin{aligned} R(n) =&cn + f^2/n + (f - 1)^2/\tau_d + (b + (f - 1)\mu_d)^2 \\ =&cn + \left(\frac{n}{n + \tau_e}\right)^2 \cdot \frac{1}{n} + \frac{\tau_e^2}{(n + \tau_e)^2} \cdot \frac{1}{\tau_d} + \left(\frac{\tau_e\mu_e}{n + \tau_e} - \frac{\tau_e\mu_d}{n + \tau_e}\right)^2 \\ =&cn + \frac{n}{(n + \tau_e)^2} + \frac{\tau_e^2}{(n + \tau_e)^2}\left[\frac{1}{\tau_d} + (\mu_e - \mu_d)^2\right] \\ =&cn + \frac{n + \tau_e - \tau_e}{(n + \tau_e)^2} + \frac{\tau_e^2}{(n + \tau_e)^2}\left[\frac{1}{\tau_d} + (\mu_e - \mu_d)^2\right] \\ =&cn + \frac{1}{n + \tau_e} + \frac{1}{(n + \tau_e)^2}\left\{\tau_e^2\left[\frac{1}{\tau_d} - \frac{1}{\tau_e} + (\mu_e - \mu_d)^2\right]\right\} \end{aligned}$$

Let $r = \tau_e^2\left[\frac{1}{\tau_d} - \frac{1}{\tau_e} + (\mu_e - \mu_d)^2\right]$.

Then

$$R(n) = cn + \frac{1}{n+\tau_e} + \frac{r}{(n+\tau_e)^2}.$$

This is a particularly convenient expression because the optimal choice of n, the one that minimizes R, is a function only of r, τ_e and c. Dan wishes to minimize R over all choices of $n \geq 0$. Although only integer values of n make sense, we consider the minimum of $R(n)$ over all non-negative numbers n. The integer minimum is then one of the two integers nearest to the optimal real number found.

Let $y = \sqrt{c}(n + \tau_e)$. Then

$$R = -\sqrt{c}\tau_e + \sqrt{c}\left\{y + \frac{1}{y} + \frac{\sqrt{c}r}{y^2}\right\}.$$

Instead of minimizing $R(n)$ over the space $n \geq 0$, we may equivalently minimize

$$g(y) = y + \frac{1}{y} + \frac{\tilde{r}}{y^2}$$

over the space $y \geq \tilde{\tau}_e$, where $\tilde{r} = \sqrt{c}r$ and $\tilde{\tau}_e = \sqrt{c}\tau_e$.

Thus only $\tilde{r}$ and $\tilde{\tau}_e$ matter for finding the optimal y, and hence, the optimal sample size.

Consider the first derivative of g:

$$g'(y) = 1 - \frac{1}{y^2} - \frac{2\tilde{r}}{y^3}.$$

Set equal to zero, this is equivalent to

$$y^3 - y - 2\tilde{r} = 0.$$

Over the range $-\infty < y < \infty$, the cubic equation has the limits as follows:

$$\lim_{y\to\infty}[y^3 - y - 2\tilde{r}] = \infty$$

and

$$\lim_{y\to-\infty}[y^3 - y - 2\tilde{r}] = -\infty.$$

Since $y^3 - y - 2\tilde{r}$ is continuous there exists at least one real solution to the equation

$$y^3 - y - 2\tilde{r} = 0.$$

Let $y(\tilde{r})$ be the largest root of this equation.

Then we can characterize the optimal choice of sample size as follows:

Theorem 11.4.1. *If*

(a) $\tilde{r} > -1/(3\sqrt{3})$ *and*

(b) $-\tilde{r}/y^2(\tilde{r}) < \tilde{\tau}_e < y(\tilde{r})$

then $y(\tilde{r})$ *minimizes* $g(y)$ *and the optimal sample size is* $(y(\tilde{r}) - \tilde{\tau}_e)/\sqrt{c}$. *Otherwise the minimum is at* $y = \tilde{\tau}_e$, *and the optimal sample size is zero.*

Proof. The function $y^3 - y = y(y-1)(y+1)$ has roots at 1, 0 and -1. On the positive axis its minimum occurs at the solution to $3y^2 = 1$, which implies $y = 1/\sqrt{3}$ and its value there is $y^3 - y = (1/\sqrt{3})^3 - (1/\sqrt{3}) = (1/\sqrt{3})(\frac{1}{3} - 1) = -2/(3\sqrt{3})$. Therefore if $\tilde{r} \leq -1/(3\sqrt{3})$, $g(y)$ increases for $y > 0$, and hence the minimum on the set $y \geq \tilde{\tau}_e$ occurs at $y = \tilde{\tau}_e$. Second, we consider $\tilde{r} \geq 0$. The second derivative of $g(y)$ is $g''(y) = \frac{2}{y^3} + \frac{6\tilde{r}}{y^4} > 0$. Then $y^3 - y - 2\tilde{r}$ has only one positive root. The optimal y is then $y(\tilde{r}) - \tilde{\tau}_e$ if $y(\tilde{r}) - \tau_e > 0$ and $\tilde{\tau}_e$ otherwise. The conditions $\tilde{r} \geq 0$ and $y(\tilde{r}) > \tau_e$ together imply condition (b) of the theorem.

Finally, we consider the case $-1/(3\sqrt{3}) < \tilde{r} < 0$. In this case there are two positive roots, of which the larger is a local minimum and the smaller a local maximum. Thus the minimum of the function g over the domain $y \geq \tilde{\tau}_e$ occurs either at $y(\tilde{r})$ or at $\tilde{\tau}_e$. There is a critical value t^* for $\tilde{\tau}_e$ such that if $\tilde{\tau}_e \leq t^*$, the minimum of g occurs at $y = \tau_e$, and the optimal sample size is zero. However, if $\tilde{\tau}_e > t^*$, then the minimum of g occurs at $y(\tilde{r})$.

The value of t^* is characterized by the equation $g(\tilde{\tau}_e) = g(y(\tilde{r}))$, together with the fact that $g(y(\tilde{r}))$ is a relative minimum of $g(y)$.

To simplify the notation for this calculation, let $y(\tilde{r}) = y$ and $\tilde{\tau}_e = x$. Then $g(y(\tilde{r})) = g(\tau_e)$ implies

$$y + \frac{1}{y} + \frac{\tilde{r}}{y^2} = x + \frac{1}{x} + \frac{\tilde{r}}{x^2}. \tag{11.41}$$

Additionally y satisfies $g'(y) = 0$, so

$$\begin{aligned} 1 - \frac{1}{y^2} - \frac{2\tilde{r}}{y^3} &= 0, \text{ hence} \\ y^3 - y - 2\tilde{r} &= 0, \text{ and so} \\ y &= y^3 - 2\tilde{r}. \end{aligned} \tag{11.42}$$

Now (11.41) is equivalent to

$$x^2(y^3 + y + \tilde{r}) = y^2(x^3 + x + \tilde{r}).$$

So

$$\begin{aligned} 0 =& x^2y^3 - y^2x^3 + x^2y - y^2x + x^2\tilde{r} - y^2\tilde{r} \\ =& x^2y^2(y - x) + xy(x - y) + \tilde{r}(x - y)(x + y) \\ =& (y - x)[x^2y^2 - xy - r(x + y)]. \end{aligned}$$

Now substitute (11.42) for y in the middle term:

$$\begin{aligned} 0 =& (y - x)[x^2y^2 - x(y^3 - 2\tilde{r}) - \tilde{r}(x + y)] \\ =& (y - x)[x^2y^2 - xy^3 + 2\tilde{r}x - \tilde{r}x - \tilde{r}y] \\ =& (y - x)[xy^2(x - y) + \tilde{r}(x - y)] \\ =& -(y - x)^2[xy^2 + r]. \end{aligned}$$

Solving for x, we have $x = -\tilde{r}/y^2$. Thus the critical value for $\tilde{\tau}_e$ is $t^* = -\tilde{r}/y^2$. If $\tilde{\tau}_e > -\tilde{r}/y^2$ then $\tilde{y}(\tilde{r})$ is the minimum of $g(y)$ over the space $y \geq \tilde{\tau}_e$. Otherwise the minimum occurs at $\tilde{\tau}_e$, and the optimal sample size is zero. This concludes the proof of the theorem. □

11.4.1 *Notes and references*

This setup and theorem are from Etzioni and Kadane (1993), who also consider a multivariate case and another loss function. Lindley and Singpurwalla (1991) consider an acceptance sampling problem from a similar viewpoint. Lodh (1993) analyzes a problem in which the variance is also uncertain.

The work of Tsai and Chaloner (Not dated) and Tsai (1999) tackles multiparty designs with a utility that focuses on Edward's utility rather than Dan's.

11.4.2 *Summary*

Dan and Edward agree on a normal likelihood with known precision (here taken to be 1). They each have conjugate normal priors on the mean θ, but have possibly different means and precisions for their priors. Edward chooses an estimator, after seeing the data, to minimize his expected squared error loss. Dan chooses a design, before seeing the data, to minimize his expected squared error loss of the decision Edward makes, plus a cost per observation, c.

The theorem gives Dan's optimal sample size.

11.4.3 *Exercises*

1. Suppose it happens that Dan and Edward have the same distribution, and in particular $\mu_d = \mu_e$ and $\tau_d = \tau_e$.
   (a) What is r? What is $\tilde{r}$?
   (b) What is $g(y)$?
   (c) What is $y(\tilde{r})$, the largest root of $g(y) = 0$?
   (d) Apply the theorem. What is the optimal sample size?
   (e) Give an intuitive explanation of your answer.

2. Consider the case $n = 0$.
   (a) What will Edward's estimate be?
   (b) What will Dan's expected loss be? Find this by evaluating $R(0)$.
   (c) Explain your answer to (b).
3. Consider the case in which $\tau_e \to 0$.
   (a) What is r?
   (b) How does the analysis compare to that found in exercise 1 above?

11.4.4 *Research problem*

Explain why $\sqrt{cr}$ and $\sqrt{c\tau_e}$ are the only functions of c, μ_d, μ_e, τ_d and τ_e that matter. I suspect that the reason has something to do with invariance.

11.4.5 *Career problem*

Recreate the theory of experimental design from a Bayesian perspective. Under what sorts of prior distributions is each of the popular designs optimal? For which designs is a two-party perspective necessary or useful? See DuMouchel and Jones (1994) for a start.

11.5 Optimal Bayesian randomization in a multiparty context

In section 7.10, we showed that a Bayesian designing an experiment for his own use would never find it strictly optimal to randomize. In this section we return to this topic in the context of several parties, and display a scenario in which randomization is a strictly optimal design strategy.

The scenario we study is phrased in terms of a clinical trial, although the conclusions are more general, as discussed in the end of this section. In addition to Dan (the designer) and Edward (the estimator), we have a third character, Phyllis (the physician), who implements Dan's design.

The purpose of this imaginary trial is to compare the efficacy of two treatments, 1 and

2. We'll suppose that the outcome of a treatment assigned to a patient is either a success or a failure. Suppose n_1 patients are assigned to treatment 1, and n_2 to treatment 2. Also let $X_i = 1$ if the ith patient's treatment is a success, and zero otherwise. Finally, let $t_i = 1$ if patient i is assigned to treatment 1 and $t_i = 2$ otherwise.

Edward, unaware of any patient covariates, views the data from the trial as two independent binomial samples. So Edward's sufficient statistics are

$$\hat{p}_1 = \left(\sum_{i:t_i=1} X_i \right) / n_1$$

and

$$\hat{p}_2 = \left(\sum_{i:t_1=2} X_i \right) / n_2.$$

As $n_j \to \infty$, $\hat{p}_j \to P\{X = 1 \mid t = j\}$ for $j = 1$ and 2, where the P is Edward's probability. We consider the case in which n_1 and n_2 are large, so Edward's prior is irrelevant.

Phyllis, the physician, assigns the patients to a treatment subject to whatever design Dan chooses. She also has information about a covariate Edward does not know about. Let $h_i = 1$ if the ith patient is healthy and $h_i = 0$ otherwise. Neither Dan nor Edward has data on the health of patients. The health of the patient may affect the probability of success of a treatment. Let p_{jk} be Dan's probability that a patient is a success under treatment j with health $h = k$, assumed to be the same for all patients with treatment j and health k. If Dan's design permits her to, Phyllis may use the health of the patient in assigning patients to treatments. It does not matter, for the analysis to follow, whether this is a conscious or subconscious choice on her part.

Dan specifies the design; that is, Dan gives rules to Phyllis for how patients are to be assigned to treatments. Dan knows that Phyllis will make allocations of patients to treatments within the context of the design he specifies, and that Edward will analyze the data. Dan is concerned about which treatment will be used after the trial is over, and therefore wants Edward's estimates to be as accurate as possible. The covariate h is assumed not to be known about future patients. The population of patients in the trial is believed to be the same as the population of future patients. Therefore he judges the effectiveness of each treatment by its effectiveness for the population as a whole. He is aware that there may be a covariate like h, but does not have data on h for individual patients. Let w be the proportion of healthy patients in the population. Dan wants Edward's estimates to converge to his view of the correct population quantities

$$\begin{aligned} p_1^* &= w p_{11} + (1 - w) p_{10} \text{ and} \\ p_2^* &= w p_{21} + (1 - w) p_{20}, \end{aligned}$$

respectively. These are the probabilities that a random member of the population would have a successful outcome if assigned to treatment 1 or 2, respectively, in Dan's opinion. If Edward were to have measurements on h_i for each patient, his estimates could possibly be made more accurate by including that information, but Dan knows that Edward will not have that information.

The result of whatever design Dan chooses, and, given that choice, whatever Phyllis does in assigning patients to treatments, can be characterized by λ_1, Dan's probability that a healthy patient is assigned to treatment 1, and λ_0, Dan's probability that an unhealthy patient is assigned to treatment 1.

Then

$$\begin{aligned} P\{X_i = 1 \mid t_i = 1\} = &P\{X_i = 1 \mid t_i = 1,\ h_i = 1\} P\{h_i = 1 \mid t_i = 1\} \\ &+ P\{X_i = 1 \mid t_i = 1,\ h_i = 0\} P\{h_i = 0 \mid t_i = 1\}. \end{aligned}$$

The term $P\{h_i = 1 \mid t_i = 1\}$ can be expressed in the notation above as

$$P\{h_i = 1 \mid t_i = 1\}$$
$$= \frac{P\{t_i = 1 \mid h_i = 1\}P\{h_i = 1\}}{P\{t_i = 1 \mid h_i = 1\}P\{h_i = 1\} + P\{t_i = 1 \mid h_i = 0\}P\{h_i = 0\}}$$
$$= \frac{w\lambda_1}{w\lambda_1 + (1-w)\lambda_0}.$$

Therefore

$$P\{X_i = 1 \mid t_i = 1\} = \frac{p_{11}w\lambda_1}{w\lambda_1 + (1-w)\lambda_0} + \frac{p_{10}(1-w)\lambda_0}{w\lambda_1 + (1-w)\lambda_0}.$$

So Dan is concerned about

$$\begin{aligned}
p_1^* - &P\{X_i = 1 \mid t_i = 1\} \\
=& wp_{11} + (1-w)p_{10} - \frac{p_{11}w\lambda_1}{w\lambda_1 + (1-w)\lambda_0} - \frac{p_{10}(1-w)\lambda_0}{w\lambda_1 + (1-w)\lambda_0} \\
=& wp_{11}\left[1 - \frac{\lambda_1}{w\lambda_1 + (1-w)\lambda_0}\right] + (1-w)p_{10}\left[1 - \frac{\lambda_0}{w\lambda_1 + (1-w)\lambda_0}\right] \\
=& wp_{11}\frac{[(1-w)(\lambda_0 - \lambda_1)]}{w\lambda_1 + (1-w)\lambda_0} + \frac{(1-w)p_{10}w(\lambda_1 - \lambda_0)}{w\lambda_1 + (1-w)\lambda_0} \\
=& \frac{w(1-w)(p_{11} - p_{10})(\lambda_0 - \lambda_1)}{w\lambda_1 + (1-w)\lambda_0}.
\end{aligned}$$

Similarly

$$p_2^* - P\{X_i = 1 \mid t_i = 2\} = \frac{w(1-w)(p_{21} - p_{20})(\lambda_0 - \lambda_1)}{w\lambda_1 + (1-w)\lambda_0}.$$

Hence for $\hat{p}_1$ to approach p_1^* and $\hat{p}_2$ to approach p_2^*, there are three cases to consider:

(a) $w(1-w) = 0$

(b) $w(1-w) \neq 0$ and $p_{11} = p_{10}$ and $p_{21} = p_{20}$

(c) $w(1-w) \neq 0$, $p_{11} \neq p_{10}, p_{21} \neq p_{20}$ and $\lambda_0 = \lambda_1$.

In case (a), there is no health covariate. Either all the patients are healthy or they all are unhealthy. In case (b), there is a health covariate, but it doesn't matter. Dan's probability of success with each treatment does not depend on the covariate. So when there is a covariate that matters, for Dan's design to succeed he must have $\lambda_0 = \lambda_1$. How can Dan arrange things so that λ_1, his probability of a patient being assigned to treatment 1 if the patient is healthy, is the same as λ_0, his probability of the patient being assigned to treatment 1 if the patient is unhealthy? If Dan's design instructs Phyllis to flip a (possibly biased) coin to decide on the treatment of each patient, independently of the other assignments of treatments to patients, then $\lambda_0 = \lambda_1$, and Dan succeeds in designing so that Edward's estimates will approach p_1^* and p_2^*, respectively. Not having individual data on the health of patients, any other design leaves Dan vulnerable to $\lambda_0 \neq \lambda_1$. Thus in this circumstance, Dan's best design is randomization.

Suppose Dan's design were to allow each patient to choose a treatment. If healthy patients have a different probability of choosing treatment 1 than do unhealthy patients, then $\lambda_1 \neq \lambda_0$, and the design is suboptimal from Dan's perspective.

Suppose instead that Dan's design were to allow Phyllis to choose a treatment for each patient. Suppose that Phyllis believes treatment 1 to be better for healthy patients and treatment 2 for unhealthy patients. Also suppose that Phyllis wishes to maximize the probability of success for each patient in the trial. Then she will choose so that $\lambda_1 = 1$ and $\lambda_0 = 0$, a suboptimal design from Dan's perspective.

Now suppose that Phyllis wants treatment 1 to look better than treatment 2 for whatever reason, financial or ideological. Knowing that healthy patients are more likely to succeed in treatment than are unhealthy ones, she assigns the healthy patients to treatment 1 and the unhealthy patients to treatment 2. Again we have $\lambda_1 = 1$ and $\lambda_0 = 0$. Thus Phyllis's motives are not at issue here.

Even when Phyllis is not explicitly measuring the health of the patients, and believes she is assigning treatments to patients in a manner unrelated to covariates, she may not be. Thus only explicit randomization guarantees $\lambda_0 = \lambda_1$, and the success of the trial, from Dan's perspective.

While the discussion above uses the scenario and language of a clinical trial, the same considerations occur in other contexts. In a sample survey, the role of Phyllis is played by an interviewer who chooses whom to interview. In an agricultural experiment, the role of Phyllis is played by the gardener, who chooses which plot of land to plant with each kind of seed.

11.5.1 Notes and references

This section is based on Berry and Kadane (1997). Previous literature on Bayesian views of randomization include Stone (1969), Lindley and Novick (1981) and Kadane and Seidenfeld (1990).

11.5.2 Summary

In contrast to the findings of section 7.10 concerning a single Bayesian decision maker, in the context of a multi-party Bayesian model randomization can be optimal.

11.5.3 Exercises

1. Prove
$$p_2^* - P\{X_i = 1 \mid t_i = 2\} = \frac{w(1-w)(p_{21} - p_{20})(\lambda_0 - \lambda_1)}{w\lambda_1 + (1-w)\lambda_0}.$$
2. Suppose that Phyllis measures the covariate h_i and reports it to Dan before assigning a treatment. Edward, however, still does not know the covariate h_i. What is the optimal design under these circumstances?

11.6 Simultaneous moves

> "I knew one [school-boy] about eight years of age, whose success at guessing in the game of 'even and odd' attracted universal admiration. This game is simple, and is played with marbles. One player holds in his hand a number of these toys and demands of another whether that number is even or odd. If the guess is right, the guesser wins one; if wrong, he loses one. The boy to whom I allude won all the marbles of the school. Of course he had some principle of guessing; and this lay in mere observation and measurement of the astuteness of his opponents. For example, an arrant simpleton is his opponent, and, holding up his closed hand, asks, 'Are they even or odd?' Our school-boy replies, 'Odd,' and loses; but upon the second trial he wins, for he then says to himself: "The simpleton had them even upon the first trial, and his amount of cunning is just sufficient to make him have them odd upon the second; I will therefore guess odd'; – he guesses odd, and wins. Now, with a simpleton

a degree above the first, he would have reasoned thus: 'This fellow finds that in the first instance I guessed odd, and, in the second, he will propose to himself, upon the first impulse, a simple variation from even to odd, as did the first simpleton; but then a second thought will suggest that this is too simple a variation, and finally he will decide upon putting it even as before. I will therefore guess even'; – he guesses even, and wins. Now this mode of reasoning in the school-boy, whom his fellows termed 'lucky,' – what, in its last analysis, is it?

'It is merely,' I said, 'an identification of the reasoner's intellect with that of his opponent.' " Edgar Allan Poe, *The Purloined Letter* (pp. 165, 166)

We now consider a different structure for the interaction of the decision-makers (we'll call them players in this section). In particular, we'll suppose that their moves are simultaneous, and thus without knowledge of what the other player (or players) do. This is the assumption of traditional game theory, although most games that people actually play more typically allow for sequential, rather than simultaneous, play.

Game theorists can claim that sequential games are a special case of simultaneous games, by the trick of having a player specify – in principle – what move they would choose in every possible situation resulting from the play up to that point. The difficulty is that in games such as chess, bridge, poker, etc., the number of possible situations is so large as to make this approach impractical.

There is a huge literature on this subject, only a small portion of which is relevant for this book. To understand how game theory and Bayesian decision-making intersect, I first rehearse a few of the most important results from game theory. Later I address the nature of the assumptions made.

11.6.1 Minimax theory for two person constant-sum games

Suppose there are two players, P1 and P2. Suppose P1 has a set of available actions $\{a_1, \ldots, a_m\}$, and P2 has a set $\{b_1, \ldots, b_n\}$. The outcome of a choice by P1 and P2 simultaneously is a pair (a_i, b_j). This has utility u_{ij} for P1, and utility $-u_{ij}$ for P2. It is because their utilities sum to zero, for each pair of choices that they might make, that these are called "zero-sum" games.

There is a more general class of games to which the results below apply. If P1 has utility u^1_{ij} if P1 chooses a_i and P2 chooses b_j, and if P2 has utility u^2_{ij} under those circumstances, then constant sum games are defined by the constraint

$$u^1_{ij} + u^2_{ij} = c \quad \text{for all } i \text{ and } j, \tag{11.43}$$

and for some c. Zero-sum games correspond to the special case $c = 0$. Since the analysis of zero-sum games is conceptually the same as constant-sum games for any fixed c, we study the zero-sum case.

We now allow for the possibility that each player may randomize his strategy. Thus let p_i be the probability that P1 chooses a_i, and similarly let q_j be the probability that P2 chooses b_j. We assume that $p_i \geq 0$ and $q_j \geq 0$ for all i and j, and $\sum_{i=1}^m p_i = \sum_{j=1}^n q_j = 1$.

Let $\mathbf{p} = (p, \ldots, p_m)$ and $\mathbf{q} = (q_1, \ldots, q_n)$. We now suppose that P1 will choose $\mathbf{p}$ from the set P of all possible probability distributions on $(a_1, \ldots, a_m)$, and similarly P2 will choose $\mathbf{q}$ from the set Q of all possible probability distributions on $(b_1, \ldots, b_n)$.

To make these choices, we imagine that P2 knows P1's probability distribution $\mathbf{p}$, but not the specific choice P1 is to make among $\{a_1, \ldots, a_m\}$ in accord with $\mathbf{p}$. Similarly we imagine that P2's probability distribution $\mathbf{q}$, but not the specific choice P2 is to make among $\{b_1, \ldots, b_n\}$, governed by $\mathbf{q}$, is known to P1.

In this case, P1's expected utility arising from the choice of a_i is

$$\sum_{j=1}^{n} u_{ij} q_j.$$

Hence P1's expected utility arising from his choice of the randomized strategy $\mathbf{p}$ is

$$M(\mathbf{p}, \mathbf{q}) = \sum_{i=1}^{m} \sum_{j=1}^{n} u_{ij} p_i q_j. \tag{11.44}$$

Reversing the role of P1 and P2, P2's expected utility arising from his choice of the randomized strategy $\mathbf{q}$ is $-M(\mathbf{p}, \mathbf{q})$.

Suppose, then, that P1 chooses $\mathbf{p}$, which is known to P2. P2, then, would choose $\mathbf{q}$ to minimize $M(\mathbf{p}, \mathbf{q})$, and the resulting utility is a function of $\mathbf{p}$, say $V_1(\mathbf{p}) = \min_{\mathbf{q} \epsilon Q} M(\mathbf{p}, \mathbf{q})$.

Now P1, in his choice of $\mathbf{p}$, is assumed to make this choice to maximize $V_1(\mathbf{p})$ over choice of $\mathbf{p} \epsilon P$, resulting in a value V_1 from the best such choice $\mathbf{p}^*$. Then

$$V_1 = V_1(\mathbf{p}^*) = \max_{\mathbf{p} \epsilon P} V_1(\mathbf{p}) = \max_{\mathbf{p} \epsilon P} \left[\min_{\mathbf{q} \epsilon Q} M(\mathbf{p}, \mathbf{q}) \right]. \tag{11.45}$$

The choice $\mathbf{p}^*$ is called the maximin strategy.

Now we do the symmetric analysis, for P2. We suppose that P2 chooses $\mathbf{q} \epsilon Q$, which is known to P1. P1 would then choose $\mathbf{p} \epsilon P$ to maximize $M(\mathbf{p}, \mathbf{q})$, and the resulting utility is a function of $\mathbf{q}$, say

$$V_2(\mathbf{q}) = \max_{\mathbf{p} \epsilon P} M(\mathbf{p}, \mathbf{q}). \tag{11.46}$$

Now P2, in his choice of $\mathbf{q}$, is assumed to make this choice to minimize $V_2(\mathbf{q})$ over choice of $\mathbf{q} \epsilon Q$, resulting in a value V_2 from the best such choice $\mathbf{q}^*$. Then

$$V_2 = V_2(\mathbf{q}^*) = \min_{\mathbf{q} \epsilon Q} V_2(\mathbf{q}) = \min_{\mathbf{q} \epsilon Q} \left[\max_{\mathbf{p} \epsilon P} M(\mathbf{p}, \mathbf{q}) \right]. \tag{11.47}$$

The choice $\mathbf{q}^*$ is called the minimax strategy.

Now

$$V_1 = V_1(\mathbf{p}^*) = \min_{\mathbf{q} \epsilon Q} M(\mathbf{p}^*, \mathbf{q}) \leq M(\mathbf{p}^*, \mathbf{q}) \text{ for all } \mathbf{q} \epsilon Q. \tag{11.48}$$

Therefore

$$V_1 \leq M(\mathbf{p}^*, \mathbf{q}^*). \tag{11.49}$$

Similarly

$$V_2 = V_2(\mathbf{q}^*) = \max_{\mathbf{p} \epsilon P} M(\mathbf{p}, \mathbf{q}^*) \geq M(\mathbf{p}, \mathbf{q}^*) \text{ for all } \mathbf{p} \epsilon P. \tag{11.50}$$

Therefore

$$V_2 \geq M(\mathbf{p}^*, \mathbf{q}^*). \tag{11.51}$$

Summarizing

$$V_1 \leq M(\mathbf{p}^*, \mathbf{q}^*) \leq V_2, \text{ so } V_1 \leq V_2. \tag{11.52}$$

That in fact $V_1 = V_2$ is the content of the famous minimax theorem of Von Neumann (von Neumann and Morgenstern (1944)). A proof of this result is given in the appendix to this chapter.

The zero-sum two person game is widely regarded as "solved" by this result. Much effort has been expended in extending this result to games involving more than two people and to non-zero-sum games.

Consider the game of "even and odd" discussed by Poe, and, suppose that we identify utilities with marbles. Then "even and odd" is a zero-sum, two-person game. The minimax strategy is to randomize, choosing independently odds with probability one-half and evens otherwise. Good advice for the simpletons, but this is bad advice for the school-boy in question.

11.6.2 Comments from a Bayesian perspective

The first thing to notice, I think, is how peculiar the assumptions are in this formulation. Each player is presumed to know the other player's utility, and that it is the exact opposite of his own. It seems to me extraordinary to have such knowledge.

A Bayesian facing such a game would be uncertain about which choice her opponent is about to make. Being a Bayesian, such a person would have probabilities, non-negative and summing to one, about what that other player will do. Then expected utility maximization can be accomplished as follows: Consider P1's decision first. By assumption, P1 has probabilities $\mathbf{q} = (q_1, \ldots, q_n)$ about the action of P2. Then P1's expected utility of choosing action a_i is

$$\sum_{j=1}^{n} u_{ij} q_j, \tag{11.53}$$

so P1's optimal choice is that value i (or any of them, in case of ties), that maximizes (11.53). By the same argument, P2's optimal choice (or choices) minimizes over index j the expected utility

$$\sum_{i=1}^{m} u_{ij} p_i \tag{11.54}$$

where $\mathbf{p} = (p_1, \ldots, p_m)$ reflect P2's opinion about the choice P1 will make.

These choices obey the principle of dominance, as follows: Consider P1's decision problem, and suppose there are actions a_i and $a_{i'}$ available to P1, satisfying the following inequality:

$$u_{ij} \geq u_{i'j} \text{ for all } j = 1, \ldots, n. \tag{11.55}$$

Choice a_i is said to dominate choice $a_{i'}$ for P1 in this case. Then whatever probabilities $\mathbf{q}$ P1 may have on P2's choice, a_i will always be at least as good a choice for P1 as will $a_{i'}$. Thus $a_{i'}$ may be eliminated from among P1's choices without loss of expected utility to P1.

Now consider P2's decision problem, and suppose there are decisions b_j and $b_{j'}$ available to P2, satisfying the inequality

$$u_{ij} \leq u_{ij'} \quad \text{for all } i = 1, \ldots, m. \tag{11.56}$$

In this case choice b_j is said to dominate choice $b_{j'}$ for P2. Then whatever probabilities $\mathbf{p}$ P2 may have on P1's choice, b_j will always be at least as good a choice for P2 as will $b_{j'}$. Thus $b_{j'}$ may be eliminated from among P2's choices without loss of expected utility to P2.

What relationship is there between the expected-utility maximizing choices in (11.53) and (11.54), and the minimax solutions $\mathbf{p}^*$ and $\mathbf{q}^*$ found above? Let's suppose P1 is sure that P2 will use his randomized minimax choice $\mathbf{q}^*$. Then the associated maximin solution $\mathbf{p}^*$ for P1 puts positive probability on a number of choices for P1. We can, without loss of generality, renumber the choices for P1 so that $\mathbf{p}^*$ is positive for choices $i = 1, \ldots, m' \leq m$. Each of the choices $a_i, i = 1, \ldots, m'$ then is utility-maximizing for P1, and each has the same expected utility, as shown in the corollary in the appendix to this chapter. Then any randomized strategy that puts positive probability only on $a_1, \ldots, a_{m'}$ will also have this same (optimal) expected utility. In particular, $\mathbf{p}^*$ is one of those randomized strategies, and therefore maximizes P1's expected utility. But this is a weak recommendation for $\mathbf{p}^*$ as a strategy for P1. P1 need not randomize among the strategies $a_1, \ldots, a_{m'}$ at all. If P1 does not have a belief about P2's strategy $\mathbf{q}$ that coincides with $\mathbf{q}^*$, then $\mathbf{p}^*$ will be suboptimal for P1 in general. Thus $\mathbf{p}^*$ is not very impressive as a utility-maximizing choice for P1.

The same can be said for P2. If P2's beliefs $\mathbf{p}$ about P1's choice coincide with $\mathbf{p}^*$ exactly, then the strategies for P2 can be renumbered so that those with indices $j = 1, 2, \ldots, n'$, and only those, have positive probability under $\mathbf{q}^*$. Then every randomization of $b_1, \ldots, b_{n'}$

maximizes P2's expected utility, including the choices $b_1, \ldots, b_{n'}$, again, as shown in the appendix. If P2's beliefs do not coincide with $\mathbf{p}^*$, then $\mathbf{q}^*$ in general is a suboptimal choice for P2. The fact that $\mathbf{q}^*$ is so weakly recommended for P2 makes $\mathbf{p}^*$ even less attractive as a belief for P1.

Game theory as developed by von Neumann and Morgenstern (1944) and their successors places great stress on two distinctions among games: whether there are two players or more than two, and whether the game is constant-sum or not. Neither of these distinctions seems critical from the Bayesian perspective. If there are $k > 2$ players, then P1 must assess his probability of the decisions of each of the other players, but again will optimally choose a_i to maximize (11.53), where now the index j ranges over the joint choices of each of the other players. Similarly (11.53) applies to P1's choice whether or not the game has constant-sum utilities. Thus these two distinctions do not affect the Bayesian theory in any conceptual way.

What does matter for the Bayesian theory, but not for classical game theory, is sequential play. For a Bayesian, previous play by an opponent or opponents is data, from which a Bayesian learns information that can be useful in predicting the future play of either those or other opponents. However, for minimax players, the previous history is not relevant; such a player continues to use the same mixed strategy regardless of the choices made by the same or other opponents in past play, of the same game or other games. In the simpler context of sequential play, as in section 11.3, we have seen that the fact that your opponent will learn about you from your play leads to major complication in the Bayesian theory.

11.6.3 An example: Bank runs

The essential problem for a traditional bank is that it accepts deposits for which repayment can be demanded in a short time, and makes loans that have a long time horizon. If everyone demands their money back from a bank at the same time, the bank cannot pay because it cannot call in its loans, and bankruptcy ensues.

The heart of this problem can be modeled by imagining two players, P1 and P2, each of whom has deposited an amount D in the bank. The bank has invested this money in a project. If the bank is forced to liquidate the project before it matures, the bank can recover $2r$, where we assume $D > r > D/2$. At maturity, the project will pay $2R$, where $R > D$. The question for the players is whether to demand their money now, that is, withdraw, or allow the project to proceed to maturity. We'll assume that the utilities of each player are linear in money.

The payoffs to the two players can be expressed in the following matrix:

		Player 2	
		W	NW
Player 1	W	r, r	$D, 2r - D$
	NW	$2r - D, D$	R, R

Here the first number gives P1's payoff, and the second is P2's payoff.

If P1 is sure that P2 will not withdraw, his optimal strategy is not to withdraw as well, since $R > D$. Similarly, if P1 is sure that P2 will withdraw, then his optimal strategy is to withdraw as well, since $2r - D < r$. In the language of traditional game theory, both (W, W) and (NW, NW) are Nash equilibria, since the knowledge of the other player's strategy would not change one's own. This fact does not help P1 determine his optimal strategy, however.

How does Bayesian theory suggest that P1 play this game? P1's uncertainty here is what P2 will do. Suppose that P1's probability that P2 will withdraw is θ. Then P1's expected utility for withdrawal is

$$r\theta + D(1 - \theta),$$

and his expected utility for not withdrawing is

$$(2r - D)\theta + R(1 - \theta).$$

Then withdrawal is strictly optimal for P1 if and only if

$$r\theta + D(1 - \theta) > (2r - D)\theta + R(1 - \theta), \text{ or}$$

$$\theta > \frac{R - D}{R - r} \quad .$$

Not withdrawing is optimal if

$$\theta < \frac{R - D}{R - r},$$

and P1 is indifferent between withdrawing and not withdrawing if

$$\theta = \frac{R - D}{R - r}.$$

It makes sense that if θ is large, P1 should withdraw, while if θ is small he should not. P2's analysis is similar (with perhaps a different θ), because his utilities are assumed to be the same as P1's.

Of course this is a highly simplified version of the actual situation. Usually there are many depositors, but their problem is captured by this simple structure: if the bank is going down, they want their money immediately. The history of banking has many instances of panics in which depositors, sometimes in response to rumors, simultaneously demand their money from a particular bank, or from many banks.

In response to the high social costs of bank runs and bank failures, governments have instituted two basic policies: regulation of banks to ensure their soundness, and governmental deposit insurance. Both of these policies aim at reassuring the public that their money is safe, thus reducing the θ's of the players. As a public policy, these measures have been quite successful.

It seems to me that the Bayesian analysis of bank runs illuminates the essential problem, which is what the depositors believe other depositors will do. Not to have room for those beliefs in the traditional theory seems to me to deprive it of insight.

It is also to be noted that there aren't useful principles in this game to tell P1 what θ to believe about P2. Sometimes bank runs occur, sometimes they do not. To hold that the payoffs to the game, plus "common knowledge" and "common priors" can resolve P1's problem seems to me to be a hopeless quest.

11.6.4 Example: Prisoner's Dilemma

This is a famous game, attributed by Luce and Raiffa (1957, p. 94) to A. W. Tucker. The story is that two persons suspected of jointly committing a crime are taken into custody and separated. They are believed to have committed a serious crime. If both confess, they will get 8 years imprisonment each. If one confesses and the other does not, the one confessing will get 3 months, and the other will get 10 years. If neither confesses, they will each get 1 year on minor charges. We'll suppose that their losses are linear functions of the time they spend in jail.

In the literature on this problem, to cooperate (C) with the other player is not to confess, while to defect (D) is to confess.

From the viewpoint of Prisoner 1 (P1), his major uncertainty is whether P2 will confess. Suppose his probability that P2 will confess is θ, $0 \leq \theta \leq 1$. Then his expected jail time if he confesses is

$$8\theta + .25(1 - \theta).$$

Similarly, if P1 does not confess, his expected jail time is

$$10\theta + (1 - \theta).$$

Since $10\theta+(1-\theta) > 8(\theta)+.25(1-\theta)$ for all θ, $0 \leq \theta \leq 1$, it follows that the optimal strategy for P1 is to confess, regardless of his probability θ on P2's behavior. By the same analysis, it is optimal for P2 to confess, regardless of his probability on P1's behavior. Some find this analysis uncomfortable, because both prisoners could do better not confessing (getting only 1 year in jail each), than confessing (8 years each).

Rapoport (1960, p. 175), for example, argues that

> Instead of taking as the basis of calculations the question "Where am I better off?," suppose each prisoner starts with the basic assumption: "My partner is like me. Therefore he is likely to act like me. If I conclude that I should confess, he will probably conclude the same. If I conclude that I should not confess, this is the way he probably thinks. In the first case, we both get (10 years); in the second case (1 year). This indicates that I personally benefit by not confessing."

Later, however, Rapoport (1966, p. 130) appears to change his position:

> If no binding agreement can be effected, the mutually advantageous choice (C, C) is impossible to rationalize by appeal to self-interest. By definition, a "rational player" looks out for his own interest only. On the one hand, this means that the rational player is not malicious – that is, he will not be motivated to make choices simply to make the other lose (if he himself gains nothing in the process). On the other hand, solidarity is utterly foreign to him. He does not have any concept of collective interest. In comparing two courses of action, he compares only the payoffs, or the expected payoffs, accruing to him personally. For this reason, the rational player in the absence of negotiation or binding agreements cannot be induced to play C in the game we are discussing. Whatever the other does, it is to his advantage to play D.

If the players had the opportunity to make an enforceable agreement, they could agree not to confess. Essentially an enforceable agreement changes the utilities of some of the choices, which of course changes the analysis.

There are people for whom "confessing" has high disutility, because it means doing something that will harm another person, perhaps a friend. For such people, their losses are not linear in the time spent in jail. For such a person, the analyses above should be redone using his or her personal loss function.

It is noteworthy that in market situations involving few players (oligopolies), the players do better cooperating (to raise prices, or constrain output). The US Antitrust laws specifically make contracts in restraint of trade unenforceable. Thus in the case of oligopolies, the public interest is served by the "confess" strategies in which the companies do not cooperate. There are other situations (such as the outbreak of World War I), in which it could be argued that whatever alliance mobilized first would have a great advantage. Since the sides were not able to make an enforceable agreement, both mobilized, war ensued, and both alliances lost utility.

Whether the advice to defect in a single play Prisoner's Dilemma is paradoxical is left to the reader. Iterated Prisoner's Dilemmas are addressed in 11.6.6.

11.6.5 Notes and references

The book of von Neumann and Morgenstern (1944) is the classic work on game theory. It expounds the minimax view explained in 11.6.1. The proof of the minimax theorem given in the appendix is based on that of Loomis (1946). The material in 11.6.2 is based on Kadane and Larkey (1982a). The example concerning bank runs in 11.6.3 is discussed in

Sanchez et al. (1996) and Gibbons (1992). It is an example of a class of games known in the literature as "stag hunts" (see Skyrms (2004)). An early game theory paper supporting the use of Bayesian decision theory is Rosenthal (1981).

The views expressed in Kadane and Larkey (1982a) have not found universal acceptance in the game theoretic community. Harsanyi (1982a) comments on the paper with essentially two arguments. The first is to present a case for a necessitarian view of prior distributions, alleging that "in some situations there is only **one** rational prior distribution" (p. 20). In particular he cites Jaynes' work in physics to support this view. While the evaluation of Jaynes' work in physics is for physicists to work out (see, for example, Shalizi (2004)), let us suppose that Jaynes' assumptions allow him successfully to re-derive thermodynamics. This would not support the proposition that in games there is only one prior probability distribution that it is rational to believe about one's opponents' moves. Harsanyi's second argument is the complaint that Kadane and Larkey offer no guidelines about "how this probability distribution is to be chosen by a rational player" (p. 121). He claims that "Most game theorists answer this question by constructing various normative 'solution concepts' based on suitable rationality postulates and by assuming that the players will act, and will also expect each other to act, in accordance with the relevant solution concept."

Let us suppose for the sake of the argument that each solution concept corresponds to some prior distribution on the other players' actions. In that case a player will not be a sure loser by acting in accord with that solution concept. And such an action is then endorsed by the subjective Bayesian viewpoint of this book. The controversy, then, is whether obedience to such solution concepts is the only rational choice a player can make. The fact that game theorists have produced many solution concepts, not all of which coincide, is a hint that this program can't succeed in uniquely defining rational play. I regard the prior distributions generated by solution concepts as interesting subjects of study, and as special cases of possible belief. Whether a particular such solution concept applies to a particular instance of a game is still, I believe, a matter of (subjective) judgment to be made by a player. Perhaps what the debate comes to is that Harsanyi's vision of game theory seeks to limit attention to mutual assumptions of rationality while my vision recognizes conflict situations in which "rationality" of an opponent need not be assumed.

The debate continued with a reply from Kadane and Larkey (1982b), and a rejoinder from Harsanyi (1982b). A longer response to Harsanyi came in Kadane and Larkey (1983). This paper discusses the distinction between "ought" and "is," that is, between recommendations of how to play the game and descriptions of how people (other players) actually do play. They write "Taking the Bayesian norm as **prescriptively** compelling for my play leads me to want the best **description** I can find of my partner/opponent's play" (p. 1376). Shubik (1983) commenting on this paper takes a middle position, writing "Those of us concerned with the applications of game theoretic methods to the social sciences are well aware of the importance and the limitations of our assumptions concerning the perception, preferences and abilities of individuals" (p. 1380).

Further contributions to the subjective view of games can be found in Wilson (1986) and Laskey (1985), concerning iterated Prisoner's Dilemma games, Kadane et al. (1992), about elicitation of probabilities in a game theoretic context, and Larkey et al. (1997) on skill in games.

There is a variety of reactions to this issue. Mariotti (1995) argues that "a divorce is required between game theory and individual theory" (p. 1108). Mariotti bases his claim on an example in which a Bayesian is required to have preferences over games, but the game description does not include the prior of the player. Hence it is not possible to compute an expected utility for the play of the game, and a contradiction to simple Bayesian principles ensues. His conclusion is that game theory should abandon trying to justify its recommended choices from the perspective of Bayesian decision theory, and instead invent some other kind of decision theory. (See also discussion on Mariotti in Aumann and Drèze (2008).)

I think Mariotti is correct in calling for greater precision by game theorists in specifying exactly what assumptions are being made in justifying the claim of Bayesianity for their recommended choices. But I think he is too pessimistic in giving up hope of a reconciliation between game theory and individually rational Bayesian behavior.

A more recent comment on the debate is by Aumann and Drèze (2008). They write "On its face, the Kadane-Larkey viewpoint seems straightforward and reasonable. But it ignores a fundamental insight of game theory: that a rational player should take into account that *all the players are rational, and reason about each other*. Let's call this 'interactive rationality' " (p. 3). Later they argue that Kadane and Larkey fail to "bear in mind that *in estimating how the others will play*, a rational player must take into account that the others are – or should be – estimating how she will play" (p. 25). The issue here is in the force of the "must" and in the distinction between "are" and "should be." To reiterate the general point, I think that "interactive rationality" is an interesting special case of coherence, but not the only one.

As emphasized above, a rational player may or may not model his counterpart as rational. He does not violate the axioms of Bayesian rationality if he models his counterpart as not completely rational. However, let's play along and suppose that he does. Then the regress cited by Aumann and Drèze occurs. The point here is that whether that regress stops at some stage or continues indefinitely, it is only the marginal distribution of what move the counterpart will make that matters. More generally, players can have whatever models they may have of the other player, with however many uncertain parameters; again, only the marginal distribution of the other player's move affects the optimal decision.

In the end, Aumann and Drèze seem not to disagree. They write "Theories of games may be roughly classified by 'strength': the fewer outcomes allowed by the theory, the stronger – more specific – it is" (p. 23) ... "Viewed thus, the Harsanyi and Selten (1987) selection theory, which specifies a single outcome for each game, is the strongest. Next come refinements of the Nash equilibrium, like Kohlberg and Mertens (1986); next, Nash equilibrium (1951) itself; next correlated equilibrium; and then interactive rationality. Weaker is rationalizability (Bernheim (1984), Pearce (1984)) and weaker still, the Kadane-Larkey 'theory' " (p. 24).

This is a very reasonable view of the situation, I think. As the theories rise in strength, they require more and more restrictive assumptions about what the players believe about each other. Thus more and more is packed into phrases like "common knowledge," "common knowledge of rationality" and "common priors." The usefulness of these special assumptions has to be determined case-by-case in application. Is the strength of the assumption justified in the application? This is also what Shubik (1983) is suggesting.

In emphasizing the general case, I would not denigrate the special cases. Rather I would simply remind the reader that in each use, the assumptions underlying a special case have to be justified.

11.6.6 Iterated Prisoner's Dilemma

Fool me once, shame on you. Fool me twice, shame on me.

Unless some day somebody trusts somebody, there'll be nothing left on earth excepting fishes.
—The King and I

Suppose now that the Prisoner's Dilemma, instead of being played once, is played n times. Does repeated play affect the player's strategies?

From the viewpoint of classical decision theory, the answer is "no." At the nth iteration, it is uniquely optimal for each player to confess, or, in other words, to defect. Under the assumption that each player knows the other player to be "rational," both players then are sure that the other will confess in the last iteration. Now consider the $(n-1)^{st}$ iteration. Knowing the outcome of the last game, it is optimal for each player to confess on the $(n-1)^{st}$ game, since there is nothing to gain by not confessing. By backward induction, both players confess in each of the n iterations.

By contrast, a Bayesian player is not so constrained. For example, suppose our Bayesian believes that if he confesses in the first game, his opponent will confess at every iteration after that, while if he does not confess at the first iteration, his opponent will never confess at all ensuing iterations. In such a circumstance, clearly not confessing is the expected-utility maximizing choice at the first game. The calculations involved in the Bayesian approach to the iterated Prisoner's Dilemma can be substantial, but see Wilson (1986).

There is a vast literature about iterated Prisoner's Dilemmas. One line of work is experimental (Rapoport and Chammah (1965)), while another involves computer experiments of strategies against each other. Axelrod (1984)'s contest among strategies was won by Rapoport's "tit for tat" strategy, which cooperates on its first iteration, and on subsequent iterations makes whatever decision the other player made on the previous iteration.

Axelrod's view of optimal play for the indexsubjectindexPrisoner's Dilemma, iterated iterated Prisoner's Dilemma straddles the two views being contrasted here. On the one hand he endorses the backward induction argument, writing "Thus two egoists [he means utility maximizers, JBK] playing the game *once* will both choose their dominant choice, defection, and each will get less than they both could have gotten if they had cooperated. If the game is played a known finite number of times, the players still have no incentive to cooperate" (1984, p. 10). On the other hand he offers this explanation for cooperation: "What makes it possible for cooperation to emerge is the fact that the players might meet again. This possibility means that the choices made today not only determine the outcome of this move, but can also influence the later choices of the players. The future can therefore cast a shadow back upon the present and thereby affect the current strategic situation" (1984, p. 12). That the number of iterations is uncertain seems irrelevant to the first argument, since however many iterations are to be played, defection is optimal from that perspective. Axelrod seems not to notice or address the apparent contradiction between these two arguments, the first based on the assumption that the other player is sure to defect, while the second does not make that assumption.

The upshot of both the experimental and simulation work is that always confessing is not what people do, and not the strategy that wins tournaments. This might be regarded as evidence that the standard of rationality proposed by classical game theory is not necessarily good advice.

11.6.7 Centipede Game

Consider the game illustrated in Figure 11.4:

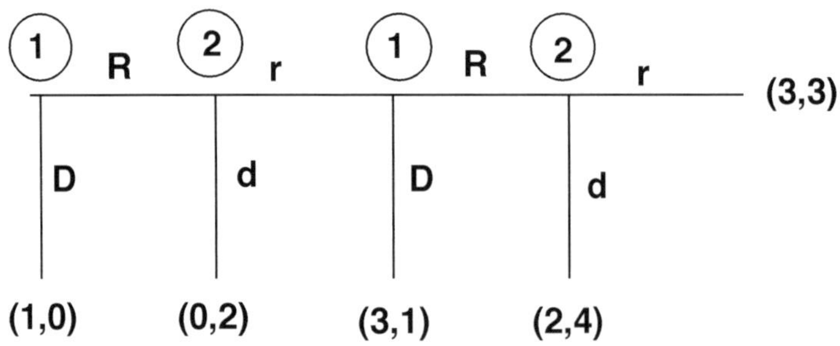

Figure 11.4: Extensive form of the Centipede Game.

To understand the diagram, player 1 decides between D and R at each stage, while player 2 decides between d and r. A choice of D or d ends the game; a choice of R or r passes the choice to the other player, except for 2's second choice, which also ends the game. The game proceeds from left to right. The payoffs (x, y) mean that player 1 gets x and player 2 gets y.

Consider first 2's second choice (if reached). Choosing r results in $(3, 3)$, which gives 3 for player 2. Choice of d results in $(2, 4)$, which means 4 for player 2. Hence, player 2 prefers d. Now consider 1's second choice (if reached). Choice of R passes the choice to player 2, leading (if player 2 accepts the analysis above) to a payoff of 2 for player 1. However, choice of D results in 3 for player 1, which she prefers. Hence choice of D is best for player 1, under these assumptions. Now consider player 2's first choice (if reached). If player 1 behaves as predicted above, the resulting payoff is 1 to player 2 from the choice of R. Otherwise he chooses d, resulting in a payoff of 2 for player 2. Hence his best choice is d, and this results in 0 for player 1. Hence at player 1's first choice, choosing D, resulting in 1, would be the best choice. If the players could make an enforceable agreement, they would both benefit from the choices R and r, leading to a $(3, 3)$ payoff. A similar game with 100 stages resulted in the name indexsubjectindexcentipede game "centipede."

The analysis here relies on each player believing that each player (including herself) will play according to backward induction. Introduced by Rosenthal (1981), the experimental results of McKelvey and Palfrey (1992) and Nagel and Tang (1998) show that the first player does not always choose D at the first choice. The message is the same: backward induction is not necessarily a good prediction of behavior.

11.6.8 Guessing a multiple of the average

Suppose there are n people in a group. Each person is to choose a number in a set S. A prize is divided among those whose guess is closest to p times the average of all the guesses, which we'll call the target. Thus this game is characterized by n, S and p.

If everyone chooses the same number $x \epsilon S$, then each person gets $1/n$ of the prize. Suppose the game is played on the set S of real numbers $[a, b]$, with $a \geq 0$, and $p < 1$. Then the average cannot be greater than b, so the target cannot be greater than pb. If everyone understands this and acts on it, there is effectively a new game played on $[a, pb]$, provided $pb > a$. If $pb \leq a$, choice of a is uniquely optimal. Successive iterations of this reasoning lead everyone to choose a. (Similarly, if $p > 1$, iteration of this reasoning would lead everyone to choose b when $a > 0$.) This "solution" requires that everyone in the group understands and acts on this induction.

When S is limited to integers between two integers a and b, and again $p < 1$, the argument is similar except that the upper limit of the new game after an iteration is the integer closest to pb, which might be b iteself. In this case the above argument does not necessarily reduce to the single point a. Thus if $p = 2/3, a = 0$ and $b = 2$, $pb = 4/3$, so the closest integer is 1. However, if $b = 1$, $pb = 2/3$ and again the closest integer is 1. Thus, if one believes that everyone else is following this argument, the set of choices reduces to $\{0, 1\}$, but is not further reduced. Again, if $p > 1$, then $a > 0$ is raised to the closest integer to ap, which again might be a itself.

What the backward induction leaves out is a description of who the members of the group are, and consequently how likely they are to follow the path outlined above. Members of the (fictional) "Society for the Propagation of Induction in Game Theory" are likely to behave differently than would a class of seventh grade students. A wise player of the game would want to know about the other players, and to think about their likely behavior in making a choice.

11.6.9 *References*

Keynes (1936) invented a game to explain his view of stock market prices. He imagines a contest in which contestants are provided photographs of women, and are asked to choose which six are most beautiful. Those that choose the most popular are eligible for a prize. The point is not to choose what you find to be the most beautiful, but what you predict others will. Or, to take it one level deeper, to predict what others will predict others will choose, etc. This is like the game of predicting p times the average, with $p = 1$. Nagel (1995) has done empirical work on how some people behave in the game with $p < 1$. Some of the literature on this game concentrates on $p = 2/3$, and call it "Guess 2/3 of the average."

11.6.10 *Summary*

There is a growing literature on Bayesian approaches to optimal decisions in the context of simultaneous move games. At this time, there appears to be a glimmer of hope of a consensus: a hierarchy of assumptions ranging at the low end from only the assumption of coherence (Kadane-Larkey) to, at the high end, the Harsanyi-Selten work that gives a single recommended strategy. The choice of which assumptions are reasonable depends on the context of a given application.

11.6.11 *Exercises*

1. Vocabulary: State in your own words the meaning of
   (a) constant-sum game
   (b) minimax strategy
   (c) maximin strategy
   (d) value of a zero-sum, 2 person game
   (e) dominance
   (f) n-person game
2. Why might a good prescriptive theory of how to play a game require a good descriptive theory of the opponents' play?
3. Suppose, in the single iteration Prisoner's Dilemma (section 11.6.4), that the prisoner's loss function is monotone but not necessarily linear in the amount of jail time they serve. This means that each prefers less jail time to more jail time. Show that under this assumption the same result applies: it is optimal for each to confess.
4. Consider the following modification of the iterated Prisoner's Dilemma problem. Instead of punishments, years in jail, suppose the problem is phrased in terms of rewards. If both cooperate, they each get 3 points. If both defect, they each get 1 point. If one cooperates and the other defects, the defector gets 5 points and the cooperator 0 points. Suppose the player to be advised wishes to maximize expected points. [By problem 3, this change does not affect the fact that defection is the optimal strategy in a single iteration situation.]
   Now suppose that the player we advise is to play a five-iteration Prisoner's Dilemma as specified in the above paragraph against each of 10 players. He will then choose one of these 10 players to play a 100-iteration Prisoner's Dilemma with. How would you advise our player to play in the first phase, which player should he choose for the 100-iteration game, and how should he play in the 100-iteration second phase?
   Explain your reasoning.
5. Recall that a median of an uncertain quantity X is a number m such that $P\{X \leq m\} \geq 1/2$ and $P\{X \geq m\} \geq 1/2$.

(a) Does the induction for the "guessing p times the mean" game also work for the "guessing p times the median" game?

(b) More generally, the qth quantile of an uncertain quantity X is a number x_q such that $P\{X \leq x_q\} \geq q$ and $P\{X \geq x_q\} \geq 1 - q$. [$x_{1/2}$ is the median.] Does the induction work for the "guessing p times the qth quantile" game?

11.7 The Allais and Ellsberg Paradoxes

You can believe nearly every word I say.

—Delaney & Bonnie & Friends

Paradoxes play an important role in a normative theory, such as Bayesian decision theory. A single unresolved paradox could lead to the abandonment of the theory, as it would have been shown to be an inadequate guide to optimal behavior. Such an example would have the character of proposing a scenario and reasonable seeming choices within it that contradict the normative theory. The two most serious challenges to Bayesian theory were proposed by Allais and by Ellsberg.

11.7.1 The Allais Paradox

Allais (1953) proposed a paradox, which is discussed extensively in Savage (1954). In situation 1, would you prefer choice A (\$500,000 for sure) to choice B (\$500,000 with probability .89, \$2,500,000 with probability 0.10 and status quo (\$0) with probability 0.01)? In situation 2, would you prefer choice C (\$500,000 with probability 0.11, and status quo otherwise) or choice D (\$2,500,000 with probability .1, status quo otherwise)? Allais proposes that many would choose A in situation 1 and D in situation 2, and that these choices, jointly, contradict expected utility theory.

Allais's argument is as follows: Your expected utility for choices A and B are respectively $U(\$500{,}000)$ and $.89\ U(\$500{,}000) + .1\ U(\$2{,}500{,}000) + .01\ U(\$0)$. Therefore you will prefer A to B in situation 1 if and only if

$$U(\$500,000) > .89\ U(\$500,000) + .1\ U(\$2,500,000) + .01\ U(\$0),$$

if and only if

$$.11\ U(\$500,000) > .1\ U(\$2,500,000) + .01\ U(\$0). \tag{11.57}$$

In situation 2, your expected utility for choice D is

$$.1\ U(\$2,500,000) + .9\ U(\$0),$$

while your expected utility for choice C is

$$.11\ U(\$500,000) + .89\ U(\$0).$$

Hence you will prefer D to C in situation 2 if and only if

$$.1\ U(\$2,500,000) + .9\ U(\$0) > .11\ U(\$500,000) + .89\ U(\$0),$$

if and only if

$$.1\ U(\$2,500,000) + .01\ U(\$0) > .11\ U(\$500,000). \tag{11.58}$$

But your utilities cannot satisfy both (11.57) and (11.58). Therefore, Allais argues, a rational Bayesian agent cannot prefer A to B in situation 1 and D to C in situation 2. Allais' argument depends on the acceptance of the proffered probabilities .89, .1 and 0.1 as your subjective probabilities.

Savage (1954) agreed that his first impulse was to choose A and D, and gave the following table:

		Ticket Number		
		1	2-11	12-100
Situation 1	Choice A	5	5	5
	Choice B	0	25	5
Situation 2	Choice C	5	5	0
	Choice D	0	25	0

Prizes, in units of $100,000

Based on this analysis, Savage used the sure-thing principle to change his choice in situation 2 from D to C. [It is not clear whether Allais's subjects would choose A and D if those choices were presented in Savage's table.]

In situation 1, if I am to contemplate choice B, I would be very curious about the random mechanism that would be used to settle the gamble, and about the incentives faced by the kind person making these offers. I might well decide that my subjective probability of getting nothing if I chose B is higher than 0.01. Thus, I might have some healthy skepticism about the offered probabilities. Simply because someone says I have a high probability of winning some fabulous prize doesn't imply that I am required to believe them.

By contrast, in situation 2, I am unlikely to win anything anyway, and hence am about equally vulnerable to being cheated whether I choose C or D.

Suppose my probability is θ that the person offering me a gamble will cheat me by giving me the lowest payoff possible in whatever gamble I choose. Also suppose, without loss of generality, that my utility function satisfies $1 = u(\$2,500,000) > u(\$500,000) = w > u(\$0) = 0$. Then choice A has expected utility w, while choice B has expected utility $(1-\theta)[.89w + .1]$. Thus subjective expected utility favors A over B if and only if

$$\begin{aligned} w >&(1-\theta)[.89w + 0.1], \text{ or} \\ w >&\frac{(0.1)(1-\theta)}{1-(1-\theta)(0.89)}. \end{aligned} \tag{11.59}$$

Similarly D is preferred over C if

$$\begin{aligned} (1-\theta)(0.1) >&(1-\theta)(0.11)w, \text{ so} \\ \frac{0.1}{0.11} >&w. \end{aligned} \tag{11.60}$$

Thus we can ask, under what conditions on θ is there a w satisfying both (11.59) and (11.60), which requires $\frac{0.1}{0.11} > \frac{(0.1)(1-\theta)}{1-(1-\theta)(0.89)}$.

But this inequality holds if and only if

$$\begin{aligned} 1-(1-\theta)(.89) >&(1-\theta)(.11), \text{ or equivalently} \\ 1 >&1-\theta, \text{ or} \\ \theta >&0. \end{aligned} \tag{11.61}$$

Thus choices A and D are compatible provided I put any positive probability θ on being cheated. If I choose A in situation 1, I can sue if I don't get paid my $500,000. With choices B, C and D, the situation is much murkier.

11.7.2 The Ellsberg Paradox

Suppose there are two urns containing red and black balls. One ball is to be drawn at random from one of the urns. To "bet on Black$_I$" means you choose to have a ball drawn from urn 1, and will win \$1 if the ball drawn is black, and nothing otherwise. To "bet on Red_I, Black_{II} or Red_{II}" are defined similarly. Urn 1 contains 100 balls, some of which are red and some black, but you do not know how many of each are in urn 1. In urn 2, you confirm that there are 50 red balls and 50 black balls.

Now consider the following questions:

#1 Do you prefer to bet on Red_I or Black_I, or are you indifferent?

#2 Do you prefer to bet on Red_{II} or Black_{II}, or are you indifferent?

#3 Do you prefer to bet on Red_I or Red_{II}, or are you indifferent?

#4 Do you prefer to bet on Black_I or Black_{II}, or are you indifferent?

Many people are indifferent in the first two choices, but prefer Red_{II} to Red_I and Black_{II} to Black_I. Suppose they are your choices. Are these choices coherent?

With slight abuse of notation, let Black_I be the event that a black ball is drawn from urn 1, and similarly for Red_I, Black_{II} and Red_{II}.

Indifference in question #1 implies that, for you,

$$P\{\text{Red}_I\} = P\{\text{Black}_I\}. \tag{11.62}$$

Since in addition

$$P\{\text{Red}_I\} + P\{\text{Black}_I\} = 1, \text{ we conclude}$$
$$P\{\text{Red}_I\} = P\{\text{Black}_I\} = 1/2. \tag{11.63}$$

Similarly indifference in question #2 implies that, for you,

$$P\{\text{Red}_{II}\} = P\{\text{Black}_{II}\} = 1/2. \tag{11.64}$$

Then it is incoherent to prefer Red_{II} to Red_I, and to prefer Black_{II} to Black_I. Thus these answers appear to be incoherent. But are they?

I note that the experimenter probably knows the content of urn 1, which is unknown to you. By deciding which bet is "on," the experimenter might choose to put you at a disadvantage. Not knowing the experimenter's utilities, you don't know if he wants to do this or not. Only by choosing Red_{II} over Red_I and Black_{II} over Black_I can you ensure yourself against such manipulation.

Suppose your probability is θ_1 that, if the proportion of red balls in urn 1 is less than 1/2 and you bet on Red_I in question #3, the experimenter will malevolently choose to enact question #3. With probability $1 - \theta_1$, the bet will be enacted without regard to the contents of urn 1. Similarly, suppose your probability is θ_2 that, if the proportion of black balls in urn 1 is less than 1/2 and you bet on Black_I in question #4, the experimenter will malevolently choose to enact question #4. With probability $1 - \theta_2$, under these conditions, the bet occurs regardless of the contents of urn 1.

Let $\tilde{P}_R$ be the proportion of red balls in urn 1. $\tilde{P}_R$ is a known constant to the experimenter, but is a random variable to you. Let P_R be the expectation, to you, of $\tilde{P}_R$. Thus P_R is your probability for a red ball being drawn from urn 1. By your answers to question 1, we know that $P_R = 1/2$. We will suppose that $\tilde{P}_R$ has positive variance for you, so you put positive probability on the event $\{\tilde{P}_R > 1/2\}$ and on the event $\{\tilde{P}_R < 1/2\}$. Let m_1 be your conditional expectation of $\tilde{P}_R$ if $\tilde{P}_R$ is less than or equal to 1/2. Similarly, let m_2 be your conditional expectation of $1 - \tilde{P}_R$ if $\tilde{P}_R$ is greater than or equal to 1/2. Then $0 \leq m_1 < 1/2$ and $0 \leq m_2 < 1/2$. With this notation, your probability of winning if you bet on Red_I in question #3 is $(1 - \theta_1)P_R + \theta_1 m_1$. Similarly if you bet Black_I in question #4, your probability of winning is $(1 - \theta_2)(1 - P_R) + \theta_2 m_2$.

So the question is whether there are values of θ_1 and θ_2 such that

$$\frac{1}{2} > \frac{(1 - \theta_1)}{2} + \theta_1 m_1 \tag{11.65}$$

and

$$\frac{1}{2} > \frac{(1-\theta_2)}{2} + \theta_2 m_2. \tag{11.66}$$

But

$$\frac{1-\theta_2}{2} + \theta_2 m_2 > 1/2 \text{ for all } \theta_2 > 0. \tag{11.67}$$

Similarly

$$\frac{1-\theta_1}{2} + \theta_1 m_1 = \frac{1}{2} - \theta_1(\frac{1}{2} - m_1) > 1/2, \text{ for all } \theta_1 > 0. \tag{11.68}$$

The final question to address is whether there is a probability distribution on $\tilde{P}_R$ satisfying the following constraints:

(a) The conditional mean of $\tilde{P}_R$ if $\tilde{P}_R \leq 1/2$ is m_1.

(b) The conditional mean of $\tilde{P}_R$ if $\tilde{P}_R \geq 1/2$ is m_2.

(c) The mean of $\tilde{P}_R$ is 1/2.

Consider the distribution for $\tilde{P}_R$ that puts all its probability on $m_1 < 1/2$ and on $1 - m_2 > 1/2$. Then (a) and (b) are automatically satisfied. Suppose m_1 has probability $(1/2 - m_2)/(1 - m_1 - m_2) > 0$ and m_2 has probability $(1/2 - m_1)/(1 - m_1 - m_2) > 0$. These probabilities sum to 1, since

$$\frac{(1/2 - m_2)}{1 - m_1 - m_2} + \frac{1/2 - m_1}{1 - m_1 - m_2} = \frac{1 - m_1 - m_2}{1 - m_1 - m_2} = 1. \tag{11.69}$$

The mean of $\tilde{P}_R$ is

$$\begin{aligned} E(\tilde{P}_R) &= \frac{m_1(1/2 - m_2)}{1 - m_1 - m_2} + \frac{(1 - m_2)(1/2 - m_1)}{1 - m_1 - m_2} \\ &= \frac{(1/2)m_1 - m_1 m_2 + 1/2 - (1/2)m_2 - m_1 + m_1 m_2}{1 - m_1 - m_2} \\ &= \frac{1/2 - m_1/2 - m_2/2}{1 - m_1 - m_2} = 1/2. \end{aligned}$$

Therefore (c) is satisfied.

Thus if a person has any suspicion (*i.e.*, $\theta_1 > 0, \theta_2 > 0$), then the common choices are coherent.

11.7.3 *What do these resolutions of the paradoxes imply for elicitation?*

My resolution of both paradoxes involves what I call healthy skepticism of the experimenter. In both cases, I would argue that the setup of the paradox enhances reasonable fear. In the Allais case, the enormous rewards involved provide the experimenter the motive to cheat. In the Ellsberg case, the mechanism, and hence the opportunity to cheat, is all too apparent.

But might not the same skepticism affect every elicitation of probability? Yes, it would, but it need not. Much depends on the circumstances of the elicitation, including anonymity, whether the person doing the elicitation has an obvious stake in the outcome, etc. Reasonable elicitations are performed without these issues apparently corrupting them. But these paradoxes serve a healthy warning that the entire circumstances of an elicitation must be thought about carefully. See Kadane and Winkler (1988) for more on the impact of incentive effects on elicitation.

11.7.4 Notes and references

The Allais Paradox first appeared in Allais (1953) and is commented on by Savage (1954, pp. 101-103). The Ellsberg Paradox is from Ellsberg (1961). They appeared at a time in which it was widely understood that utilities might differ from person to person, but many still held the idea that probabilities were interpersonal. The discussion in this section is based on Kadane (1992).

The Allais Paradox led Machina (1982, 2005) and others to explore the consequences to expected utility theory of abandoning the sure-thing principle.

11.7.5 Summary

This section shows how the paradoxes of Allais and Ellsberg can be explained by "healthy skepticism," which essentially asks "what's in it for the other guy?" In this sense, it is an explanation with a game-theoretical flavor.

11.7.6 Exercises

1. Vocabulary: Explain in your own words
   (a) Allais Paradox
   (b) Ellsberg Paradox
   (c) Healthy skepticism
2. Do you think healthy skepticism offers a good explanation of why coherent actors would make the choices prescribed by the Allais and Ellsberg Paradoxes? Why or why not?

11.8 Forming a Bayesian group

> Make of our hands, one hand
> Make of our hearts, one heart
>
> —West Side Story

> Can we all get along?
>
> —Rodney King

This section concerns the conditions under which two Bayesian agents can find a Bayesian compromise, that is, find a probability and utility that represents them together. The Bayesian agents are each assumed to have probabilities in the sense of Chapter 1 and utilities in the sense of Chapter 7. I do not assume interpersonal utility comparisons, the idea that one person cares more than another about a particular choice (see Arrow (1978), Elster and Roemer (1991), Harsanyi (1955), and Hausman (1995) for commentary).

I also must specify the sense in which I use the word "compromise." What I mean is the satisfaction of a **weak Pareto condition**, which says that if each of the agents strictly prefer one option to another, then so must the compromise. I seek conditions under which there is a probability and utility for the agents jointly, so that they can be modeled as a Bayesian group.

One solution to this problem is **autocratic solutions**, that is, to choose one individual and adopt that person's probabilities and utilities. While such a solution satisfies the weak Pareto condition, it does not comport with what in ordinary language might be thought of as a compromise. To introduce this result, recall from section 7.3 that a consequence c_{ij} is the outcome if you decide to do decision d_i and that state-of-the-world θ_j ensues. Each of the two Bayesians, whom we'll call Dick and Jane, have utility functions over the set of consequences. These utility functions are denoted $U_D(\cdot)$ and $U_J(\cdot)$, respectively.

Case	Utility	Probability	Result	Lemma
1	lin[a], $r > 0$	yes	o.a.c.[b]	11.8.8
2	lin[a], $r > 0$	no	compromises	11.8.2
3	lin[a], $r < 0$	yes	compromises	11.8.2
4	lin[a], $r < 0$	no	o.a.c.[b]	11.8.7
5	nonlin[c]	yes	compromises	11.8.2
6	nonlin[c]	no	o.a.c[b]	11.8.6

[a]lin: $U_J(\cdot) = rU_D(\cdot) + s, r \neq 0$
[b]only autocratic compromises
[c]nonlin: a fails for all $r \neq 0, s$

Table 11.1: Cases for Theorem 11.8.1.

Additionally, Dick and Jane are assumed to have probability distributions on Θ, denoted respectively $p_D(\cdot)$ and $p_J(\cdot)$. The following definitions recur repeatedly:

1. $p_J(\cdot) \equiv p_D(\cdot)$. [Dick and Jane are said to agree in probability.]
2. $U_J(\cdot) \equiv rU_D(\cdot) + s$ for some constants $r > 0$ and s. [Dick and Jane are said to agree in utility.]

If Dick and Jane are not distinct, there are no compromises that need to be made.

Structurally, I assume that Dick and Jane agree about the distribution of one uniformly distributed random variable. Thus the set of prizes may be taken to be a convex set, because mixtures of prizes are available, and mean the same to both parties. This device is equivalent to the "horse lotteries" of Anscombe and Aumann (1963).

With these remarks as introduction, I can now state the result to be proved in this section:

Theorem 11.8.1. *There exist non-autocratic, weak Pareto compromises for two Bayesians if and only if they either agree in probability, but not in utility, or agree in utility, but not in probability.*

The proof of Theorem 11.8.1 divides into six cases, as shown in Table 11.1. There are two important facts to be gleaned from this table. First, the six cases are disjoint and do not omit possibilities claimed by the theorem. Second, if each of the results stated in the table is proved by the related lemma as claimed, then the theorem is established.

It is relatively simple to prove the existence of non-autocratic Pareto-respecting compromises when they exist, so cases 2, 3 and 5 are dealt with in the following lemma.

Lemma 11.8.2 (Existence of Compromises). *In cases 2, 3 and 5, there are non-autocratic compromises.*

Proof. Case 2: Here the parties agree in utility, but not in probability. Let the consensus utility U be $U(\cdot) = U_D(\cdot)$, (U_J would do as well), and let the consensus probability $P(\cdot)$ satisfy

$$P(\cdot) = \alpha P_D(\cdot) + (1 - \alpha) P_J(\cdot)$$

for some $\alpha, 0 < \alpha < 1$. Suppose both Dick and Jane strictly prefer decision d_1 to decision d_2, which means

$$U_D(d_1) > U_D(d_2) \text{ and } U_J(d_1) > U_J(d_2).$$

Then

$$\begin{aligned}
U(d_1) - U(d_2) = &\int [U(d_1,\theta) - U(d_2,\theta)]\, d\,[\alpha P_D(\theta) + (1-\alpha)P_J(\theta)] \\
=&\alpha \int [U_D(d_1,\theta) - U_D(d_2,\theta)]\, dP_D(\theta) \\
&+ (1-\alpha) \int [U_D(d_1,\theta) - U_D(d_2,\theta)]\, dP_J(\theta) \\
=&\alpha\, [U_D(d_1) - U_D(d_2)] \\
&+ (1-\alpha) \int [(rU_J(d_1,\theta) + s) - (rU_J(d_2,\theta) + s)]\, dP_J(\theta) \\
=&\alpha\, [U_D(d_1) - U_D(d_2)] + (1-\alpha) r\, [U_J(d_1) - U_J(d_2)] \\
>&0.
\end{aligned}$$

Thus (U, P) respects the Pareto condition. Since P does not coincide with either P_D or P_J, the pair (U, P) is a non-autocratic, Pareto-respecting compromise.

Case 3: In this case, because $r < 0$, their utilities are directly opposed, but their probabilities agree. Suppose Dick strictly prefers d_1 to d_2, so $U_D(d_1) > U_D(d_2)$.

Then

$$\begin{aligned}
U_J(d_1) - U_J(d_2) = &\int [U_J(d_1,\theta) - U_J(d_2,\theta)]\, dP_J(\theta) \\
=&\int [(rU_D(d_1,\theta) + s) - (rU_D(d_2,\theta) + s)]\, dP_J(\theta) \\
=&r \int [U_D(d_1,\theta) - U_D(d_2,\theta)]\, dP_D(\theta) \\
=&r\, [U_D(d_1) - U_D(d_2)] \\
&< 0.
\end{aligned}$$

Hence whenever Dick strictly prefers d_1 to d_2, Jane strictly prefers d_2 to d_1. Then the Pareto condition is vacuous, and every pair (U, P) is a Pareto-respecting compromise.

Case 5: In this case, the utilities of the parties are not linearly related, so there do not exist $r \neq 0$ and s such that $U_J(\cdot) = rU_D(\cdot) + s$, and $P_D(\cdot) \equiv P_J(\cdot)$. In this case let $P = P_D(\cdot) = P_J(\cdot)$, and let $U(\cdot) = \alpha U_D(\cdot) + (1-\alpha)U_J(\cdot)$ for some α, $0 < \alpha < 1$. Suppose that Dick and Jane both strictly prefer d_1 to d_2, or, in notation, $U_D(d_1) > U_D(d_2)$ and $U_J(d_1) > U_J(d_2)$.

Then

$$\begin{aligned}
U(d_1) - U(d_2) = &\int \big(\big[\alpha U_D(d_1,\theta) + (1-\alpha)U_J(d_1,\theta)\big] - \\
&\big[\alpha U_D(d_2,\theta) + (1-\alpha)U_J(d_2,\theta)\big]\big) dP(\theta) \\
=&\alpha \int \big[U_D(d_1,\theta) - U_D(d_2,\theta)\big] dP_D(\theta) \\
&+ (1-\alpha) \int \big[U_J(d_1,\theta) - U_J(d_2,\theta)\big] dP_J(\theta) \\
=&\alpha[U_D(d_1) - U_D(d_2)] + (1-\alpha)\big[U_J(d_1) - U_J(d_2)\big] \\
&> 0.
\end{aligned}$$

Thus the (U, P) pair is Pareto-respecting. Since in addition it is not autocratic, this pair satisfies the conditions of the lemma. □

Lemma 11.8.3. *In Case 6, reversing the roles of Dick and Jane if necessary, there exists an event F and consequences $r*$, r_* and c satisfying the following:*

a) $P_D(F) < P_J(F)$

b) $U_J(r^*) = U_D(r^*) = 1$

$U_J(r_*) = U_J(r_*) = 0$

$U_D(c) < U_J(c)$

Proof. In this case, $U_J(\cdot) \neq rU_D(\cdot) + s$ for all $r \neq 0$ and s, and $P_D(\cdot) \neq P_J(\cdot)$.

Dick is assumed to have some strict preferences (so his utility is not constant). Thus there exist consequences r^* and r_* satisfying $U_D(r^*) > U_D(r_*)$. There are now three cases to consider:

Case A: For all consequences such that $U_D(r^*) > U_D(r_*)$, Jane's preferences are such that $U_J(r^*) < U_J(r_*)$.

Case B: There are consequences such that $U_D(r^*) > U_D(r^*)$ and $U_J(r^*) = U_J(r_*)$. Similarly there are consequences t^* and t_* such that $U_J(t^*) > U_J(t_*)$ and $U_D(t^*) = U_D(t_*)$.

Case C: There are consequences r^* and r_* such that $U_D(r^*) > U_D(r_*)$ and $U_J(r^*) > U_J(r_*)$.

Below Case A is shown to contradict the utility assumption, Case B is shown to reduce to Case C, and Case C leads to the conclusion of the lemma.

Case A: If $U_D(r^*) > U_D(r_*)$ implies $U_J(r^*) < U_J(r_*)$. then we must have

$$U_J(\cdot) = rU_D(\cdot) + s \text{ with } r < 0$$

and $P_D(\cdot) = P_J(\cdot)$, both of which contradict the assumptions of the case. For more on this, see Kadane (1985).

Case B:

Let

$$r_{**} = \begin{cases} r_* & \text{with probability } 1/2 \\ t_* & \text{with probability } 1/2 \end{cases}$$

and

$$r^{**} = \begin{cases} r^* & \text{with probability } 1/2 \\ t^* & \text{with probability } 1/2 \end{cases}.$$

Then

$$\begin{aligned} E_D U_D(r_{**}) =& \frac{1}{2} U_D(r_*) + \frac{1}{2} U_D(t_*) \\ E_D U_D(r^{**}) =& \frac{1}{2} U_D(r^*) + \frac{1}{2} U_D(t*) \\ E_J U_J(r_{**}) =& \frac{1}{2} U_J(r_*) + \frac{1}{2} U_J(t_*) \\ E_J U_J(r^{**}) =& \frac{1}{2} U_J(r^*) + \frac{1}{2} U_J(t^*). \end{aligned}$$

Both parties prefer r^{**} to r_{**}. Since r^{**} and r_{**} are in the convex set of rewards, they are legitimate rewards themselves, and hence satisfy case C.

Case C: In this case, there are r^* and r_* such that $U_D(r^*) > U_D(r_*)$ and $U_J(r^*) > U_J(r_*)$. Without loss of generality, we may normalize Dick and Jane's utilities so that

$$\begin{aligned} U_D(r^*) =& U_J(r^*) = 1 \\ U_D(r_*) =& U_J(r_*) = 0. \end{aligned}$$

If there were no reward such that $U_D(c) \neq U_J(c)$, then we would have $U_D(\cdot) = U_J(\cdot)$, which would contradict the utility assumption of case 6. Hence there is some c such that $U_D(c) \neq U_J(c)$. We may identify Dick as the party such that $U_D(c) < U_J(c)$. This shows part b) of the lemma.

To show part a), since $P_D(\ ) \neq P_J(\ )$, there is some event G such that $P_D(G) \neq P_J(G)$. If $P_D(G) < P_J(G)$, let $F = G$. If $P_D(G) > P_J(G)$, then $P_D(\overline{G}) < P_J(\overline{G})$, so let $F = \overline{G}$. In both cases, $P_D(F) < P_J(F)$, which is part a). □

The strategy we now pursue is to see what utilities $U(c)$ and probabilities $P(F)$, candidates for the compromise utility and probability of Dick and Jane, are compatible with Pareto optimality. There may be many choices of c satisfying condition b of Lemma 11.8.3.

Let $Z_1 = U_D(c)$ and $Z_2 = U_J(c)$.

Lemma 11.8.4. *Under the conditions of Lemma 11.8.3, there exists a choice of c such that* $0 < Z_1 < Z_2 < 1$.

The interpretation of Lemma 11.8.4 is that this new consequence lies strictly between r_* and r^* in utility for both parties.

Proof. To this end, choose $0 < \alpha < 1/2$ (there are further constraints on α imposed later) and let $c^N = \alpha r_* + \alpha r^* + (1 - 2\alpha)c$.

Then

$$\begin{aligned} U_D(c^N) &= \alpha + (1 - 2\alpha)Z_1 \\ U_J(c^N) &= \alpha + (1 - 2\alpha)Z_2. \end{aligned} \tag{11.70}$$

Since $Z_1 < Z_2, U_D(c^N) < U_J(c^N)$, so condition b) is satisfied. What remains to be shown is that α can be chosen so that

$$0 < U_D(c^N) < 1$$

and

$$0 < U_J(c^N) < 1.$$

To that end,

$$\begin{aligned} &\alpha + (1 - 2\alpha)Z_i < 1 \\ \text{iff } &(1 - 2\alpha)Z_i < 1 - \alpha, \text{ or} \\ &Z_i < \frac{1 - \alpha}{1 - 2\alpha} \quad i = 1, 2. \end{aligned}$$

Similarly $0 < \alpha + (1 - 2\alpha)Z_i$ iff

$$(1 - 2\alpha)Z_i > -\alpha$$

or

$$Z_i > -\alpha/(1 - 2\alpha) \quad , i = 1, 2.$$

These are both satisfied if

$$\frac{-\alpha}{1 - 2\alpha} < Z_i < \frac{1 - \alpha}{1 - 2\alpha} \text{ for } i = 1, 2. \tag{11.71}$$

Now if $\alpha \to 1/2$ from below, $\frac{-\alpha}{1-2\alpha} \to -\infty$ and $\frac{1-\alpha}{1-2\alpha} \to \infty$. Thus for fixed Z_1 and Z_2, there are values of α, less than but sufficiently close to 1/2, so that (11.71) is satisfied.

Indeed an inspection of equation (11.70) shows that choosing α close to 1/2 arbitrarily diminishes the influence of the term $(1-2\alpha)Z_i$ on the sum, which is why this works. For this argument to work, it is necessary that utility be finite, so that $Z_i \neq \infty$ or $-\infty$ in (11.71).

Recalling (11.70), we now have, without loss of generality,

$$0 < Z_1 < Z_2 < 1. \tag{11.72}$$

This completes the proof of Lemma 11.8.4. □

We now suppose that there may be a probability p and a utility U satisfying the Pareto principle. It will turn out that the only such p and U are autocratic, that is, identical to those of one of the parties. The technique is to use the Pareto condition repeatedly. To do so, I can choose decisions to compare. When I choose decisions such that both Dick and Jane prefer one to the other, then so must the consensus. This gives me control over what the consensus utility and probability can be. The choice of which decisions to compare is not always obvious.

The first step is to normalize U. Both parties prefer r^* to r_*. Therefore the consensus utility U must also prefer r^* to r_*. Consequently we may normalize U so that $U(r^*) = 1$ and $U(r_*) = 0$. With U so normalized, we may state the next lemma.

Lemma 11.8.5. *If r_*, r^*, c and F satisfy the conditions of Lemma 11.8.4, then there is one of the parties (either Dick or Jane), whose utilities and probabilities will be subscripted with a $*$, such that $p_*(F) = p(F)$ and $U_*(c) = U(c)$.*

[The party denoted $*$ turns out to be the autocrat.]

Proof. I first show that the consensus utility $U(c)$, (if it exists) must satisfy $Z_1 \leq U(c) \leq Z_2$.

To show $Z_1 \leq U(c)$, consider the decision $d_1(\epsilon)$ that yields r^* with probability $Z_1 - \epsilon$ and r_* otherwise, with $0 < \epsilon < Z_1$. Also let decision d_2 yield c with probability 1.

For both parties

$$\begin{aligned} U_D(d_1(\epsilon)) &= Z_1 - \epsilon = U_J(d_1(\epsilon)). \\ U_D(d_2) &= Z_1 \text{ and } U_J(d_2) = Z_2. \end{aligned}$$

Thus both parties prefer d_2 to $d_1(\epsilon)$, for all $\epsilon > 0$.

Therefore so must the consensus utility. Hence we must have

$$Z_1 - \epsilon < U(c) \text{ for all } \epsilon, \quad 0 < \epsilon < Z_1.$$

Therefore $Z_1 \leq U(c)$.

Similarly consider the decision $d_3(\epsilon)$ that yields r^* with probability $Z_2+\epsilon$, and r_* otherwise, where $0 < \epsilon < 1-Z_2$. To both parties the expected utility of $d_3(\epsilon)$ is $Z_2+\epsilon$, larger than that of d_2. Therefore the consensus utility must prefer $d_3(\epsilon)$ to d_2 for all $\epsilon, 0 < \epsilon < 1 - Z_2$.

Thus we must have

$$U(c) < Z_2 + \epsilon \text{ for all } \epsilon, 0 < \epsilon < 1 - Z_2,$$

i.e., $U(c) \leq Z_2$.

Hence we have

$$Z_1 \leq U(c) \leq Z_2. \tag{11.73}$$

I now show that the Pareto condition implies that $p_D(F) \leq p(F)$, where $p(F)$ is the compromise probability. If $p_D(F) = 0$ there is nothing to prove. Then suppose that $p_D(F) > 0$. Let ϵ be chosen so that $0 < \epsilon < p_D(F)$. Consider the decision d_4 that yields r^* if F occurs and r_* if F does not occur. Consider also the family of decisions $d_5(\epsilon)$ that yields r^* with probability $p_D(F) - \epsilon$ and r_* otherwise. The specified d_4 has expected utility $p_D(F)$ to Dick and $p_J(F)$ to Jane. So the expected utility to each is $p_D(F)$ or higher. The expected utility of $d_5(\epsilon)$ is $p_D(F) - \epsilon$ to both. Therefore for each ϵ, they prefer d_4 to $d_5(\epsilon)$. Therefore, by the

Pareto condition, so must the consensus. The consensus expected utility of d_4 is $p(F)$, and of $d_5(\epsilon)$ is again $p_D(F) - \epsilon$. Therefore, we must have, for each ϵ satisfying $p_D(F) > \epsilon > 0$,

$$p_D(F) - \epsilon < p(F).$$

Therefore we have

$$p_D(F) \leq p(F). \tag{11.74}$$

To show that $p(F) \leq p_J(F)$, if $p_J(F) = 1$ there is nothing to prove. So suppose $p_J(F) < 1$ and choose $\epsilon > 0$ so that $p_J(F) < 1 - \epsilon < 1$. Now consider decisions $d_6(\epsilon)$ yielding r^* with probability $p_J(F) + \epsilon$, and r_* otherwise. Then $d_6(\epsilon)$ has expected utility $p_J(F) + \epsilon$. Since d_4 has expected utility no higher than $p_J(F)$ to both Dick and Jane, they both prefer $d_6(\epsilon)$ to d_4 for all ϵ satisfying $0 < \epsilon < 1 - p_J(F)$. Therefore so must the consensus, using the Pareto condition. Hence we must have

$$p_J(F) + \epsilon > p(F) \text{ for all } \epsilon, \quad 0 < \epsilon < 1 - p_J(F).$$

Consequently

$$p_J(F) \geq p(F). \tag{11.75}$$

We may summarize (11.74) and (11.75) by stating that the consensus probability $p(F)$ must satisfy

$$p_D(F) \leq p(F) \leq p_J(F). \tag{11.76}$$

Equations (11.73) and (11.76) show that the consensus $p(F)$ and $U(c)$, if they exist, are constrained to lie in a rectangle whose lower left corner is $(p_D(F), U_D(c))$ and whose upper right corner is $(p_J(F), U_J(c))$. The last part of this argument shows that only those two corner points, which correspond to autocratic solutions, are possible. This is shown using decisions that are somewhat more complicated than those we studied above.

To simplify the notation, let $x_1 = p_D(F)$, $x_2 = p_J(F)$, and $x_0 = \frac{x_1+x_2}{2}$. Similarly, recalling $Z_1 = U_D(c)$ and $Z_2 = U_J(c)$, let $Z_0 = \frac{Z_1+Z_2}{2}$.

Now let $d_7(\epsilon)$ be a decision with the following consequences:

If F occurs, c has probability $1 - x_0$. Otherwise r_* has probability x_0.

If $\overline{F}$ occurs, r^* has probability $\frac{x_1Z_2+x_2Z_1}{2} + \epsilon$ and r_* has probability $1 - \frac{x_1Z_2+x_2Z_1}{2} - \epsilon$.

Also let $d_8(\epsilon)$ be a decision with these consequences:

If F occurs, r^* happens with probability $\frac{Z_2(1-x_1)+Z_1(1-x_2)}{2} - \epsilon$, and otherwise r_* happens.

If $\overline{F}$ occurs, c happens with probability x_0, and r_* happens otherwise.

Obviously $\epsilon > 0$ can be chosen small enough that all the probabilities above involving ϵ are positive and less than 1.

Now I compute the expected utility of the difference between $d_7(\epsilon)$ and $d_8(\epsilon)$ for each of the parties.

For $i = D, J$,

$$E_i[U_i(d_7(\epsilon)) - U_i(d_8(\epsilon))] = p_i(F)\left\{ -\frac{Z_2(1-x_1) - Z_1(1-x_2)}{2} - \epsilon \right\} \tag{11.77}$$

$$+p_i(F)U_i(c)\{1 - x_0\} + (1 - p_i(F))\left\{\frac{x_1Z_2 + x_2Z_1}{2}\right.$$

$$\left. + \epsilon\right\} + (1 - p_i(F))U_i(c)\{-x_0\} \tag{11.78}$$

$$= \epsilon + p_i(F)U_1(c) - p_i(F)Z_0 - U_i(c)x_0 + \frac{x_1Z_2 + x_2Z_1}{2} \tag{11.79}$$

$$= (p_i(F) - x_0)(U_i(c) - Z_0) + \frac{x_1Z_2 + x_2Z_1}{2} - x_0Z_0 + \epsilon. \tag{11.80}$$

We now re-express the constant

$$\begin{aligned}
&\frac{x_1Z_2 + x_2Z_1}{2} - x_0Z_0 \\
=&\frac{x_1Z_2 + x_2Z_1}{2} - \left(\frac{x_1 + x_2}{2}\right)\left(\frac{Z_1 + Z_2}{2}\right) \\
=&\frac{1}{4}\left\{2x_1Z_2 + 2x_2Z_1 - x_1Z_1 - x_1Z_2 - x_2Z_1 - x_2Z_2\right\} \\
=&\frac{1}{4}\left\{x_1Z_2 + x_2Z_1 - x_1Z_1 - x_2Z_2\right\} \\
=& -\frac{1}{4}(x_2 - x_1)(Z_2 - Z_1).
\end{aligned} \tag{11.81}$$

Hence we have, substituting (11.81) into (11.77),

$$\begin{gathered}
E_i[U_i(d_7(\epsilon)) - U_i(d_8(\epsilon)] = \\
(p_i(F) - x_0)(U_i(F) - Z_0) - \frac{1}{4}(x_2 - x_1)(Z_2 - Z_1) + \epsilon.
\end{gathered} \tag{11.82}$$

From Dick's point of view, this comes to

$$\begin{gathered}
E_D[U_D(d_7(\epsilon) - U_D(d_8(\epsilon))] = \\
(x_1 - x_0)(Z_1 - Z_0) - \frac{1}{4}(x_2 - x_1)(Z_2 - Z_1) + \epsilon.
\end{gathered}$$

Now

$$x_1 - x_0 = x_1 - \left(\frac{x_1 + x_2}{2}\right) = \frac{x_1 - x_2}{2}.$$

Similarly

$$Z_1 - Z_0 = \frac{Z_1 - Z_2}{2}.$$

Therefore

$$\begin{aligned}
&E_D[U_D(d_7(\epsilon)) - U_D(d_8(\epsilon))] \\
=&\frac{(x_1 - x_2)}{2}\frac{(Z_1 - Z_2)}{2} - \frac{1}{4}(x_2 - x_1)(Z_2 - Z_1) + \epsilon = \epsilon.
\end{aligned} \tag{11.83}$$

Therefore, for all sufficiently small $\epsilon > 0$, Dick prefers $d_7(\epsilon)$ to $d_8(\epsilon)$.

Now we examine the same utility difference from Jane's perspective, as follows:

$$\begin{gathered}
E_J[U_J(d_7(\epsilon)) - U_J(d_8(\epsilon))] = \\
(x_2 - x_0)(Z_2 - Z_0) - \frac{1}{4}(x_2 - x_1)(Z_2 - Z_1) + \epsilon.
\end{gathered} \tag{11.84}$$

Now

$$x_2 - x_0 = \frac{x_2 - x_1}{2}$$

and

$$Z_2 - Z_0 = \frac{Z_2 - Z_1}{2}.$$

Therefore

$$\begin{gathered}
E_J[U_J(d_7(\epsilon)) - U_J(d_8(\epsilon))] = \\
\left(\frac{x_2 - x_1}{2}\right)\left(\frac{Z_2 - Z_1}{2}\right) - \frac{1}{4}(x_2 - x_1)(Z_2 - Z_1) + \epsilon = \epsilon.
\end{gathered} \tag{11.85}$$

Therefore for each sufficiently small $\epsilon > 0$, Jane also prefers $d_7(\epsilon)$ to $d_8(\epsilon)$. By the weak Pareto principle, we then require that, for all sufficiently small $\epsilon > 0$, the compromise $U(c)$ and $p(F)$ also prefer $d_7(\epsilon)$ to $d_8(\epsilon)$. Thus $U(c)$ and $p(F)$ must satisfy

$$(p(F) - x_0)(U(c) - Z_0) - \frac{1}{4}(x_2 - x_1)(Z_2 - Z_1) + \epsilon > 0 \tag{11.86}$$

for all small $\epsilon > 0$, so

$$(p(F) - x_0)(U(c) - Z_0) \geq \frac{1}{4}(x_2 - x_1)(Z_2 - Z_1). \tag{11.87}$$

To appreciate (11.87), it is useful to rewrite it as follows:

$$\left[\frac{2(p(F) - x_0)}{x_2 - x_1}\right]\left[\frac{2(U(c) - Z_0)}{Z_2 - Z_1}\right] \geq 1. \tag{11.88}$$

Let $r = \frac{2(p(F)-x_0)}{x_2-x_1}$ and $s = \frac{2(U(c)-Z_0)}{Z_2-Z_1}$.
Then (11.88) can be rewritten as

$$rs \geq 1. \tag{11.89}$$

In this notation, the constraint (11.73) can be written as

$$-1 \leq s \leq 1. \tag{11.90}$$

Similarly the constraint (11.76) can be written as

$$-1 \leq r \leq 1. \tag{11.91}$$

It is obvious that there are only two solutions to equations (11.89), (11.90) and (11.91): $(r, s) = (-1, -1)$, corresponding to $p(F) = x_1 = p_D(F)$, and $U(c) = Z_1 = U_D(c)$, and $(r, s) = (1, 1)$, corresponding to $p(F) = x_2 = p_J(F)$ and $U(c) = Z_2 = U_J(c)$.

This completes the proof of Lemma 11.8.5. □

It remains to show that the party identified in Lemma 11.8.5 is an autocrat, that is, that the consensus probability p and utility U are identical with those of $*$, whichever party that may identify.

First I consider probabilities. Let G be an arbitrary event. If $p_D(G) \neq p_J(G)$, then Lemma 11.8.5 applies to the pair (G, c), [relying on Lemma 11.8.4 for the existence of such a c]. We denote the autocrat found in this application of Lemma 11.8.5 with a double star $**$.

Then we have

$$p_{**}(G) = p(G) \text{ and } U_{**}(c) = U(c).$$

But $U(c) = U_*(c)$ and $Z_1 \neq Z_2$. Therefore $**$ and $*$ are the same party, and

$$p_*(G) = p(G). \tag{11.92}$$

Now suppose $p_D(G) = p_J(G) \neq p(G)$. In this case, $p_*(G) = p_D(G) = p_J(G)$. In particular, suppose that $p_D(G) = p_J(G) < p(G)$. Then there is a real number x such that

$$p_D(G) = p_J(G) < x < p(G).$$

Let $d_9 = \begin{cases} r^* & \text{with probability } x \\ r_* & \text{with probability } 1 - x \end{cases}$

and let $d_{10} = \begin{cases} r^* & \text{if } G \text{ occurs} \\ r_* & \text{if } G^c \text{ occurs} \end{cases}$.

Then, for both parties

$$U_i(d_9) = x > U_i(d_{10}) = p_i(G) \quad i = D, J$$

so by the Pareto principle, the consensus utility must prefer d_9 to d_{10}. But

$$U(d_9) = x < U(d_{10}) = p(G),$$

a contradiction.

Similarly, if $p_D(G) = p_J(G) > p(G)$, there is a real number y such that $p_D(G) = p_J(G) > y > p(G)$. Then comparing

$$d_{11} = \begin{cases} r^* & \text{with probability } y \\ r_* & \text{with probability } 1 - y \end{cases}$$

to d_{10}, we find that both parties prefer d_{10} to d_{11}, but the consensus prefers d_{11} to d_{10}, again a contradiction. Thus we must have

$$p_*(G) = p(G),$$

when $p_d(G) = p_J(G)$. Combining this result with that in equation (11.92), we have $p_*(G) = p(G)$ for all events G.

Finally, it remains to show that $U_*(g) = U(g)$ for all consequences g. First suppose that $U_J(g) \neq U_D(g)$. Using the event E whose existence is proved in Lemma 11.8.3, there is a consequnce g' whose existence is proved in Lemma 11.8.4 satisfying the conditions of Lemma 11.8.4. Then Lemma 11.8.5 applies to the pair (E, g') and there is an autocrat (again denoted **) such that

$$p_{**}(E) = p(E) \text{ and } U_{**}(g') = U(g).$$

Now using the fact that $p_J(E) \neq p_D(E)$, we again find that the autocrat $**$ is the same party as the autocrat $*$, so $U_*(g') = U(g')$. The construction in Lemma 11.8.4 shows that g and g' are related, for some $0 < \alpha < 1/2$, by

$$g = \alpha r^* + \alpha r_* + (1 - 2\alpha)g'.$$

Thus $U(g) = \alpha + (1 - 2\alpha)U(g') = \alpha + (1 - 2\alpha)U_*(g') = U_*(g)$.

Now suppose $U_J(g) = U_D(g) \neq U(g)$.

Again we apply Lemma 11.8.4 to the two utilities, and find that we may assume, without loss of generality that $0 < U_J(g) = U_D(g) < U_*(g) < 1$. Then there is some real number z such that $0 < U_J(g) - U_D(g) < z < U(g) < 1$.

Now let

$$d_{12} \text{ have consequence } g \text{ with probability 1, and}$$

$$d_{13} = \begin{cases} r^* & \text{with probability } z \\ r_* & \text{with probability } 1 - z \end{cases}.$$

Then for both parties

$$U_i(d_{12}) = U_i(g) < z = U_i(d_{13}), \; i = D, J$$

so by the Pareto principle, the consensus utility must prefer d_{13} to d_{12}. But

$$U(d_{12}) = U(g) > z = U(d_{13}),$$

a contradiction. Therefore we must have $U_J(g) = U_D(g) = U(g)$ and hence $U_*(g) = U(g)$ for all g.

Hence the party $*$, whichever it may be, has

$$p_*(g) = p(g) \text{ for events } G$$

and $U_*(g) = U(g)$ for all consequences g. Therefore, the party $*$ is an autocrat.

This proves the following:

Lemma 11.8.6. *In Case 6, the only Pareto-respecting compromises are autocratic.*

Lemma 11.8.7. *In Case 4, the Pareto-respecting compromises are autocratic.*

Proof. In Case 4, Dick and Jane's utilities satisfy $U_J(\cdot) = rU_D(\cdot) + s$ with $r < 0$. Thus if ℓ and m are prizes such that $U_D(\ell) > U_D(m)$, then we have $U_J(\ell) < U_J(m)$. Without loss of generality, we may normalize so that $U_D(\ell) = U_J(m) = 1$ and $U_D(m) = U_J(\ell) = 0$. Thus $s = 0$, $r = -1$ and $U_D + U_J = 1$. Also, since $P_D(\cdot) \neq P_J(\cdot)$, there is an event C such that $P_D(C) \neq P_J(C)$.

The consensus utility U must satisfy either $U(\ell) \geq U(m)$ or $U(\ell) < U(m)$. By reversing the identities of Dick and Jane if needed we may suppose that $U_D(\ell) > U_D(m)$ and $U(\ell) \geq U(m)$.

In addition, we may assume that there is an event F such that $P_D(F) < P_J(F)$: if $P_D(C) < P_J(C)$, let $F = C$. Otherwise let $F = \overline{C}$.

Now let $G = \begin{cases} \ell & \text{if } F \text{ occurs} \\ m & \text{if } \overline{F} \text{ occurs} \end{cases}$.

Then $E_DU_D(G) = P_D(F)$ and $E_JU_J(G) = 1 - P_J(F)$.

Let ϵ be chosen so that $P_D(F) - P_J(F) > \epsilon > 0$, and let

$$G^*(\epsilon) = \begin{cases} \ell & \text{with probability } P_D(F) - \epsilon \\ m & \text{with probability } 1 - P_D(F) + \epsilon \end{cases}.$$

Then $E_DU_D(G^*) = P_D(F) - \epsilon$ and $E_JU_J(G^*) = 1 - P_D(F) + \epsilon$.

Thus both Dick and Jane prefer G to G^*.

Therefore the consensus utility and probability (U, P) must also strictly prefer G to G^*.

This implies first that $U(\ell) \neq U(m)$, since if $U(\ell) = U(m)$, the consensus would be indifferent between G and G^*. Hence we must have $U(\ell) > U(m)$. Now we may normalize U so that $U(\ell) = 1$ and $U(m) = 0$.

With respect to the consensus (U, P),

$$E_PU(G) = P(F) \text{ and } E_P(U(G^*)) = P_D(F) - \epsilon.$$

Therefore $P(F) > P_D(F) - \epsilon$ for all ϵ in the range specified, and hence,

$$P(F) \geq P_D(F). \tag{11.93}$$

Now let

$$H = \begin{cases} \ell & \text{if } \overline{F} \text{ occurs} \\ b & \text{if } F \text{ occurs} \end{cases}.$$

$E_DU_D(H) = 1 - P_D(F)$; $E_JU_J(H) = P_J(F)$.

Also let

$$H^*(\epsilon) = \begin{cases} \ell & \text{with probability } 1 - P_D(F) + \epsilon \\ m & \text{with probability } P_D(F) - \epsilon \end{cases}.$$

$E_DU_D(H^*) = 1 - P_D(F) + \epsilon$; $E_JU_J(H^*) = P_D(F) - \epsilon$.

Both parties prefer H^* to H for all ϵ in the designated range. Evaluated at the consensus, (U, P),

$$E_P U(H^*) = 1 - P_D(F) + \epsilon \text{ and } E_P U(H) = 1 - P(F).$$

Then

$$1 - P_D(F) + \epsilon > 1 - P(F)$$

so

$$1 - P_D(F) \geq 1 - P(F).$$

Consequently

$$P(F) \leq P_D(F). \tag{11.94}$$

Combining (11.93) and (11.94), we have $P(F) = P_D(F)$.

Now we deal with the case in which F satisfies $P_J(F) = P_D(F)$. By assumption of the lemma, there is at least one event F^* such that $P_J(F^*) \neq P_D(F^*)$, and by the analysis above, $P_D(F*) = P(F^*)$. Also $\overline{F}^*$ satisfies the same equations.

Now $F^* = (F^* \cap F) \cup (F^* \cap \overline{F})$ and this is a disjoint union.

Then

$$\begin{aligned} P_D(F^*) &= P_D(F^* \cap F) + P_D(F^* \cap \overline{F}) \\ P_J(F^*) &= P_J(F^* \cap F) + P_D(F^* \cap \overline{F}). \end{aligned}$$

Now $P_D(F^*) \neq P_J(F^*)$ implies one of the following:

a) $P_D(F^* \cap F) \neq P_J(F^* \cap F)$

b) $P_D(F^* \cap \overline{F}) \neq P_J(F^* \cap \overline{F})$

c) both a) and b).

If a) holds, then $P_D(F^* \cap F) = P(F^* \cap F)$. which implies $P_D(F^* \cap \overline{F}) = P(F^* \cap \overline{F})$. Similarly, if b) holds, then $P_D(F^* \cap \overline{F}) \neq P_J(F^* \cap \overline{F})$, so $P_D(F^* \cap F) = P(F^* \cap F)$. So in all cases $P_D(F^* \cap \overline{F}) = P(F^* \cap \overline{F})$ and $P_D(F^* \cap F) = P(F^* \cap F)$.

Now

$$P(F) = P(F \cap F^*)_P(F \cap \overline{F}^*) = P_D(F \cap F^*) + P_D(F \cap \overline{F}^*) = P_D(F).$$

Hence in all cases $P \equiv P_D$.

It remains to show that $U(c) = U_D(c)$ for all consequences c. Choose c different from ℓ and m. Redefining them as needed, we may assume without loss of generality, that Dick prefers ℓ to c to m, so $U_D(\ell) \geq U_D(c) \geq U_D(m)$. Then there is some probability p such that Dick is indifferent between the gamble

$$X = \begin{cases} \ell & \text{with probability } p \\ m & \text{with probability } 1 - p \end{cases}$$

and the gamble $Y_{\{}c$ with probability $1\}$. By this construction, $p = U_D(c)$.

To say that Dick is indifferent between X and Y is equivalent to saying that Dick strictly prefers Y to all gambles of the form

$$X(\epsilon) = \begin{cases} \ell & \text{with probability } p - \epsilon \\ m & \text{with probability } 1 - p + \epsilon \end{cases}$$

and strictly prefers $X^*(\epsilon)$ to Y, where

$$X^*(\epsilon) = \begin{cases} \ell & \text{with probability } p + \epsilon \\ m & \text{with probability } 1 - p - \epsilon \end{cases}.$$

Since Jane's preferences are opposite, she strictly prefers $X(\epsilon)$ to Y for all $\epsilon > 0$ and Y to $X*(\epsilon)$ for all $\epsilon > 0$. Then Jane is also indifferent between X and Y.

Let $R = \frac{1}{2}G + \frac{1}{2}X$ and $R^* = \frac{1}{2}G^* + \frac{1}{2}Y$.

Since both parties are indifferent between X and Y and both prefer G to G^*, both prefer R to R^*.

$$E_P U(R) = \frac{1}{2}E_P U(G) + \frac{1}{2}EU(x) = \frac{1}{2}P(F) + \frac{1}{2}P$$

$$E_P U(R^*) = \frac{1}{2}E_p U(G^*) + \frac{1}{2}EU(Y) = \frac{1}{2}[P_D(F) - \epsilon] + \frac{1}{2}U(c).$$

Then

$$\frac{1}{2}P(F) + \frac{1}{2}P > \frac{1}{2}[P_D(F) - \epsilon] + \frac{1}{2}U(c),$$

so $\frac{1}{2}P(F) + \frac{1}{2}P \geq \frac{1}{2}P_D(F) + \frac{1}{2}U(c)$.

Recalling $P(F) = P_D(F)$. we have

$$p \geq U(c). \tag{11.95}$$

Similarly let

$$T = \frac{1}{2}H + \frac{1}{2}X \text{ and } T^* = \frac{1}{2}H^* + \frac{1}{2}Y.$$

Both parties prefer H^* to H, so both prefer T^* to T.

$$E_p(T) = \frac{1}{2}(1 - P(F)) + \frac{1}{2}p$$

$$E_p(T^*) = \frac{1}{2}(1 - P_D(F) + \epsilon) + \frac{1}{2}U(c).$$

Therefore $\frac{1}{2}(1 - P_D(F) + \epsilon) + \frac{1}{2}U(c) > \frac{1}{2}(1 - P(F) + \frac{1}{2}p$. So $\frac{1}{2}(1 - P_D(F) + \frac{1}{2}U(c) \geq \frac{1}{2}(1 - P(F)) + \frac{1}{2}P$. Hence

$$U(c) \geq p. \tag{11.96}$$

Combining (11.95) and (11.96), we have

$$U(c) = p = U_D(c)$$

so Dick is an autocrat. □

Lemma 11.8.8. *In Case 1, the only Pareto-respecting compromise is autocratic.*

Proof. In this case there is an obvious compromise: choose $P(\cdot) = P_D(\cdot) - P_J(\cdot)$ and choosing any $a > 0$ and b, $U = aU_D(\cdot) + b$. Suppose both Dick and Jane prefer d_1 to d_2. Then the expected utility of d_1 is greater than that of d_2 for both, and hence also under the compromise (U, P). It may be peculiar to think of this compromise as autocratic, but it is, because it coincides with one (here both) of the party's utilities and probabilities.

The heart of this lemma, then, is to prove that only the choice above respects the Pareto condition. Therefore, we suppose that (U, P) is any other choice of utility and probability, and show that it violates the Pareto condition. For clear notation, let $P^*(\cdot) \equiv P_J(\cdot) \equiv P_D(\cdot)$ and $U^*(\cdot) \equiv aU_D(\cdot) + b$ for some $a > 0$ and b.

Since U^* is the utility of both parties, there must be prizes $r*$ and r_* such that $U^*(r^*) > U^*(r_*)$. If d_1 yields r^* with probability 1, and d_2 yields r_* with probability 1, both parties prefer d_1 to d_2. Therefore, by Pareto, so does (U, P). Therefore, we must have $U(r^*) > U(r_*)$ Now we can normalize U and U^* so that

$$U^*(r^*) = U(r^*) = 1$$

and

$$U^*(r_*) = U(r_*) = 0.$$

Suppose there is a consequence c such that $U(c) < U^*(c)$. Then there is some x such that $U(c) < x < U^*(c)$.

Let

$$d_3 = \begin{cases} r^* & \text{with probability } x \\ r_* & \text{with probability } 1 - x \end{cases}$$

and let $d_4 = \{c \text{ with probability } 1\}$.

Then

$$\begin{aligned} E_{P^*}U^*(d_3) &= E_P U(d_3) = x. \\ E_{P^*}U^*(d_4) &= U^*(c) \text{ and } E_P U_P(d_4) = U(c). \end{aligned}$$

Hence both parties strictly prefer d_4 to d_3. However, under (U, P), the purported compromise strictly prefers d_3 to d_4, which violates the Pareto condition.

Now suppose there is a consequence c such that $U(c) > U^*(c)$. The same argument as above applies, reversing (U, P) and (U^*, P^*). Therefore, we must have $U(c) = U^*(c)$ for all c.

Finally, suppose that there is an event F such that $P(F) < P^*(F)$. Then there is a y such that $P(F) < y < P^*(F)$.

Let

$$d_5 = \begin{cases} r^* & \text{with probability } y \\ r_* & \text{with probability } 1 - y \end{cases}$$

$$d_6 = \begin{cases} r^* & \text{if } F \text{ occurs} \\ r_* & \text{if } \overline{F} \text{ occurs.} \end{cases}.$$

Then

$$\begin{aligned} E_{P^*}U^*(d_5) &= E_P(U(d_5) = y \\ E_{P^*}(U^*(d_6)) &= P^*(F) \text{ and } E_P(U_P(d_6)) = P(F). \end{aligned}$$

Hence both parties prefer d_6 to d_5, but the purported compromise prefers d_5 to d_6, violating the Pareto condition.

If $P^*(F) < P(F)$, the same argument applies, again reversing (U, P) and (U^*, P^*). Hence we have $P^*(\cdot) = P(\cdot)$.

Therefore the only Pareto-respecting compromise is autocratic. □

11.8.1 Summary

We may summarize the results of this section with the following theorem:

There exist non-autocratic, weak Pareto compromises for two Bayesians if and only if they either agree in probability, but not in utility, or agree in utility, but not in probability.

11.8.2 Notes and references

Case 6 is discussed in Seidenfeld et al. (1989). Goodman (1988) gives an extensive discussion of the relationship of hyperbolas to differences in the utilities of different decisions. He also explores generalization of the result here to more than two decision makers. Many cases ensue, in some of which there are non-trivial, non-autocratic, weak Pareto compromises.

Case 4 emerged from consideration of an error pointed out to us by Dennis Lindley in a previous "proof" of this theorem.

Where there is a weak Pareto condition, there must also be a strong one. The **strong Pareto condition** says that if A_1 is preferred to or is indifferent to A_2 for all agents, and at least one agent prefers A_1 to A_2 (strictly), then the compromise prefers A_1 to A_2. Seidenfeld et al. (1989) show that the strong Pareto condition eliminates the autocratic solutions, leaving none in the interesting Case 3.

Earlier literature on this problem include important papers by Hylland and Zeckhauser (1979) and by Hammond (1981). Those papers restrict the group amalgamation to separately amalgamating probabilities and utilities, which the work described in 11.8 does not. Additionally, the results described here apply to all agents whose probabilities and utilities differ, while Hylland and Zeckhauer's and Hammond's are restricted to showing that there is some configuration of probabilities and utilities that causes difficulty for amalgamation.

There is an extensive literature on the amalgamation of probabilities, and a somewhat less extensive literature on the amalgamation of utilities. (See Genest and Zidek (1986), French (1985) and the discussion of Kadane (1993).) The results of this section make pressing the question of what meaning these amalgamations may have.

This result is different from Arrow's famous impossibility theorem (Arrow (1951)) in that he finds, under general conditions, that there is no non-dictatorial social utility function. His result requires only an ordering of alternatives from each participant, and aims to return a social ordering. It is a generalization of the observation that three voters, with preferences among alternatives A, B and C satisfying:

$$\begin{array}{ll} \text{Voter 1:} & A > B > C \\ \text{Voter 2:} & B > C > A \\ \text{Voter 3:} & C > A > B \end{array}$$

will have intransitive pairwise majority votes:

$$A > B > C > A.$$

By contrast the result discussed in this section requires more of the participants, that their preferences be coherent, and hopes to deliver more, that their consensus be coherent, under the weak Pareto condition.

One interpretation of the result given here is that it emphasizes how profoundly personal the theory of maximization of expected utility is. Perhaps a satisfactory decision theory should be required to address both individuals and groups, where the group decision relates gracefully to its constituent individuals. If so, the result of this section suggests that Bayesian decision theory that has separate probabilities and utilities fails to meet this criterion.

Alternatively, we may work with the perspective of Rubin (1987) mentioned in section 7.3. Then we may work with the functions

$$h_i(\theta, d) = U_i(\theta, d)p_i(\theta) \quad \text{for } i = J, D.$$

If both Dick and Jane strictly prefer decision d_1 to d_2, then

$$\int h_i(\theta, d_1)d\theta > \int h_i(\theta, d_2)d\theta \quad \text{for } i = J, D.$$

Consider the function

$$h(\theta, d) = \alpha h_D(\theta, d) + (1 - \alpha)h_J(\theta, d) \quad \text{for some } \alpha, 0 < \alpha < 1.$$

Then

$$\begin{aligned} \int h(\theta, d_1)d\theta &= \int [\alpha h_D(\theta, d_1) + (1 - \alpha)h_J(\theta, d_1)]d\theta \\ &= \alpha \int h_D(\theta, d_1)d\theta + (1 - \alpha) \int h_J(\theta, d_1)d\theta \\ &> \alpha \int h_D(\theta, d_2)d\theta + (1 - \alpha) \int h_J(\theta, d_2)d\theta \\ &= \int h(\theta, d_2)d\theta. \end{aligned}$$

Hence the function $h(\theta, d)$ can be regarded as a compromise decision function for Dick and Jane that respects the Pareto condition.

11.8.3 Exercises

1. State in your own words what is meant by
   (a) weak Pareto condition
   (b) strong Pareto condition
   (c) Bayesian group
   (d) autocratic compromise
2. Show that if one Bayesian is indifferent, say $U_1(d,\theta) = b$ for all $d\epsilon\mathcal{D}$ and $\theta\epsilon\Omega$, then the Pareto condition is vacuous regardless of whether $p_1(\theta) = p_2(\theta)$ for all $\theta\epsilon\Omega$, or whether $p_1(\theta) \neq p_2(\theta)$.
3. Show that, in Case 2(b), if $p_1(\theta) \neq p_2(\theta)$ and $U_D(d,\theta) = aU_J(d,\theta) + b$, where $a > 0$, then $p(\theta) = \alpha p_D(\theta) + (1-\alpha)p_J(\theta)(0 \leq \alpha \leq 1)$ and $U(d,\theta) = U_D(d,\theta)$ satisfy the Pareto condition.
4. Verify that (11.73) implies (11.90).
5. Verify that (11.76) implies (11.90).
6. Show that $r = -1, s = -1$ implies $p(E) = p_D(E)$ and $U(c) = U_D(c)$.
7. Show that $r = 1, s = 1$ implies $p(E) = p_J(E)$ and $U(c) = U_J(c)$.

Appendix A: The minImax theorem

Let $U = (u_{ij})$ be an $m \times n$ matrix. Let $p = \{(p_1, \ldots, p_m) \mid p_i \geq 0, \sum_{i=1}^m p_i = 1\}$ be the set of all m-dimensional probability vectors $\mathbf{p}$ and similarly let Q be the set of all n-dimensional probability vectors $\mathbf{q}$.

Theorem 11.1. *There exists a unique λ, and (not necessarily unique) vectors $\mathbf{p}\epsilon p$ and $\mathbf{q}\epsilon Q$ such that*

$$\lambda \geq \sum_{j=1}^{n} u_{ij}q_j \quad \textit{for } i = 1, \ldots, m \tag{11.A.1}$$

and

$$\lambda \leq \sum_{i=1}^{m} p_i u_{ij} \quad \textit{for } j = 1, \ldots n. \tag{11.A.2}$$

Proof. To show this, we introduce a different symbol for λ in (11.A.2):

$$\mu \leq \sum_{i=1}^{m} p_i u_{ij} \quad \text{for } j = 1, \ldots, n. \tag{11.A.3}$$

There are λ's and $\mathbf{q}$'s satisfying (11.A.1), and μ's and $\mathbf{p}$'s satisfying (11.A.3). Also (11.A.2) and (11.A.3) yield

$$\mu \leq \sum_i \sum_j p_i u_{ij} q_j \leq \lambda, \tag{11.A.4}$$

so $\mu \leq \lambda$. The values of λ satisfying (11.A.1) are bounded below. Since Q is compact (closed and bounded), the greatest lower bound λ_0 can be used in (11.A.1) for some vector $\mathbf{q}^0\epsilon Q$. Similarly the least upper bound, μ_0, for μ can be used for some vector $\mathbf{p}^0\epsilon p$. Since (11.A.4) holds for all $p\epsilon p$ and $q\epsilon Q$, we have $\mu_0 \leq \lambda_0$. We wish to show $\mu_0 = \lambda_0$.

The proof is by induction on $m+n$. If $m+n=2$, then $m=n=1$ and the theorem is trivial. If equality occurs in (11.A.1) for all $i = 1, \ldots, m$ when $\lambda = \lambda_0$ and $\mathbf{q} = \mathbf{q}^0$ then

$$\sum_j u_{ij} q_j^0 = \lambda_0 \text{ for all } i = 1, \ldots, m.$$

Let $e^i = (0, 0\ 0, 1\ 0, \ldots, 0)$ where the 1 occurs at the ith coordinate. Then

$$\sum_k \sum_j e^i_k u_{kj} q^0_j = \lambda_0 \quad \text{for } i = 1, \ldots, m.$$

Hence $\mu_0 \geq \lambda_0$. But since $\mu_0 \leq \lambda_0$, we have $\mu_0 = \lambda_0$ and the result is proved.

Now consider the case in which strict inequality holds at least once in (11.A.1). Renumbering if necessary, we have

$$\begin{aligned} \lambda_0 &= \sum_j u_{ij} p^0_j \qquad i = 1, \ldots, m_1 \\ \lambda_0 &> \sum_j u_{ij} p^0_j \qquad i = m_1 + 1, \ldots, m. \end{aligned} \tag{11.A.5}$$

Consider now the reduced matrix $U^* = (u_{ij})$, which is an $m_1 \times n$ matrix, and let λ_1 and μ_1 be the greatest lower bound in (11.A.1) and the least upper bound in (11.A.3) for U^*. Then we claim

$$\lambda_1 \leq \lambda_0 \text{ and } \mu_1 \leq \mu_0. \tag{11.A.6}$$

The first inequality follows from the observation that every λ and $\mathbf{q}$ satisfying (11.A.1) for $i = 1, \ldots, m$ also satisfies (11.A.1) for the reduced set $i = 1, \ldots, m_1$. The second inequality is shown because every μ and $p\epsilon p$ satisfying the reduced (11.A.3) (with m_1 replacing m in the sums) also satisfies the original (11.A.3) if $\mathbf{p}$ is extended to $p = (p_1, \ldots, p_{m_1}, 0, \ldots, 0)$.

Now we assert $\lambda_1 = \lambda_0$. Suppose to the contrary that $\lambda_1 < \lambda_0$, and that λ_1 is associated with $\mathbf{p} = p'$, so

$$\lambda_1 \geq \sum_{j=1}^{n} u_{ij} p'_j \text{ for } i = 1, \ldots, m_1. \tag{11.A.7}$$

Let $\mathbf{p} = \alpha \mathbf{p}^0 + (1 - \alpha)\mathbf{p}'$, where $0 < \alpha < 1$. Then $\mathbf{p}\epsilon p$.

Using both (11.A.5) and (11.A.7), for $i = 1, \ldots, m_1$

$$\begin{aligned} \sum_{j=1}^{n} u_{ij} p_j =& \sum_j u_{ij} (\alpha p^0_j + (1 - \alpha) p'_0) \\ =& \alpha \sum_j u_{ij} p^0_j + (1 - \alpha) \sum_j u_{ij} p'_j \\ \geq& \alpha \lambda_0 + (1 - \alpha) \lambda_1 > \lambda_0. \end{aligned} \tag{11.A.8}$$

It follows from the second set of equations in (11.A.5) and the continuity of the linear function that

$$\sum_{j=1}^{n} u_{ij} x_j > \lambda_0$$

for $i = m_1 + 1, \ldots, m$ if α is small enough. Hence λ_0 is not the greatest lower bound in (11.A.1), a contradiction. Therefore $\lambda_1 = \lambda_0$. Noting that $\lambda_1 = \mu_1$ by the inductive hypothesis, we have

$$\lambda_0 = \lambda_1 = \mu_1 \leq \mu_0 \leq \lambda_0.$$

Hence $\lambda_0 = \mu_0$ and the theorem is proved. □

Corollary 11.8.9. *Without loss of generality, renumber the m choices available to P1 so that $p_i > 0$ for $i = 1, \ldots, m'$ and $p_i = 0$ for $i = m' + 1, \ldots, m$. Here $1 \leq m' \leq m$. Similarly renumber the n choices available to P2 so that $q_j > 0$ for $j = 1, \ldots, n'$ and $q_j = 0$ for $j = n' + 1, \ldots, n$. Here $1 < n' \leq n$.*

(a) Every pure strategy $a_1, \ldots, a_{m'}$ maximizes P1's expected utility, as does every randomized strategy that puts probability 1 on $a_1, \ldots, a_{m'}$.

(b) Every pure strategy $b_1, \ldots, b_{n'}$ maximizes P2's expected utility, as does every randomized strategy that puts probability 1 on $b_1, \ldots, b_{n'}$.

Proof. The minimax theorem shows that for all $i, i = 1, \ldots, m$

$$\sum_{j=1}^{n} u_{ij} q_j \leq \lambda.$$

Therefore

$$\sum_{i=1}^{m} p_i \sum_{j=1}^{n} u_{ij} q_j \leq \lambda \sum_{i=1}^{m} p_i = \lambda. \tag{11.A.9}$$

Similarly the minimax theorem says that for all $j, j = 1, \ldots, n$,

$$\sum_{i=1}^{m} p_i u_{ij} \geq \lambda.$$

Therefore

$$\sum_{j=1}^{n} q_j \sum_{i=1}^{m} u_{ij} p_i \geq \lambda \sum_{j=1}^{n} q_j = \lambda. \tag{11.A.10}$$

Putting (11.A.9) and (11.A.10) together yields

$$\lambda \leq \sum_{i=1}^{m} p_i \sum_{j=1}^{n} u_{ij} q_j = \sum_{j=1}^{n} q_j \sum_{i=1}^{m} u_{ij} p_i \leq \lambda.$$

Therefore

$$\lambda = \sum_{i=1}^{m} p_i \sum_{j=1}^{n} u_{ij} q_j = \sum_{j=1}^{n} q_j \sum_{i=1}^{m} u_{ij} p_i. \tag{11.A.11}$$

I now proceed to prove (a). Let, for $i = 1, \ldots, m$

$$x_i = \sum_{j=1}^{n} u_{ij} q_j.$$

Then x_i is the expected utility of P1's choice a_i. From the minimax theorem, we have, for $i = 1, \ldots, m$

$$x_i \leq \lambda. \tag{11.A.12}$$

From (11.A.11) we have

$$\sum_{i=1}^{m} p_i x_i = \lambda. \tag{11.A.13}$$

Now (11.A.13) can be rewritten as

$$\sum_{i=1}^{m} p_i (x_i - \lambda) = 0. \tag{11.A.14}$$

Since $p_i \geq 0$, in view of (11.A.12), (11.A.14) is a sum of non-positive terms which sums to zero. Therefore we must have

$$p_i (x_i - \lambda) = 0 \quad \text{for } i = 1, \ldots, m. \tag{11.A.15}$$

Since $p_i > 0$ for $i = 1, \ldots, m'$, we have

$$x_i = \lambda \qquad \text{for } i = 1, \ldots, m'. \tag{11.A.16}$$

If $\mathbf{p}' = (p'_1, \ldots, p'_{m'}, 0, \ldots, 0)$ is an arbitrary mixture of the strategies $a_1, \ldots, a_m$ that puts probability 1 on $a_1, \ldots, a'_m$, then (11.A.16) implies

$$\sum_{i=1}^{m'} p'_i x_i = \lambda$$

which completes the proof of (a).

The proof of (b) is similar. Let, for $j = 1, \ldots, n$,

$$y_j = \sum_{i=1}^{m} u_{ij} p_i.$$

Here y_j is the expected loss of P2's choice of b_j. From the minimax theorem, we have, for $j = 1, \ldots, n$,

$$y_j \geq \lambda. \tag{11.A.17}$$

From (11.A.11) we have

$$\sum_{j=1}^{n} q_j y_j = \lambda. \tag{11.A.18}$$

Again we have, rewriting (11.A.11),

$$\sum_{j=1}^{n} q_j (y_j - \lambda) = 0. \tag{11.A.19}$$

In view of (11.A.17), and the fact that $q_j \geq 0$, for $j = 1, \ldots, n$, we know that (11.A.19) is the sum of non-negative quantities that sum to zero. Therefore we must have

$$q_j (y_j - \lambda) = 0 \qquad \text{for } j = 1, \ldots, n. \tag{11.A.20}$$

Because $q_j > 0$ for $j = 1, \ldots, n'$, we have

$$y_j = \lambda \qquad \text{for } j = 1, \ldots, n'. \tag{11.A.21}$$

Again, if $\mathbf{q}' = (q'_1, q'_2, \ldots, q'_{n'}, 0, 0, \ldots, 0)$ is an arbitrary mixture of the strategies $b_1, \ldots, b_n$ that puts probability 1 on $b_1, \ldots, b_{n'}$, then (11.A.21) implies

$$\sum_{j=1}^{n'} q'_j y_j = \lambda, \tag{11.A.22}$$

which proves (b). □

11.A.1 Notes and references

The proof here follows that of Loomis (1946). Other proofs use the Brouwer fixed point theorem, separating hyperplanes, or duality theory in linear programming. The λ, $\mathbf{p}$ and $\mathbf{q}$ can be computed for any matrix B using linear programming.

Chapter 12

Exploration of Old Ideas: A Critique of Classical Statistics

I can see clearly now, the pain is gone
I can see all obstacles in my way
Gone are the dark clouds that had me blind
Gonna be a bright (bright), bright (bright) sun-shiny day!

—Johnny Nash

12.1 Introduction

A volume on statistics would be remiss if it failed to comment on sampling theories, since they occupy so much space in many statistics books and journals. The distinction between the sampling theory and Bayesian viewpoint is stark. It comes down to the issue of what is to be considered fixed and what is to be considered random.

The Bayesian viewpoint is quite simple. All the quantities of interest in a problem are tied together by a joint probability distribution. (Often this joint probability distribution is expressed as a likelihood (*i.e.*, a probability distribution of the data given the parameters) times a prior distribution on the parameters.) This probability distribution reflects the beliefs of the person doing the analysis. Since these beliefs are not necessarily shared by the intended readers, the reasoning behind the beliefs should be explained and defended. Any decisions that are to be made before new data are available are made by maximizing expected utility, where the expectation is taken with respect to the probability distribution specified.

When new information becomes available, in the form of data or otherwise, that new information is conditioned upon, leading to a posterior distribution. And that posterior distribution is used as the distribution with respect to decisions that are made after the data become available. Thus, the probability distributions reflect the uncertainty of the author, both before and after data are observed.

Sampling theory reverses what is random and what is fixed. The parameter is taken to be fixed but unknown (whatever that might mean). The data are taken to be random, and comparisons are made between the distribution of a statistic before the data are observed, and the observed value of the statistic. It further assumes that likelihoods are known (because they are objective or by consensus) while priors are highly suspect (because they are subjective). In the sections below we'll look at examples of reasoning of this kind.

Chapter 9 discusses how to handle missing data in a Bayesian framework. From a sampling theory framework, it is unclear whether missing data are *(i)* fixed parameters that become random when they are observed, *(ii)* "data" that are to be treated as random when they are observed, or *(iii)* a third kind of quantity with its own set of rules.

A simple example can show how difficult it is to adhere to sampling theory. Suppose I observe a random sample of n from a normal distribution with mean μ and variance σ^2. Then the likelihood function is

$$
\begin{aligned}
f(x, \ldots, x_n \mid \mu, \sigma^2) &= \prod_{i=1}^{n} \frac{1}{\sigma\sqrt{2\pi}} \exp\left\{-\frac{1}{2}\left(\frac{x_i - \mu}{\sigma}\right)^2\right\} \\
&= \frac{1}{(2\pi)^{n/2}} \cdot \frac{1}{\sigma^n} \exp\left\{-\frac{1}{2}\sum_{i=1}^{n}\left(\frac{x_i - \mu}{\sigma}\right)^2\right\}. \qquad (12.1)
\end{aligned}
$$

We'll adopt the popular sampling theory estimation using maximum likelihood. It is easily shown that the maximum likelihood estimators are

$$
\hat{\mu} = \bar{X} = \sum_{i=1}^{n} \frac{x_i}{n} \text{ and } \hat{\sigma}^2 = \frac{\sum_{i=1}^{n}(x_i - \bar{x})^2}{n}. \qquad (12.2)
$$

Now suppose that in fact there were originally $2n$ observations, of which the n observed are chosen at random. How should $X_{n+1}, \ldots, X_{2n}$ be treated?

It would seem legitimate to think that from $x_1 \ldots, x_n$ I have learned something about $X_{n+1}, \ldots, X_{2n}$. So perhaps I can treat them as parameters. If I do, the maximum likelihood estimates are now

$$
\begin{aligned}
\hat{\mu} &= \bar{X} = \sum_{i=1}^{n} x_i/n, \\
\hat{X}_{n+1} &= \hat{X}_{n+2} = \ldots = \hat{X}_{2n} = \bar{X} \qquad (12.3) \\
\hat{\sigma}^2 &= \frac{\sum_{i=1}^{2n}(x_i - \bar{x})^2}{2n} = \frac{\sum_{i=1}^{n}(x_i - \bar{x})^2}{2n}.
\end{aligned}
$$

Hence by imagining another n data points I never saw, the estimate of the variance is now half of what it was. And of course if I imagine kn normal random variables, I get

$$
\hat{\sigma}^2 = \frac{\sum_{i=1}^{n}(x_i - \bar{x})^2}{kn} \qquad (12.4)
$$

so by dint of a great imagination ($k \to \infty$), the maximum likelihood estimate of the variance vanishes!

Of course what should be done with $X_{n+1}, \ldots, X_{2n}$ is to integrate them out. But there is no real distinction between unobserved data and a parameter. And to integrate out a parameter means it must have a distribution, which is where Bayesians were to begin with.

For more on this, see Bayarri et al. (1988).

Some hint of the havoc caused by the doctrine that parameters do not have distributions can be seen in the classical treatment of fixed and random effects, as in Scheffé (1999). Take for example an examination given to each child in a class, and several classes in a school, as discussed in Chapter 9. If you are interested in each individual child's performance, classical doctrine says to use a fixed effect model. However, if you are interested in how the classes compare, you should use a random effects model. This is a puzzle on several grounds:

1) According to the classical theory, the model represents how the data were in fact generated. But the above account has the model dependent on the interest of the investigator, the children or the classes, which is a utility matter.
2) To have a random effects model means to use a prior on the parameters for each child, and to integrate those parameters out of the likelihood. So here a classical statistician apparently feels OK about using such a prior. If that's OK for random effects, why not elsewhere?

3) An investigator might be interested in both each child and the classes. What model would classical statistics recommend then?

Because of this fundamental difference about what is fixed and what is random, attempts to find compromises or middle grounds between Bayesian and sampling theory statistics have failed. For example, Fisher (1935) proposed something he called fiducial inference. An instance of it looks like this:

$$X \sim N(\theta, 1) \tag{12.5}$$

$$X - \theta \sim N(0, 1) \tag{12.6}$$

$$\theta - X \sim N(0, 1) \tag{12.7}$$

$$\theta \sim N(X, 1) \tag{12.8}$$

This looks plausible if one isn't too precise about what $\sim$ means. A more careful version would write

$$X \mid \theta \sim N(\theta, 1). \tag{12.9}$$

Then one can proceed through analogs of (12.6) to get an analog of (12.7),

$$\theta - X \mid \theta \sim N(0, 1), \tag{12.10}$$

from which (12.8) does not follow. Barnard (1985) also attempted to find compromises essentially having to do with what he, following Fisher, called pivotals, like $X - \theta$ above, which have the distribution $N(0, 1)$ whether regarding X or θ as random. Fraser's structural inference (1968, 1979) is yet another attempt to find cases that can be interpreted either way. But at best these are examples of a coincidence that holds only in special cases. As soon as there is divergence, the issue must be addressed of which is fundamental and which is not. Hence, each reader has to decide for themselves what path to take.

12.1.1 Summary

The key distinction between Bayesian and sampling theory statistics is the issue of what is to be regarded as random and what is to be regarded as fixed. To a Bayesian, parameters are random and data, once observed, are fixed. To a sampling theorist, data are random even after being observed, but parameters are fixed. Whether missing data are a third kind of object, neither data nor parameters, is a puzzle for sampling theorists, but not an issue for Bayesians.

Some standard modern references for sampling theory statistics are Casella and Berger (1990) and Cox and Hinkley (1974).

12.1.2 Exercises

1. Show that $\hat{\mu}$ and $\hat{\sigma}^2$ given in (12.2) maximize (12.1) with respect to μ and σ^2. HINT: maximize the log of f, first with respect to μ, and then substitute that answer in and then maximize with respect to σ.
2. Show that $\hat{\mu}, \hat{\sigma}^2$ and $\hat{X}_{n+1}, \ldots, \hat{X}_{2n}$ given in (12.3) maximize the analogue of (12.1) with respect to μ, σ^2 and $X_{n+1}, \ldots, X_{2n}$. Same hint.
3. Explain what is random and what is fixed to (a) a Bayesian and (b) a sampling theorist.

12.2 Testing

There are two flavors of testing that are part of sampling theory, Fisher's significance testing and the Neyman-Pearson testing of hypotheses. We'll consider them in that order.

Suppose that $X_1, \ldots, X_n$ are a random sample (*i.e.*, independently and identically distributed, given the parameter) from a normal distribution with mean μ and variance 1, which can be written

$$X_1 \ldots, X_n \sim N(\mu, 1). \tag{12.11}$$

Then we know that

$$\frac{(\bar{X} - \mu)}{\sqrt{n}} \sim N(0, 1). \tag{12.12}$$

Suppose that we wish to test the hypothesis that $\mu = 0$. (Such a hypothesis is called "simple," reflecting the fact that it consists of a single point in parameter space. A "composite" hypothesis consists of at least two points.) If $\mu = 0$,

$$P\{| \bar{X}_n |> 1.96/\sqrt{n}\} = 0.05, \tag{12.13}$$

so Fisher would say that the hypothesis that $\mu = 0$ is rejected at the .05 level. This is (more's the pity) the most common form of statistical inference used today.

Of course, the number 0.05 (called the size of the test) is arbitrary and conventional, but that's not the heart of the difficulties with this procedure.

What does it mean to reject such a hypothesis? Fisher (1959a, p. 39) says that it means that either the null hypothesis is false or something unusual has happened. However this theory does not permit one to say which of the above is the case, nor even to give a probability for which is the case. If the null hypothesis is not rejected, nothing can be said. Furthermore, one may reject a true null hypothesis, or fail to reject when the null hypothesis is false.

The biggest issue with significance testing, however, is a practical one. It is easy to see (and many users of these methods have observed) that when the sample size is small, very few null hypotheses are rejected, while when the sample size is large, almost all are rejected. This is because of the $\sqrt{n}$ behavior in (12.12). Thus, while significance testing purports to be addressing (in some sense) whether $\mu = 0$, in fact the acceptance or rejection of the null hypothesis has far more to do with the sample size than it does with the extent to which the null hypothesis is a good reflection of the truth.

This lesson was driven home to me by some experiences I had early in my career. I was coauthor of a study of participation in small groups (Kadane et al. (1969)). There was a simple theory we were testing. The theory was rejected at the .05 level, the .01 level, indeed at the 10^{-6} level. I had to think about whether I would be more impressed if it were rejected at say the 10^{-13} level, and decided not. The issue was that we had a very large data set, so that any theory that isn't exactly correct (and nobody's theory is *exactly* correct) will be rejected at conventional levels of significance. A simple plot showed that the theory was pretty good, in fact.

Sometime later I was working at the Center for Naval Analyses. A study had been done comparing the laboratory to the field experience on a new piece of equipment. The draft report said that there was no significant difference. On further scrutiny, it turned out that, while the test was correctly done, there were only five field-data points (which cost a million dollars apiece to collect). Indeed, the machine was working roughly 75% as well in the field, which seemed a far more useful summary for the Navy.

These experiences taught me that with a large sample size virtually every null hypothesis is rejected, while with a small sample size, virtually no null hypothesis is rejected. And we generally have very accurate estimates of the sample size available without having to use significance testing at all!

Significance testing has been criticized for years because of its binary character (a data set, model and null hypothesis are either "significant" or "not significant".) Most recently, Amrhein, Greenland and McShane, together with over 800 signatures have called for "it's time for statistical significance to go" (Amrhein et al., 2019).

A more general view of significance testing relies on p-values., the probability under the null hypothesis of seeing data as or more discrepant than that observed. P-values have become so widely misused that the American Statistical Association, for the first time in its history, undertook to issue a statement on a methodological matter. The ASA's six points are:

"1. P-values can indicate how incompatible the data are with a specified statistical model.

2. P-values do not measure the probability that the studied hypothesis is true, or the probability that the data were produced by random chance alone.

3. Scientific conclusions and business or policy decisions should not be based only on whether a p-value passes a specific threshold.

4. Proper inference requires full reporting and transparency.

5. A p-value, or statistical significance, does not measure the size of an effect or the importance of a result.

6. By itself, a p-value does not provide a good measure of evidence regarding a model or hypothesis.."

Significance testing violates the likelihood principle, which states that, having observed the data, inference must rely only on what happened, and not on what might have happened but did not. The Bayesian methods explored in this book obey this principle. But the probability statement in (12.13) is a statement about $\bar{X}_n$ before it is observed. After it is observed, the event $\mid \bar{X}_n \mid > 1.96/\sqrt{n}$ either happened or did not happen, and hence has probability either one or zero.

There's one other general point to make about significance testing. As discussed in section 1.1.2, it is based on a limiting relative frequency view of statistics. The interpretation is that if μ were zero and $\bar{X}_n$ were computed from many samples of size n, the proportion of instances in which $\mid \bar{X}_n \mid$ would exceed $1.96/\sqrt{n}$ would approach .05. But the application of this method is to a single instance of $\bar{X}_n$. Thus, a theory that relies on an arbitrarily large sample for its justification is being applied to a single instance.

Consider, for example, the following trivial test. Flip a biased coin that comes up heads with probability 0.95, and tails with probability 0.05. If the coin comes up tails, reject the null hypothesis. Since the probability of rejecting the null hypothesis if it is true is 0.05, this is a valid 5% level test. It is also very robust against data errors; indeed it does not depend on the data at all. It is also nonsense, of course, but nonsense allowed by the rules of significance testing.

A Bayesian with a continuous prior on μ (any continuous prior) puts probability zero on the event $\mu = 0$, and hence is sure, both prior and posterior, that the null hypothesis is false. It is an unusual situation in which a hypothesis of lower dimension than the general setting (here the point $\mu = 0$ on the real line for μ) is so plausible as to have a positive lump of probability on exactly that value.

Neyman and Pearson (1967) modify significance testing by specifying an alternative distribution, that is, an alternative value (or space of values) for the parameter. Thus, they would test (using (12.11) again) the null hypothesis $H_0 : \mu = 0$ against an alternative hypothesis, like $H_a : \mu = \mu_0 > 0$, for a specific chosen value of μ. In this case Neyman and Pearson would choose to see whether the event

$$\bar{X}_n > 1.68/\sqrt{n} \tag{12.14}$$

occurs, because, under the null hypothesis, this event has probability 0.05 and, under the alternative hypothesis, it has maximum probability. The emphasis on this probability, which they call the power of the test, is what distinguishes the Neyman-Pearson theory of testing hypotheses from Fisher's tests of significance.

Neyman and Pearson use language different from Fisher's to explain the consequences of such a test. If the event (12.14) occurs, they would reject the null hypothesis and accept the alternative. Conversely, if it does not they would accept the null hypothesis and reject the alternative. The probability of rejecting the null hypothesis if it is true is called the type 1 error rate; the probability of rejecting the alternative if it is true is called the type 2 error rate.

Again, Neyman-Pearson hypothesis testing violates the likelihood principle, because the event (12.14) either happens or does not, and hence has probability one or zero. Again, the behavior of the test depends critically on the sample size, particularly when it is used with a fixed type 1 error rate, as it most typically is. And again, a single instance of $\bar{X}_n$ is being compared to a long-run relative frequency.

The trivial test that relies on the flip of a biased coin that comes up heads with probability 0.95 is again a valid test of the null hypothesis within the Neyman-Pearson framework, but it has disappointingly low power.

Often in practice the Neyman-Pearson idea is used, not with a simple alternative (like $\mu = \mu_0$) in mind, but with a whole space of alternatives instead. This leads to power (one minus the type 2 error) that is a function of just where in the alternative space the power is evaluated.

From a Bayesian perspective, it would make more sense to ask for the posterior probability of the null hypothesis, as a substitute for significance testing, or for the conditional posterior probability of the null hypothesis given that either the null or alternative hypothesis is correct, as a substitute for the testing of hypotheses.

12.2.1 Further reading

The classic book on testing hypotheses is Lehmann and Romano (2005). More recent developments have centered on the issue of maintaining a fixed size of test when simultaneously testing many hypotheses (see, for instance, Miller (1981)). Still more recently, literature has sprung up concerning limiting the false discovery rate (Benjamini and Hochberg (1995)). A recent defense is Mayo (2018).

For a detailed comparison of methods in the context of an application, see Kadane (1990).

There have been various attempts to square testing with the Bayesian framework. For example, Jeffreys (1961) proposes to put probability 1/2 on the null hypothesis. This is unobjectionable if it is an honest opinion an author is prepared to defend, but Jeffreys presents it as an automatic prior to use in a testing problem. Thus, Jeffreys would change his prior depending on what question is asked, which is incoherent. Bayes factors are yet another way to try to fit testing hypotheses into the Bayesian framework. See (2.14).

For the context of the ASA statement, see Wasserstein and Lazar (2016). For the statement itself, and the commentary on the six points, see American Statistical Association (2016).

12.2.2 Summary

Although widely used in statistical practice, testing, whether done using the Fisher or the Neyman-Pearson approach, rests on shaky foundations.

12.2.3 Exercises

1. Vocabulary. State in your own words the meaning of:

(a) test of significance
(b) test of a hypothesis
(c) null hypothesis
(d) alternative hypothesis
(e) type I and type II error
(f) size of a test
(g) power of a test
(h) the likelihood principle

12.3 Confidence intervals and sets

The rough idea of a confidence interval or, more generally, a confidence set, is to give an interval in which the parameter is likely to be. However the fine print that goes with such a statement is crucial.

There is a close relationship between testing and confidence intervals. Indeed a confidence set can be regarded as the set of simple null hypotheses which, had they been tested, would not have been rejected at the (say) 0.05 level. More formally, it is a procedure (*i.e.*, an algorithm) for producing an interval or set having the property that (say) 95% of the time it is used it will contain the parameter value. Recall, however, that this is part of sampling theory, in which the data are random and the parameters fixed but unknown. Therefore, what is random about a confidence interval (or set) is the interval, not the parameter.

It is appealing, but wrong, to interpret such an interval as a probability statement about the parameter, because that would require a Bayesian framework in which parameters have distributions. There are such intervals and sets, called credible intervals and credible sets, which contain, say, 95% of the (prior or posterior) probability.

Like their testing cousins, confidence intervals and sets violate the likelihood principle. Also, like them, such sets rely on a single instance in a hypothetical infinite sequence of like uses for their justification. The trivial flip-of-a-biased coin example of the preceding section has the following confidence set equivalent: if the coin comes up heads (which it will with 95% probability) take the whole real line. Otherwise (with probability 5%) take the empty set. Such a random interval has the advertised property, namely that 95% of the time it will contain the true value of the parameter, whatever that happens to be. Therefore, this is a valid confidence interval. It is also useless, since we know immediately whether this is one of the favorable instances (the 95% of the time we get the whole real line), or one of the 5% of the time we get the empty set.

While such an example is extreme, the same kind of thing happens in more real settings. Consider a random sample of size two, Y_1 and Y_2, from a distribution that is uniform on the set $(\theta - 1/2, \theta + 1/2)$ for some θ (fixed but unknown). First, we do some calculations:

$$P\{\min(Y_1, Y_2) > \theta \mid \theta\} = P\{Y_1 > \theta \text{ and } Y_2 > \theta \mid \theta\} = \\ P\{Y_1 > \theta \mid \theta\}P\{Y_2 > \theta \mid \theta\} = 1/2 \cdot 1/2 = 1/4. \quad (12.15)$$

Similarly,

$$P\{\max(Y_1, Y_2) < \theta \mid \theta\} = P\{Y_1 < \theta \text{ and } Y_2 < \theta \mid \theta\} = \\ P\{Y_1 < \theta \mid \theta\}P\{Y_2 < \theta \mid \theta\} = 1/2 \cdot 1/2 = 1/4. \quad (12.16)$$

Therefore

$$P\{\min(Y_1, Y_2) < \theta < \max(Y_1, Y_2) \mid \theta\} = 1/2 \text{ for all } \theta, \quad (12.17)$$

so the interval $(\min(Y_1, Y_2), \max(Y_1, Y_2))$ is a valid 50% confidence interval for θ. If the length of this interval is small, however, it is less likely to contain θ than if the interval has length approaching one. Indeed if the interval has length one, we would know that θ lies within the interval, and, even more, we would know that θ is the midpoint of that interval. Thus, in this case the length of the interval gives us a very good hint about whether this is one of the favorable or unfavorable cases for the confidence interval, which is like the previous example. Because whether a procedure yields a valid confidence interval is a matter of its coverage over many (a limiting infinite number!) uses and not its character in this particular use, examples like this cause embarrassment. (This example is discussed in Welch (1939) and DeGroot and Schervish (2002, pp. 412-414).)

What property might make a particular confidence interval desirable among confidence intervals? Presumably one would like it to be short if it contains the point of interest, and wide otherwise. The standard general method is to minimize the expected length of the interval, where the expectation is taken with respect to the distribution of possible samples at a fixed value of the parameter. However this criterion is challenged by Cox (1958), who discusses the following example: Suppose the data consist of the flip of a fair coin, which is coded as $X = 0$ for heads and $X = 1$ for tails.

If $X = 0$, we see data

$$Y \sim N(\theta, \sigma_0).$$

If $X = 1$, however, we see data

$$Y \sim N(\theta, 100\sigma_0).$$

In this case, urges Cox, doesn't it make sense to offer two confidence intervals, one of $X = 0$ and a different one of $X = 1$, each having the standard structure? An interval with shorter average length can be found by making the interval, conditional on $X = 1$, a lot shorter at the cost of making the interval conditional on $X = 0$ a bit longer. See also the discussion in Fraser (2004) and in Lehmann (1986, Chapter 10). A statistic such as X is called ancillary, because its distribution is independent of the parameter. Cox and Fraser advocate conditioning on the ancillary statistic. However, Basu (1959) shows that ancillary statistics are not unique, which calls into question the general program of conditioning on ancillary statistics.

As teachers of statistics, it is common that, no matter how carefully one explains what a confidence interval is, many students misinterpret a confidence interval as if it were a (Bayesian) credible interval, that the probability is α that the parameter lies in the interval specified, where what is random and hence uncertain, is the parameter. Credible intervals and sets can be seen as a part of descriptive statistics, that is, as a quick way of conveying where the center of a distribution, prior or posterior, lies.

A confidence interval (say a 95% confidence interval) is precisely the set of null hypotheses which, had they been tested, would not have been rejected at the 5% level. Consequently, these are exactly the null hypotheses about which nothing can be said, according to Fisher. Thus, the Fisherian and Neyman-Pearson viewpoints are far apart on how to interpret a confidence interval.

An attempt to square confidence intervals with Bayesian ideas is through the concept of highest posterior density regions. Such regions depend critically on the parameterization, unlike Bayesian decision theory, which does not. I see highest posterior density regions as an attempt by early Bayesian writers not to seem too radical to their frequentist colleagues. But Bayesian thought is radically different from frequentism.

12.3.1 Summary

Like the theory of testing, the basis of confidence intervals is weak.

12.4 Estimation

An estimator of a real-valued parameter is a real-valued function of the data hoped to be close, in some sense, to the value of the parameter. As such, it is an invitation to certainty-equivalence thinking, neglecting the uncertainty about the value of the parameter inherent in the situation. Sometimes certainty-equivalence is a useful heuristic, simplifying a problem so that its essential characteristics become clearer. But sometimes, when parameter uncertainty is crucial, such thinking can lead to poor decisions. Thus, estimation is a tool worth having, but not one to be used automatically.

In order to think about which estimators might be good ones to use, it is natural to have a measure of how close the estimator $\hat{\theta}(\mathbf{x})$ is to the value of the parameter. The most commonly used measure of loss (*i.e.*, negative utility) is squared error,

$$(\hat{\theta}(\mathbf{x}) - \theta)^2. \tag{12.18}$$

When uncertainty is taken with respect to a distribution on θ (prior or posterior) the optimal estimator is

$$\hat{\theta}(\mathbf{x}) = E(\theta) \tag{12.19}$$

and the variance of θ (prior or posterior) is the resulting loss. (Indeed this estimator is called in some literature "the Bayes estimate," as if squared error were a law of nature, rather than a statement of the user's subjective utility function.)

However, when (12.18) is viewed from a sampling theory point of view, the expectation must be taken over $\mathbf{x}$ with θ regarded as fixed. The result is an expected loss that depends, with rare exceptions, on the value of θ. The two candidate estimators can have expected loss functions that cross, meaning that for certain values of the parameter one would be preferred, and for other values of the parameter, a different one would be preferred. Since the sampling theory paradigm has no language to express the idea that certain parts of the parameter space are more likely (and hence more important) than others, an impasse results. A plethora of principles then ensues, with no guidance of how to choose among them except for the injunction to use something "sensible," whatever that might mean.

One criterion often used by sampling theory statisticians is unbiasedness, which requires that

$$E(\hat{\theta}(\mathbf{X})) = \theta \tag{12.20}$$

for all θ, where the expectation is taken with respect to the sampling distribution of $\mathbf{X}$. And among unbiased estimators, one with minimum (sampling) variance is to be preferred. Of course this violates the likelihood principle, since it depends on all the samples that might have been observed but were not. Nonetheless, I can see some attractiveness to this idea in the case in which the same commercial entities do business with each other repetitively. Each can figure that whatever such a rule may cost them today will be balanced out over the long run. And here there is a valid long run to consider, unlike most other applications of statistics.

However, unbiased estimates don't always exist, and many times minimum-variance unbiased estimates exist only when unbiased estimates are unique. Consider estimating the function $e^{-2\lambda}$ where X has a Poisson distribution with parameter λ.

An unbiased estimate is

$$I\{X \text{is even}\} - I\{X \text{ is odd}\},$$

which has expectation

$$E(I\{X\text{is even}\} - I\{X\text{is odd}\}) =$$
$$e^{-\lambda}(1 + \frac{\lambda^2}{2!} + \frac{\lambda^4}{4!} + \ldots) - e^{-\lambda}\left(\frac{\lambda}{1!} + \frac{\lambda^3}{3!} + \ldots\right) = e^{-2\lambda}, \tag{12.21}$$

and indeed it can be shown (Lehmann (1983, p. 114)), that this is the only unbiased estimator, and hence a fortiori the minimum variance unbiased estimator. But this estimator is either +1 or −1. Of course −1 is surely too small since $e^{-2\lambda}$ is always positive, and +1 is too big, since $e^{-2\lambda}$ is always less than 1.

Another popular method is maximum likelihood estimation. Were the likelihood multiplied by the prior, what would be found is the mode of the posterior distribution. Under some circumstances, maximum likelihood estimation can thus be a reasonable general method for finding an estimate, if it is necessary to find one. However, the example discussed in section 12.1 shows that even maximum likelihood estimates can have problems when the parameter space is unclear.

That's not all of the story, however. Consider the following example: with probability p, we observe a normal distribution with mean μ and variance 1; with probability $1-p$, we observe a normal distribution with mean μ and variance σ^2. Thus, the likelihood is

$$p\phi(x-\mu) + \left(\frac{1-p}{\sigma}\right)\phi\left(\frac{x-\mu}{\sigma}\right) \tag{12.22}$$

for a single observation x, and the product of these for a sample of size n is:

$$f(x \mid \mu, \sigma, p) = \prod_{i=1}^{n}\left[p\phi(x_i-\mu) + \frac{1-p}{\sigma}\phi\left(\frac{x_i-\mu}{\sigma}\right)\right]. \tag{12.23}$$

Maximizing (12.23) with respect to μ, σ and p yields the following: if $\hat{\mu} = x_i$ for some i, $\sigma \to 0$ and $\hat{p} = 1/2$, the likelihood goes to infinity! Thus, for a sample of size n there are n maximum likelihood estimates for μ. And this example has only 3 parameters and independent and identically distributed observations.

Another example shows just how unintuitive maximum likelihood estimation can be. An urn has 1000 tickets, 980 of which are marked 10θ and the remaining 20 are marked θ, where θ is the parameter of interest. One ticket is drawn at random, and the number x on the ticket is recorded. The maximum likelihood estimate $\hat{\theta}$ of θ is $\hat{\theta} = x/10$, and this has 98% probability of being correct.

Now choose an $\epsilon > 0$; think of ϵ as positive but small. Let $a_1, \ldots, a_{980}$ be 980 distinct constants in the interval $(10-\epsilon, 10+\epsilon)$. Suppose now that the first 980 tickets in the urn are marked $\theta a_1, \ldots, \theta a_{980}$, while the last 20 continue to be marked θ. Again, we choose one ticket chosen at random, and observe the number x marked. Then the likelihood is

$$L(\theta|x) = \begin{cases} .02 & \theta = x \\ .001 & \theta = x/a_i \; i = 1, 2, \ldots, 980 \\ 0 & \text{otherwise} \end{cases}.$$

Hence the maximum likelihood estimator in this revised problem is $\hat{\theta} = x$, which has only a 2% probability of being correct. We know that there is a 98% probability that θ is in the interval $(x/(10+\epsilon), x/(10-\epsilon))$, but maximum likelihood estimation is indifferent to this knowledge.

12.4.1 Further reading

The classic book on estimation is Lehmann (1983). An excellent critique of estimation from a Bayesian perspective is given by Box and Tiao (1973, pp. 304-315). The second example of peculiar behavior of a maximum likelihood estimate is a modification of one given in Basu (1975).

12.4.2 Summary

Estimation is useful (sometimes) as a way of describing a prior or posterior distribution, particularly when it is concentrated around a particular value. As such, for Bayesians it is part of descriptive statistics.

12.4.3 Exercise

1. Let $\epsilon > 0$. Show that there are 980 distinct numbers between $10 - \epsilon$ and $10 + \epsilon$.

12.5 Choosing among models

Model choice is estimation applied to the highest level in the hierarchical model specified in section 9.4. Under what circumstances is it useful to choose one particular model and neglect the others? One circumstance might be if one model had all but a negligible amount of the probability. This case corresponds to estimation where a posterior distribution is concentrated around a particular value.

As a general matter, I would think it is sounder practice to keep all plausible models in one's calculations, and hence not to select one and exclude the others.

12.6 Goodness of fit

There is a burgeoning literature in classical statistics examining whether a particular model fits the data well. However, the assumptions underlying goodness of fit are rarely questioned.

Typically, fit is measured by the probability of the data if the model were true. As such, the best fitting model is one that says that whatever happened had to happen. Such a model is useless for prediction of course, but fits the data excellently. Why do we reject such a model out of hand? Because it fails to express our beliefs about the process generating the data. Also it is operational only after seeing the data, and hence is prone to hindsight bias (see section 1.1.1).

Generally, goodness of fit has to do with how regular (or well-understood) the process under study is, compared to some, often unexpressed, independence model. I think a better procedure is to be explicit about what alternative is contemplated, and then use the methods outlined in section 9.4.

12.7 Sampling theory statistics

A general issue for sampling theory statistics goes under the name of "nuisance parameters," which roughly are parameters not of interest, those that do not appear in a utility or loss function. But "nuisance" hardly describes the havoc such parameters wreak on many sampling theory methods. Bayesian analyses are undisturbed by nuisance parameters: you can integrate them out and deal only with the marginal distribution of the parameters of interest, or you can leave them in. Either way the expected utility of each decision, and hence the expected utility of the optimal decision, will be the same.

As you can see, I find serious foundational problems with each of these methods. But to voice these concerns is not to denigrate the authors cited or the many others who have contributed to sampling theory. Quite the contrary: I stand in awe and dismay at the enormous amount of statistical talent that has been devoted to work within, and try to make sense of, a paradigm with such weak foundations.

To find that the foundations of frequentism are weak is not the same as to hold that the inferences drawn using it are necessarily incorrect. Indeed, it may be that the very messiness of frequentism is a strength, in that it can encourage an adventurous attitude, in which some justification (and frequentism offers many competing justifications) can be found for a method that seems intuitively reasonable. Perhaps there's something to be learned from the history of shrinkage. It was originally approached from a frequentist perspective, but later was found to have a simple Bayesian interpretation (see section 8.2.1).

12.8 "Objective" Bayesian methods

> The notion of a reasonable degree of belief must be brought in before we can speak of a probability.
>
> —H. Jeffreys (1963, p. 402)

This volume would also be incomplete if it failed to address "Objective Bayesian" views (Bernardo (1979); Berger and Bernardo (1992)). For example, suppose a Bayesian wants to report his posterior to fellow scientists who share his model and hence his likelihood. Objective Bayesians search for priors that have a minimal effect on the posterior, in some sense. Some comments are in order:

1. It is not an accident that this hypothetical framework is exactly that of classical, sampling theoretical statistics. From the viewpoint of this book, this framework exaggerates the general acceptability of the model, and also exaggerates the lack of general acceptability of the prior. The likelihood is rarely so universally acclaimed, and often there is useful prior information to be gleaned. If you accept the argument of this book, likelihoods are just as subjective as priors, and there is no reason to expect scientists to agree on them in the context of an applied problem. Yet another difficulty with this program is ambiguity in hierarchical models of just where the likelihood ends and the prior begins.
2. The purpose of an algorithmic prior is to escape from the responsibility to give an opinion and justify it. At the same time, it cuts off a useful discussion about what is reasonable to believe about the parameters. Without such a discussion, appreciation of the posterior distribution on the parameters is likely to be less full, and important scientific information may be neglected.
3. The literature is replete with various attempts to find a unifying way to produce "low information" priors. Often these depend on the data, and violate the likelihood principle. Some make distinctions between parameters of interest and nuisance parameters, which implicitly depends on the utility function of an unstated decision problem. Some are disturbed by transformation: if a uniform distribution on $[0, 1]$ is ok for p, is the consequent for the distribution of $1/p$ also ok? Jeffreys' (Jeffreys (1939, 1961)) priors do not suffer from this, but do violate the likelihood principle. The fact that there are many contenders for "the" objective prior suggests that the choice among them is to be made subjectively. If the proponents of this view thought their choice of a canonical prior were intellectually compelling, they would not feel attracted to a call for an internationally agreed convention on the subject, as have Berger and Bernardo (1992, p. 57) and Jeffreys (1955, p. 277). For a general review of this area, see Kass and Wasserman (1996), and, on Jeffreys' developing views, *ibid.* (pp. 1344 and 1345).

And finally, there is the issue of the name. A claim of possession of the objective truth has been a familiar rhetorical move of elites, whether political, social, religious, scientific, or economic. Such a claim is useful to intimidate those who might doubt, challenge or debate the "objective" conclusions reached. History is replete with the unfortunate consequences, nay disasters, that have ensued. To assert the possession of an objective method of analyzing data is to make a claim to extraordinary power in our society. Of course it is annoyingly arrogant, but, much worse, it has no basis in the theory it purports to implement.

I write sometimes about legal problems, in which it is critical that both the likelihood and the prior are fair to the parties involved. I think of myself as modeling an impartial arbitrator. This is different from scientific reporting, in which I am comfortable with likelihoods and priors that convey real information (provided I explain the considerations that led to the modeling choices made).

12.9 The Frequentist Guarantee

This is a slogan used to justify procedures from a frequentist viewpoint. It says that, with specified long-run frequency, the procedure in question will be "correct." It says nothing about the specific use of the procedure on a specific dataset, however.

Suppose there is such a procedure P with a frequentist guarantee. Then I can create a new procedure P' as follows: Let A be an alternative procedure, and let N be a (large) integer. The procedure P' will use A for the first N times this model is encountered, and will use procedure P, starting a time $N + 1$. The procedure P' has the same frequentist guarantee as does P. Note that procedure A is completely arbitrary, and N can be, say, a billion. Thus, I can do anything with the data before me, and claim a frequentist guarantee. In this sense, the frequentist guarantee is meaningless.

12.10 Final word about foundations

Debate about foundations became very fraught and personal, particularly in the 1930's. Since then, as a reaction, I believe, a modern consensus developed in which foundations are not discussed in polite company. Everybody does their thing, and nobody is to raise questions about what it means.

I think that's a mistake for the profession. For statistics to thrive, there should be no questions disallowed. We can discuss foundations without getting nasty, and we should. I feel that I would be doing my readers a disservice if I failed to give my take on foundational questions.

It is no secret that my view is that of a personalistic Bayesian. A more balanced view, if you want one, can be found in Barnett (2009). When I started, the most bothersome issues in Bayesian thought were identification and randomization. I think those have now been answered. The looming question for me is to find an adequate theory for group decision-making, given the result of Theorem 11.8.1.

Chapter 13

Epilogue: Applications

"In theory, there is no difference between theory and practice. In practice, there is."*

A centipede has sore feet. Slowly, painfully, he climbs the tree to see the owl, and explains his problem. "Oh, I see," says the owl. "Then walk three inches above the forest floor and your feet won't hurt." "Thank you, Owl," says the centipede, as he starts, slowly, painfully, to descend the tree. Suddenly he reverses, and comes back to see the owl. "Owl, how do I do that?" asks the centipede. The owl replies "I've solved the problem in principle. The implementation is up to you."

It may come as a surprise that after what may seem like endless mathematics, I now take the position that the material discussed in this book is only a prelude to the most important aspects of the subject of uncertainty. As mathematics, probability theory has some charms, but certainly lacks the elegance of other branches of mathematics. Much of statistics has to do with special functions and other topics that cannot be regarded as fundamental to further mathematical development.

The reason to study these subjects then, is that they are useful. If our justification is to be that we are useful, we had better attend to being useful. Applications of statistics and probability is where the center of the subject is.

In my view, probability is like a language. Just as grammar specifies what expressions follow the rules that make thoughts intelligible, the rules of coherence specify what probability statements are intelligible. That sentences are grammatical says nothing about the wisdom of what is expressed. Similarly beliefs expressed in terms of probability may or may not be acceptable or interesting to a reader. That is a different discussion, one having to do with rhetoric, with persuading a reader of the reasonableness of the beliefs expressed.

The ideas expressed in this book introduce probability as a disciplined way of conveying beliefs about uncertain quantities, and utilities (losses) as a disciplined way of expressing values. Here "disciplined" means only "free of certain internal contradictions." As I have stressed, the theory here places no other constraints on the content of those beliefs and values. Thus it is possible, using Bayesian methods, to express beliefs and values that are wise or foolish, meritorious or evil. What is offered here is a common language that

*I have seen this attributed both to Jan van de Snepscheut and to Yogi Berra. I have not been able to verify to whom it should be attributed.

encourages being explicit about assumptions, beliefs and values. The hope is that the use of this language will encourage better communication. As such, it may help contending interpreters of data only to understand more precisely where they disagree. But that, in itself, can be a step toward progress.

In doing applied work, the focus has to be on the applied problem. In addressing it, one uses all the tools at one's disposal. I have had, more than once, the experience of doing applied work in a way that did not satisfy me, and only later seeing my way to doing the problem "right." And for me, doing it right means expressing it in the way outlined in this book.

"Statistics is never having to say you're certain."

13.1 Computation

> For better an approximate answer to the *right* question, which is often vague, than an *exact* answer to the wrong question, which can always be made precise.
>
> —John Tukey (1962, pp. 13,14)

An attentive reader will have noticed, and perhaps been disturbed to realize, that in the main I have made no concessions to pleas that what I propose is difficult to compute (or would take billions of years, or whatever). My reason is precisely that stated by Tukey: until the right question is identified, it is hopeless to rush to the computer. Thus the emphasis here is on a framework for posing questions. How to find approximate answers to these questions is a continuously unfolding story, on which there has been, and continues to be, dramatic progress. I have reflected what I take to be the most important computational developments to date, particularly in Chapter 10, but expect more progress to be made.

13.2 A final thought

The perspective of this book is to honor the possibility of alternative points of view about the assumptions: prior, likelihood and utility, that go into the analysis of data. There is no claim that I can sustain, that another person is obligated to agree with my specifications of these objects. Rather, it is my obligation, as author, to explain the considerations that led to my choices, in the hope that a reader may find them acceptable. But I have no right to pretend that my views have *per se* authority, no right to claim that these views are "objective," and hence no basis for a claim that my assumptions live on some mystical higher plane than those of the reader.

What about the thought that at a higher methodological level, this book is rather opinionated about appropriate methodology? It is precisely to explain the reasons why I find certain methodologies appropriate, and others less so, that I undertook to write this book.

> As the bird far flies
> as stars are above skies
> statistics applies.
>
> —Doha Akad

Bibliography

Akaike, H. (1973). "Information Theory and an Extension of the Maximum Likelihood Principle." In *2nd International Symposium on Information Theory*, 267–281. Budapest: Akademia Kiado.

— (1974). "A new look at the statistical model identification." *IEEE Transactions on Automatic Control*, 19, 6, 716–723.

Allais, M. (1953). "Le comportement de l'homme rationel devant le resque: Critique des postulats et axioms de l'ecole Americane." *Econometrica*, 21, 503–546.

American Statistical Association (2016). "Statement on Statistical Significance and *P*-values." *The American Statistician*, 70, 2, 131–133.

Amrhein, W., Greenland, S., and McShane, B. (2019). "Scientists rise up against statistical significance." *Nature*, March 20, 2019.

Andel, J. (2001). *Mathematics of Chance*. New York: J. Wiley & Sons.

Anscombe, F. J. and Aumann, R. J. (1963). "A definition of subjective probability." *Annals of Mathematical Statistics*, 34, 199–205.

Appleton, D. R., French, J. M., and Vanderpump, M. P. J. (1996). "Ignoring a covariate: An example of Simpson's Paradox." *The American Statistician*, 50, 340–341.

Arntzenius, F. and McCarty, D. (1997). "The two envelopes paradox and infinite expectations." *Analysis*, 57, 42–50.

Arrow, K. J. (1951). *Social Choice and Individual Values*. New York: John Wiley & Sons.

— (1971). *Essays in the Theory of Risk-Bearing*. Chicago: Markham Publishing.

— (1978). "Extended sympathy and the possibility of social choice." *Philosophia*, 7, 223–237.

Artin, E. (1964). *The Gamma Function*. New York: Holt, Rinehart and Winston.

Asimov, I. (1977). *On Numbers*. Garden City, NY: Doubleday.

Aumann, R. J. (1987). "Correlated equilibrium as an expression of Bayesian rationality." *Econometrica*, 55, 1–18.

Aumann, R. J. and Drèze, J. H. (2008). "Rational expectations in games." *American Economic Review*, 98, 1, 72–86.

Axelrod, R. (1984). *Evolution of Cooperation*. New York: Basic Books.

Barnard, G. A. (1985). "Pivotal inference." In *Encyclopedia of Statistical Sciences*, vol. VI, 743–747. New York: J. Wiley & Sons. N. L. Johnson and S. Kotz, eds.

Barnett, A. (2009). *Comparative Statistical Inference*. 3rd ed. New York: J. Wiley & Sons.

Barone, L. (2006). "Translation of Bruno DeFinetti's 'The Problem of Full-Risk Insurances'." *Journal of Investment Management*, 4, 3, 19–43.

Barron, A. R., Schervish, M. J., and Wasserman, L. (1999). "The consistency of posterior distributions in non-parametric problems." *Annals of Statistics*, 536–561.

Bartle, R., Henstock, R., Kurzweil, J., Schechter, E., Schwabik, S., and Vyborny, R. (1997).

"An Open Letter." www.math.vanderbilt.edu/~schectex/ccc/gauge/letter/.

Basu, D. (1959). "The family of ancillary statistics." *Sankhya*, Series A, 21, 247–256.

— (1975). "Statistical information and likelihood." *Sankhya*, 37, Series A, 1–71.

Bayarri, M. J. and Berger, J. (2004). "The interplay between Bayesian and frequentist analysis." *Statistical Science*, 19, 58–80.

Bayarri, M. J., DeGroot, M. H., and Kadane, J. B. (1988). "What is the Likelihood Function?" In *Proceedings of the Fourth Purdue Symposium on Decision Theory and Related Topics*, 3–27 (with discussion). New York: Springer-Verlag. S. Gupta and J. Berger, eds.

Beam, J. (2007). "Unfair gambles in probability." *Statistics and Probability Letters*, 77, 7, 681–686.

Benjamini, Y. and Hochberg, Y. (1995). "Controlling the false discovery rate: A practical and powerful approach to multiple testing." *Journal of the Royal Statistical Society, Series B*, 57, 289–300.

Bentencourt, M. (2018). *A Conceptual Introduction to Hamiltonian Monte Carlo*. arXiv: 1701.02434.

Berger, J. and Bernardo, J. (1992). "On the development of reference priors." In *Bayesian Statistics 4*, 35–60. Oxford: Oxford University Press. J. M. Bernardo, J. O. Berger, A. P. Dawid, and A. F. M. Smith, eds.

Berger, J. O. and Berry, D. A. (1988). "Statistical analysis and the illusion of objectivity." *American Scientist*, 76, 159–165.

Berkson, J. (1946). "Limitations of the application of fourfold table analysis to hospital data." *Biometrics Bulletin*, 2, 47–53.

Bernardo, J. M. (1979). "Reference posterior distributions for Bayesian inference." *Journal of the Royal Statistical Society*, 41, 113–147 (with discussion).

Bernheim, B. D. (1984). "Rationalizable strategic behavior." *Econometrica*, 52, 1007–1028.

Bernoulli, D. (1954). "Exposition of a new theory on the measurement of risk." *Econometrica*, 22, 23–36. Translation of his 1738 article.

Berry, D. and Fristedt, B. (1985). *Bandit Problems: Sequential Allocation of Experiments*. New York: Chapman & Hall.

Berry, S. M. and Kadane, J. B. (1997). "Optimal Bayesian randomization." *Journal of the Royal Statistical Society, Series B*, 59, 813–819.

Bhaskara Rao, K. and Bhaskara Rao, M. (1983). *Theory of Charges: A Study of Finitely Additive Measures*. London: Academic Press.

Bickel, P. J., Hammel, E. A., and O'Connell, J. W. (1975). "Sex bias in graduate admissions: Data from Berkeley." *Science*, 187, 398–404.

Billingsley, P. (1995). *Probability and Measure*. Wiley Series in Probability and Mathematical Statistics, 3rd ed. John Wiley & Sons.

Blackwell, D. and MacQueen, J. (1973). "Ferguson distributions via polya urn schemes." *Annals of Statistics*, 1, 2, 353–355.

Blei, D., Kucukelbir, A., and McAuliffe, J. (2017). "Variational Inference: A review for statisticians." *Journal of the American Statistical Association*, 112, 859–877.

Blythe, C. R. (1972). "On Simpson's Paradox and the sure thing principle." *Journal of the American Statistical Association*, 67, 364–366.

— (1973). "Simpson's Paradox and mutually favorable events." *Journal of the American*

Statistical Association, 68, 746.

Box, G. E. P. (1980). "Sampling and Bayes inference in scientific modeling and robustness." *Journal of the Royal Statistical Society, Series A*, 143, 383–430 (with discussion).

— (1980a). "There's no Theorem like Bayes Theorem." In *Bayesian Statistics, Proceedings of the First International Meeting Held in Valencia (Spain)*. University Press. J. M. Bernardo, M. H. De Groot, D. V. Lindley and A. F. M. Smith, eds., `http://www.biostat.umn.edu/~brad/cabaret.html`.

Box, G. E. P. and Tiao, G. (1973). *Bayesian Inference in Statistical Analysis*. Reading, Mass.: Addison-Wesley. Reprinted (1992) as a Wiley Classic, John Wiley & Sons: New York.

Breiman, L. (1961). "Optimal gambling systems for favorable games." In *Fourth Berkeley Symposium on Probability and Statistics*, vol. I, 65–78.

Bremaud, P. (1999). *Markov Chains: Gibbs Fields, Monte Carlo Simulation, and Queues*. New York: Springer-Verlag.

Brier, G. W. (1950). "Verification of forecasts expressed in terms of probability." *Monthly Weather Review*, 78, 1–3.

Bright, J. C., Kadane, J. B., and Nagin, D. S. (1988). "Statistical sampling in tax audits." *Journal of Law and Social Inquiry*, 13, 305–338.

Brockwell, A. and Kadane, J. B. (2003). "A gridding method for sequential analysis problems." *Journal of Computational and Statistical Graphics*, 12, 3, 566–584.

Bullen, P. S. and Vyborny, R. (1996). "Arzela's dominated convergence theorem for the Riemann integral." *Bollettino della Unione Mathematica Italiana*, 10-A, 347–353.

Buzoianu, M. and Kadane, J. B. (2008). "Adjusting for verification bias in diagnostic test evaluation: A Bayesian approach." *Statistics in Medicine*, 27, 13, 2453–73.

Campbell, J. Y. and Viceira, L. M. (2002). *Strategic Asset Allocation: Portfolio Choice for Long-Term Investors*. Oxford: Oxford University Press.

Casella, G. and Berger, R. (1990). *Statistical Inference*. Pacific Grove, California: Wasdsworth and Brooks/Cole.

Casella, G. and Robert, C., eds. (2004). *Monte Carlo Statistical Methods*. 2nd ed. New York: Springer-Verlag.

Chalmers, D. J. (2002). "The St. Petersburg Two-Envelope Paradox." *Analysis*, 62, 155–157.

Chaloner, K. and Verdinelli, I. (1995). "Bayesian experimental design: A review." *Statistical Science*, 10, 237–304.

Chen, L. (1975). "Poisson approximation for dependent trials." *Annals of Probability*, 3, 3, 534–545.

Chen, L., Goldstein, L., and Shao, Q. (2011). *Normal Approximation by Stein's Method*. Berlin Heidelberg: Springer-Verlag.

Chernoff, H. and Moses, L. (1959). *Elementary Decision Theory*. New York: J. Wiley & Sons. Reprinted in paperback by Dover Publications, New York.

Chipman, J. S. (1960). "The foundations of utility." *Econometrica*, 28, 193–224.

Chow, Y. S. and Teicher, H. (1997). *Probability Theory: Independence, Interchangeability, Martingales*. 3rd ed. New York: Springer.

Church, A. (1940). "On the concept of a random sequence." *Bulletin of the American Mathematical Society*, 46, 130–135.

Cleveland, W. S. (1993). *Visualizing Data*. Summit, NJ: Hobart Press.

— (1994). *The Elements of Graphing Data.* New York: Chapman & Hall.

Cochran, W. (1977). *Sampling Techniques.* 3rd ed. New York: John Wiley & Sons.

Cohen, M. and Nagel, E. (1934). *An Introduction to Logic and Scientific Method.* New York: Harcourt, Brace and Company.

Coletti, G. and Scozzafava, R. (2002). *Probabilistic Logic in a Coherent Setting.* Dordrecht: Kluwer Academic Publishers.

Cornford, J. (1965). "A note on the likelihood function generated by randominzation over a finite set." *35th Session of the International Statistics Institute.* Beograd.

Courant, R. (1937). *Differential and Integral Calculus*, vol. I & II. New York: Wiley Interscience.

Courant, R. and Hilbert, D. (1989). *Methods of Mathematical Physics*, vol. 1. New York: J. Wiley & Sons.

Courant, R. and Robbins, R. (1958). *What Is Mathematics? An Elementary Approach to Ideas and Methods.* Oxford University Press. I. Steward, ed.

Cox, D. and Hinkley, D. (1974). *Theoretical Statistics.* Boca Raton: Chapman & Hall.

Cox, D. R. (1958). "Some problems connected with statistical inference." *Annals of Mathematical Statistics*, 29, 357–372.

Cox, R. T. (1946). "Probability, frequency and reasonable expectation." *American Journal of Physics*, 14, 1–13.

— (1961). *The Algebra of Probable Inference.* Baltimore, MD: Johns Hopkins University Press.

Crane, J. and Kadane, J. (2008). "Seeing things: The Internet, the Talmud and Anais Nin." *The Review of Rabbinical Judaism*, 342–345. Koninklijke Brill NV, Leiden. Also available online at `www.brill.nl`.

— (2013). "Seeing Things." *Review of Rabbinic Judaism*, 16, 88.

Cunningham, F. J. (1967). "Taking limits under the integral sign." *Mathematics Magazine*, 40, 179–186.

Cyert, R. M. and DeGroot, M. H. (1970). "Multiperiod decision models with alternating choice as a solution to a duopoly problem." *Quarterly Journal of Economics*, 84, 410–429.

— (1977). "Sequential strategies in dual control problems." *Theory and Decision*, 8, 173–192.

Dagpunar, V. (2007). *Simulation and Monte Carlo, with Applications in Finance and MCMC.* Chichester: J. Wiley & Sons.

Darmois, G. (1935). "Sur les lois de probabilitíes à estimation exhaustive." *C.R. Acad. Sci., Paris*, 200, 1265–1266.

Dawid, A. (2000). "Causal inference without counterfactuals." *Journal of the American Statistical Association*, 95, 407–424 (with discussion).

Deeley, J. J. and Lindley, D. (1981). "Bayes empirical Bayes." *Journal of the American Statistical Association*, 76, 833–841.

Deemer, W. L. and Olkin, I. (1951). "The Jacobians of certain matrix transformations useful in multivariate analysis." *Biometrika*, 38, 345–367.

DeFinetti, B. (1937). "Foresight: Its Logical Laws, Its Subjective Sources." In *Studies in Subjective Probability*, eds. H. E. Kyburg and H. E. Smokler, Translated by H. Kyburg. Huntington, NY: Robert E. Kreiger Publishing Company (1980).

— (1940). "Il problema dei pieni." *Giornale dell'Istituto Italiano degli Attuari*, 18, 1, 1–88.

— (1952). "Sulla preferibilita." *Giornale degli Economisti e Annali di Economia*, 11, 685–709.

— (1974). *Theory of Probability*. London: J. Wiley & Sons. Translated from Italian (1974), 2 volumes.

— (1981). "The role of 'Dutch books' and 'proper scoring rules.'." *British Journal of the Philosophy of Science*, 32, 55–56.

DeGroot, M. H. (1970). *Optimal Statistical Decisions*. New York: McGraw-Hill. Reprinted (2004) by J. Wiley & Sons, Hoboken, in the Wiley Classics Series.

— (1987). "The use of peremptory challenges in jury selection." In *Contributions to the Theory and Applications of Statistics*, 243–271. New York: Academic Press. A. Gelfand, ed.

DeGroot, M. H. and Kadane, J. B. (1980). "Optimal challenges for selection." *Operations Research*, 28, 952–968.

— (1983). "Optimal Sequential Decisions in Problems Involving More than One Decision Maker." In *1982 Proceedings of the ASA Business and Economics Section*, 10–14. And in *Recent Advances in Statistics–Papers Submitted in Honor of Herman Chernoff's Sixtieth Birthday*, Academic Press, 197-210, H. Rizvi, J. S. Rustagi and D. Siegmund, eds.

DeGroot, M. H. and Schervish, M. (2002). *Probability and Statistics*. Boston: Addison-Wesley.

Devroye, L. (1985). *Non-uniform Random Variate Generation*. New York: Springer-Verlag.

Doyle, S. A. (1981). *The Penguin Complete Sherlock Holmes*. New York: Viking Penguin.

Draper, D. (1995). "Assessment and propagation of model uncertainty." *Journal of the Royal Statistical Society, Series B*, 57, 45–97 (with discussion).

Dresher, M. (1981). *The Mathematics of Games of Strategy: Theory and Applications*. New York: Dover Publishing.

Drèze, J. (1974). "Bayesian theory of identification in simultaneous equation models." In *Studies in Bayesian Econometrics and Statistics*, 159–174. Amsterdam: North Holland. S. E. Fienberg and A. Zellner, eds.

Dubins, L. (1968). "A simpler proof of Smith's roulette theorem." *Annals of Mathematical Statistics*, 39, 390–393.

Dubins, L. and Savage, L. J. (1965). *How to Gamble If You Must: Inequalities for Stochastic Processes*. New York: McGraw-Hill.

DuMouchel, W. and Harris, J. (1983). "Bayes methods for combining the results of cancer studies in humans and other species." *Journal of the American Statistical Association*, 78, 293–308.

DuMouchel, W. and Jones, B. (1994). "A simple Bayesian modification of D-optimal designs to reduce dependence on an assumed model." *Technometrics*, 36, 37–47.

Dunford, N. and Schwartz, J. T. (1988). *Linear Operators Part II: Spectral Theory: Self-Adjoint Operators in Hilbert Space*. New York: J. Wiley & Sons.

Dunn, M., Kadane, J. B., and Garrow, J. (2003). "Comparing harm done by mobility and class absence: Missing students and missing data." *Journal of Education and Behavioral Statistics*, 28, 3, 269–288.

Efron, B. and Morris, C. (1977). "Stein's Paradox in statistics." *Scientific American*, 236, 5, 119–127.

Ekstrom, C. (2012). *The R Primer*. Boca Raton: CRC Press, Chapman and Hall.

Ellsberg, D. (1961). "Risk, ambiguity and the Savage axioms." *Quarterly Journal of Economics*, 75, 643–699.

Elster, J. and Roemer, J., eds. (1991). *Interpersonal Comparisons of Well-Being*. Cambridge: Cambridge University Press.

Etzioni, R. and Kadane, J. B. (1993). "Optimal experimental design for another's analysis." *Journal of the American Statistical Association*, 88, 1404–1411.

Feller, W. (1957). *An Introduction to Probability Theory and Its Applications*, vol. 1. 2nd ed. John Wiley & Sons.

Ferguson, T. (1973). "A Bayesian analysis of some non-parametric problems." *Annals of Statistics*, 1, 2, 209–230.

Fienberg, S. and Haviland, A. (2003). "Discussion of Pearl (2003)." *TEST*, 12, 319–327.

Fischhoff, B. (1982). "For those condemned to study the past: Heuristics and biases in hindsight." In *Judgment under Uncertainty: Heuristics and Biases*, chap. 23, 335–351. Cambridge University Press. D. Kahneman, P. Slovic and A. Tversky, eds.

Fishburn, P. C. (1970). *Utility Theory for Decision Making*. New York: John Wiley & Sons.

— (1988). *Nonlinear Preferences and Utility Theory*. Baltimore: Johns Hopkins Press.

Fisher, R. and Ury, W. (1981). *Getting to Yes: Negotiating Agreement Without Giving In*. Penquin Publishing Group.

Fisher, R. A. (1922). "On the Mathematical Foundations of Theoretical Statistics." *Philosophical Transactions of the Royal Society A*, 222, 309–368.

— (1925). "Theory of statistical estimation." In *Proceedings of the Cambridge Philosophical Society*, vol. 22, 700–725.

— (1935). "The fiducial argument in statistical inference." *Annals of Eugenics*, 6, 391–398.

— (1959a). "Mathematical probability in the natural sciences." *Technometrics*, 1, 21–29.

— (1959b). *Statistical Methods and Scientific Inference*. 2nd ed. Edinburgh and London: Oliver and Boyd.

Fraser, D. A. S. (1968). *The Structure of Inference*. New York: J. Wiley & Sons.

— (1979). *Inference and Linear Models*. New York: McGraw Hill.

— (2004). "Ancillaries and conditional inference." *Statistical Science*, 19, 333–369 (with discussion).

French, S. (1985). "Group consensus probability distributions: A critical survey." In *Bayesian Statistics 2*, 183–201 (with discussion). North-Holland Publishing. J. M. Bernardo, M. H. DeGroot, D. V. Lindley and A. F. M. Smith, eds.

Gelman, A., Carlin, J. B., Stern, H. S., and Rubin, D. B. (1995). *Bayesian Data Analysis*. London: Chapman & Hall.

Gelman, A. and Hill, J. (2007). *Data Analysis Using Regression and Multilevel/Hierarchical Models*. Cambridge: Cambridge University Press.

Gelman, A. and Rubin, D. (1992). "Inference from iterative simulation using multiple sequences." *Statistical Science*, 7, 457–472 (with discussion).

Genest, C. and Zidek, J. (1986). "Combining probability distributions: A critique and an annotated bibliography." *Statistical Science*, 1, 114–148 (with discussion).

Geyer, C. (1992). "Practical Markov chain Monte Carlo." *Statistical Science*, 7, 473–483 (with discussion).

Ghahramani, Z. (2005). "Nonparametric Bayesian methods, Uncertainty in Artificial Intelligence Tutorial." Unpublished Notes.

Gibbons, R. (1992). *Game Theory for Applied Economists.* Princeton, NJ: Princeton University Press.

Goldstein, M. (1983). "The prevision of a prevision." *Journal of the American Statistical Association*, 78, 817–819.

Good, I. J. and Mittal, Y. (1987). "The amalgamation and geometry of two-by-two contingency tables." *Annals of Statistics*, 15, 694–711.

Goodman, J. H. (1988). "Existence of Compromises in Simple Group Decisions." Unpublished Ph.D. dissertation, Carnegie Mellon University, Department of Statistics.

Goodman, N. (1965). *Fact, Fiction and Forecast.* Indianapolis: Bobbs-Merrill.

Grimmett, G. and Stirzaker, D. (2001). *Probability and Random Processes.* 3rd ed. Oxford: Oxford University Press.

Halmos, P. R. (1958). *Finite-Dimensional Vector Spaces.* 2nd ed. Princeton, NJ: D. Van Nostrand.

— (1985). *I Want to Be a Mathematician: An Automathography.* New York: Springer-Verlag.

Halperin, J. Y. (1999a). "A counterexample to theorems of Cox and Fine." *Journal of Artificial Intelligence Research*, 10, 67–85.

— (1999b). "Cox's theorem revisited." *Journal of Artificial Intelligence Research*, 11, 429–435.

Hammond, P. (1981). "Ex-ante and ex-post welfare optimality under uncertainty." *Economica*, 48, 235–250.

Hardy, G. H. (1955). *A Course in Pure Mathematics.* 10th ed. Cambridge: Cambridge University Press.

Harsanyi, J. C. (1955). "Cardinal welfare, individualistic ethics, and interpersonal comparisons of utility." *Journal of Political Economy*, 63, 309–321.

— (1982a). "Subjective probability and the theory of games: Comments on Kadane and Larkey's paper." *Management Science*, 28, 120–124.

— (1982b). "Rejoinder to Professors Kadane and Larkey." *Management Science*, 28, 124–125.

Harsanyi, J. C. and Selten, R. (1987). *A General Theory of Equilibrium in Games.* Cambridge, MA: MIT Press.

Hausman, D. M. (1995). "The impossibility of interpersonal utility comparisons." *Mind*, 104, 473–490.

Heath, J. D. and Sudderth, W. (1978). "On finitely additive priors, coherence, and extended admissibility." *Annals of Statistics*, 6, 333–345.

Heckerman, D. (1999). *Learning with Graphical Models*, chap. A Tutorial on Learning with Bayesian Networks. Cambridge, MA: MIT Press. M. Jordan, ed.

Henstock, R. (1963). *Theory of Integration.* London: Butterworths.

Hewitt, E. and Savage, L. J. (1955). "Symmetric Measures on Cartesian Products." *Transmissions of the American Mathematical Society*, 80, 470–501.

Heyde, C. C. and Johnstone, I. M. (1979). "On asymptotic posterior normality for stochastic processes." *Journal of the Royal Statistical Society, Series B*, 41, 184–189.

Hoeting, J. A., Madigan, D., Raftery, A. E., and Volinsky, C. T. (1999). "Bayesian model averaging: A tutorial." *Statistical Science*, 14, 382–521 (with discussion).

Holland, P. (1986). "Statistics and causal inference." *Journal of the American Statistical Association*, 81, 945–960 (with discussion).

Hylland, A. and Zeckhauser, R. (1979). "The impossibility of Bayesian group decisions with separate aggregation of beliefs and values." *Econometrica*, 47, 1321–1336.

Iyengar, S. (2010). *The Art of Choosing*. New York: Twelve Books.

James, W. and Stein, C. (1961). "Estimation with quadratic loss function." In *Proceedings of the 4th Berkeley Symposium*, vol. 1, 361–380. University of California Press.

Jaynes, E. T. (2003). *Probability Theory: The Logic of Science*. Cambridge: Cambridge University Press. G. L. Brethurst, ed.

Jeffreys, H. (1939). *Theory of Probability*. Oxford: Clarendon Press.

— (1955). "The present position in probability theory." *The British Journal for the Philosophy of Science*, 5, 275–289.

— (1961). *Theory of Probability*. 3rd ed. Oxford: Clarendon Press.

Jeffreys, H. and Jeffreys, B. (1950). *Methods of Mathematical Physics*. 2nd ed. Cambridge: Cambridge University Press.

Johnson, R. A. (1967). "An asymptotic expansion for posterior distributions." *Annals of Mathematical Statistics*, 38, 1899–1906.

— (1970). "Asymptotic expansions associated with posterior distributions." *Annals of Mathematical Statistics*, 41, 857–864.

Joseph, V. R. (2006). "A Bayesian approach to the design and analysis of fractionated experiments." *Technometrics*, 48, 219–229.

Kadane, J. (2008). "Load factors, crowdedness and the Cauchy-Schwarz Inequality." *Chance*, 21, 1, 33–34.

— (2017). "Optimal sample size for risk-adjusted survey assessment." *Law, Probability and Risk*, 16, 4, 151–162.

Kadane, J. and Bellone, G. (2009). "DeFinetti on risk aversion." *Economics and Philosophy*, 25, 2, 153–159.

Kadane, J. B. (1974). "The role of identification in Bayesian theory." In *Studies in Bayesian Econometrics and Statistics*, 175–191. Amsterdam: North Holland. S. E. Fienberg and A. Zellner, eds.

— (1985). "Opposition of interest in subjective Bayesian theory." *Management Science*, 31, 1586–1588.

— (1990). "A statistical analysis of adverse impact of employer decisions." *Journal of the American Statistical Association*, 85, 925–933.

— (1992). "Healthy scepticism as an expected-utility explanation of the phenomena of Allais and Ellsberg." *Theory and Decision*, 32, 57–64.

— (1993). "Several Bayesians: A review." *TEST*, 2, 1-2, 1–32 (with discussion).

Kadane, J. B. and Hastorf, C. (1988). "Bayesian paleoethnobotany." In *Bayesian Statistics III*, 243–259. Oxford University Press. J. Bernard, M. DeGroot, D. V. Lindley and A. F. M. Smith, eds.

Kadane, J. B. and Larkey, P. (1982a). "Subjective probability and the theory of games." *Management Science*, 28, 113–129.

— (1982b). "Reply to Professor Harsanyi." *Management Science*, 28, 124.

— (1983). "The confusion of is and ought in game theoretic contexts." *Management Science*, 29, 1365–1379.

Kadane, J. B. and Lazar, N. (2004). "Methods and criteria for model selection." *Journal of the American Statistical Association*, 99, 465, 279–290.

Kadane, J. B., Levi, I., and Seidenfeld, T. (1992). *Elicitation for Games*, 21–26. Knowledge,

Belief, and Strategic Interaction. Cambridge University Press. C. Bicchieri and M. L. Dalla Chiara, eds.

Kadane, J. B., Lewis, G., and Ramage, J. (1969). "Horvath's theory of participation in group discussion." *Sociometry*, 32, 348–361.

Kadane, J. B. and O'Hagan, A. (1995). "Using finitely additive probability: Uniform distributions on the natural numbers." *Journal of the American Statistical Association*, 626–631.

Kadane, J. B., Schervish, M. J., and Seidenfeld, T. (1986). *Bayesian Inference and Decision Techniques: Essays in Honor of Bruno deFinetti*, P. K. Goel and A. Zellner, eds. Statistical Implications of Finitely Additive Probability, 59–76. Elsevier Science Publishers. Reprinted in *Rethinking the Foundations of Statistics*, Cambridge University Press, Cambridge, 1999, pp. 211–232, J. B. Kadane, M. J. Schervish and T. Seidenfeld, eds.

— (1996). "Reasoning to a foregone conclusion." *Journal of the American Statistical Association*, 91, 1228–1235.

— (2001). "Goldstein's dilemma: Abandon finite additivity or abandon 'Prevision of prevision,'." *Journal of Statistical Planning and Inference*, 94, 89–91.

— (2008). "Is ignorance bliss?" *Journal of Philosophy*, 60, 1, 5–36.

Kadane, J. B. and Seidenfeld, T. (1990). "Randomization in a Bayesian perspective." *Journal of Statistical Planning and Inference*, 25, 329–345.

— (1992). "Equilibrium, common knowledge and optimal sequential decisions." In *Knowledge, Belief and Strategic Interaction*, 27–45. Cambridge: Cambridge University Press. C. Bicchieri, C. and M. L. Dalla Chiara, eds.

Kadane, J. B., Stone, C., and Wallstrom, G. (1999). "Donation paradox for peremptory challenges." *Theory and Decision*, 47, 47, 139–151.

Kadane, J. B. and Terrin, N. (1997). "Missing data in the forensic context." *Journal of the Royal Statistical Society, Series A*, 160, 351–357.

Kadane, J. B. and Winkler, R. L. (1988). "Separating probability elicitation from utilities." *Journal of the American Statistical Association*, 83, 357–363.

Kahneman, D. (2011). *Thinking, Fast and Slow*. New York: Farrar, Straus and Giroux.

Kahneman, D., Slovic, P., and Tversky, A., eds. (1982). *Judgment under Uncertainty: Heuristics and Biases*. Cambridge University Press.

Kallenberg, O. (2005). *Probabalistic Symmetries and Invariance Properties*. New York: Springer-Verlag.

Kass, R. and Steffey, D. (1989). "Approximate Bayesian inference in conditionally independent hierarchical models (parametric empirical Bayes)." *Journal of the American Statistical Association*, 84, 717–726.

Kass, R. and Wasserman, L. (1996). "The selection of prior distributions by formal rules." *Journal of the American Statistical Association*, 91, 1343–1377.

Kass, R. E., Tierney, L., and Kadane, J. B. (1988). "Asymptotics in Bayesian computation." In *Bayesian Statistics*, 261–278. Oxford University Press. J. Bernardo, M. DeGroot, A. F. M. Smith and D. V. Lindley, eds.

— (1989a). "Approximate marginal densities of nonlinear functions." *Biometrika*, 76, 425–433. Correction: 78, 233–234.

Kaufman, G. (2001). "Statistical identification and estimability." In *The International Encyclopedia of the Behavioral and Social Sciences*, 15025–15031. Amsterdam: Elsevier. N. Smelser and P. Baltes, eds.

Kelly, Jr., J. L. (1956). "A new interpretation of information rate." *Bell System Technical*

Journal, 917–926.

Kempthorne, O. (1971). *Foundations of Statistical Inference*, chap. Discussion of Lindley's (1961), 451–453. Toronto: Holt, Reinhart and Winston. V. P. Godambe and D. A. Sprott, eds.

Kestelman, H. (1970). "Riemann integration of limit functions." *American Mathematical Monthly*, 77, 182–187.

Keynes, J. (1936). *The General Theory of Employment, Interest and Money*. New York: Harcourt Brace and Co.

Keynes, J. M. (1937). "The general theory of employment." *Quarterly Journal of Economics*, 51, 2, 209–223.

Khuri, A. I. (2003). *Advanced Calculus with Applications in Statistics*. 2nd ed. Hoboken, NJ: J. Wiley & Sons. Theorem 5.4.4, pp. 176–177.

Kingman, J. (1978). "Uses of Exchangeability." *Annals of Probability*, 6, 83–197.

Knapp, T. R. (1985). "Instances of Simpson's Paradox." *College Mathematics Journal*, 16, 209–211.

Knight, F. H. (1921). *Risk, Uncertainty and Profit*. Boston: Houghton-Mifflin.

Kohlberg, E. and Mertens, J.-F. (1986). "On the strategic stability of equilibria." *Econometrica*, 54, 1003–1037.

Kolmogorov, A. N. (1933). *Foundations of the Theory of Probability*. New York: Chelsey. Translated from German (1950).

Koopman, B. (1936). "On Distributions Admitting a Sufficient Statistic." *Transactions of the American Mathematical Society*, 39, 3, 399–409.

Kosinski, A. S. and Barnhard, H. X. (2003). "Accounting for nonignorable verification bias in assessment of a diagnostic test." *Biometrics*, 59, 163–171.

Kosslyn, S. M. (1985). "Graphics and human information processing: A review of five books." *Journal of the American Statistical Association*, 80, 499–512.

Krause, A. and Olson, M. (1997). *The Basics of S and S-Plus*. New York: Springer.

Kullback, S. and Liebler, R. (1951). "On information and sufficiency." *Annals of Mathematical Statistics*, 22, 79–86.

Kyburg, H. E. and Smokler, H. E., eds. (1964). *Studies in Subjective Probability*. New York: J. Wiley & Sons.

Larkey, P., Kadane, J. B., Austin, R., and Zamir, S. (1997). "Skill in games." *Management Science*, 43, 596–609.

Laskey, K. B. (1985). "Bayesian Models of Strategic Interactions." Ph.D. thesis, Carnegie Mellon University.

Lauritzen, S. (1996). *Graphical Models*. Oxford: Clarenden Press.

— (2004). "Discussion on causality." *Scandinavian Journal of Statistics*, 31, 189–192.

L'Ecuyer, P. (2002). *Encyclopedia of the Social and Behavioral Sciences*, chap. Random Numbers, 12735–12738. Amsterdam: Elsevier. N. J. Smelser and P. B. Baltes, eds.

Lehmann, E. L. (1983). *Theory of Point Estimation*. New York: J. Wiley & Sons.

— (1986). *Testing Statistical Hypotheses*. 2nd ed. New York: J. Wiley & Sons.

Lehmann, E. L. and Romano, J. (2005). *Testing Statistical Hypotheses*. 3rd ed. Springer.

LeRoy, S. F. and Singell, L. D. (1987). "Knight on risk and uncertainty." *Journal of Political Economy*, 95, 2, 394–406.

Lewin, J. W. (1986). "A truly elementary approach to the bounded convergence theorem."

American Mathematical Monthly, 93, 395–397.

Lewis, D. (1973). *Counterfactuals.* Cambridge: Harvard University Press.

Li, M. and Vitanyi, P. (1993). *An Introduction to Kolmogorov Complexity and Its Applications.* 2nd ed. New York: Springer.

Liesenfeld, R. and Richard, J.-F. (2001). *Encyclopedia of the Social and Behavioral Sciences*, chap. Monte Carlo Methods and Bayesian Computation, 10000–10004. Amsterdam: Elsevier. N. J. Smelser and P. B. Baltes, eds.

Lindeberg, J. (1922). "Eine neue Herleitung des Exponentialgesetzes in der Wahrscheinlichkeitsrechnung." *Math. Z.*, 15, 211–225.

Lindley, D. (1971). *Foundations of Statistical Inference*, chap. The estimation of many parameters, 435–455 (with discussion). Toronto: Holt, Reinhart and Winston. V. P. Godambe and D. A. Sprott, eds.

— (1976). "A class of utility functions." *Annals of Statistics*, 4, 1–10.

— (1985). *Making Decisions.* 2nd ed. Chichester: J. Wiley & Sons.

— (2002). "Seeing and doing: The concept of causation." *International Statistical Review*, 70, 191–197 (with discussion).

— (2006). *Understanding Uncertainty.* Hoboken, NJ: J. Wiley & Sons.

Lindley, D. and Smith, A. (1972). "Bayes estimates for a linear model." *Journal of the Royal Statistical Society, Series B*, 34, 1–41 (with discussion).

Lindley, D. V. (1982). "Scoring rules and the inevitability of probability." *International Statistical Review*, 50, 1–26.

Lindley, D. V. and Novick, M. R. (1981). "The role of exchangeability in inference." *Annals of Statistics*, 9, 45–58.

Lindley, D. V. and Singpurwalla, N. D. (1991). "On the evidence needed to reach action between adversaries, with application to acceptance sampling." *Journal of the American Statistical Association*, 86, 933–937.

Little, R. J. A. and Rubin, D. B. (2003). *Statistical Analysis with Missing Data.* 2nd ed. New York: J. Wiley & Sons.

Lodh, M. (1993). "Experimental Studies by Distinct Designer and Estimators." Ph.D. dissertation, Carnegie Mellon University, Pittsburgh.

Lohr, S. (1995). "Optimal Bayesian design of experiments for the one-way random effects model." *Biometrika*, 82, 175–186.

Loomis, L. H. (1946). "On a Theorem of von Neumann." In *Proceedings of the National Academy of Science*, vol. 32, 213–215.

Luce, R. D. (2000). *Utility of Gains and Losses: Measurement-Theoretical and Experimental Approaches.* Mahwah, NJ: Lawrence Erlbaum Associates.

Luce, R. D. and Raiffa, H. (1957). *Games and Decisions: Introduction and Critical Survey.* New York: John Wiley & Sons.

Luxemburg, W. A. J. (1971). "Arzela's dominated convergence theorem for the Riemann integral." *American Mathematical Monthly*, 78, 970–979.

Machina, M. (1982). "Expected utility: Analysis without the independence axiom." *Econometrica*, 50, 277–323.

— (2005). "Expected utility/subjective probability analysis without the sure-thing principle or probabilistic sophistication." *Economic Theory*, 26, 1–62.

Mariano, L. T. and Kadane, J. B. (2001). "The effect of intensity of effort to reach survey respondents: A Toronto smoking survey." *Survey Methodology*, 27, 2, 131–142.

Mariotti, M. (1995). "Is Bayesian rationality compatible with strategic rationality?" *The Economic Journal*, 105, 1099–1109.

Markowitz, H. M. (1959). *Portfolio Selection: Efficient Diversification of Investments*. New York: John Wiley & Sons.

— (2006). "DeFinetti scoops Markowitz." *Journal of Investment Management*, 4, 3, 5–18.

Martin-Lof, P. (1970). "On the notion of randomness." In *Intuitionism and Proof Theory*, 73–78. North-Holland. A. Kino, et al., eds.

Mayo, D. (2018). *Statistical Inference as Severe Testing: How to Get Beyond the Statistics Wars*. Cambridge: Cambridge University Press.

McKelvey, R. and Palfrey, T. (1992). "An experimental study of the centipede game." *Econometrica*, 60, 803–836.

McShane, E. J. (1983). *Unified Integration*. Orlando: Academic Press.

Metropolis, N., Rosenbluth, A., Rosenbluth, M., Teller, A., and Teller, E. (1953). "Equations of state calculations by fast computing machines." *J. Chem. Phys.*, 21, 1087–1092.

Meyn, S., Tweedie, R., and Glynn, P. (2009). *Markov Chains and Stochastic Stability*. 2nd ed. Cambridge University Press.

Miller, R. G. (1981). *Simultaneous Statistical Inference*. 2nd ed. New York: Springer-Verlag.

Mirsky, L. (1990). *An Introduction to Linear Algebra*. New York: Dover Publications.

Mitchell, T. (1997). *Machine Learning*, vol. 17. New York: McGraw Hill.

Morrell, C. H. (1999). "Simpson's Paradox: An example for a longitudinal study in South Africa." *Journal of Statistics Education*, 7, 3.

Mosteller, F. (1962). "Understanding the birthday problem." *The Mathematics Teacher*, 55, 322–325. Reprinted in *Selected Papers of Frederick Mosteller*, New York: Springer 2006, 349–353.

Murphy, K. (2012). *Machine Learning, A Probabalistic Approach*. Cambridge, MA: MIT Press.

Nagel, R. (1995). "Unraveling in guessing games: An experimental study." *American Economic Review*, 85, 1313–1326.

Nagel, R. and Tang, F. (1998). "An experimental study of the Centipede Game in normal form – An investigation on learning." *Journal of Mathematical Psychology*, 42, 356–384.

Nash, J. F. (1951). "Non-cooperative games." *Annals of Mathematics*, 54, 286–295.

Natarajan, R. and McCulloch, C. (1998). "Gibbs sampling with diffuse proper priors: A valid approach to data-driven inference?" *Journal of Computational Statistics and Graphics*, 7, 267–277.

Neal, R. (2011). *MCMC Using Hamiltonian Dynamics*. Boca Raton: Chapman & Hall.

Neyman, J. (1923). "On the application of probability theory to agricultural experiments: Essay on principles." *Roczniki Nauk Rolniczych*, X, 1–51 (in Polish). English translation by D.M. Dabrowska and T.P. Speed (1990) *Statistical Science* 9, 465–480.

Neyman, J. and Pearson, E. S. (1967). *Joint Statistical Papers*. Cambridge, U.K.: Cambridge University Press.

Novick, M. (1972). "Discussion of the paper of Lindley and Smith." *Journal of the Royal Statistical Society, Series B*, 34, 24–25.

Nummelin, E. (1984). *General Irreducible Markov Chains and Non-negative Operators*. Cambridge: Cambridge University Press.

— (2002). "MC's for MCMC'ists." *International Statistical Rview*, 70, 215–240.

O'Hagan, T. and Foster, A. (2004). "Bayesian inference." In *Kendall's Advanced Theory of Statistics*, vol. 2B. London: Arnold Publishers.

Paisley, J. (2010). "A simple proof of the stick breaking construction of the Dirichlet Process." Tech. rep., Princeton University.

Pascal, B. (1958). *Pascal's Pensees.* With an introduction by T. S. Eliot. New York: E. P. Dutton.

Patrick, G. (1890). "The psychology of prejudice." *Popular Science Monthly*, 36, 633–643.

Pearce, D. G. (1984). "Rationalizable strategic behavior and the problem of perfection." *Econometrica*, 52, 1029–1050.

Pearl, J. (2000). *Causality: Models, Reasoning and Inference.* Cambridge: Cambridge University Press.

— (2003). "Statistics and causal inference: A review." *TEST*, 12, 281–345 (with discussion).

Pearson, K., Lee, A., and Bramley-Moore, L. (1899). "Mathematical contributions to the theory of fertility in man, and of fecundity in thoroughbred racehorses." *Philosophical Transactions of the Royal Society of London, Series A*, 192, 257–330.

Pfeffer, W. (1993). *The Riemann Approach to Integration: Local Geometric Theory.* Cambridge: Cambridge University Press.

Pippenger, N. (2012). "Elementary Proofs of the Main Limit Theorems of Probability." *arXiv: 1207.6078v1*.

Pitman, E. (1936). "Sufficient statistics and intrinsic accuracy." *P. Cam. Phil. Soc.*, 32, 567–579.

Plummer, M., Best, N., Cowles, K., and Vines, K. (2006). "CODA: Convergence Diagnosis and Output Analysis for MCMC." *R News*, 6, 1, 7–11.

Poskitt, O. S. (1987). "Precision, complexity and Bayesian model determination." *Journal of the Royal Statistical Society, Series B*, 49, 199–208.

Poundstone, W. (2005). *Fortune's Formula: The Untold Story of the Scientific Betting System That Beat the Casinos and Wall Street.* New York: Hill and Wang.

Pratt, J. W. (1964). "Risk aversion in the small and in the large." *Econometrica*, 32, 112–136.

Predd, J., Seiringer, R., Lieb, E., Osherson, D., Poor, H., and Kulkarni, S. (2009). "Probabalistic coherence and proper scoring rules." *IEEE Transactions on Information Theory*, 55, 10, 4786–4792.

Press, S. J. (1985). "Multivariate group assessment of probabilities of nuclear war." In *Bayesian Statistics 2*, 425–462. Amsterdam: Elsevier Science Publishers B.V. (North Holland). J. M. Bernardo, M. H. DeGroot, D. V. Lindley and A. F. M. Smith, eds.

Press, S. J. and Tanur, J. M. (2001). *The Subjectivity of Scientists and the Bayesian Approach.* New York: J. Wiley & Sons.

Propp, J. and Wilson, D. (1996). "Exact sampling with coupled Markov chains and applications to statistical mechanics." *Random Structures and Algorithms*, 9, 1 and 2, 223–252.

Raiffa, H. (1985). *The Art and Science of Negotiation: How to Resolve Conflicts and Get the Best Out of Bargaining.* Harvard University Press.

— (2007). *Negotiation Analysis: The Science and Art of Collaborative Decision Making.* Harvard University Press.

Raiffa, H. and Schlaifer, R. (1961). "Applied Statistical Decision Theory." Tech. rep., Division of Research, Graduate School of Business Administration, Harvard University.,

Boston. Reprinted in the Wiley Classics Series.

Ramsey, F. P. (1926). *Truth and Probability*. Reprinted in Kyberg and Smokler, *Studies in Subjective Probability*. New York: J. Wiley & Sons.

Rapoport, A. (1960). *Fights, Games and Debates*. Ann Arbor: University of Michigan Press.

— (1966). *Two Person Game Theory*. Ann Arbor: University of Michigan Press.

Rapoport, A. and Chammah, A. M. (1965). *Prisoner's Dilemma: A Study in Conflict and Cooperation*. Ann Arbor: University of Michigan Press.

Rasmussen, E. and Williams, C. (2006). *Gaussian Processes for Machine Learning*. ISBN 026218253X: MIT Press.

Richenbach, H. (1948). *The Theory of Probability, an Inquiry into the Logical and Mathematical Foundations of the Calculus of Probability*. University of California Press.

Robbins, H. (1956). "An Empirical Bayes Approach to Statistics." In *Proceedings of the Third Berkeley Symposium on Statistics*, vol. 1, 157–163.

Roberts, G., Gelman, A., and Gilks, W. (1997). "Weak convergence and optimal scaling of random walk Metropolis algorithms." *Annals of Applied Probability*, 7, 110–120.

Robins, J. (1986). "A new approach to causal inference in mortality studies with sustained exposure periods – Application to control of the healthy worker survivor effect." *Mathematical Modeling*, 7, 1393–1512.

— (1987). "Addendum to 'A new approach to causal inference in mortality studies with sustained exposure periods – Application to control of the healthy worker survivor effect'." *Computers and Mathematics with Applications*, 14, 923–945.

Robins, J. and Greenland, S. (1989). "The probability of causation under a stochastic model for individual risk." *Biometrics*, 45, 1125–1138.

Rosenthal, R. (1981). "Games of perfect information, predatory pricing and the chain-store paradox." *Journal of Economic Theory*, 25, 92–100.

Ross, N. (2011). "Fundamentals of Stein's Method." *Probability Surveys*, 8, 210–293.

Rotando, L. M. and Thorp, E. O. (1992). "The Kelly criterion and the stock market." *American Math Monthly*, 922–31.

Roth, A., Kadane, J. B., and DeGroot, M. H. (1977). "Optimal peremptory challenges in trial by juries: A bilateral sequential process." *Operations Research*, 25, 901–19.

Rubin, D. (1974). "Estimating causal effects of treatments in randomized and non-randomized studies." *Journal of Educational Psychology*, 66, 688–701.

— (1976). "Inference and missing data." *Biometrika*, 63, 581–592 (with discussion).

— (1978). "Bayesian inference for causal effects: The role of randomization." *Annals of Statistics*, 6, 34–58.

— (1980). "Comment on 'Randomization analysis of experimental data: The Fisher randomization test' by D. Basu." *Journal of the American Statistical Association*, 75, 591–593.

— (1986). "Which ifs have causal answers? (comment on 'Statistics and causal inference' by P.W. Holland)." *Journal of the American Statistical Association*, 81, 961–962.

— (1988). "Using the SIR algorithm to simulate posterior distributions." In *Bayesian Statistics 3*, eds. J. M. Bernardo et al., 395–402. Oxford: Oxford University Press.

— (2004). "Direct and indirect causal effects via potential outcomes." *Scandinavian Journal of Statistics*, 31, 161–170 (with discussion).

Rubin, H. (1987). "A weak system of axioms for 'rational' behavior and the nonseparability of utility from prior." *Statistics and Decisions*, 5, 47–58.

Rubinstein, M. (2006). "Bruno DeFinetti and mean-variance portfolio selection." *Journal of Investment Management*, 4, 3, 3–4.

Rubinstein, R. and Kroese, D. (2008). *Simulation and the Monte Carlo Method*. 2nd ed. Hoboken: J. Wiley & Sons.

Rudin, W. (1976). *Principles of Mathematical Analysis*. 3rd ed. New York: McGraw-Hill.

Samuelson, P. A. (1971). "The 'fallacy' of maximizing the geometric mean in long sequences of investing or gambling." *Proc. Nat. Acad. of Sci.*, 68, 2493–96.

— (1973). "Mathematics of speculative price." *SIAM Review*, 15, 1–42.

— (1979). "Why we should not make mean log of wealth big though years to act are long." *Journal of Banking and Finance*, 3, 305–307.

Sanchez, J., Kadane, J. B., and Candel, A. (1996). "Multiagent Bayesian theory and economic models of Duopoly, R. & D., and bank runs." In *Advances in Econometrics*, vol. II, Part A. Greenwich, CT: JAI Press. T. C. Fomby and R. C. Hills, eds.

Savage, L. J. (1954). *The Foundations of Statistics*. New York: J. Wiley & Sons. Reprinted by Dover Publications, 1972.

— (1971). "Elicitation of personal probability and expectations." *Journal of the American Statistical Association*, 66, 783–801.

Scheffé, H. (1959). *The Analysis of Variance*. John Wiley & Sons.

— (1999). *The Analysis of Variance*. John Wiley & Sons. Reprinted as a Wiley Classic.

Schelling, T. (1960). *The Strategy of Conflict*. Harvard University Press.

Schervish, M., Seidenfeld, T., and Kadane, J. B. (1984). "The extent of non-conglomerability in finitely additive probabilities." *Zeitschrift fur Wahrscheinlictkeitstheorie und verwandte Gebiete*, 65, 205–226.

Schervish, M. J., Seidenfeld, T., and Kadane, J. B. (2009). "Proper scoring rules, dominated forecasts, and coherence." *Decision Analysis*, 6, 4. Doi: 10.1287/deca.1090.0153.

Schirokauer, O. and Kadane, J. B. (2007). "Uniform distributions on the natural numbers." *Journal of Theoretical Probability*, 20, 429–441.

Schott, J. R. (2005). *Matrix Algebra for Statistics*. Hoboken: J. Wiley & Sons.

Schwarz, G. (1978). "Estimating the dimension of a model." *Annals of Statistics*, 6, 2, 461–464.

Seidenfeld, T. (1979). "Why I am not a subjective Bayesian: Some reflections prompted by Rosenkrantz." *Theory and Decision*, 11, 413–440.

— (1987). *Foundations of Statistical Inference*, chap. Entropy and Uncertainty, 259–287. D. Reidel Publishing Co. I. B. MacNeill and G. J. Umphrey, eds.

Seidenfeld, T., Kadane, J. B., and Schervish, M. (1989). "On the shared preferences of two Bayesian decision-makers." *Journal of Philosophy*, 5, 225–244. Reprinted in *The Philosopher's Annual*, Vol. XII, (1989), 243–262.

Seidenfeld, T., Schervish, M. J., and Kadane, J. (2009). "Preference for equivalent random variables: A price for unbounded utilities." *Journal of Mathematical Economics*, 45, 5-6, 329–340.

Sethuraman, J. (1994). "A constructive definition of Dirichlet priors." *Statistica Sinica*, 4, 639–650.

Shafer, G. (2000). "Comment on 'Causal Inference without Counterfactuals' by A. P. Dawid." *Journal of the American Statistical Association*, 95, 438–442.

Shalizi, C. (2004). "The Backwards Arrow of Time of the Coherently Bayesian Statistical

Mechanics." Unpublished.

Shannon, C. E. (1948). "A mathematical theory of communication." *Bell System Technical Journal*, 27, 379–423, 623–656.

Shubik, M. (1983). "Comment on 'The confusion of is and ought in game theoretic contexts.'." *Management Science*, 29, 1380–1383.

Simpson, E. H. (1951). "The interpretation of interaction in contingency tables." *Journal of the Royal Statistical Society, Series B*, 13, 238–41.

Skyrms, B. (2004). *The Stag Hunt and the Evolution of Social Structure*. Cambridge: Cambridge University Press.

Smith, B. (2005). "Bayesian Output Analysis Program (OA) Version 1.1.5." The University of Texas. `http://www.public-health.uiowa.edu/boa`.

Smith, G. (1967). "Optimal strategy for roulette." *Z. Wahrscheinlichkeitstheorie Verw. Gebeite*, 8, 9–100.

Spiegelhalter, D., Best, N. G., Carlin, B., and van der Linde, A. (2002). "Bayesian measures of model complexity and fit." *Journal of the Royal Statistical Society, Series B*, 64, 583–639 (with discussion).

Spiegelhalter, D., Thomas, A., Best, N., and Lunn, D. (2003). "WinBugs User Manual." MRC Biostatistics Unit; Cambridge. `http://www.mrc-bsu.cam.ac.uk/bugs/`.

Spirtes, P., Glymour, C., and Scheines, R. (1993). *Causation, Prediction and Search*. Cambridge: MIT Press.

— (2000). *Causation, Prediction and Search*. 2nd ed. Cambridge: MIT Press.

State of New Jersey vs. Pedro Soto et al. (1996). 324 NJ Super 66: 734 A. 2d 350. Superior Court of New Jersey. Law Division, Gloucester County. Decided March 4, 1996. Approved for publication July 15, 1999.

Stein, C. (1956). "Inadmissibility of the usual estimator for the mean of a multivariate normal distribution." In *Proceedings of the 3rd Berkeley Symposium*, vol. 1, 197–206. University of California Press.

— (1962). "Confidence sets for the mean of a multivariate normal distribution." *Journal of the Royal Statistical Society, Series B*, 24, 2, 265–296 (with discussion).

— (1972). "A bound for the error in the normal approximation to the distribution of a sum of independent random variables." In *Proceedings of the Sixth Berkeley Symposium on Mathematical Statistics and Probability*, vol. 2, 583–602. University of California Press.

Stern, R. B. and Kadane, J. (2015). "Coherence of countably many bets." *Journal of Theoretical Probability*, 28, 2, 520–538.

Stigler, S. M. (1980). "Stigler's law of eponomy." *Trans. New York Academy of Science*, Series 2, 39, 147–158.

Stone, M. (1969). "The role of experimental randomization in Bayesian statistics: Finite sampling and two Bayesians." *Biometrika*, 56, 681–683.

Taylor, A. E. (1955). *Advanced Calculus*. Boston: Ginn and Company.

Tierney, J. (1991). "Behind Monty Hall's doors: Puzzle, debate and answer." *New York Times*, July 21, page 1.

Tierney, L. (1992). "Practical Markov chain Monte Carlo, comment." *Statistical Science*, 7, 499–501.

— (1994). "Exploring posterior distributions." *Annals of Statistics*, 22, 1701–1734 (with discussion).

Tierney, L. and Kadane, J. (1986). "Accurate approximations for posterior moments and marginal densities." *Journal of the American Statistical Association*, 81, 82–86.

Tierney, L., Kass, R. E., and Kadane, J. B. (1989). "Fully exponential Laplace's approximations to expectations and variances of non-positive functions." *Journal of the American Statistical Association*, 84, 710–716.

Trotter, H. (1959). "An Elementary Proof of the Central Limit Theorem." *Arch. Math*, 10, 226–234.

Tsai, C. (1999). "Bayesian Experimental Design with Multiple Prior Distributions." Ph.D. thesis, School of Statistics, University of Minnesota.

Tsai, C. and Chaloner, K. (Not dated). "Bayesian design for another Bayesian's analysis (or a frequentist's)." Tech. rep., School of Statistics, University of Minnesota.

Tufte, E. (1990). *Envisioning Information*. Cheshire, CT: Graphics Press, LLC.

— (1997). *Visual Explanations: Images and Quantities, Evidence and Narrative*. Cheshire, CT: Graphics Press, LLC.

— (2001). *The Visual Display of Quantitative Information*. 2nd ed. Cheshire, CT: Graphics Press, LLC.

— (2006). *Beautiful Evidence*. Cheshire, CT: Graphics Press, LLC.

van der Vaart, A. (1998). *Asymptotic Statistics*. Cambridge: Cambridge University Press.

van Fraassen, B. (1977). "Relative Frequencies." *Synthese*, 34, 2, 133–166.

Venables, W. N. and Ripley, B. D. (2002). *Modern Applied Statistics with S*. 4th ed. New York: Springer.

Verdinelli, I. (2000). "Bayesian design for the normal linear model with unknown error variance." *Biometrika*, 87, 222–227.

Ville, J. (1936). "Sur la notion de collectif." *C.R. Acad. Sci Paris*, 203, 26–27.

— (1939). *Etude Critique de la Notion de Collectif*. Gauthier-Villars.

von Mises, R. (1939). *Probability, Statistics and Truth*. Macmillan. Dover reprint (1981).

von Neumann, J. and Morgenstern, O. (1944). *The Theory of Games and Economic Behavior*. Princeton, NJ: Princeton University Press.

von Winterfeld, D. and Edwards, W. (1986). *Decision Analysis and Behavioral Research*. Cambridge University Press.

Wagner, C. H. (1982). "Simpson's Paradox in real life." *The American Statistician*, 35, 46–48.

Wald, A. (1950). *Statistical Decision Functions*. New York: J. Wiley & Sons.

Walker, A. M. (1969). "On the asymptotic behavior of posterior distributions." *Journal of the Royal Statistical Society, Series B*, 31, 80–88.

Walley, P. (1990). *Statistical Reasoning with Imprecise Probabilities*. No. 42 in Monographs on Statistics and Applied Probability. Chapman & Hall.

Wasserstein, R. and Lazar, N. (2016). "The ASA's statement on p-values: Context, Process and Purpose." *The American Statistician*, 70, 2, 129–131.

Weil, A. (1992). *The Apprenticeship of a Mathematician*. Basel: Berkhaeuser Verlag. Translation by Jennifer Gage.

Weisstein, E. W. (2005). "Birthday Problem." From Mathworld: A Wolfram Web Resource, `http://mathworld.wolfram.com/BirthdayProblem.html`.

Welch, B. L. (1939). "On confidence limits and sufficiency, with particular reference to parameters of location." *Annals of Mathematical Statistics*, 39, 58–69.

Westbrooke, I. (1998). "Simpson's Paradox: An example in a New Zealand survey of jury composition." *Chance*, 11, 2, 40–42.

Wickham, H. and Grolemund, G. (2016). *R for Data Science*. O'Reilly Media, Inc. (also free on the web).

Wilson, J. (1986). "Subjective probability and the Prisoner's Dilemma." *Management Science*, 32, 45–55.

Wright, G. and Ayton, P., eds. (1994). *Subjective Probability*. Chichester: J. Wiley & Sons.

Yee, L. P. and Vyborny, R. (2000). *The Integral: An Easy Approach after Kurzweil and Henstock*. Cambridge: Cambridge University Press.

Yule, G. U. (1903). "Notes on the theory of association of attributes in statistics." *Biometrica*, 2, 121–134.

Zellner, A. (1971). *An Introduction to Bayesian Inference in Econometrics*. New York: J. Wiley & Sons. Reprinted in the Wiley Classics Series.

Subject Index

Person Index

Made in the USA
Columbia, SC
17 September 2026

9d8f9a31-4b3f-4ab6-9cba-14ca97e112dfR01